Praise for *No G...*
(Volume Two of C...)

"*Carbon Ideologies* is an almanac of global energy use . . . a travelogue to natural landscapes riven by energy production . . . a compassionate work of anthropology that tries to make sense of man's inability to weigh future cataclysm against short-term comfort. . . . One of the most honest books yet written on climate change." —Nathaniel Rich, *The Atlantic*

"In the face of complex, contested data, Vollmann is a diligent and perceptive guide. He's also deeply mindful of those who've been sacrificed in the name of profits and political expediency. Amid the Trump administration's rollbacks of environment protections, these are incontestably important books." —*San Francisco Chronicle*

"Vollmann's many fans . . . will not be disappointed. . . . He packs research and voice into his impassioned works. . . . Reading these two books did have an effect on me; I became even more conscious of the resources I waste in my own life." —John Schwartz, *The New York Times Book Review*

"One of the enjoyable things about this massive work is the way Vollmann employs irony, and that bluntest of irony called sarcasm, throughout the volume. He can be quite humorous. You might even call this the *Infinite Jest* of climate books. . . . There's something admirable, even noble, about the sheer time and effort—and sheer *humanity*—that went into these volumes." —*The Baffler*

"Equal parts gonzo journalism, hand-wringing confessional, and one hot mess . . . The books document Vollmann's quest to understand how capitalism, consumerism, and fossil fuels are ruining the planet." —*Sierra*

"The best parts of the books [are] the conversations Vollmann had during his travels, the sensitive histories he gives of the places he visited, and the moral impressions those conversations and places have made on him. It's these parts that made *Carbon Ideologies* a unique, lasting, definitive contribution to the global warming literature." —*The Humanist*

"An elegy to our damned epoch that's also a work of enlightenment and education . . . The book is a performance of the vexations involved in trying to understand our energy reality. . . . [Vollmann's] project—not unlike that of his historical fiction—is to show with utmost fidelity what it was like to be a human involved in terrible things."
—*Los Angeles Review of Books*

"Vigilant in his precision, open-mindedness, and candor, Vollmann takes on global warming. . . . [His] careful descriptions, touching humility, molten irony, and rueful wit, combined with his addressing readers in 'the hot dark future,' make this compendium of statistics, oral history, and reportage elucidating, compelling, and profoundly disquieting."
—*ALA Booklist* (starred review)

"[A] rewarding, impeccably researched narrative . . . Vollmann apologizes to the future that we've ruined, charting how our choices of energy sources made the planet scarcely inhabitable."
—*Kirkus Reviews*

PENGUIN BOOKS

NO GOOD ALTERNATIVE:
VOLUME TWO OF CARBON IDEOLOGIES

William T. Vollmann is the author of ten novels, including *Europe Central*, which won the National Book Award. He has also written four collections of stories, including *The Atlas*, which won the PEN Center USA Award for fiction; a memoir; and eight works of nonfiction, including *Rising Up and Rising Down* and *Imperial*, both of which were finalists for the National Book Critics Circle Award, and *No Immediate Danger* (Volume One of *Carbon Ideologies*). He lives in California.

CARBON
IDEOLOGIES

VOLUME II

NO
GOOD
ALTERNATIVE

WILLIAM T. VOLLMANN

PENGUIN BOOKS

PENGUIN BOOKS
An imprint of Penguin Random House LLC
penguinrandomhouse.com

First published in the United States of America by Viking Penguin,
an imprint of Penguin Random House LLC, 2018
Published in Penguin Books 2019

"I Am Here Only for Working" first appeared in *Harper's* in 2017.

Map illustrations and photographs by the author.

ISBN 9780525558514 (paperback)

THE LIBRARY OF CONGRESS HAS CATALOGED THE HARDCOVER EDITION AS FOLLOWS:
Names: Vollmann, William T., author.
Title: Carbon ideologies / William T. Vollmann.
Description: New York City : Viking, 2018- |
Identifiers: LCCN 2018013219 (print) | LCCN 2018017544 (ebook) |
ISBN 9780525558507 (ebook) | ISBN 9780525558491 (hardcover : v. 2) |
ISBN 9780525558507 (ebook : v. 2)
Subjects: LCSH: Carbon—Environmental aspects. |
Energy development—Environmental aspects. |
Atmospheric carbon dioxide—Environmental aspects. | Coal mines and mining—
Environmental aspects.
Classification: LCC QD181.C1 (ebook) | LCC QD181.C1 V648 2018 (print) |
DDC 363.738/74—dc23
LC record available at https://lccn.loc.gov/2018013219

Printed in the United States of America
1 3 5 7 9 10 8 6 4 2

Set in Garamond Premier Pro
Designed by Nancy Resnick

Contents: VOLUME II

COAL

For source notes to both volumes of *Carbon Ideologies,* please see:
https://www.penguinrandomhouse.com/carbonideologies

List of Maps and Illustrations

MAPS

ILLUSTRATIONS

Coal: Appalachia

Coal: Bangladesh

Fracking and Natural Gas: Colorado, West Virginia, United Arab Emirates

Oil: Mexico, California, Oklahoma

Oil: United Arab Emirates

Happily Ever After

And So, and Then

CARBON IDEOLOGIES

VOLUME II

NO
GOOD
ALTERNATIVE

This book does not stand alone. For quantitative fundamentals on greenhouse emissions, fuel efficiencies and other matters, please see **Volume I,** *No Immediate Danger.*

Readers beginning with this volume should be advised that fuel efficiency is fundamental to *Carbon Ideologies*.

What Should Be Measured?

*. . . wherefore the best means that I could imagine to wake him out of his
trance was to cry loud in his ear 'Hough host! What's to pay? Will no man
look to the reckoning here?'*

Thomas Nashe, *The Unfortunate Traveller,* 1594

I n a boggy meadow a few paces to the side of a former logging road in West
Virginia, a pipe assemblage crouched hissing. Bubbles rushed up from the pool
from which rose one of its tubular columns; there was a reek of natural gas,
and from the side of that pipe came a steady outrush of gas, comparable in force
to the jet of compressed air from a motorized tire pump, which when I was alive
could blow a penny across a parking lot. Since methane is a primary constituent
of natural gas, I suspected its presence in this petrochemical breeze that gushed

so continually, the product of negligent waste. Until meeting the retired mining inspector Stanley
Sturgill I used to believe that what I smelled in
places such as this was methane, which actually
bears no scent; the odor was crude oil.—In its first
20 years, methane, as you may recall, is at least 86
times more effective at trapping heat than carbon
dioxide.* The environmental activist who was
guiding us here remarked that he had already informed the state Department of Environmental
Protection of this gas rig's ongoing crime. Of
course nothing was done, here and in thousands
and millions of other sites.

Holding the pancake frisker directly in the gas jet, I took a one-minute timed
count, and obtained a reading of 15 counts per minute, 0.06 microsieverts per
hour—the most modest I had ever obtained in West Virginia, and indeed pretty
close to the lowest possible measurement.

---

* See "About Methane," p. 307, and the table of Comparative One-Century Global Warming Potentials, I:178. (Henceforth, cross-references will be as follows: A page number preceded by a roman nu-
meral I, e.g., I:178, directs you to Volume I of *Carbon Ideologies*—that is, *No Immediate Danger.* A page
number by itself cites the volume that you hold.)

Coal-fired power plants give off poisonous emissions, some of which are even said to be radioactive*—but my superficial measurements never found that to be so. The air dose by the John E. Amos Power Plant in Nitro, West Virginia (as measured by poking the frisker out the window of a moving vehicle, a procedure which should have increased the number of encountered particles), was a trifling 39 counts per minute, 0.12 micros per hour. (I neither dug in the ground nor frisked anybody's tomatoes.)

In order to accurately inform you to what extent the four modes of resource extraction and utilization considered in *Carbon Ideologies* were or were not harmful, I would have needed to measure at least the following: carbon dioxide, methane, chlorinated fluorocarbons, aerial particulate sizes, concentrations and compositions; acidity, turbidity, conductivity and metal content of water, the latter subcategorized into selenium, cadmium, aluminum, etcetera; I also ought to have sampled the density of specific microorganisms, amphibians and crustaceans. Accomplishing this was impossible for me, and I am sorry. At a manageable price I could indeed have bought a meter to monitor certain emissions such as carbon dioxide, but the salesman sadly explained that because those gases mix so instantaneously with air, a useful measurement could only have been obtained right up on the smokestack—from which I was excluded by the so-called *regulated community*.

That environmental activist in West Virginia (his name was Chad Cordell) carried a device to sample water for pH and conductivity; maybe I should have bought one. But more often than not, I found myself in waterless places.

So please consider the pancake frisker readings in the remainder of this book as placeholders for the measurements that I would have taken had I possessed more money and power.

Perhaps it was just as well. My readings in the red zones scored sufficient drama in their way; and only one variable—radioactivity—needed to be considered. The manifold effects of fracking, coal-burning, oil refining and kindred operations augmented one other variable of yet more crucial interest: the warming of our planetary home. Since I could frisk neither a smokestack nor a barrel of oil for greenhouse emissions, there was nothing for it but to let this book grow inchoate. I could hint at the villainous parts played by this heavy metal and that gas; there were so many villains! I could portray well-meaning ignorance, mercenary dishonesty and ruthlessness, indifference, useless heroism and sensible

---

* *Scientific American* reported that a coal plant's fly ash, thanks to uranium and thorium, "carries into the surrounding environment 100 times more radiation than a nuclear plant producing the same amount of energy." However, the amount was a trivial 19 microsieverts per year, or 0.53% of the natural background dose—two extra days' worth.

accommodation. Among the tales of coal and oil and natural gas I never heard of accidents comparable to the reactor failures at Fukushima. What then could my narratives be?

Although they do not speak directly to climate change, I ask you to consider the preceding nuclear section as a concentrated relation of this book's theme, which runs like this:

Once upon a time there was something dangerous that could not be seen, felt, heard or smelled. ("Because it's invisible . . . ," sighed my Japanese taxi drivers.) Making use of its associated fuel had been convenient, but terribly mistaken. The best plan would have been to get away from this nearly unknowable thing, but such a course of action appeared so utterly inconvenient that we preferred to continue on as before, which might entail killing our children. Then again, maybe our children would be lucky enough to die from their own stupidity instead of ours. As *The Wall Street Journal* reminded us, I think by way of reassurance:

> It's easy enough to drive out to the country and find somebody in overalls willing to blame the latest flood, drought, windstorm or six-legged pest outbreak on the increased carbon in the atmosphere.

About Permissible Limits

For every human presupposition and every enunciation has as much authority as another, unless reason shows the difference between them.

Montaigne, "Apology for Raymond Sebond," 1575–80

Assuming that our generation had in fact stood ready to measure, record and publicly share local and planetary levels of hydrofluorocarbons, nitrous oxides, carbon dioxide and all those other invisible analogues of radiocontamination, the next and still more contentious step must have been to establish legal ceilings for each, along with procedures for addressing violations.

At the end of Volume I, I compared Japan's and Ukraine's statutory limits for cesium-137 in various foods.* Their variability of categorizations, as in the case of milk, did not entirely obscure consistent ways of thinking. For example, the Ukrainian allowance for higher radioactivity in dried than in fresh foods presumably derived from a supposition that before consumption the dried items would get rehydrated in a significant bulk of (I hope less radioactive) liquid; therefore, one would ingest a smaller quantity of dried than fresh milk, even if the respective liquid volumes were the same. This is arguable, but plausible. And although the Ukrainian allowance of 2,500 becquerels per kilo for dried berries did seem awfully lenient, in fact the two nations' chains of reasoning ran somewhat parallel:

The Ministry of Health of Ukraine set its standards *based* (I quote) *on the fact that the content of Cs-137 and Sr-90 in food and drinking water* should respect *the accepted boundaries of the annual effective exposure of 1 mSv.*† And the Japanese Ministry of the Environment had likewise said: *To achieve further food safety and consumer confidence, Japan is planning to reduce [the] maximum permissible dose from 5 mSv/year to 1 mSv/year.*

Yes, permissible limits would inevitably be arbitrary, like speed limits, felony

---

* See I:506.

† As for iodine-131 and the rest, unfortunately, "the exposure due to intake of other man-made and natural nuclides are [*sic*] excluded."

charges and rules of war, but that alone could not invalidate them, because the *lack* of limits was more perilous.

And we needed to draft them for greenhouse gases. In addition to that annual per capita food radiation limit of 1 millisievert, we should have established annual per person emissions ceilings for, at the very least, the dozen most pernicious heat-trapping agents (doubtless you from the future have a longer retrospective wish list).—But whatever those ceilings were, we would have hated them as infringements of our freedoms. And so when I was alive we wasted years arguing about national and international carbon budgets. It was as if Japan and Ukraine had agreed to disagree as to whether they needed to safeguard their citizens against unregulated intake of cesium-137.

That was one reason we kept burning and selling coal when I was alive.

COAL

West Virginia, U.S.A. (2012–15)
Kentucky, U.S.A. (2015)
Barapukuria, Phulbari and Dhaka, Bangladesh (2015)

Overleaf: Roadside view of a West Virginia coal mine

Coal Ideology

Assertions

UNIQUE OR INTRINSIC BENEFIT

"The lowest cost, most dependable form of energy available."

West Virginia Coal Association, 2013

"Coal is our most abundant fossil fuel source."

George A. Olah, Alain Goeppert and G. K. Surya Prakash, 2009

"Our coal is a basic feedstock of our nation's chemical industry . . ."

West Virginia Coal Association, 2013

"Coal has also contributed to the steady, longterm progress achieved in reducing atmospheric pollutant levels, as well as other improvements."

National Coal Association, 1993

"A secure domestic fuel, unaffected by the politics and instabilities of the Middle East."

National Coal Association, 1993

"The energy density of coal is almost double that of firewood with otherwise similar properties."

Rolf Peter Sieferle, 1982*

"Renewable fuel sources . . . cannot do what coal does. They cannot power America 24 hours a day, 7 days a week, rain or shine. And they don't produce steel!"

West Virginia Coal Association, 2012

---

* Maybe so, for anthracite. To evaluate this assertion, see the table of Calorific Efficiencies beginning on I:208.

"Kentucky's elk are all but parked on their surface coal mine reclaims for its [*sic*] tremendous food and cover value . . . We've never advocated surface mining for the creation of wildlife habitat, but it does what it does for certain species with proper reclamation."

Bob Fala, West Virginian "outdoors columnist," 2014

Among the fruits of the "West Virginia Coal Tree," which occasionally appears as a newspaper or brochure illustration: billiard balls, disinfectant, perfumes, fertilizer, laughing gas, baking powder, medicines, paint pigments, sugar substitute, food preservatives, linoleum, lipstick, batteries, varnish, soda water, roofing and paving.

> *There is a similar tree for petroleum, whose branches are not nearly so diverse, but do happily lead to munitions and insecticides. The fracking tree, which you will find described in the appropriate place, much resembles the trees for coal and petroleum.*

A University of Kentucky experiment concluded that certain coal mine microbes might "help fight disease."

Embellishments

ECONOMIC APPEAL

"Provides small businesses and seniors on fixed incomes with affordable electricity rates."

Joshua Nelson, Representative, West Virginia House of Delegates, 23rd District, 2015

"We are the Saudi Arabia of America . . . When we still have so much coal, so much opportunities [*sic*] in coal, we just can't turn our back on that."

Bill Cole (R-Mercer), West Virginia Senate President, 2015*

"From my point . . . the coal [is] very important in China. *It needs the coal!* The coal is very important. China should develop another coal more, for productivity—but only one, since there are many environmental harms. So only one coal more will not do so much harm."

Mr. Zhang Wen, Chinese guest worker at the Barapukuria coal mine, Bangladesh, 2015

---

* See p. 108 for "The Bill Cole Plan."

"This is black gold money and we all should benefit."
Mr. Rabiul Islam Rabi, President of the Barapukuria Workers' Union, 2015

"The one saving grace for coal in the US has been users' memory of natural gas prices spiking . . . in the mid-2000s . . ."
Chris Faulkner, CEO, Breitling Energy Corporation, 2014

"Coal matters. Vote for jobs."
Banner at West Virginia Coal Festival, Madison, West Virginia, 2013

EXHORTATIONS AND IDENTIFICATIONS

"We support the veterans. Keep the coal burning."
Painted on American Legion building, Williamson, West Virginia [seen 2015]

"Remember—Coal is West Virginia!"
Friends of Coal slogan, 2014

"Proud To Be A Coal Miner."
In display inside restaurant, Northfork, West Virginia, 2014

"YES COAL."
Fire department sign display at West Virginia Coal Festival, 2013

LOCAL APPEAL

"As sources of energy, oil and gas are better and cleaner and their prices are lower. But here in Bangladesh we don't have oil or gas, so obviously coal is better."
Administrator at the Barapukuria coal mine, 2015

"There's no alternative to coal, since we've run out of gas here."
Security guard at the same place, 2015

"There's no substitute for coal in Bangladesh, and we have really good bituminous coal."

Mr. Rabiul Islam Rabi, 2015

"Coal mining is the tie that binds our generation. It is our heritage and has shaped our culture."

West Virginia Coal Association, 2013

"One of West Virginia's leading industries. For decades, it has been providing thousands of good-paying jobs, infusing millions of dollars into local and state economies and providing low-cost energy for its industries and residents."

West Virginia Coal Association, 2013

"KEEP WV ALIVE. SUPPORT COAL!"

Decal on car in Madison, West Virginia, 2013

In McDowell County "you're either a coal miner, you're a prison guard or you're in prison."

Diner waitress, Charleston, West Virginia, 2014

NATIONALISTIC APPEAL

"America runs on coal."

West Virginia Coal Festival slogan, 2014

"Our role is critically important to our nation's quest to become energy independent and break that unholy grip of our dependence on foreign oil."

West Virginia Coal Association, 2013

"Coal is the answer for powering America . . . Coal remains the least expensive fuel source . . ."

Joy Mining Machinery, Pennsylvania, 2009

"We Salute the Coal Industry. It's America's Power."

James A. Redding Company, 2014

"When America runs out of oil, it won't run out of energy thanks to our vast reserves of COAL."

Alliance Resource Partners, L.P., 2009

"Coal is expected to play a major role in addressing some of India's energy challenges."

Australian government report, 2015

"The role of coal, and especially of black coal, is ... of overwhelming importance to the future development of the energy economy and hence to the social and economic growth of the nation."

Dr. Ing. Zymunt Falecki, Scientific and Development Centre for Energy Problems [Poland], 1980

"Harnessing Bangladesh's natural gas reserves and vast quantities of coal in the Phulbari region could improve the lives of 150 million Bangladeshis. U.S. energy sector cooperation also offers the prospect of commercial benefit."

U.S. Ambassador James Moriarty, confidential cable, 2009

"Coal put 'great' into Great Britain—it's as simple as that."

Chris Kitchen, president of the National Union of Mineworkers [U.K.], 2015

RELIGIOUS APPEAL

"Each new day is a blessing working as a coal miner."

Homemade sign in front of house, Madison, West Virginia, 2014

TAUTOLOGICAL TWADDLE

"Coal lights our nights and brightens our future."

Utah Power and Light, before 1985

"Coal, the overwhelming energy choice for today's living."

West Virginia Coal Association, 2009

"It's A Matter of Pride."

Slogan at the Pinnacle Mine, near Welch, West Virginia, 2014

"IF YOU DON'T LIKE COAL DON'T USE ELECTRICITY."

Bumper sticker in Harlan, Kentucky, 2015

"KEN AND PAUL'S FAMILY STEAKHOUSE. MOODY'S 4 PARTS. COAL Keeps the Lights On!"

Bumper sticker in Harlan, Kentucky, 2015

"Coal is not a four-letter word."

Fred Tucker, UMWA (ret.) and co-chair of [West Virginia] Coal Forum, 2015

"As a governor, I never shall apologize for encouraging economic development in this state."

West Virginia Governor Cecil Underwood, defending mountaintop removal at a "Coalfield Justice Rally," 1999

Defenses Against Criticisms

"Let me assure you, the 'War on Coal' is real."

West Virginia Coal Association, 2013

"Coal is under attack, and I shall not stand by and allow federal bureaucrats and others to dictate our future. Air quality issues, unfair and unreasonable, threaten the coal industry . . . Now many mining families face still another challenge— the legal entanglements that jeopardize the future of mountaintop mining."*

Governor Underwood, at a "Coal Miners' Memorial Rally," 1999

"What would you call a radical organization that threatens to shut down 25% of our electric grid?

~~Anarchist~~
~~Militia~~
~~Terrorist~~
Obama's EPA

. . . With radicals like this in power, who needs enemies?"

The Environmental Policy Alliance, 2014

---

* The term preferred by the industry, since mountaintop *removal* might trouble the squeamish. As you read this chapter you will find that the retention or deletion of that word will almost infallibly show the speaker's bias.

"For every one mining job lost, seven related jobs would be impacted."

West Virginia Attorney General Patrick Morrisey, 2015

"For each job in the coal industry, an additional seven jobs are created throughout the economy—in electric utilities, transportation, manufacturing and other sectors."

National Coal Association, 1993

"As a coal miner with over four years in the industry, I believe it is an outrage we have laws that restrict the coal market."

Josh Nelson, West Virginia House of Representatives, 23rd District, 2015

"Promoting the idea that oil and gas was better for the climate than coal was 'a marketing ploy' by the oil and gas industry . . . Come on—the last time I looked there was plenty of carbon in methane and . . . in oil."

Andres Mackenzie, CEO of BHP Billton, "the world's most valuable listed mining group," 2015

"I am for one thing—our coal mining jobs . . . The Sierra Club . . . and many of these other groups have gone on record opposing coal mining. They have gone on record supporting the Obama Administration . . . And they can't have it both ways. They can't pretend to be my friend and cut my throat at the same time."

Roger Horton, President, United Citizens for Coal, West Virginia, 2015

Ukrainian coal miners, memorialized in the Moscow subway

"God Bless the West Virginia Coal Miner!"

About Coal

The brain has accomplished what the arm and hand never attempted, converting the power of the sunshine into blessings for the human race.

L. W. Ellis and Edward A. Rumely, *Power and the Plow,* 1911

Like petroleum, that stuff consisted of crushed plant and animal matter from long ago.* It surprised me that such great quantities lay underground for our convenience: *proved reserves,* we called them, *currently sufficient to meet 153 years of global production!* So we dug, blasted and burned. In 2012 the International Energy Agency announced: *Examining CO_2 emissions in 2009, it is apparent that coal remains the largest source of anthropogenic CO_2 emissions at 12.5 gigatons . . . [C]oal emissions remain 1.86 gigatons greater than those from oil . . . and more than twice those of natural gas.* In 2014, the year of my second West Virginia Coal Festival, coal made up merely 29% of our planet's total *primary energy supply,* but rewarded us with *46% of . . . global CO_2 emissions due to its heavy carbon content per unit of energy released . . .* So half of all our carbon dioxide came from coal! On the other hand, we didn't need to "enrich" it like uranium, in order to shield it like plutonium, and as for the waste . . . Well, you will see.

We bulldozed contour strips along mountainsides, until we got still more ambitious and took the mountaintops right off! That was called *removal of overburden.* Sometimes we mined in open pits, or went underground, hollowing away the coal seam in "room-and-pillar" configurations. Profit-squeezers such as the West Virginia coal baron Don Blankenship made those pillars thinner,† which might have endangered miners but certainly reduced waste! Meanwhile, higher-tech longwall machines allowed us to slice back and forth along a seam without leaving any pillars at all. As we retreated from our accomplishment, the ceiling caved in zone by zone.

Deep mining never became truly safe. At concentrations of 5 to 15%, methane, which seeped out of most mines, could react explosively with air. "Black

---

* Perhaps the plant-stuff predominated; hence the reference to sunshine in that rather poetic epigraph.
† See below, p. 63.

damp" suffocated the unwary. Miners died in collapses.* They ruined their lungs with silicosis. Well, they got the coal out.

The next step, since what we had taken might consist of 25% "mineral matter," was mechanical separation. Then came the motorized cleaning stage, which most frequently consisted of "jig washing," making coal grains dance up and down in water or, as may be, "a suspension of pulverized magnetite"; the lighter particles came to the top; those were the good clean coal. Because we were geniuses, we knew how to get them even cleaner. Our method was froth flotation. Here again we got help from fossil-fueled electricity. Air came magically spurting through a black stew of oil, water and powdered coal; the coal affixed itself to the bubbles. Sometimes we added flocculants and other chemicals.† Finally, at the expense of a few more BTUs, vibrating screens and centrifuges accomplished dewatering—because our customers expected a good dry weight of coal.

In storage, coal lost calorific efficiency at a rate of slightly under 1% per year. From time to time it spontaneously combusted. All, in all, as was the case for that other hurry-up-and-use-it fuel, plutonium-241,‡ it was better not to keep coal around too long. Burn it up, and we'd sell you more!

Montcoal, the site of the Upper Big Branch accident

It came in four basic categories (see the following pages):

---

* "For it must be admitted," wrote Upton Sinclair, "that a man buried under a mountain is as well buried as a company could be expected to arrange it."

† The disaster caused by one of these is described on p. 170.

‡ See I:233.

Coal Types, Simplified,

if not quite enough

These definitions change over places and years. For a hint of the glorious confusion they evoke, see I:553–56. For units of calorific efficiency (HHVs and LHVs), see I:534–35.

All HHVs expressed in BTUs per pound, converted when needed from International Energy Agency round-number figures of kilocalories per kilogram. As such, their precision may be spurious.* Thus anthracite's 10,243 BTUs/lb might better be read as a cool 10,000. But such decisions are not for this humble number-cruncher.

Boldfaced HHVs are from 2012 IEA category definitions. As you will immediately see, they guarantee nothing.

STEAM COAL

Hard coal

Anthracite

"A high rank, non-agglomerating coal." Also called hard, black or stone coal. Often distinguished from bituminous, but not always by the IEA. Prized above other steam coals for its HHV. Sometimes called "metallurgical" coal since it is used in steelmaking, but "met coal" may actually be a blend of high-carbon coals. Rare.

> HHV: **10,243** *minimum.* [The U.S. National Coal Association pegged the stuff at 15,000.] One Pennsylvanian sample came in at 11,980, while a Peruvian hunk was only 4,059, and a Vietnamese specimen measured 14,776. So much for that "minimum" HHV.

Soft coal

Low-volatile bituminous

"An agglomerating coal." [Good bituminous coal of the coking type was also sometimes called "metallurgical."] The coal I heard about in West Virginia and Harlan County, Kentucky, was mostly of this type; likewise the substance in the profiled Bangladeshi chapter (which might or might not have been high-volatile). Bituminous coals vary widely in carbon content, but usually contain less carbon than anthracite and more than lignite.

> HHV: **10,243** *minimum.* By this criterion, West Virginia low-volatile bituminous coal from Elkhorn, Pocahontas No. 3 bed,

---

* Some of the lower HHVs may in fact be LHVs. The sources are not always clear.

burned smashingly at 14,550. Some other samples: Taiwanese, 13,710; and Hungarian, 12,562.

Subbituminous

"A non-agglomerating coal . . . containing more than 31% volatile matter." HHV: **7,492–10,243.** The National Coal Association ranked it higher than bituminous. A Kyrgyzstani piece dating from the Lower Jurassic burned well at 11,758, and a Nigerian specimen read 12,958.

BROWN COAL (or THERMAL COAL, a category which sometimes includes subbituminous)

Lignite

"A non-agglomerating coal" with a low carbon content and more than 31% volatile matter. Oil shale sometimes included in this category.
HHV: *Less than* **7,492.** One Pliocene-era coal from Afghanistan weighed in at a mere 81.685—even less than the feeble stuff called blast furnace gas. A Hungarian variety read 1,995, a Pliocene-era Greek sample 2,859. The South African, Indonesian and Indian coals mentioned in that chapter must have been lignite, and their inefficiencies approached the Greek sample's.

"Methane emissions are normally a function of coal rank (a classification related to the percentage of carbon in the coal) . . ."*

Sources: U.S. Geological Survey, 2006; International Energy Agency, 2012; U.S. Environmental Protection Agency, 2016.

---

* The aphorism continues: "And depth" of the coal mine.

Most of us burned bituminous coal when we could afford it—anthracite being out of reach. Poorer Earthlings burned lignite. Of course a lot depended on which subtype lay buried nearest. Eastern Europeans found much lignite at hand, while Appalachians, by so many other definitions impoverished, received bituminous-flavored electricity, and occasionally even went out to some highway cut to chip out a bucket or two of bituminous coal to keep their families warm.*

---

* See p. 99.

West Virginian toy cast from coal dust and resin

And for a long time coal seemed as cheap and easy, if not quite as magically fabulous, as nuclear power.

If its energy could all be utilized in a steam tractor, the authors of our epigraph calculated, *one pound of coal, costing two fifths of a cent and having 14,000 B.t.u., would produce about eleven million foot-pounds of work, or approximately the useful work of one horse for one day.*

Yes, it heated our homes (and, proximately and unintentionally, the atmosphere), made steam which turned electric turbines, and employed miners. In Appalachia the longterm experience of digging and using coal became part of the culture.

Almost all of New Mexico's electrical production emissions came from coal in 2000. Rolling through that state in 2016, I found coal memorialized in thoroughfares and eateries:

Gallup, New Mexico

When I was alive my old chemistry textbooks mostly roosted in a spider-webbed niche which I could reach only by ladder. What did I care about chemicals? You in your starveling future existence might feel the same—for the productive surplus enabling some level of universal education must have burned and flooded away, in which case please skip the next two pages; they list certain industrial reasons that we loved coal. Why trouble yourself? When I got out my acrylic colors and the pretty model assumed her pose, neither one of us had to think about the coal tar derivatives on my paintbrush. Afterward she took a shower, lathering up her hair with coal. I could go on and on about this fossil fuel (in which, as a matter of fact, one sometimes discovered petrified ferns and other extinct wonders). But all I really need to say is: *no good alternative.*

Sometimes we burned coal only partially, into coke,* with which we made steel. And when coked, each ton of soft coal gave up 120 pounds of coal tar, which in turn produced two pounds of benzene (which we once upon a time had to extract from whale oil), half a pound of toluene, 0.1 pound of xylenes (*starting materials for the production of certain synthetic fibers and films*), half a pound of phenol, two pounds of cresols and five pounds of naphthalene, which was perfect for mothballs and dyes.—Oh, but benzene, that was the cat's pajamas! Without it, we would long since have used up every rubber tree on earth—because we had to keep manufacturing tires for cars and trucks; and benzene satisfied that demand.† (We had learned how to pull it out of oil just as easily as we took it from coal, but don't let me distract you from America's best friend.) Benzene offered up *a strong, yet fairly pleasant aromatic odor.* (According to environmentalists, the benzene that came out of a single fracked well could contaminate more than 100 billion gallons of drinking water.)—In 1960 we made 150 million gallons of benzene alone! Since the density of that compound is 7.73 pounds per gallon, calculating backward from the above recipe reveals that we must have coked 579,750,000 tons (1.159 trillion pounds) of soft coal that year. Marrying it to alcohol, we compelled benzene to become high-performance racing fuel. It could increase the octane rating of gasolines. Upon benzene we founded insecticides, detergents, plastics and aniline dyes. People with time on their hands employed it to dissolve rubber. Mixed with chlorine, ethanol and sulfuric acid, it became that ghastly wonder, DDT. (*"Slow knock-down, sure kill" is the outstanding characteristic of DDT towards most insects. Considerable quantities are used in [India's] malaria eradication programme.*) Hence in 1966 two organic chemists enthused:

> It would be hard to exaggerate the importance to the chemical industry
> and to our entire economy of the large-scale production of benzene and
> the alkylbenzenes . . . Just where do the enormous reservoirs of simple

---

* The grey solid (almost pure carbon) remaining after the volatiles and the coal tar are heated out of bituminous coal in an oxygen-poor vessel. [See I:556.]

† For artificial rubber synthesis, see "About Oil," p. 408.

aromatic compounds come from? There are two large reservoirs of organic material, coal and petroleum, and aromatic compounds are obtained from both. Aromatic compounds are separated as such from coal tar, and are synthesized from the alkanes of petroleum.

A byproduct of coking was ammonium sulfate, which we loved to spread on our fields in order to fertilize the crops and warm the atmosphere.[*]

So kindly refrain from pretending that all we did with coal was burn it. It served us so delightfully as to leave us no good alternative.

—Oh, and that superb welding fuel and plastic-maker known as acetylene, which our chemists used to call *the alkyne of chief industrial importance,* could be *obtained by a few steps from three abundant, cheap raw materials: water, coal, limestone.*

By now we had become whizzes at magicking coal into methanol, which was *one of the 10 largest volume organic chemicals produced in the world.* (The Sierra Club called it one of *the 12 most commonly used air toxics* from fracking *in the Los Angeles basin.*) My high school chemistry book introduced it thus: *Methanol is a colorless liquid with a rather pleasant odor. It is very poisonous*—a quality we took advantage of when we "denatured" alcohol with it, so that winos would die from drinking what was sold only as fuel.[†] It could dissolve any number of substances, and it served as a versatile feedstock. For instance, we blew aerated methanol vapor across hot copper, and here came formaldehyde! Morticians loved that, and so did plastic-makers. Methanol made antifreeze; methanol made shellac. Sometimes we carbonylated it into acetic acid, which offered many seductive industrial applications. (I used the latter in my darkroom, under the name of "stop bath.") Good old methanol![†]—We could also dehydrate it into dimethyl ether, which lies halfway to gasoline. (In the 1980s New Zealanders followed a comparable process to turn their offshore natural gas into motor fuel.)

Eastman Kodak turned coal into acetic anhydride, which was a perfect starting point for medicines, sweeteners and cellulosic plastics. The South African Sasol plant improved it into propylene and polypropylene. In Beulah, North

---

[*] See "About Agriculture," I:114.

[†] "In almost all countries heavy taxes are levied on manufactured alcohol mainly as a source of revenue . . . The great importance of alcohol in the [technocratic] arts has necessitated the introduction of a duty-free product which is suitable for most industrial processes, and at the same time is perfectly unfit for beverages."

[†] One popular science book from the 20th century described it as *an alcohol made from coal, wood, natural gas, or garbage.* If they took all those feedstocks away from us, we could still have pulled it out of oil shale or tar sands. It had 35% fewer "smog-producing hydrocarbons" and 20–40% fewer "airborne toxics" than gasoline, 20% greater horsepower and acceleration, but only half the mileage, and it gave off formaldehyde. Well, nobody's perfect.

Dakota, lignite entered the Dakota Gasification Company Great Plains, and came out ammonia, sulfur or synthetic natural gas!

Our women wore coal on their legs, in the form of nylon stockings. They dabbed it on their faces as perfumes. We killed weeds and fueled rockets with it. When I got a headache, I swallowed two coal pills*—although I readily confess that we could and did also make these products out of natural gas.

But in 1980 a talking head at an energy symposium concluded: *The single largest end use for coal in recent years has been electric power generation . . .*

No wonder we loved coal! (We especially loved selling it.) From 1971 through 2009, carbon emissions from coal increased by an ecstatic 140%.

Coal consumption for electricity generation increased by 248% in China between 2000 and 2009, or 569% when compared to 1990 . . . In 2009, China contributed 5.7 gigatons . . . which was 46% of global coal emissions.

On that happy subject, a scientist observed: *Burning coal is a primary source of sulfur gas. Yet the use of coal is rapidly growing . . . The sulfur pollution produced by burning coal in China contributes to 400,000 premature deaths per year.*

---

* Or, if you like, they contained a handy coal tar derivative, descended from benzene through the promiscuously soluble, almond-smelling, pale yellow liquid called nitrobenzene, and midwifed by sulfuric acid, which the more knowing sorts called 4-acetamidophenol, while the rest of us bought it over the counter as acetaminophen. For another fossil fuel source of this ubiquitous drug, see the risqué epigraph to "About Industrial Chemicals," I:124.

About Sulfur

And now, once again, I bid my hideous progeny go forth and prosper. I have an affection for it, for it was the offspring of happy days, when death and grief were but words which found no true echo in my heart.

Mary Shelley, 1831 introduction to *Frankenstein,* 1818

Although this book has privileged carbon, a case could be made for a study entitled *Sulfur Ideologies*—which offers another reason for me to hope that you from the future might pity as well as blame us. When we exterminated the buffalo from the open plains, we knew what we were doing; General Sherman made that clear. But when we meddled with our atmosphere, so many subtle complexities began to drift discreetly out of kilter that we might as well have been children.

Regarded by the alchemists, on account of its inflammable nature, as the principle of combustion, our yellow protagonist, which combusted with a suitably eerie blue flame, distinguished itself from the whites and greys of most other elements. And since it was so special, what if its most famous combustion product, sulfur dioxide, had been the primary cause of global warming?

When I was alive, the prestigious Intergovernmental Panel on Climate Change did not think so. Carbon dioxide was the ace of spades! The panelists allowed that *pollutants such as . . . volatile organic compounds* . . . and sulphur dioxide . . . , which by themselves are negligible G[reen] H[ouse] G[ase]s, have an indirect effect . . . by altering, through atmospheric chemical reactions, the abundance of important gases . . . such as [methane] and ozone . . . , and/or by acting as precursors of secondary aerosols.* Let this indirect effect now inspire our musings.

Scientists measured the sulfate in Greenlandic ice cores. As might be expected, concentrations fluctuated over years and centuries. Long before humans began burning coal, volcanoes had done their mite to spice up the air, and the ice cores showed it. A man named Peter L. Ward made certain correlations and concluded: *If large amounts of SO_2 are erupted frequently enough . . . the atmosphere loses its oxidizing capacity . . . Greenhouse gases such as water vapor, methane and carbon monoxide, and other pollutants increase, causing global warming.*

---

* See p. 352.

26

Here was how it worked: Occasional large volcanic eruptions actually cooled the planet for about three years each, because their upflung droplets of sulfuric acid that soon reached the upper atmosphere blocked some absorption of the sun's rays.* But they simultaneously depleted the supply of oxidizers—namely, ozone and its two derivatives, hydrogen peroxide and the hydroxyl radical. These compounds normally broke down carbon monoxide, methane and other green-house gases. Hydroxyl radicals could even decompose such nasty humanmade greenhouse gases as hydrofluoroethers, which we had invented as substitutes for ozone-killing CFCs. *All three oxidants are highly reactive, short lived, and in limited supply, especially at night, at high latitudes, and during the winter.* Accordingly, after any heavy eruption greenhouse gases began to build up—but if several years elapsed between eruptions, the oxidizers returned to their prior concentrations, even while the sulfuric acid droplets continued to shield the earth from heat. Hence a happily spaced out continuation of these sporadic eruptions could bring about a cool drought, or even an Ice Age. Ward claimed that extreme declines in sulfate levels in those Greenlandic ice cores correlated with *the demise of major civilizations.*

But more frequent volcanic activity—say, an eruption each year for at least 10 years—blocked the reformation of oxidants long enough for methane and carbon dioxide to reach dangerous levels, thereby precipitating warming. At intervals of around 20 million years, super-eruptions defeated *the oxidizing capacity of the atmosphere for tens of thousands of years or more, causing mass extinctions.*

Ward now grew alarmingly specific. He wrote: *Note that climate change appears to be initiated by levels of sulfate far less than one "large" volcanic eruption per year,* by which he meant one producing at least 50 parts per billion of sulfates.

Come the Industrial Revolution, anthropogenic sulfur caught fire. In Sicily we used to refine that ore by burning it in a hollow in the ground, after which we ladled out the molten remains. *This exceptionally wasteful process, in which only one-third of the sulphur is recovered, has been improved by conducting the fusion in a sort of kiln.* Meanwhile the other two-thirds had already ascended into the air. Well, there was plenty more to burn!

We learned to love sulfur nearly as much as coal. You see, it was *of prime importance to the fertilizer industry.* In the early 20th century we put sulfur dioxide to work in our refrigerators, but like ammonia it proved inconveniently toxic and corrosive. Fortunately, it was magnificent for "preserving" vegetables, wine and dried fruits.

---

* It was also blocking the light lower down. One popular history of coal assured me that sulfur dioxide "is the principal reason people can't see great distances in the eastern United States." It supposedly reduced visibility "on an average day" from as much as 45 to 90 miles down to 14 miles.

Often we transformed it into that *colorless, viscous liquid,* sulfuric acid, which my 1911 *Britannica* called *perhaps the most important of all chemicals, both on account of the large quantities made in all industrial countries and of the multifarious uses to which it is put.* Here were two: making hydrogen to buoy up balloons, and manufacturing artificial indigo. Weren't those worth warming the planet for? Not that sulfuric acid manufacture warmed it all that much, compared to what we did in steelmaking. *Where spent acid is used as raw material, it usually is decomposed in furnaces fired by gas, oil or other fuels . . . and the high temperature gas from such furnaces can also generate steam or power.* And we caught more than 98% of the sulfur dioxide produced, thanks to four catalytic filtering stages. So we didn't stint ourselves. *The largest sulfur burning plants as of the mid-1930s produce approximately 3300 metric tons of acid per day.* By then it was already producing papers, paints, detergents, rayon, phosphate fertilizers . . .

By then (as you might have heard), we'd found ever more cause to burn fossil fuels, some of which contained sulfur as impurities.* And presently Ward advised us: *The amount of sulfate deposited in each layer (1.7 years)* of Greenland ice *had risen by 1962 to 50 ppb,* his trigger point. Then it kept rising.

For the last three decades of the 20th century, Euro-American pollution controls beat out Asian progress, so that total SO_2 emissions actually declined by 15%. By 2000, progress had won as usual. In the next five years alone, we added 9 gigagrams (19.8 million pounds) to our generous annual donation.

For every 2,494 pounds of aviation fuel we burned, we offered the atmosphere a pound of sulfur. Gasoline and diesel likewise did what they could.—I am proud to say that coal made its own contribution. In 2014, Americans sent upward just under 10 billion pounds of sulfur dioxide. Most of it derived *from coal combustion for electric power generation and the metals industry.*

What if Ward were correct, and the widespread combustion of sulfur-laced coal were more doom-inducing than all our carboniferous follies? I could, of course, have asked other experts, such as those on the Intergovernmental Panel on Climate Change. If they'd agreed with Ward,[†] I might have started believing him. If they'd disagreed, I could have believed *them* or else thrown up my hands. Maybe nobody knew. In 2016 the U.S. Environmental Protection Agency decided that *the radiative forcing [global warming] estimates from both the first and*

---

* Torgny Schütt, 1980: "Acidification of lakes and land ecosystems by emitted sulphur dioxide . . . from oil combustion is at present the most severe environmental problem in Sweden. More than 10,000 lakes are already 'dead' or badly disturbed . . . And this process is accelerating. At present about 800,000 [metric] tonnes of SO_2 are annually emitted and also deposited in Sweden."

† In fact their position was simpler: "A reduction in sulphur dioxide (SO_2) emissions leads to more warming."

the second indirect effect of sulfur dioxide on its surroundings *are believed to be negative, as is the combined radiative forcing of the two. However, because SO₂ is short-lived and unevenly distributed in the atmosphere, its radiative forcing impacts are highly uncertain.* Moreover, the sulfate concentrations reported by some authorities differed from Ward's.

Lacking both the intellect of a Galileo and the wealth of a corporate politician, I could hardly fly to Greenland, launch probes into the troposphere and sample volcanic effluvia. What about spot-checking Ward's chain of logic? Well, that sounded awfully laborious. I preferred to follow some plausible authority somewhere. And if that defined the scientific method even of your well-meaning, hardworking author of *Carbon Ideologies,* how could I fault underprivileged people for following, say, the local preacher or the state coal association? Thus the story of this chapter.

2

America's Best Friend

June 2013, April 2014, June 2014, July 2015

COAL. It is America's best friend.

West Virginia Coal Association, *Coal Facts,* 2014

APPALACHIAN PLACES MENTIONED

A	Acidified tap water
F	Buffalo Creek flood (coal slurry dam failure)
M	Mountaintop removal mine
P	Polluted residential wells (due to coal mining)
R	Radioactive groundwater contamination (from fracking)
S	Elk River chemical spill (MCHM, a coal-cleaning substance)
U	Upper Big Branch accident

"THE ONLY THING WE'VE EVER HAD WHERE YOU COULD MAKE A LIVING"

Well," said Pastor Bob Blevins, "I was born in a coal company house, right down in Bradshaw, just off the road you come up on. My father was a coal miner, and he worked for an old time coal company, Pocahontas Coal, and 12 or 13 years later, we got high waters when they put the coal in the tipple,* and all the residue was dumped along the river, and it begin to push the river toward the road, and the level of the road was just slightly above the high water mark. I was born in '36. In '47, heavy rains and high water comin' down that river had nowhere to go, and the high water washed both of those houses away. By that time we were living on the mountain out here. The coal company my Dad worked for, there was a coal mining explosion that happened in 1941 at that Bartley Mine— killed 91 miners. My Dad worked on the evening shift, and that mine blew up in the day. So my father, he escaped that explosion. We lived up by the Davis Baptist Church, but two mile down that holler was where my home was, until I went in the Army in 1953. The Korea conflict ended on July 8, and I went in on August 20. I wanted to get out of the mine. I was in the mine, I was only 17 year old, and the mine was hard and low pay, and I hadn't worked then but a little while but I knew I didn't want to work in the mine. I come out a year later and married my wife, whose father was also a coal miner. We were goin' 47 years, and we retired in Africa."

He was a greyhaired fellow with a bald spot and narrow-framed spectacles. In 2014, when I met him, coal-burning plants still generated almost half the power in America. I would call him a pretty solid pro-coal man. He was kind, success-fully self-educated, effective and tolerant. As I listened to him giving instructions to workmen for a funeral I could see that he had a very practical bent. I liked Pastor Blevins, and I hope that you will like him, too.

He said: "We neither one were raised in Christian homes. My folks, they went to Bradshaw Church of God, and they were baptized in the creek back there. But we moved on the mountain there and they didn't go to church. They

---

* The spot outside a mine where coal was tipped out of the cars.

didn't have a car. Church of God just taught, if you quit going to church you were back in the world. They were a lot like Freewill Baptists.*

"We became Christian after I finished the Army and moved to Dayton, Ohio. In the subdivision, my little daughter, she was four years old, her friends invited her to Sunday school, and in about 1960, my wife was going to have a thyroid problem, and she began to think, maybe my mother and daddy may not have been right. So maybe I ought to see for myself. What about if I take Bobbie Lee and I go to Sunday school? So I said, fine, you can take the child to school, but you've never driven. She said, well, I'll just drive about a mile an hour through this subdivision. And they prayed over her and she received Christ in her heart in our home. She did officially on Christmas, and I did in April.

Pastor Bob Blevins

---

* "In this county, the main denomination is gonna be Pentecostal. We Baptists are a minority. Freewill Baptists do not believe in the eternal security of the believer. Southern Baptists, we believe that when you receive Jesus as your Savior and you show repentance of your sin, and you have the spirit of Christ living in your heart, you have eternal security in the believer. Now, the Freewill, they believe, you have to remain faithful; you're saved for as only as long as you want to be. And if you die in that condition where you have given up your faith, you're eternally lost. Primitive Baptists in this area belong to an association called Universalism. They believe that everybody is saved anyway. Sounds good, don't it? You can join if you believe in God or if you don't believe in it. Old Regular Baptists are Calvinists. Now, the Freewill broke off years and years ago, because of this Old Regular belief in unlimited grace. Now the American Baptists here, that is pretty strong in this county, and their doctrine is pretty much the same as Southern. The strongest thing about Southern Baptism is evangelism and mission."

"And we were both young, I think maybe 24 or 25 years old, and nine years later, I felt called to preach, and my wife and I went to three-year Bible college. The women didn't have to attend as many years as the men did.

"I became a Christian in '61, and in '64 I left the National Cash Register Company, and went to work for a Ford dealer. Myself and another fella got a job there because they had so many customers they couldn't keep up. So we became salesmen for the Ford Mustang for the next six years; that's the most money I ever made. And in 1970 we left them to go to Bible college. Later my wife worked part time in a shoe store and I worked part time as a bus driver.

"My wife and me, we were first volunteer missionaries in 1979 and 1980.

"Our church is still small. We had 20 in congregation last Sunday.

"Now this place here was called the Acapulco Club, this whole complex. They had a restaurant, dancehall, go-go stage, three bars, a pool room, a mechanical stage, a large room and three or four bedrooms for *whatever*. And the original owner was a machinist that made mining equipment. He had a small plane, flew into Mexico for six weeks at a time, and one time he come in with a young Mexican girl. People assumed he just had a live-in girlfriend, and then not so long after, his plane crashed, and it turned out he really had married her. So she married a local guy, turned it into the Acapulco Club, and with her contacts, she brought the boys and girls here. She would only keep 'em here about six months. Anyway, the new husband got it all converted, and they ran it all full blast for many years. This property at one time had a $420,000 mortgage, so somebody thought it would be worth a lot of money. But he was into drugs big time. Some way or other, he satisfied all those mortgages but he got caught the third time for those drugs, and went to prison. He'd had a $10,000 mortgage, and that was where he had borrowed money from a drugstore operator in Welch. And that fella then filed a claim, and this was all sold then at a sheriff's sale. And a young man bought it in 1999, because he thought it would be a good deal. We come along in 2001, and later bought it for a church.* And we have renovated it top and bottom. The go-go stage is our altar. We have good parking spaces and a little more room, being on top of the mountain. We have a lot going on here with funerals. The one we're having this weekend, she's being buried in one of the largest graveyards, in a mountain cemetery.

"Her husband that was killed in the plane crash, he built what was called small mining equipment. These hills were full of small mines. This equipment would hold about a ton of coal, operated with electricity through a cable. Before

---

* "The husband then was in prison, and we bought property then next door, and that's when we got acquainted with her, and somewhere about 2005, or '06, she died, down in Tennessee. He then got out of prison about three years ago, and he went down in Tennessee where his wife died. He's got diabetes; his life is a wreck, it came to a terrible end. He was smart enough to know that he himself couldn't use the drug he sold."

that, they used two small ponies. They loaded up the face inside, and a guy outside dumped the coal into a place, and the ponies were trained to turn around and go straight back in. Later they had some of those devices that were operated by battery power. Her husband took car differentials and cut them in half to fabricate that stuff, and he made a good profit out of it. He would take some of the people that didn't have the money to open up the mine and they'd go to the landowner and get the permit, and this operator would pay that man, in them early days, 25 cents a ton, and they would pay him 20 cents a ton until they got that equipment paid for. My wife's brother was an operator and he did that; he had two or three small mines going. He didn't make a lot of money, but he made more than he could have made working for the big operators.

"In 1953 when I was working in the mine, they sold the coal for $4.50 a ton. They paid the landowner 25 cents a ton, and they paid the coal miners a dollar each. And he would pay his electric bills, and buy the dynamite he needed to shoot. So he could end up making maybe 50 or 60 dollars a day.

"I was 17 years old, so I had to work with an experienced miner and we got 90 cents a car* for drill, shoot and load, and the two of us together would get about 10 cars, so we divided $4.50 apiece.

"My Dad worked for a larger company and was a union miner. I would say probably my Dad made 12 to 15 dollars a day. Dad's company was part of Pocahontas called Pond Creek Pocahontas Coal Company; it was later called Allen Creek and eventually consolidated with others and took the name of Console, which is probably the third or fourth largest in the country. They have several coal operations that are now shut down, mainly due to low demand.

"In my growing up years, there were three things you could do to make a living. One, you could work in the forest to cut timber. Second woulda been, you coulda worked on the sawmills. We had probably, reasonably close to us here, five or six sawmills. Next would have been the coal. That's the only thing we've ever had in this area where you could make a living. Nearly everybody had a small farm; most of the coal miners had a large garden crop. Where we lived on this ridge, we had seven acres and seven children. We had large good crops of potatoes, tomatoes, good green beans and cucumbers. We sold those. That's where in the summer we would make enough money for school supplies in the fall, clothing and shoes and whatnot. What we would call farming and raising was mostly for the benefit of feeding our cows, our chickens, and then we used corn for our cornmeal. We had it ground for our bread.

---

* Carload of coal.

"Looking back now, it would look like we were poor, but the way we worked and had that small farm . . .—but my Dad's daily wage was misleading, because he hardly got more than two or three days' work in a week, because of slack runs on the railroad, and a lot of downtime because of union strikes. Back in those days they were just getting the unions organized good, so they were often on strike for safety reasons."

"On that subject, why are all the UMWA* offices closed around here?"

"I would say that that the biggest reason is, they have depleted a lot of the coal, and others are down because they don't have many men working. Back in the early days, they used to have so many men working that they used to keep the office open all the time. They'd keep a secretary . . . Now they have so few members that they don't pay the dues and they don't have the offices open. Now, there's a union office in Buchanan County, next county over, and they've rented out the top floor for a dentist office and part of the bottom floor for a drugstore."

"When you were working in that mine, how did you feel about the unions?"

"My Dad was a strong union man, helped organize a lot of what they call the local. The company where I worked was not union, so I didn't really have to say about it."

"Was he disappointed that you did non-union mining work?"

"Well, he was, but it was a small mine, and they never could have paid union wages. Now, those larger companies probably got a lot more than that $4.50 a ton for steel coal or for steam coal in the power plants, and we have both types of coal here, and the best coal here is small seams, and back in that day it was really hard to mine it, but now they only need about 36 inches to get a machine in, and if the coal is about 20 inches, they cut it all together. That's the best coal; it's metallurgical coal; it has low sulfur, high BTU. The coal we were mining, it was metallurgical coal, and the small operator we worked for, he sold it to a broker, and then they sold it to a larger steel company."

"What proportion of the miners are small operators today?"

"I don't know of any. With the regulations and the laws that were passed, the small operator just couldn't cope. My wife's brother, he had a good small mine going, and in the '70s, when they made some new laws, one was that you had to pipe water into the mine and keep water spray on it as it was being mined, to keep the dust away, and to run those pipes in, and where were you gonna get that water? If you had a small stream, you could only take only so much, because the people downstream, you might cut them off. So that was a problem, and the other was, you had to be free to hire both men and women in the mine, and not discriminate,

---

* United Mine Workers of America.

and the law said, you had to provide toilet facilities for both, and even if you didn't have a lady working, you had to have provisions for a lady. Well, this scared my wife's brother for good. So he sold his equipment and his mine, and he went out and took retirement. A lot of people followed that same kind of pattern.

"A man, he really works more closer to Welch, named Eddy Asbury, he was a small operator, and with his management and skills, he got some leases on a small seam of coal, and then a larger one, and he's still in business today, and he's known to be a good businessman, and he's non-union.* Matter of fact, I'm not familiar with any union operation in this county. In this state we've got right-to-work laws, because in those old days, they were violating the law. Some people were referring to them as bootleg mines. If I'm not mistaken, when you're going out of Welch toward Bluefield on U.S. 52, one of the two large operations is his, and the other is a Russian owned company that had got behind on their taxes, and the sheriff in McDowell County had to file suit against them, and had already issued a summons to that company to confiscate their equipment and sell it. That was just a week ago."

"Were miners better off in old days?"

"The owner probably was better off. But the people working for him was just by the skin of their teeth. Now I was not married and only 17, and neither of us partners was married, so we thought we made good money, and what we had in the mine, we did have an electric drill; that was a step up from what they had recently; they used to have a breastplate with a big hand drill. So the electric drill was so much better. Still though, we would drill, and would shoot the coal, and we had to go to another place we could go when the smoke was coming out when we had blasted the coal with that dynamite, because that smoke had a strong odor. Well, the two of us, when we would get 10 cars of that coal, far as we were concerned, that was about the end of the day. But some guys were much better coal loaders than we were. It was just a hard job. That's why when I went in the Army, it was a good thing for me. I got 72 dollars a month."

"Did the darkness bother you?"

"Not much. I was never fearful in the mine, but we didn't have a lot of light. We had a small light that fit up on our cap. They charged it all night, so we could use it through the day. As long as you were busy and working, you seemed to not pay a lot of attention to it. My wife's other brother, he never got used to it. He would have nightmares about the dark closing in on him."

"And you thought the owner did better than you?"

"Well, we were so few working where I worked; the owner, he was just like one of us; he didn't ever do anything that would make it hard on us."

---

* Mr. Asbury declined to be interviewed for *Carbon Ideologies*.

John E. Amos coal-fired power plant, Nitro, West Virginia

I'M NOT DENYING THAT WE DON'T HAVE A PROBLEM

1: The Unconscionable Omission

I n January 2014, responding to our President's State of the Union address, West Virginia Senator Joe Manchin, best friend of America's best friend, *pointed out that Obama did not mention coal when addressing energy, which Manchin described as "unconscionable."* This word must have rung harsh from the mouth of one whom *The New Yorker* described as *an affable, drawling figure in his sixties with a thick brush of gray hair;* for in due time he added: *I'm not denying that we don't [sic] have a climate problem that we've all been a part of contributing to or that we don't have a responsibility to make it better and cleaner.*

But, reader, if *you* denied that we have a climate problem, how could I have blamed you? Maybe in 2042 or even 2029 I might have sadly shaken my head, but back in 2014, a jubilee sort of year—for American energy consumption had nearly tripled since 1949—how could we have begun to see climate change for itself, much less perceive the degree of coal's agency in our undoing? My friend Priscilla informed me that on her property in Maine she was discovering plants she had never before seen at that high latitude, but couldn't their presence indicate some temporary and local "natural" trend? In Nitro, West Virginia, a plump middle-aged woman stood on her lawn while the John E. Amos Power Plant gushed cotton-white coal-smoke right behind her, and she assured me: "Well, I been here all my life and it never bothered me." I had expected that smoke to reek, but I could barely taste it on my tongue. Where then lay verifiable danger? Down in Becco I passed a great cylindrical vinyl pool and three blond children splashing in it, a coal elevator across the road behind them, and their summer went on with creaking and coal-smells, clattering on, that smell of fresh coal almost like asphalt but not quite; so many times I smelled coal from those long squeaking trains that were heaped up so high with the black stuff which sometimes went invisible in the forest shade, and what I mostly seemed to see was *life!* On a washed-out road in Man, looking up at the newly blasted cliffs (for a road, not for coal), I watched buttercups and ivy under the clouds of evening, and nothing seemed unnerving.—That night in Miyako Oji when the taxi finally arrived at the illuminated warning signs on the dark and empty highway whose median line curved a few paces farther ahead into the forbidden zone, up to the place where the glowsticks and that one streetlight and our headlights lost the power to make it visible, I could not resist believing in the boundary before me, however arbitrary it might be: This marked the inner ring. Over the next 20 kilometers, radioactivity would increase to dangerous and then lethal levels in the central ruin where the fuel rods fiercely decayed; after that, it would fall off for another 20. Were Reactor No. 1 to melt down right now, would we know, and could we escape in time? These considerations, compounded by the loneliness of the abandoned village not far behind, made us believe in our peril with nearly the same conviction as if we had come upon a scattering of corpses. The dosimeter turned over another digit. But without a dosimeter and warning signs, and *with* lights and people back in Miyako Oji, what problem would anyone need to deny? And in West Virginia what problem could there possibly be, but not enough coal?

Once upon a time, a panel on "environmentally benign manufacturing" dispatched members to three different sectors of the world, where three moderately different carbon ideologies operated. I summarize the panel's findings in the following table:

COMPARATIVE CARBON EMISSIONS, ENERGY USE, MANUFACTURING EFFICIENCIES AND ATTITUDES ABOUT INDUSTRIAL REGULATION:

Japan, Germany and the United States, *ca.* 1995–97

in multiples of Japan's per capita CO_2 emissions

1

Japan

"A greater sense of shared values concerning the environment . . . compared to the United States." "Of the three regions studied, Japan appears to have the greatest concern with CO_2 emissions and global warming." "Emphasis on recycling . . . between that of the U.S. and the EU."

Per capita carbon dioxide emissions:	**9.3 metric tons**
	(10.25 U.S. tons)

Per capita commercial energy use: 166.58 million BTUs
Per capita manufacturing energy use: not given
Per capita industrial water use: 578 cubic meters
(20,412 cu ft)
Per capita municipal waste: 400 kg (880 lbs)

1.13

Germany

"More collaborative relationships between governments, industries and universities ... than in either Japan or the United States." Some Green Party politicians in power. "A world leader in life cycle assessment" of manufactured products "and the integration of LCA into business practices." "New environmental directives were not met with the same level of skepticism that one would see in the U.S."*

Per capita carbon dioxide emissions:	**10.5 metric tons**
	(11.57 U.S. tons)

Per capita commercial energy use: 172.57 million BTUs
Per capita manufacturing energy use: 35.07 billion BTUs
Per capita industrial water use: 1,865 cubic meters
(65,863 cu ft)
Per capita municipal waste: 400 kg (880 lbs)

2.15

United States

A "traditionally adversarial relationship between U.S. government regulators and industry," compounded by "the litigious nature of the U.S. society." [In this connection you will presently meet upstanding members of West Virginia's defiant "regulated community."] Environmental advances often "linked to concern over future liability." "Protection of ... air and water ... appears to be equal to or better than Japan and Europe." But generally worse than they on government, industry, R & D and educational actions.

---

* These assessments all referred to the European Union countries visited by the research panel, not to Germany specifically—although Germany was among the EU countries studied; and these were described as if their carbon ideologies were relatively homogenous.

Per capita carbon dioxide emissions:	**20.0 metric tons** (22.05 U.S. tons)
Per capita commercial energy use:	329.40 million BTUs
Per capita manufacturing energy use:	50.23 billion BTUs
Per capita industrial water use:	5,959 cubic meters (210,442 cu ft)
Per capita municipal waste:	720 kg (1.584 lbs)

Source: *Journal of Cleaner Production,* citing various organizations, with calculations by WTV.

Yes, in comparison to Japan the United States emitted more than twice as much carbon dioxide *per capita;* excreted almost twice as much municipal waste *per capita;* used more than 10 times as much industrial water *per capita;* I could not refrain from those triple italics of disgusted hopelessness once an intimation of my far more populous nation's *absolute* ecocidal accomplishments occurred to me.

On the other hand, the fact that the U.S. per capita consumption of commercial energy was nearly twice as high as Japan's[*] could have been interpreted optimistically. What if that meant we lived twice as well? Bleed this question of any sarcasm: Our houses were bigger; our food and transportation were cheaper. Best of all, we kept the lights on.

THE ART OF LIVING WELL: AN AMERICAN EXAMPLE (1994)

From an industrial encyclopaedia: *In the United States, ca 600 kg of coke[†] is used to produce a metric ton of steel. Japanese equipment and practice reduce the requirement to 400–450 kg.*

On this subject I turned again to Mrs. Keiko Golden, intelligent, first-generation Japanese-American, and frequent traveller between San Francisco and Tokyo. She told me: "In Japan, you have to heat each room separately. You

---

[*] In 2010, which is to say in the year of the Upper Big Branch accident and the year before the Fukushima accident, Japan's per capita energy-related emissions remained nearly the same as in 1995, at 8.97 metric tons, Germany's had fallen slightly to 9.32, and America's had declined the most (but not enough) to 17.31—still about the double of Japan's.

[†] Depending on the amount of volatile matter in it, 1,000 kilograms of coal yielded between 620 and 750 kilograms of coke. Therefore, the American steelmaking process required *fully twice as much coal as its Japanese equivalent*—which of course entailed releasing twice as much greenhouse gas.

actually heat up only the room you are in. Air conditioners also. Material life in Japan, people try to be modest, and they have not so many goods, but much *quality* goods. Just even the nail clipper, it's much much better! You can keep it a long time, so they don't waste resources making another one. Vacuum cleaner, here it's so heavy and suction isn't really good. In Japan it's two pounds, not 20, and the price is half the price here. So, in order to make a Japanese vacuum, you use less resources. And power the vacuum needs is less in Japan. In Japan, they don't use dishwasher. Plates are small, so they're easy to wash. Here it takes more time, because American plates are larger."

"Then you wish that you had a dishwasher?"

She giggled and said: "I wish I do!"

Who could blame her? She cooked for three . . .

Now let me bring you to the very American place called West Virginia, so that you can decide precisely how good those people had it.

COMPARATIVE USE OF COAL FOR ELECTRICITY GENERATION, 2013–15

Japan (*ca.* 2013)	28%*
EU-15 nations (*ca.* 2013)	46%
U.S.A. (2014)	39%[†]
U.S.A. (2015, projected)	36%
West Virginia (2015)	95%

Sources: *Greenhouse Gas Inventory, Japan*, 2015; EU greenhouse report, 2014; *Coal Facts*, 2012; *Charleston Gazette*, 2014, 2015.

* Of fossil fuels only—all that was stated (and very likely almost all the fuels in Japan in 2013, with the nuclear plants shut down).

[†] In 2014, 82% of all BTUs spent in America derived from fossil fuels.

Yes, I had felt uneasy in Miyako Oji. But on the equally deserted road between Van and Twilight, after the other bristling self-announcements of the Talon Load-out Company—seven or eight miles from Montcoal's Upper Big Branch mine, in the depths of which, 11 months before Fukushima, 29 miners had perished in what investigators called not only *the most deadly US mining disaster in 40 years* but also an entirely preventable explosion—the weedy railroad ballast offered for contemplation a **VENDOR ENTRANCE** that appeared to also be an **AMBULANCE ENTRANCE**; and this seemed less sinister than merely thought-provoking. After all, the 29 had not died exactly here at Talon; nor did their calamity endanger me here and now; the methane fireball had burned out; as for climate change, *well, I been here all my life and it never bothered me.*—So why compare Upper Big Branch to the continuing Japanese catastrophe?*—Because those who sacrifice people now will squander a planet later.—Japan had seen fit to shut down her nuclear reactors, if only for awhile†—while Upper Big Branch's closure was also temporary—for there were lights to keep on, its Eagle Seam coal being of the highest quality "met-allurgical" type. A security chief and a superintendent went to jail; strange to say, a subsidiary's President eventually served time (and published a booklet naming himself a "political prisoner"); an assistant underground foreman surrendered his license for three years without admitting to anything—and five years after those 29 men died, a special panel determined that conditions in many mines remained the same! One inspector proposed that this fact ought to "really send chills up peoples' spines," but it didn't.

---

* If you would rather, please feel free to compare it to the Deepwater Horizon offshore oil disaster, which occurred in the same year as Upper Big Branch—although it took "only" 11 lives. See p. 171.

† The carbon ideologues at the International Energy Agency saw an opportunity, and their report from the first year after the accident concluded: "Japan has 40 GW [gigawatts] of coal-fired capacity, which operated with a 75% load factor in 2010. In addition to destroying the Fukushima Dai-Ichi nuclear plant, the Tohoku earthquake and subsequent tsunami caused substantial damage to around 7.5 GW of coal-fired capacities, of which 3.2 GW recovered by the end of July . . . Restoration of the damaged conventional plants is proceeding at an impressive speed. The coal-fired units damaged by the earth-quake are expected to be operational at some point. At current prices, their marginal cost will be substantially below *L[iquefied] N[atural] G[as]* or especially oil fired units . . . This could mean 3–5 Mt [million metric tons] of additional coal demand for partial recovery or 10–13 Mt with the whole fleet."

On a windy morning in 1849, Thoreau strolled past a horde of Cape Cod beachcombers who were *often obliged to separate fragments of clothing from* the seaweed which they were gathering, *and they might at any moment have found a human body under it. Drown who might, they did not forget that this weed was a valuable manure. This shipwreck had not produced a visible vibration in the fabric of society.* And why should it? Inertia is the universal rule. However many coal miners have been killed in West Virginia, which according to the Upper Big Branch official report boasted *the highest rates of fatal accidents and injuries in the country,* however many coal strikers and union men have been shot down by the big boys, however many dwellers in shady "hollers" have been drowned by the failures of coal slurry dams,* however many residents of Northfork, Welch, Bim, Charleston and Beckley might have sickened, lost their teeth or developed ulcers in consequence of drinking mine-polluted tapwater (for this last cause-and-effect chain, see pp. 152–53), I fear that in the long run none of these various known, forgotten, unrecorded and hypothetical casualties *produced a visible vibration in the fabric of society.* Consider the Jed Mine coal dust explosion of March 26, 1912; it destroyed 81 miners.† No doubt there was horror, sympathy and talk in plenty then—a true "vibration," if you will. But *many of the victims were foreign-born, and when no one came to claim their bodies, they were hastily buried in a potter's field that the gravediggers called "Little Egypt."* I have never been there; evidently it lies in what is now Havaco. *The potter's field remains 100 years later, unmarked, unattended, and essentially unknown to most people in McDowell County.*

In McDowell County alone, how many miners have been killed, and who remembers? Between 1902 and 1958 the catalogue of catastrophes reads Algoma, Lick Branch, Second Lick Branch, Bottom Creek, Jed, Yukon, Second Yukon, then another Yukon which inexplicably is not called Third, followed by Bartley, the above-mentioned Havaco, Bishop and Second Bishop: by my count, 441 lives smashed, crushed or suffocated away.

I have read that more than 100,000 American coal miners were killed on the job in the 20th century. (As you will see, a prominent strategy of West Virginia's carbon ideology was not to hide such casualties, but rather treasure them up into flashes of triumphalism.) How many miners' neighbors also became coal's victims is unknown to me; perhaps it was never totted up. One instance: In

---

* Slurry (or sludge) is the thick dark liquid created when coal is washed in order to reduce the pollutants it will give off when burned. See the next page.

† Coal dust appears in an Army list of "commodities useful for making improvised explosives" (but so do powdered coffee and confectioners' sugar). "A glance at this list will show the great complexity of the problem to security forces who must deny these to insurgents. The strictest possible control of their purchase and sale is imperative at the earliest possible moment."

Logan County, on February 26, 1972, two coal sludge dams* constructed by Pittston Coal gave way on Buffalo Creek. Of the 5,000 residents of that area, 4,000 became homeless, 1,121 were injured, not to mention 125 killed.—A retired mining inspector told me: "It could have been prevented, in my opinion, if they'd had stricter regulations. When they come up with this 1977 Coal Mine Act, that made it a lot stricter. But as far as surface impoundments and these things, a lot more could be done. We got 'em here in Harlan, Kentucky, and they ever get turned loose, it's gonna wipe out everything down the river."

Wondering how Pittston's little mistake might be remembered, I decided to go and ask. They'd named that narrow tree-hung road the Buffalo Creek Memorial Highway, and under a wooden awning in the town of Kistler stood a lectern or pulpit with a white ceramic pair of praying hands lying atop it; just outside was the granite slab IN MEMORY OF THOSE WHO DIED IN THE BUFFALO CREEK FLOOD FEBRUARY 26, 1972. So it was a flood; an act of God; that was how they'd pitched it.

* As one survivor put it, "they had built two more dams in behind the first dam . . . They say these two dams let go first and the dam in front could not hold the water."

Mrs. Glenna Wiley, aged 89, said: "I was right here. Well, I was going to work, and one of the girls who was going, she stopped me and said, you can't go because we had a flood. I said you don't know what you're talking about, but I went back to the house, and people were calling and trying to get ahold of their loved ones. I hope I never see anything like it again. It got out in my yard, but it never did get up here, thank goodness. I just opened my doors to anyone that wanted. We just cried and prayed about it and then made some food . . ."

"I'm sorry," I said.

Mrs. Glenna Wiley

"You got to have strength to go on, just not to give up. We lived up on the hill there. I thought, you know, if I go down to the store and call the owner, I would go up on the hill and beat the flood. Right out here it didn't come to too much force. We saw people's refrigerators, stoves, all kinds of things. People didn't know where to stay. They were looking for their families. They were scared to death."

"What do you think caused it?"

"Honey, I guess it was the dams they turned loose from coal mines. People had predicted a long time ago, and they just didn't pay attention to what they heard."

"Was it the coal company's fault?"

"Well, they say it wasn't but we don't know what to believe. Somebody said the company added too much water to the dam. I just know that it was a lot of water and a lot of people killed. My husband's father, he washed all the way to Accoville, which is about five miles . . ."*

"How would you sum up the pros and cons of coal?"

"I am all for coal, 100 percent, because there are people living here that have given their young lives to coal, 100 percent, *amen*! It's dirty, but it's powered our country for ever so long."

"Do you believe in global warming?"

"Honey, I don't know."

"Some people say that fossil fuels are making the planet warm up, and some say there isn't any problem."

"You know," said the sweet old lady, "I just really don't know. Just don't know what it's all about."

Her mother used to cook on a coal stove, which would get so hot that she had to put a screen around it to keep the children from getting burned. No one she knew burned coal at home anymore.

Coal stove, Kentucky Coal Miners' Museum

Her late husband had been a good Christian man. He'd worked with heavy equipment for the coal company; the widow said there'd never been a machine

---

* Man lies at the intersection of Highway 10 and the Buffalo Creek Memorial Highway (CR 16). Kistler, where I met Glenna Wiley, is about two miles upstream from there, and Accoville about the same distance farther, with the tinier town of Crown in between them. Upstream from Accoville are Braeholm and Becco, with maybe a mile between each, and then Amherstdale, which is as large as Man and Kistler, then Robinette, Latrobe, Crites and Stowe, all little places, then Man-sized Lundale, then Craneco, Man-sized Lorado where the flood supposedly happened, where I interviewed Karen Elkins; actually, as you will see, the coal slurry dam was a trifle upstream of there. If Mrs. Wiley's husband's father "washed all the way to Accoville," he must have been taken nearly in Lorado.

that he couldn't operate. Someone had stolen his photograph, but she'd gained it back, and it hung on the wall.

Farther up the road, with Buffalo Creek still on the right and Buffalo Mountain on the other side of that, the last northeast tip of Logan County narrowing between Wyoming County on the south and Boone County on the north, one reached Lorado, which lay very near the exact place where that coal slurry had exploded out of the hills, and here Ms. Karen Elkins was sitting on her porch. She said: "It happened in February and I turned 13 in March. We lived up here, the second house up, so we were in the flood: my Dad, my mother, and I had two brothers and a sister . . . I can remember rainy nights when they would holler the dam was gonna break, and a lot of people just got to the point they didn't believe it. But my Dad would come home many a morning from work (he worked for the Amherst Coal company); and he would have to get us and carry us away up in the holler just to keep us safe. The night before it broke, we had went to visit my grandmother. Then we heard on the news that something had happened on Buffalo Creek. So we tried to come around Kelly Mountain another way. We went to Sharless* and saw my Dad's family that was there. We went back to my grandmother's, and I went to a school at my cousin's for a couple days, because they wouldn't let us back here no more. We stayed at Man Hospital for 21 days; there was 21 of us in one room. We kept watching the deceased list. We had three of our relatives on it: the baby, a little boy, and the mother; they never found the little boy; they buried the mother and daughter together. It was weeks before I would come back here. We lost several neighbors that were around us, and we lost our cousins; we lost everything we had; our house just splintered. Our neighbor, they found him in the mud face down; they called him the Wonder Child: face down in the mud, and they heard him cry! Every time it rains, even now, and I'm 57 years old, I want to get under something, and I'm scared to death."

"What caused the accident?"

"I think they blew the dam up because they was trying to release some water."

"What did they say about it?"

"I didn't have much dealing with 'em afterward, being just a kid; I think Dad got some compensation, but just barely got enough to put our house back.† All the terror and the horror that we went through, it was awful."

"Are any of those dams left?"

---

* Local spelling; not verified.

† She said that her father would know more and remember more; his was the white house past the cross-street. Slowly, weary and pallid, in suspenders and white clothes he came to the door and said he didn't want to talk about it.

"Not that I know of. I think they quit putting them up here."

(But as that retired inspector in Kentucky had intimated, the dams in fact remained in many parts of Appalachia, and they could fail at any minute.)

"Ours was so big," she said then, "that people could drive a boat on top of it. I never did; I was too scared."

I told her: "I'm very sorry for what you've suffered."

She said: "The pain that hurts me the worst is the friends that I lost. I can look up the road and tell you who each one of 'em were. You can't put that back."

"How do you feel about coal?"

"Coal is a wonderful thing. In my heart I don't think we could make it without coal. What happened here, it wasn't the coal's fault. It was the people trying to mine coal; I think they just didn't know what they were dealing with."

"And whose responsibility was Upper Big Branch?" I asked then.

"I couldn't really comment on that. I just remember what I lived through up here."

(On that subject, here is the title of a certain newspaper editorial written by two officials with expert knowledge: **Three years after UBB: More miners have died, promising tech is unused, safety ignored.*** Indeed, the official report on the explosion contains this bitter aside: *We have seen similar reports, written with the same good intent, gathering dust on the bookshelves of the national Mine Health and Safety Academy.*)

One of Karen Elkins's relatives had now come out to the porch, and instead of sitting down beside me he stood looking me over, so it was near about time to shove off. Instead of raising the subject of global warming, I asked what the Pittston dam looked like nowadays.

"You see," she said, "there's a big coal company up there now, so it's closed off. They just put one I & R† in four or five months ago. You can see the loadout, and then the big mines."

And she was right; just up the road was the Powelton mine (Greenbrier Minerals, Buffalo Energy Division), and as I listened to the steady machine moan and the hissing of water and coal there (an extremely safe 52 radioactive counts per minute, 0.18 microsieverts per hour), with cumuli in the soft blue sky, and my sweaty bluejeans sticking to my thighs, here came a long, long train heaped with open cars of steam coal.

---

* One of the piece's findings was shocking. Although black lung disease has been understood for decades, and can be prevented with masks, on this subject the dead men at Upper Big Branch testified as follows: "Autopsies showed that, of the 24 victims who had sufficient lung tissue to examine, 71 percent had evidence of black lung, including men in their early 20s . . ."

† I never learned what this meant.

The Powelton mine

Thus Buffalo Creek, whose vibration of grief and terror continued to hum in the hearts of those two women, but whose *visible vibration in the fabric of society* had not sufficed to make coal slurry dams safe.*

2: "I Can't Figure Out What Happened That Day"

Since that disaster had happened 43 years ago, while the explosion at Upper Big Branch dated back only five years, I thought to learn whether the latter *visible vibration* might still be thrumming to any good purpose. Thrumming it

---

* See below, pp. 129, 162–63.

certainly was; I heard it mentioned again and again by West Virginians. Outside of Appalachia, of course, it remained as forgotten as Fukushima.

"I guess you all heard about them 29 guys that got blew up," said a former coal miner named Mark Mooney. (You will read more about him later.) "Happened right here at Bandytown. Preacher's the one who said, that smells like the blasting agent we used to use in the mines! And we went by there and there was that black smoke. One of my best friends was in there: Mark Elwick.* It was his second day. That's why I think the union mines is a lot safer. Well, the young guys working non-union, they won't say nothing, so they can keep their jobs."

There were several monuments to victims of Upper Big Branch. One was a dead son's shirt, flown like a flag by the resolutely grieving father. I have read about it but have not seen it. Another was the site in Whitesville. *The roadside memorial plaza includes three distinct points of interest and stands as a solemn reminder of the human cost that West Virginians have so dearly paid to power this great nation.* In other words, they kept the lights on.†

To see that brand new memorial one took Highway 3 through Racine, Bloomingrose, Maxine, Comfort, then Seth, whose pastor wrote my favorite anti-humanist column in the *Coal Valley News;* winding through the lush hollers ("some of the people who live in these wretched little trailers," said my local driver, "still have a view you'd pay a thousand dollars for"), one reached the hamlets of Kirbyton, Fosterville and Coopertown; Orgas, whose name was printed larger in the atlas than the preceding three but didn't look bigger in real life; minuscule Keith, then Sylvester with its many **God Bless America** and **Heart of Coal Country** banners hung from power poles, not to mention the Round Bottom Mine and the Church of God of Prophecy; on the edge of that town there also happened to be a long elevator crossing three chutes which slanted in parallels against the summer trees; this was the Elk Run Mine, owned by Alpha Natural Resources, which had bought Massey Energy shortly after Upper Big Branch, and was now itself about to file for bankruptcy because, as the newspaper explained, *utilities are switching to natural gas and coal prices are plummeting;* and after flashing through Elk Run Junction and Janie at intervals of a sneeze apiece, one came into Whitesville, still accompanying that same winding sluggish green-brownish jelly of a river, the Big Coal.

---

* Mr. Mooney's dialect and mine were sufficiently different that I heard "Elliott." According to the stone memorial at Upper Big Branch, Michael Lee Elwick was 56 years old when he died.

† Who do you think paid for the monument, and why? I quote Sheila Combs, President of the Upper Big Branch Mining Memorial Group, Incorporated: "We are committed to the preservation of the region's coal heritage and the further development of coal related tourism in southern West Virginia." Hopefully those dead men would earn her incorporated group a few bucks.

Among other grand things, the memorial said: Tuesday, April 13. By early morning, the recovery effort is complete. The last mine rescuers on site cover the entrance to the UBB mine in a large American flag—a final homage to the men and the sacrifice they made for the country. That almost implied that their sacrifice accomplished something.

It was a hot Sunday afternoon, and on the riverward side of the wall where the dead men's names were stood a quiet couple, the wife appearing older than her husband, who was a slender, trim, greyhaired man of apparent middle age. Learning that he was retired, I told him that he looked good, to which his wife said: "He's fine on the outside, not on the inside. That coal mine just breaks you down."

He had worked at Upper Big Branch some years before the accident. As he put it, "I opened that mine." I think he was another of those people who loved coal because they had lived it.

Why did coal extraction retain its defiant cachet in Appalachia? The loyalty that coal excited was unparalleled among the various carbon ideologies.*—No Japanese reactor employees ever in my hearing described nuclear labor as their *heritage.* When I asked those Tepco public relations men, "If there were only one energy source in the world, what would it be?," I imagined that they if anyone would be single-resource enthusiasts.—Not hardly!—"In order to minimize the risk of increased prices," they replied, "we have to procure different types of energy from different countries. So there is no one answer." (The West Virginia Coal Association would almost certainly have proffered a unitary solution.)† You have seen how those atomic workers often lived in the same village, be it Kawauchi, Tomioka or Iitate, and even in the same house where their parents and grandparents were born—but none of their grandparents had devoted their labor to reactors—whereas coal goes back and back and back, like a dark seam disappearing into the mountain! Thomas Jefferson wrote in his "Notes on the State of Virginia," *circa* 1781: *In the western country, coal is known to be in so many places, as to have induced an opinion, that the whole tract between the Laurel mountain,*

---

* Harold Bonnett, 1965: "If there is a Heaven there will be English fields of long ago in it, and a little steam plowing done here and there." His little steam plow was powered by coal.

† But the Vice-President of the West Virginia Coal Association told me: "I don't see where we are as dependent on coal as a lot of people would think. Regionally, in three or four of our major coal-producing areas, since as Mingo, Boone and Logan, we're going to find a dependency, but we're about 15% of the state's gross product, not 70 or 80%. We only mine coal in half of our 55 counties. The Eastern Panhandle, when you go over there and talk about coal, or the north central part of our state, there are no coal jobs over there. Now they get coal severance money over there. We've become quite dependent on gambling in the last 10 years or so. I think proceeds and revenues are dropping off a little bit there . . ."—His name was Chris Hamilton. He will come in and out of this chapter.

Missisipi [sic], and the Ohio, yields coal. That opinion was acted on, generation after generation. When I was alive, a Congressman invoked his grandfather who had perished in a coal mine, and the Secretary of State *spoke of her roots in the coalfields of Northern West Virginia.* Good old Governor Tomblin liked to say in just about every speech (in one address he said it twice): *Growing up in the heart of coal country as a proud son of the West Virginia coalfields . . .* The President of the West Virginia Coal Association was correct when he said: *Not only does our coal make the steel and "keep the lights on . . ."[;] we turned them on in the first place.* Coal grew comforting and good. And if it sometimes seemed to me that West Virginians' elegiac memories of coal got conflated with memories of what coal had taken from them, well, could I call that a mistake or indeed anything but human?

Coal House, Williamson, West Virginia. It is built mostly out of coal.

"I opened that mine," that retired miner proudly said, and when I asked him how he would characterize the management of the company—Massey Energy, which the United Mine Workers accused of "industrial homicide"—he replied: "I had no complaints."

The Mine Health and Safety Administration felt otherwise. The President of Massey's Green Valley Resource Group was one of several who tipped off the mining crews whenever inspectors showed up; they played it safe until the inspectors went away. (He finally pled guilty to two counts of conspiracy.) In the year before the explosion, Upper Big Branch got written up for 515 safety violations, which I am told was *nearly twice the national average.*

Nor did the retired miner fault Massey's chief executive officer, Mr. Don Blankenship, who was finally about to go on trial. But a certain lawyer *who has fought him in court for years* asserted: "His legacy is a Mingo County with its coal ripped out, its mountains destroyed, and the people without jobs. He's walked away with a quarter of a billion." Blankenship's money helped elect a West Virginia Supreme Court justice who then participated in a judgment against a small coal operator whom Blankenship had ruined. The U.S. Supreme Court faulted that judgment for denial of due process; as for the loser, he dryly said: "Having two of the votes that go against you come from, one, a close personal friend of Don Blankenship's, and two, a justice who received the benefit of Don Blankenship's $3 million spending spree on the Supreme Court race certainly gives me pause."

Blankenship himself took pride in his political insertions. In his 67-page memoir he wrote: *In 2004 I began using my own money on efforts to improve the political and economic landscape . . . West Virginia Supreme Court Justice Warren McGraw was my first target because he was an activist anti-business Judge.* Well, we couldn't have that! *I also involved myself in other state elections. In 2006 I led an effort to put a Republican on every ticket in the West Virginia House of Delegates races. Previously, many Democrats ran unopposed.*

And so Blankenship dug one success after another out of the mines and the electorate of the Mountain State, right up until 2010, the year before Fukushima.

In 2010, Massey Energy produced 34 million tons of coal—better, to be sure, than the 28 million tons mined by semi-moribund Patriot Coal (whose callous villainies to its retirees will be mentioned in this volume), but considerably below the 77 million tons of Alpha Natural Resources, which soon would acquire Massey, and far below the 198 million tons of Patriot's parent, Peabody Coal. In fact Massey was a big fish in the American pond, but small in comparison to many other players in the international market. The world leader, Coal India, dug 431 million tons that year—13 times more than Massey. In the International Energy Agency's list of the 30 "leading coal companies," ranked in decreasing order of productivity, Massey was only 28th. Therefore, much more money could be made!—And Blankenship was determined to make it. Although he did sometimes invest in safety equipment—one director of mining engineering credited Blankenship for the development *many years ago* of *a reliable proximity detection device,* which *MSHA has only recently required*—at Upper Big Branch, he allegedly insisted on productivity over safety, even when the mine had a deficit of air, a condition which apparently led to the fatal methane explosion. The official report concluded that *illegal ventilation changes . . . became the norm at UBB*.

After the fact, Blankenship tried to turn the tables:

> MSHA ventilation specialist Joe Mackowiak . . . bragged in an email to his superiors at MSHA and the Department of Labor that he had picked a fight with the company by denying their ventilation plan . . . If MSHA had told the truth about the 3.5 million cubic feet of natural gas exiting the mine after the explosion, the press would likely have challenged their dust explosion story and uncovered the fact MSHA had reduced the airflow by 50%.

Why the MSHA would deliberately stifle coal miners I never quite figured out. After all, as a famous muckraker once concluded: *The business of a coal-operator was to buy his labour cheap, to turn out the maximum product in the shortest time, and to sell the product at the market price to parties whose credit was satisfactory. If a concern was doing that, it was a successful concern; for any one to mention that it was making wrecks of the people who dug the coal, was to be guilty of sentimentality and impertinence.*—Ninety-nine years passed, and here I stood, admiring the memorial's pretense that the miners had died for something wonderful.

Blankenship wrote bitterly: *This Justice Department is intent on turning industrial accidents into crimes . . .* I grant that he had not caused the explosion on purpose. But he required production reports every half-hour. He fired one employee who refused to narrow the mine's support pillars below standard industry thickness. After the explosion, he and 17 other executives from Massey *or its subsidiary that ran Upper Big Branch* exercised their Fifth Amendment rights against self-incrimination, refusing to answer questions of *several state and federal investigators.* In 2010,* Alpha Natural Resources (*one of the largest and most regionally diversified coal suppliers in the United States*) took over Massey, and Blankenship got forcibly retired. Alpha presently established a "Running Right Leadership Academy" of safety. The boss of Running Right was a certain Chris Adkins, about whom the special report on those 29 dead men concluded: *History makes him a questionable choice . . . One need look no further than UBB . . .* Indeed, he was one of those cautious souls who took the Fifth and refused to be

---

* The transfer was completed in 2011.

interviewed about the explosion.—Well, well; innocent until proven guilty!—
Every time I went from Charleston to Madison, with dark green hills swelling
around the Big Coal and the Little Coal Rivers, each tree like the lobe of an
immense cumulus, I used to roll past Running Right Way and wonder which
safety tidbit might be on the curriculum that day. Of course Alpha ignored all
my requests for interviews. At that time they seemed to be doing pretty well;
their announcement of insolvency would be typically sudden. Just after Upper
Big Branch, West Virginia still produced half the nation's coal exports, and Alpha was running right, without a doubt: *With mining operations in West Virginia, Virginia, Kentucky, Pennsylvania, and Wyoming, Alpha supplies
metallurgical coal to the steel industry and thermal coal to generate power to customers on five continents . . .* In 2012, Alpha produced the third largest tonnage
in the nation. The brave investigative journalist whose book on Upper Big
Branch informs much of this paragraph concluded that through canny campaign
contributions and cunning lobbying, Alpha's management *have so far beaten
back the Robert C. Byrd Mine Safety Protection Act . . .* But don't worry about
that: *"The success of our company and industry starts with safety," said Kevin
Crutchfield, chairman and CEO of Alpha Natural Resources.* (He was the one
who when Alpha went bankrupt would speak of *transformational opportunity.*
His golden parachute approached $8 million. The plan was this: Alpha's corpse
would keep the marginal old mines, and a new company, coincidentally staffed
by many Alpha managers, would get the profitable mines.† The Interior Secretary
now *expressed concern* that such settlements as this might force taxpayers to pay
for remediation of abandoned coal mines, but an Interior Department spokeswoman consoled us very sweetly that Alpha's deal would constitute a *managed
route.*) And the investigative journalist continued: The moguls of Alpha *have also
fought* granting the MSHA *the ability to subpoena witnesses if a mine accident kills
three or more people. Most federal agencies already have such subpoena power.*—In
2014 a district court judge *ruled that two Alpha Natural Resources mountaintop
removal mines in southern West Virginia illegally polluted streams.*‡ In other

---

* Moreover, "our local rifle range at the Chief Logan Wildlife Management area recently got a tune up
of its own thanks to some used mining belt that was recently donated by Hobet Mining and Alpha
Natural Resources . . . It will make for some great backstop material to hold your targets."

† We will see this trick again (pp. 176ff.) after the spill of the coal cleaning chemical MCHM, when the
company responsible declared bankruptcy, ensuring that the fine it owed would never be paid, after a
which a new company sprang up with two of the same chief executives, on the old company's subsidiary's premises. Patriot Coal would pull a similar bankruptcy-and-switch on the pensioners it left in
the lurch (pp. 83–84).

‡ Promising to appeal, Alpha spokesman Ted Pile said: "Understandably we're disappointed that the
court disregarded the research of highly qualified experts."

words: *"They have been a great corporate neighbor and partner," said Boone County Economic and [sic] Development Director Larry Lodato.*

Alpha paid *"the largest fine ever for violations of water pollution permits."** Not long after that, the U.S. Interior Department bestowed on Alpha the "Good Neighbor Award."

Don Blankenship did finally get indicted—but his trial was postponed, first to January 2015, then to July. That was when I flew to West Virginia, expecting to see him in the courthouse in Beckley, a fine town whose Mexican restaurant served potato chips with mayonnaise in place of corn chips with guacamole; but the trial got postponed yet again, and so I went to Upper Big Branch just as Alpha was delisted from the Stock Exchange; a month later it declared bankruptcy.

In 2016, Blankenship began serving a one-year sentence at Taft Correctional Institution in California. In the memoir he issued from there, *An American Political Prisoner,* he offered the following proof that he had been railroaded: *It's*

noteworthy that the lead prosecutor, Booth Goodwin, is the son of one of the five Federal District Judges, Judge Joe Bob Goodwin. Judge Goodwin is also a former Chairman of the West Virginia Democratic Party. I guess that about sews it up.

The former mining inspector who had called the Buffalo Creek accident preventable was named Stanley Sturgill, and he lived in Lynch, Kentucky.

When I requested his summation of Upper Big Branch, he said: "I think it's terrible. I wrote that little letter to the editor about that.[†] What really gets me on that, some of the things that they were doing, they tried to blame it on MSHA.

---

* The judgment was "$27.5 million . . . in fines, $200 million for the cleanup," about which an Associated Press newswoman remarked, overlooking Alpha's antecedents: "So, obviously, other companies have paid big fines in the past. In 2008, the EPA settled with Massey Energy, another coal company, for $20 million."

† I think he was referring to his 2012 letter in *The New York Times,* part of which ran: "There may never be justice for the U.B.B. miners and their families. There will be whitewash and wrist slaps with short jail terms, but no real justice."

But it was just the way that they were ventilatin'. They weren't ventilatin' the way their plan was approved. They didn't build those seals the way they were to have been built. Blankenship, from what I read (I wasn't with MSHA at that time), the philosophy was: any *way*, any *how* they can mine coal, you gotta mine that way."

The official report said: *Extremely low airflow was a chronic problem in some parts of the mine . . .*

For his part, the retired miner at the Upper Big Branch memorial insisted that there had never been any problem with air, at least as far as safety was concerned—"but as for your own air, you have to take care of that."

He said that when the inspectors used to ask him who set the quota, he told them: "You see this section here? It depends on my section. If it's dry and the roof don't cave in, you can't hardly stop a miner! But if it's wet and the rock keeps falling down, well, that might be how your section is for two months.—And Blankenship understood that! That's why I can't figure out what happened that day.—I knew *him;* I knew *him,* and *him* . . ."—touching various of those 29 names on the memorial's black stone. (Ten of them died from blast injuries; carbon monoxide did for the rest.)

He used to make a hundred grand a year. His son, a fourth generation coal miner, of whom he was obviously proud, might hope to make 80 if the price of coal held up.* A melancholy smile came to him when he was telling me about coal.

He thought that coal would keep going away until two things happened. First, we had to get a better President into the White House—a President who would ask China to help us by buying a guaranteed amount of coal. (According to the International Energy Agency, *since 2002,* our planet's CO_2 *emissions from coal have grown by 4.7% on an average annual basis (or 492 million tons per year), 89.4% of which has come from rapid growth in the People's Republic of China.* But those happy words had appeared back in 2012. By 2014 China was already buying less than poor Alpha Resources, for instance, had projected.)—Second, he said, we'd need to accept the fact that it didn't advantage our planet one bit if we cut emissions when China and the other countries didn't do the same.

I did not inquire into his feelings about global warming. I declined to abrade, annoy or challenge him in any way, on account of his pain. After we had chatted a good 20 minutes his wife came to him; they said goodbye and went away.

---

* According to the West Virginia Coal Association, the average coal wage *ca.* 2011 was $68,500.

AND I WILL GIVE YOU REST

Detail from the memorial

That memorial was pretty showy, all right. I looked at it for awhile. It commemorated the sacrifice without laying any blame.

Farther up the road, just before the Montcoal sign, with the Big Coal River still on the right, and on the left the forest so thick over the dark and trickling wall of rock along the roadside that no one who had not seen it for real or on a map could have sensed Coal River Mountain beyond the lushness, there was another, simpler, more genuine monument: dark-striped orange miners' helmets* screwed into the tops of red crosses to a great metal lattice tower that supported the elevator. I am guessing that these helmets had never been worn by the men they signified; they were too new. Each brim was lettered with a name. On one of the lower crosses hung a helmet in memory of Cory Davis, and just below it something had been neatly knotted in a sash that again read **CORY**. What was it and why? I left it alone among cicadas and birds, humidity and the slow flowing of that translucent greenish river seen through the great green leaves; unseen beyond it rose Cherry Pond Mountain, whose divide demarcated the edge of Raleigh County. I saw a shoal of white stones, a concrete bridge labeled **AMBU-LANCE ENTRANCE BRUSHY EAGLE #1**, the great mine elevator crossing high over the bridge and then entering the forehead of a forest cliff, and not a single car on the road, maybe because it was Sunday.

---

* The official colors of Massey Energy were orange and silver.

HIGHLIGHTS OF DON BLANKENSHIP'S TRIAL,

as told [in italics] by media sources, 2014–16

- **November 13, 2014:** Blankenship is indicted.

    Jami Cash, daughter of a miner who was killed at UBB: *"It's the best news I've had in four years. I thought he was going to walk away with blood on his hands."*

- **November 18, 2014:** *Blankenship, 64, faces a total statutory maximum of 31 years in prison if convicted on all four counts.*

- **March 11, 2015:** *Under the new indictment, Blankenship faces three felony counts, and if convicted on all three would face a statutory maximum of 30 years . . .*

- **July 24, 2015:** Blankenship's *attorneys filed a motion . . . asking a federal judge to instruct jurors that the trial does not concern the explosion that killed 29 miners, its cause or who was responsible.*

- **December 3, 2015:** *Jury finds Don Blankenship guilty on misdemeanor charge, not guilty on felony charges.*

  > Judy Peterson, "family member of UBB victim": *"To me it didn't matter what they charged him to be guilty of . . . I just wanted somebody to stand up and say he is guilty and I know that history will link his name forever to what happened in that mine and that's the loss of 29 innocent souls out of his greed, his pure greed."*

- **2016:** Blankenship serves a one-year prison sentence. He was exempted from paying any restitution. *"You don't even feel like you're in prison when you're there,"* said Larry Levine, who . . . *now advises on how to serve time behind bars . . . "It truly meets the stereotype of a club fed."*

  Blankenship appealed. To Judge Irene Berger he said: "It's important to me that everyone knows that I am not guilty of a crime."

3: "What Would We Have Done Without It?"

Patricia Ann McNeely Wheeler, coal miner's daughter, replied slowly and steadily: "I think that it was not safe enough. They did not do what they should have done."

Thus Upper Big Branch. About Buffalo Creek she said to me: "Well, I was in Huntington at that time. We felt that the mines was not up on things. They did not do what they should."

She then summed up mining regulation as follows: "Even today with things that happen, as long as things are okay, if it ain't broke don't fix it."

"How often did serious accidents occur in your father's time?"

"Now, his best friend was killed in the coal mines, while Dad was over-cautious about everything including his family: Make sure to watch for cars when you cross the road—and back then we didn't get too many cars! Well, his friend went back into the coal mine, moseying around—he had no reason to be there—and a slate fall killed him; I was about 17 at the time."

"What was the general attitude about the big bosses?"

"I never heard my Dad say anything negative about the management. I would say it was a decent wage he got; he didn't make a lot of money, and at the end he would only work maybe two or three days a week, because the mine was playing out, and those times were rough, but when I was growing up he was making $125 a week. That was pretty good for the '50s. When the mine shut down, we left Logan and moved to a farm because my Dad did not want to be somewhere where he could not work."

"How do you feel about coal generally?"

"My first thought of coal has to be good because that was our life and it had been from my grandfather and my father, and I think my great-grandfather who came from Ireland worked in coal mines. My grandfather was the electrician for the coal mine, so they treated him like royalty, and I guess he really did a lot in the mine! He had to have a car and a telephone and everything; they really treated him well. My father, I don't know if he ever went really deep in the coal mines but he dug coal, which was a very, very dirty job, and from time to time he had to go into the coal dumping pit and clean it without a mask, and he did have black lung. He would come home at night and all you could see was his eyes and where he would lick his lips. During the war they worked twenty-four seven with a lot of money and no place to spend it; you couldn't buy tires for your car or anything because they needed it in the military. My parents lived in the coal camp, in Holden, West Virginia, Mine No. 7; my brother was born there; they had neighbors that were German, Polish, Hungarian. It was opportunity; those immigrants could make money here; they would have a cow and hog and chickens in their back yard and put away enough food for the winter.* When my parents moved out, the camp was still there; the mine closed in 1960.

"The company would come in once a year and whitewash all the houses, and they would turn the electricity on for so many hours a day, and they would dump a load of coal for them to burn in their back yard. They were very good providers. The community was fantastic. Even after we left and moved into civilization, we would still go back into the town, into Holden, because that's where the company store was, the doctor was, the dentist; they had a movie theater, a school, a swimming pool; they had everything you needed.

"They had all the good stuff at the company store. It wasn't like going to Walmart.

"They had a pension in those days, and they had insurance you could not beat. It was unionized: United Mine Workers.

"Now that camp, where Mom and Dad went, it was called Frogtown Hill, and the neighbors, it was just like family,† and the neighbor across the street helped deliver my Mom's first baby.

"I was born in Chapmanville in '43. It was only about 15, 20 miles, not far. Our neighbor had a boat; they had to cross the Guyandot River to get to

---

* About these immigrants and their opportunities a high school teacher near Welch told me: "What would you rather have? Freedom with a little bit of money. Well, when they got here they was paid in scrip. Later on they changed the law so they had to pay them half real money, so the men hid the silver coin and had the women take the scrip to the company store."

† "Well, there is one little thing. The people that did not work in the coal mines always, they would say on Friday evening, you'd better be careful because the coal miners got paid, and they're gonna be partying! They were known for having two-dollar bills."

work; so all the miners would go across with him and take a truck to the mine. Where we lived in Chapmanville, we had double tracks; the coal trains was constant, twenty-four seven, it was constant, one going in, one going out. Well, you know, that was our life."

"Did your mother ever worry about your father being hurt?"

"No, she never showed any worry, because at that time Dad was outside the mine; he dumped the coal after they brought it out. When he would come home, he would talk sometimes about things that had

Mrs. Patricia Wheeler

happened, but the exciting thing was to look in his lunchbox because he would pick up little pieces of wire and bring them home; we called it *shootin' wire;* it was colored; we would make things out of it."

"Do you think that anything could have been done to protect your father from black lung?"

"I know with my father there was nothing; there was no respirator. Back in Dad's day, the most that they could have done would have been to tie a bandana or some kind of rag around their face. I don't remember him having a really bad cough, but he did have trouble breathing."

"Did you ever have any desire to go inside the mine?"

"Well, I would go there and I would see it, but he never would have let us get close enough to . . ."

Hesitating as I always did when broaching this topic in West Virginia, because when I was alive I hated to make anyone sad, I finally asked: "Could you please tell me how you think coal and other fossil fuels relate to global warming? Or would you say that there is no such thing as global warming?"

Mrs. Wheeler answered at once: "If I knew that they were able to clean the air with using coal, I just have to think that . . . well, coal is bad, but I'm very, very mixed on that, because the coal made a very good living for so many people, because what would we have done without it? But I think there's plenty of things they can do without using coal. They need an alternative."

That satisfied me; we had gotten it out of the way; and so, smiling her way back into the past, she opened her mother's strongbox and very carefully unfolded first some World War II ration booklets imprinted some with tanks and some with howitzers, then a pay stub of her father's, and other such documents from the days of Mine No. 7, Island Creek Coal Company. She fondly said: "That particular mine, I guess they'd been at it since the turn of the century . . ."

4: "And It Does Have Heritage"

Her daughter Jackie, strong and pretty, with long dark hair, ran the family recy-cling business, collecting the brown paper that other businesses threw away, spooling it and cutting it to sell back as interleaving. She said: "At first the men used to laugh at us. They used to spit on what we were doing. I reckon we've saved a few thousand trees."

She said: "I think that coal is a good energy source for some people under some conditions and it does have heritage that people really value. As alternative sources have become more available, like solar energy, wind energy and other types, then people should try to redirect and use less coal. I did some research on the environment and the use of fossil fuels, and I believe that it is not the direct cause of global warming—if global warming is even real. However, I am not say-ing that it does not contribute to the changes, because I think it may. If there is or if there is not global warming, that does not change the fact that the use of coal and the use of fossil fuels is damaging to the environment. What happened in Charleston with that chemical spill,* those things have to stop, because that's damaging people from the inside out. The question is, what is global warming? Is it this really entirely caused by the use of fossil fuels? Because when you think of the ocean and how much the ocean purifies what's in the area, you have to say, is whatever's happening just a planetary life cycle?"

(How can I say? Perhaps that's all it was. On the next day, the Charleston news-paper ran a piece about the drought in Oregon, Idaho, Montana and Washington, *the region that produces a fifth of the U.S. harvest* of winter wheat. The Chief Execu-tive Officer of the Oregon Wheat Commission was quoted as follows: "We've had dry years in the past, but if anything is different this year, it's been warmer and longer. Growers feel like this is a little out of the norm for a hot, dry year.")

Jackie said: "Like anything that becomes history, in order for it to become preserved, it either has to be written or told verbally in stories or legends. Every-body has to have their own story, and if coal goes away and there's no story, then the coal heritage will not be preserved.

---

* Of the coal cleaning chemical MCHM, which had occurred in the previous year. See below, pp. 170–78.

"You have these people who say, coal's bad, so close every coal mine—no more coal at all. I don't agree with that. But maybe they should be teaching the alternatives at an early age instead of just teaching the heritage and saying, don't let them take this away from us. I saw one time when they were passing out toy bears to little kids and training their little minds to just dig in and fight for coal."*

"When did you first start to hear about the War on Coal?"

"Probably most of my life I've heard something. I was born in '66, so as a teenager I was living with a lot of activism, so you began to listen, and there was activism on both sides."

She then said: "My feeling on it is that coal pollutes. Coal pollutes, and coal mining is dangerous. Because I wasn't dependent on coal mining, I don't associate coal with good.

"It seems me to me that the people who own the oil and the oil rights are the ones who make the largest profit. That's not the case with coal. I know somebody who mined and he made huge money even though he grew up super poor. He would make all this money, and then the tradeoff was being tired and having accidents."

5: The Punch Mine

What Jackie had said got to the heart of something: Coal was, or at least had been or could be, the poor man's resource. I could not well imagine somebody mining uranium in his back yard and then powering his house with a homemade nuclear reactor. But on the property where those three generations of the Wheeler family dwelled there was a punch mine.—Jackie led the way up the steep mucky hillside, with the two dogs following and the afternoon sweltering hazily as she said: "Of course we have ferns galore and lots of sassafras. We have tons of dogwood trees and roses; we used to have more chestnut trees but for some reason they have come to an end," and through the kudzu we went, over downed trees, watching for poison ivy as best we could; I saw May apples and huge shelf-fungi in the green darkness beneath that calm white sky, and finally we were looking down into a pine-needle-paved concavity in the hillside; it could have been a sinkhole, and in a way it was; in the middle of it was a tiny dark slit maybe two hands wide; Jackie said: "Don't step in it, or it might cave in." Until then I had been hoping to explore it. ("You don't want to go into an abandoned coal mine," said a lawyer who occasionally had to do that sort of thing. "The longer it sits, the worse it gets. Most of the big operators at least consider liability. For the small operators, they're as

---

* Her teenaged daughter Tatiana Castro put it: "If they did stuff like have kids build windmills and stuff like that for science, that would kind of plant the seed like, there's other ways to go."

dangerous as a cocked gun. I been in some dog holes, man! I been in one where you're driving along and have to swerve your head so you miss the roof bolts. There's some small mines up in Summerville...")*—Jackie's stick went all the way in and vanished silently. In this place a man had dug coal for himself until one day the mine exploded and killed him.

Here I feel obligated to remark that after our coal mines were closed—or in this case closed themselves—their noble work of global warming continued. This punch mine was likely a casualty of methane ignition. Meanwhile, more methane kept coming invisibly out. (I would not have liked to light a cigarette anywhere near that hole.) As the German government once explained: *Emissions from decommissioned hard-coal mines play a significant role in this sub-source category. As well as active mines, decommissioned hard-coal mines (degassing) represent another relevant source of fugitive CH_4 emissions.*[†]

I never heard West Virginians talk about methane. As for accidents, their attitude appeared to be that, yes, those took place, yes, but wasn't that the case for all energy extraction and production? (Was that why West Virginians so often seemed resigned?) It did seem a shame to leave coal in the ground, when there remained so much left to burn. *On a uniform calorific value basis,* ran a technical encyclopaedia, *coal constitutes 69% of the total recoverable resources of fossil fuels in the United States. Petroleum and natural gas are about 7% and oil in oil shale, which is not as of this writing* [1993] *used as a fuel, is about 23%.* Nowadays we had learned to extract oil shale, and we were fracking like crazy... but think of all that coal still waiting to be used...!

And you from the future who hate us, you who watch the dead seas in fear of methane fireballs and murderous hurricanes, tell me you're not making matters worse! If you're not, maybe you're not human.

6: "But He Really Liked Working There"

April Mounts was a coal miner's wife who worked at the Hope Chest, a charity thrift store in Welch. She said: "It's scary having a loved one underground. But he really liked working there. The drugs down there are really bad and the drug testing is not so good, but he worked in Mine Rescue down there. The people are

---

* That sad miner at the Upper Big Branch memorial had been more encouraging. He said I ought to take a bug light, because otherwise that black damp might get me, and I'd never know it. I'd start walking like I was drunk, then I'd sit down to take a break, and it would be my last break. If I could feel cold air blowing out, then it was probably okay to go in a little bit. If my bug light's flame went out, then I had to go back. It might have been just as well that I never fell into the punch mine.

† See pp. 310–11.

unreliable: a lot of oxy.* He got pinned once. There was a guy on oxy who had pinned him. I don't know; I guess it was more about providing for us. There's less mine safety now. He was laid off about a year ago. He worked down there for about seven years. He was in Southern."†

"What do you think of coal as an energy source?"

"I think it's an excellent source. God provided it for us, for us to use."

"Does it cause global warming?"

"Anything we do is gonna have an effect."

"How much more coal is left?"

"Plenty. Hidden seams everywhere . . ."‡

7: "You Better Not Talk About Climate Change"

And so for most of the West Virginians I met coal really did still seem to be America's best friend. Upper Big Branch and Buffalo Creek were sad episodes, but not coal's fault; climate change remained an awkward subject. But what did a regulator say?

A Kentucky "stripper job"

---

* The prescription narcotic OxyContin, used illegally.

† One of the coal companies owned by the subsequent Governor, Jim Justice. For details on this scofflaw, see pp. 99–109.

‡ My guide wished to insert here: "What I remember most about her was how I wanted to buy that pink knitted baby sweater and she smiled so sweetly, and wouldn't let me pay for it."

To interview that retired miner and inspector, Stanley Sturgill, one passed the exit for Holden, West Virginia, where Mrs. Wheeler's coal camp and Mine No. 7 used to be; specifically, one rolled down Highway 119 through Lincoln County, the lovely waves of deciduous trees so summery and dark all around (from the highway everything seemed quite pure); and then quickly into Kanawha* County, almost immediately into Boone, and after a moderately longer while of occasional billboards into Logan County, one of whose commissioners explained for you of the future: "I couldn't run this courthouse—I couldn't run this county—without the tax money we get from coal"—after which one descended the almost empty road into a deep cut of blue-shadowed trees, and in Mingo County (formerly called Bloody Mingo, back in the decades when the unions remained sufficiently powerful to cause "labor unrest") a long Norfolk Southern coal train came rolling empty, with the locomotive's eyes shining as it rattled up from Williamson; and that was how one got to Pike County, Kentucky, **AMERICA'S ENERGY CAPITAL**. Highway 119 kept going. Maybe fifteen miles before Kingdom Come State Park there was a ridge, and everything below it stripped to grey and heaps of boulders and gravel beneath a sign for some coal company ("Blue Ridge Coal Company is the name of that stripper job," said Mr. Sturgill). And right around here one was in Letcher County, whose coal mining job average had decreased by more than 70% in the last three years; now came Harlan County, where Cumberland, Lynch and Benham sat hollowed out like those other dying coal towns in southern West Virginia.

In 1935, West Virginia produced 25.7% ($169,254,622) of the total U.S. value of coal mining products. Harlan County generated only 3.2% of that national wealth, but an impressive 32.3% of Kentucky's coal revenues. I would have liked to see Cumberland, Lynch and Benham in 1935.

Mr. Sturgill was waiting in Benham.

As that strong, plump, patient old fighter explained: "I'm a much hated person."

I freely confess to have always enjoyed the company of strippers in or out of clothes. Mr. Sturgill used the word in a different way.† He would often say something like: "This stripper job here, mountaintop removal, was A & G." A & G was one of a horde of companies. The kind of stripping to which Mr. Sturgill referred was the kind that denuded a hill of trees, grass and most everything but gravel and mud—which would keep washing down into the hollers forever. It

---

* This place-name, which you will also see applied to a creek and a state forest, was pronounced "kan-*aw*." The West Virginia Coal Association gave its Native American meaning as "place of the white rock."

† Some people said "strip job."

was as if one of the strippers whose smiles and wiles I liked were to be skinned, partially defleshed, then left out to stink. As the West Virginia Coal Association explained: *Mountaintop mining is the most efficient and environmentally responsible type of surface mining over the long term.*

Mr. Sturgill liked stripper jobs no better than I.

He'd say: "It was Cumberland Mineral Coal Company, and, gosh, I can't think of the exact name of the strip mine part, and all that used to be Scotia. The strip mine's been done in the last 10 years. It don't take long to do a stripper job."

To get out of the sunlight we sat in a gazebo down the hill from the Kentucky Coal Museum, and I said: "Can you tell me what it was like for you when you first went into a coal mine?"

"The very first time, I'll be honest with you, right up here, you can't see it but there was a portal up here and a fan; when we was kids we'd peep in. When I first went to work, the thing that fascinated me was the conveyor belt that rode the coal out; we rode the conveyor in and we rode it out. I took to it, right off the bat. Now I'll tell you what; at one time I had a fear of heights, but I got over that, because of the metal- and non-metal mines I inspected, where I had to walk the highest places.

"I was born in 1945, June the 11th. Well, when I first left home I ended up in Columbus, Indiana, and then my wife and I came back to Kentucky. In December of 1968 I went to work in an underground coal mine. That was the farthest thing from my mind that I would ever do that. I wanted to be a schoolteacher; I wanted to teach English. In order to make a living here at one time, and I mean a good decent living, coal mining was it. At that point in time, surface mining and strip mining, they weren't into it that much. But many of the companies, they would leave gashes in the mountain and it was unbelievable. I used to like to hunt and fish. We would go huntin'; I would go huntin' with my Dad, and just to see the devastation that was starting at that time, I really didn't like it, but there was nothing I could do about it. That was my first notice of how the mountains been treated.

"When I first hired in, I worked at United States Steel Corporation. We produced metallurgical coal. So I hired in as a roof bolter,* worked on that a few months, and then bid off. It was a UMWA mine, and we had a good strong union. After a few months, I bid to be a scoop operator. That was when the Unitrax scoops were first comin' in. I bid on one of those jobs and got it. I operated that scoop, keeping the face area clean from loose coal, coal dust, and so on. It was similar to an end loader. I stayed on that for awhile and then decided to bid off as a mine operator. I ran a continuous miner until 1974 and then took a job as a foreman. I was a foreman for eight years to the day until I got laid off. They decided coal was not as good at that time for the market, and steel was not in that much of a demand, so they decided to close one of their mines down. They laid off 21 foremen and I was one. One thing that really ticked me off, we were putting up some steel beams, and they had the USS logo on 'em, made in Japan, and that really ticked me off. Even your own company that makes steel, they're gettin' it from Japan.

"Then I applied to Mine Safety and Health Administration. And I worked at that job out of Hazard, Kentucky,† for close to two years. I had little kids back then and MSHA didn't pay much. I got an offer to go with the Blue Diamond Coal Company to be a manager of safety. And it paid more than I'd ever made with any place I'd been. And I stayed there until 1991 when Arch Mineral had already moved in here. They decided to acquire Blue Diamond Coal Company and Scotia Mines. Their safety manager used to run a continuous miner for me at U.S. Steel. He called me and said, we're gonna take over and come in and pick everybody's brain and get rid of 'em, but I'm gonna work it out to try to keep you if will stay on. So I said, sounds good to me, and I stayed. They ran it for 14 months and then shut it down, lock, stock and barrel, to get rid of what union Scotia had."

---

* This operation is currently said to entail "pounding metal bolts the size of car hubcaps into the mine ceiling to hold up the roof." For one roof bolter's story, see p. 182.

† Yet another thriving coal town, in Perry County. "The first stop on Pres. Bill Clinton's tour of poverty-stricken communities that had failed to share in the boom of the 1990s."

"How many union mines are left?"

"Right now there are no union mines in Kentucky."

"What did you do after that?"

"From that, I went to work as a janitor at a school. Times was tough; you couldn't hardly get a job; my kids were in high school. I told my wife, I said, I have never in my life worked harder for five dollars an hour than I am now! I stayed on it for four or five months and called up Great Western Coal, which eventually became New Horizons Coal. I got a job as a fire safety boss and made early morning inspections, and stayed with them, until they sold out to Good Coal Company, which transferred me to Bell County, where I was a safety inspector. They decided to sell out, too, to Addington Industries, and Addington came in and talked to us, saying we don't need a union, and I got laid off again, and went to MSHA again.

"They sent me to Arkansas. I inspected gravel pits, dredges on the Mississippi, all kinds of things; we inspected all of southern Missouri, northeast Texas, Louisiana and Mississippi; we had a pretty huge territory. I wouldn't trade nothing for the experience, but you would leave home on Monday and stay in a motel every night, and not come home till Friday. Had to follow my paycheck! I kept puttin' in for transfer back to coal and finally got in back in 2000. I inspected in Harlan County, Pike County, Bell County, some in Wise County, Virginia, some in Tennessee because that was national mines.

"Well, in 2006 I had started an inspection at some mines on Highway 38 going from Harlan back toward Evarts*; the mines was called the Kentucky Darby Mines. I'd been there three days, I'd issued 26 violations in three days; and I was in bed asleep, and they called me up, said Kentucky Darby just blew up and six people unaccounted for. There was actually five people killed and one who survived. It's ironic; one of the guys that I worked with at Great Western; they called him Cotton†; he was the supervisor, and at times I had even shot Cotton down when I would inspect for things that had been done wrong. I am not puttin' any blame on anybody. I had never gotten to the set of seals on my

---

* Another prosperous Harlan County town, 37.3% of whose citizens lived below the poverty line. Between 1950 and 2015 the population declined from 1,937 to 925.

† Amon "Cotton" Brock, formally listed as "Afternoon Shift Foreman." As for the survivor, Paul Ledford, he would live out the rest of his life with impaired lungs. A report on the 10th anniversary of the accident claimed that "mine owner Ralph Napier continued in the mining industry with an interest in nine delinquent mines owing nearly $3 million in fines as of 2014, $500,000 of it related to the Darby disaster." If this is so, then Mr. Napier's approach to paying fines would have resembled that of his colleague Jim Justice, whose handiwork Stanley Sturgill will soon show you (pp. 100ff.).

inspection, and for some reason they had some roof straps that was hanging down. Cotton decided they had to come down; they didn't take a gas check; when they lit a light, it went boom.*

"I continued to inspect coal mines until 2007, and then I wanted to make a different move and applied for a special investigations job. I did that until I retired in 2009. I investigated discrimination cases.

"You can kinda see me on the picket line in that movie 'Harlan County, USA.' I flew to Denver to testify to the EPA. I went to the climate justice march in New York and we rode a train up there. There were seven of us that got to talk, and Robert Kennedy, he was standin' there talkin'. Bernie Sanders, that guy runnin' for President, and David Letterman, he was lookin' for us, and I got to meet the first lady President of Ireland, and she wanted to march with us. Our leader was Sarah Pennington . . ."

"How would you compare surface mining to underground mining?"

"I'll be honest with you. I don't have much of a problem at all with underground coal mining. But as far as this surface mining, it's just total destruction. I'll take you to a place today, this James Justice, he's running for governor, I'll show you what he left us. If you're downstream from one of those places, you're kind of damned. I'm totally against it. I went to jail for it.†

"First you look at the mountains as beauty; to see 'em gutted and scarred is one thing, and then you get to noticin' the people with the health surveys,‡ and then you see it brings in a whole 'nother situation."

"What do you think about coal in general?"

"Well, coal at one time fed everybody around here. For many years we didn't know what the effects of it was. Right now I fight against all coal. I don't think it'll ever come back here as far as that goes, because it cost too much. I think because of our state bein' a coal sanctuary state, which has actually been enacted in legislation, and that's hurt everybody. We can't bring other types of jobs here,

---

* The "light" was an acetylene cutting torch. Just as at Upper Big Branch, the gas was methane.

† From a blog by Jeff Biggers, 2012: "As . . . sit-ins against mountaintop removal spread to offices of central Appalachian members of Congress today, one arrest stands out in my mind as a litmus test for Kentuckians—and all Americans, for that matter. Despite three years of requests, Rep. Hal Rogers refused to simply meet with[,] . . . and instead, had Capitol Hill police arrest[,] Stanley Sturgill, among six other Kentucky constituents . . . Rogers has been hailed as the most 'corrupt member of Congress' by . . . Citizens for Responsibility and Ethics in Washington."

† A study "published in the peer-reviewed journal *Environmental Research*" found that central Appalachian counties in which mountaintop removal occurred suffered from an astounding 42% increase in the rate of birth defects. Moreover, MTR in one county raised the incidence of birth defects in neighboring counties.—But what could you expect from a source called *Environmental Research*? People like that might even believe in global warming.

Mr. Stanley Sturgill

it's coal or nothing, and people really believe that. You have people like Mitch McConnell, Rand Paul, Hal Rogers, and that's all they know, is coal, coal, coal. They say, we're for the coal miners, but they've voted down every health and safety regulation since Upper Big Branch. I'm a strong advocate for coal miner safety. They still need to be protected. I'll hang in there until I die, trying to protect 'em, but anymore I can't see any good out of coal.*

---

* He couldn't see any good out of it, but even he had his loving coal memories, because coal had been his life. Here is my favorite: "Back when I was a foreman I always tried to treat those guys fair and square. We would have big Thanksgiving dinners underground, and we'd have turkey, green beans, you name it . . ."

"We're over here in southeastern Kentucky, and over to the west of us I think there's 21 coal-fired power plants. We've got one of the highest cancer rates in the United States."

"And you have black lung?"

"The first time I knew it was when I went to work for Blue Diamond Coal Company. They put me to work, and I had to take a physical and everything, and he left me a message to say, stop by the office. He said, buddy, I can't hire you. I said, what do you mean? He said, you got black lung. I said, I only been working a week, Richard! He said really? Then go on back. Other jobs, they told me that, and it started to bother me. My symptoms is coughin', wheezin', real shortness of breath. Of course I don't have near the energy I used to, but mostly what annoys me is the coughin'. Sometimes you can't lie down; you have to sit up. These new most recent laws, they might make a difference . . ."

"Do you believe in climate change?"

"I think we definitely got a problem. That's the reason I still harp about it every day. If some of these people don't give in and change the ways of doin' things, I really fear for my future generations, I really do.

"This place, it's one of the last places where you can get good clean air. If they don't quit burnin' all this coal and dumpin' on us, if they don't change, we're in a real hurt.

"As far as the people around here, coal companies has beat it into 'em: If you want your job, you better not talk about climate change."

So people didn't. That was the unconscionable omission.

"I'M NOT GOING TO SAY IT BECAUSE I'M WITH THE COMPANY"

The day before the parade that winds up the West Virginia Coal Festival in Madison, there was about as much going on in that town as if three ants were crawling on a rock, so just after getting rear-ended at a railroad crossing and watching the guilty party speed away it came time to check into the Madison Hotel (40 dollars a night and dead insects in the bathtub). The year was 2013—memorable for the photographed smile of Ms. Kendra Ball, high school senior, who was from Madison and got to be one of the Arch Coal Scholars; I saw her in the *Coal Valley News*.*—That year the residents of Earth contributed from fossil fuels alone another 32,190 million metric tons of carbon dioxide to the atmosphere, America's share being 16%. But that depresses me; I'd rather chat about Kendra Ball!—In our day, carbon-selling companies loved to attach their brands to whatever crumbs of largesse they felt like dropping. (Kendra Ball would get a munificent $2,000 a year for up to four years; let's hope she picked a cheap college.) I remember one Fourth of July in Greeley, Colorado, when a fracking concern called Noble Energy deployed a balloon-studded float, a Food Bank truck, a school bus and a stagecoach, all conspicuously eponymous; thus also the tallest billboard at the rodeo called the Greeley Stampede.† In Madison, Massey Energy had insinuated its name onto the overhanging sign for the Madison Little League, while the railroad tracks were decorated by the following grand announcement: BUILDING A STRONG FUTURE: PATRIOT COAL: BE PROUD OF WHERE YOU WORK. (From the *Coal Valley News* for that year: *Patriot Coal, which was spun-off from Peabody Energy and Arch Coal, has filed for bankruptcy and could shed its obligations to retirees.* That little trick might disencumber 12,000 broken-down coal miners of their benefits.‡ And hallelujah! The federal judge went along!

---

* Her benefactors were equally memorable for *a far ranging scheme orchestrated by Arch employees, including the former Mountain Laurel General Manager . . . , to receive cash kickbacks from certain vendors in exchange for receiving work.*

† See the illustrations on pp. 371–72.

‡ UMWA International President Cecil E. Roberts: "Peabody has spent years trying to get rid of its obligations to the thousands of retirees who made it the richest coal company in the world."

Meanwhile, *bankrupt Patriot Coal this week sought court approval to distribute more than $6 million in bonuses to corporate executives and salaried employees . . .*) The following year's Coal Festival would present a sound stage from which hung a banner honoring PATRIOT COAL FIREWORKS, Sponsored By: Hobet Mining, LLC.

In retrospect, I wonder if I should apologize for the previous paragraph. Why should I have singled out Patriot so cruelly? For in 2016, Boone County's *great corporate neighbor and partner* Alpha Natural Resources, the inheritor of Upper Big Branch, set out to follow the same canny business model: *A bankrupt coal company last month unveiled a plan to pay top executives up to $11.9 million in bonuses . . . Alpha also proposed to eliminate health insurance, disability, and other benefits for mine workers.*

But excuse that interruption; let us return first to 2012, when a coal miner's son* wrote that *Patriot . . . is currently the company dismantling my ancestral home of Cook Mountain,* and then to 2013, when, although I had not yet heard about Cook Mountain, the nearby hamlet called Twilight achieved mention in the paper: another sacrifice to coal. If I wanted to see Twilight, Madison would be the nearest place to stay.

It took awhile for the old lady to come to the front desk. In the lobby, carved and painted to match the horse-carved newel post, crouched a wooden horse which a certain little girl sometimes rode; the child called it Butterscotch.

"How far is it to Twilight?"

"Oh," said the proprietress, "I don't even know if there's still anything out there."—What an epitaph for so many small towns in West Virginia!—"Can't be more than a half hour," she said.

"It's just an old hotel," she said. "I'm sorry there's no TV in the room."

"That's all right."

"The TV's been stolen out of that room three times."

There was an ashtray next to the bed, and then brown-stained or perhaps iron-burned floral-printed curtains. I studied the grime and the upturned thumbtack on the top of the air conditioner. Then I gazed out at strips of pale light through the curtains.

When it cooled down a trifle, I set out to re-sample the extreme humidity of Madison, the perfect spiderwebs on the footbridge, clouds and vines and lovely water (it was only 90°). The Coal Festival was slow—but, after all, it felt awfully hot, and, besides, from 2011 through 2014, Boone County had lost 2,689 coal mining jobs, *a 58% decrease from the last quarter 2011 average,*† so why wouldn't

---

* Mr. Dustin White, who reappears later, on pp. 93 and 213.

† During that same time, more than 10,000 coal mining jobs went away from West Virginia. And so did many people. As an anti-mountaintop-removal activist said: "That's one of the huge issues in West Virginia: Everybody leaves. Everyone, almost everyone that I grew up with, left."

it be slow? About that county a certain extremely unbiased well-wisher warned: *Take away mountaintop mining and it will simply revert to another Appalachian tragedy . . .*

Just past the library, Ralph and his wife sought to save my soul in their booth,* Ralph showing me underlined and highlighted lines of the Bible, and at the end, when I shook his hand and thanked him, while his wife stood forlornly by the sidewalk, he advised me to at least get a true King James Bible, not a revised version but a real one, his eyes steadily watching me from beneath the cap; he reminded me of a policeman just now as he regarded me and inquired whether I too had not sinned. I said that it seemed a trifle strict to punish all the rest of us for what Adam did, to which he replied: "If Adam hadn't done it, you think you wouldn't?"—Slightly sorry not to please him, I turned back to American flags, National Guardswomen, the Friends of Coal† and the golden gleam of the courthouse dome, until another freight train came singing, slowly drawing its bulk in along the brown-green river, and stopped. Boys frisked and clambered across the couplings; I hoped they would not get hurt. After awhile there came the air-hiss which presaged motion; then the train began to creep away, first humming, next clattering, then making a sound like a broom sweeping gravel, after which the noise dwindled into something like a broom sweeping dust.

Dusk dragged the temperature down 70°, and there were clouds like purple-grey smoke-plumes above the river gap; the forest was somewhere between green and black, the air something like tropical. Long before nightfall we could all make out the backward silver S of cloud across the moon, whose so-called mountains and seas, as is often the case in humid places, were beautifully distinct. The yellow light of carnival engraved itself on whirling amusement-wheel cars, some of which were clanking slowly back down to rest while others spun their riders in devilish ways. Sedate families fanned themselves; their young girls were nicely tarted up with eyeshadow. Children kept running off the platform of the whirly ride; teenagers hesitated over the speed ball con (guess the speed of the third ball). All five slot machines were in play inside Melissa's (and who Melissa might have been not even Tiffany who presided over the bar could say); it was very smoky in there; one had to play if one wished to drink, and the only thing to drink was watery American beer.

Next morning my driver took me down the road all the way from Madison through South Madison to Uneeda, Quinland, Lanta, Bigson, to the junction at Van, where Kendra Ball, that Arch Coal Scholar, had graduated from the

---

* Depicted on p. 227.

† "Friends of Coal was born out of a desire to correct the impression that coal's time has passed in West Virginia . . . The Friends of Coal 'puts a face on the industry . . .'"

Christian Faith Academy. From Van the proverbial carrion-crow could have flown to Upper Big Branch, while the motorist had to bear left to Marnie, Bandytown and finally the bright green winding ravine past Twilight, where puddly potholes dimpled the road and water dripped down the limestone at the roadside, with Lindytown not much more than a memory at the end of the road in Ducky Ferrell Hollow. There were cicadas, weeds so green, then suddenly a vastness of grey scree. I heard birdsong. By a dry creek the road narrowed, trees practically brushing the car, with Peabody-Harris No. 1 Mine ahead. Here the road ended in a sort of parking lot with abandoned bits of something mechanical on it. Not knowing what else to do, we drove back to Twilight.

I remember the lush sound of water in the shade and two raptors slowly circling high up over the steep forest-wall which rose a hundred feet and more; here as in so many parts of West Virginia a steep wall of limestone lay exposed in the hill. Creepers grew on the telephone poles as in Tomioka.

In Twilight I visited the residence of Mr. and Mrs. Avril Richmond. Mr. Richmond was a whitehaired man, somewhat deaf, who when I interrupted him was standing on a bed of bark chips and carefully watering each plant in his garden.

"Was Cook Mountain much higher in the old days?"

"Oh, Lord!" he said. "Used to be, it took you 30 minutes to go in a four-wheeler from the top over to Bandy Springs. Now it's five minutes."

"Is Bandy Springs worth hiking up to?"

"Well," he said, "it's part of that strip job."—I took that as a negative.

"I was borned up there near Number Eight," he told me. He had been a coal miner for more than 40 years. "Not a bad life. Always had money; we took a vacation every year."

"Has Twilight changed much?"

"Twilight used to be a lot better. Houses went way up the side of the holler, but the coal company bought most of 'em out."

In Twilight I also met Mark Mooney,* aged 43. You have already heard his opinions on Upper Big Branch. His father was a disabled retired coal miner about whom the son said: "He loved to work."

Mr. Mooney said: "As the technology gets better the men get laid off."

He said: "When I was up here in the '70s, this place boomed! It was all deep mines, not strip when I grew up here in Twilight."

He led the way to the vandalized family cemetery, where a wild tiger lily grew through the thick grass by the wrought iron gate. He said: "This is my great-great-

---

* I spell this as I heard it. I did not find him again.

grandpeople and someone went and destroyed it. Watch out for snakes. This is a nice place for snakes.

"This is all family," he said. "Now this tombstone, back then those guys, my people had some money. I got some old pictures. There's one of a lady standing here in a big black gown, kind of spooky looking."

I admired that lovely orange tiger lily.

"Twilight got its name in 1948 from the post office because the sun never hit it. Well, that's an exaggeration. Round about three o'clock it gets shaded up. I don't know what it was called before that. Sometimes the names get different. See, we always called Lindytown *Robin Hood*.

"I hauled coal for a number of years. Now I work for the state.

"This holler up here ain't a holler no more; it's a slate dump for the coal. My folks, when the coal companies came in, they offered 'em a little bit of nothing, and they didn't know better so they took it. When they leave here, their stock-holders don't care. They spend more on a state dinner on the weekend than any of us ever make.

"This is called Coal Mountain Strip," he said, pointing. "A lot of the boys, they live right here, so they reclaim it good. Ain't never what it was, but it's a good reclaim. That other bunch out there"—pointing toward a gate whose security lady had refused to let me see beyond her; she kept sitting in her car with the windows rolled down, guarding the dozers and such; she would be there until seven and it was only around eleven just then—"they don't care," he said. "It looks like the moon."

"Now the Lindytown road will take you all the way to the top of the strip. There used to be a cemetery up there before they stripped it. There's a cemetery for the Cooks, but about 50 yards from that there's two little girls who came down with some disease, so they buried them separate, out of superstition, I guess. Terrible how they mined all around that . . ."

Pointing down away from the family cemetery back toward Twilight, he said: "Everything used to be down real low. The deep mine come in and they filled this up, raised the baseball field up about eight feet.

"Mountaintop removal right now is down so low, it coulda been a deep mine.* They got a high dragline here as big as that ballfield. You should see them take the mountain down. Why not wait until this nation gets desperate for coal before they do that? It's really good coal, but they could've deep mined a lot of it. Now we're fightin' dust. You can wash your vehicle at night, and next morning it's deep in dust.

---

* A traditional underground mine. See I:559–61 for deep mining categories.

"Sometimes the creek dries up when the country's been so damaged that it[*] falls into the mines beneath.

"It ain't hardly home no more. They'll never put it back the way it was.

"Now up here, this is a pretty good reclaim, but over there, they don't care.

"Bolt Mountain is still real pretty.[†] Over there, they used to live up there on the mountain. But then they had to move down when the company came in. And now it ain't really home no more.

"There used to be a lot of other houses here. You know, when you cross that bridge toward Lindytown. One of 'em wouldn't sell out. They're original people.

"My mother owns six or eight acres of the land up here. One side of it's Pocahontas Land and Fuel. Mother, she lives in the log house there, she owns the timber and the mineral rights.

"My coal miner dad, he loved working. He was a neat freak. Never left home before everything was all tidied up . . ."

"What do you think about global warming?"

"I don't know if I believe in that, Bill. You know the way NASA says we're closer to the moon than when the dinosaurs were here. However it was, that was the way God intended. It's designed for perfection, and He's gonna do what He's gonna do."

Why did he reject the idea of climate change? It wasn't simply because he was Christian. Consider this credo from the organization called Interfaith Power and Light: *As communities of faith organizing a religious response to global warming, we believe that climate disruption is among the greatest challenges that humanity has ever encountered . . . We know that all forms of coal mining are dangerous . . . Particularly egregious is mountaintop removal mining . . .*—Ah, but Interfaith Power and Light identified itself as *based outside the coalfields region.*

We thanked him and went our way. I decided to walk up Cook Mountain and see those graves, not to mention the mountaintop removal, which has been defined as follows: *Serving as an excellent alternative to contour mining in hilly and mountainous areas, while using the same equipment, this method makes possible 100% coal recovery.* The most practical day would be Sunday, when security was loosest.

On Sunday a reddish doe and her fawn crossed the highway. The morning was not yet hot. I remember the lovely light, and mist through the dark green and light green trees. Toward Bandytown ran a powerful creek, with huge sugar maple leaves over it (some of them insect-eaten), and then clover and daisy fleabane in the sunshine; that brown creek went greenish-black in the shade. I

---

[*] Evidently the creek.

[†] See p. 600.

remember the sun on the greenish-brown water, cirrus clouds, bright lawns, rusty train tracks and a huge Jesus statue on a front lawn. Driving around the corner one reached a hill's brow shaved down to rock. Passing the American flags in Bandytown, and entering the green gloom with the creek trickling, we took the Lindytown road as far as we could, then parked in the grass.

Where was Lindytown? As the *Coal Valley News* once explained: *The Twilight Surface Mine Complex, once owned by Massey Energy and now owned by Alpha Natural Resources, is thought to be one of the largest surface mines east of the Mississippi River. As the Twilight operation grew bigger, the community of Lindytown vanished.**

By now it was late morning, with people at church, sweat on my chest, dark fern-edges catching the light, and mosquitoes and no-see-ums everywhere at the end of the little town where that puddly road began, there in birdsongs and shade, in sun and sweet fresh wet air. It would not be right to dramatize this walk as if it somehow resembled a visit to one of the Japanese red zones—and yet this place would certainly be poisoned and therefore dangerous, less immediately but possibly for a longer time than Tomioka or even Okuma. The heavy metals leaching down into the hollers, the health issues, well, it might have been fair to draw up an itemized bill for those mostly unpaid costs of doing business with Cook Mountain.—In Welch, while long white strings of boxcars went crawling among white houses, the coal miner's wife April Mounts had said about mountaintop removal: "I think it's uncalled for. We flooded last year from it."—And here is how a prominent Charleston lawyer named Tim Bailey made the calculation: "If you look at the mechanization of coal, underground, you don't need as many men, and a lot of these men, for mountaintop removal, some folks don't like what it does to the environment, but the other thing about it, you mine so much more coal with fewer men and take it quicker."

I listened to this and I asked myself: Whom would *that* be good for?—For the coal mining executives, no doubt, and hopefully for buyers of coal: the power plants and their ratepayers, the steelmakers and steel's primary and proximate users, which must have included nearly all of us who were alive when I was: passengers in buses, cars and jetliners, inmates of houses, citizens on whose supposed behalf wars got enacted . . .—But by the time Tim Bailey said this, I had already begun to suspect that most of us would be better off if coal were mined more slowly and if it cost more.

Mr. Bailey continued: "But if you put that flat mountain in the middle of

---

* In 2013, the "Twilight M[ountain] T[op] R[emoval]/Progress" mine (employment: 302) was the second most productive surface mine in West Virginia at 2,677,876 tons—courtesy of Alpha Natural Resources.

nowhere, then it doesn't much do anything, now does it? At the end of the day, I believe that if you're going to do something like mountaintop removal, well, you can't employ very many people in golf courses! I wouldn't mind it if the state and the companies could work together, and talk about tax credits. If you take something away from me, give me something back. Once it's gone, it's gone. The longterm giveback ought to be, can we have longterm infrastructure for that property so that we can employ people with longterm benefit? That to me would be something that we ought to look at."

What should the longterm benefit to our longterm infrastructure have been?—Why not an atmosphere containing fewer carbon compounds?—As it happened, Mr. Bailey did not seem to accept what he called *the argument that everyone in the world seems to believe,* namely *that burning any carbon fuel causes problems.* You will hear more from him from time to time, and then you can decide what you think of him. As for me, well, he gave me an hour of his time, and he litigated on behalf of injured coal miners, so of course I liked him.

There were coal fragments on the muddy road; I measured one at a measly 0.18 micros—slightly less than my average reading for Iwaki, which for that matter was the same as for all West Virginia: an eminently safe 0.20 micros. Water kept trickling down limestone walls, and a large animal was treading through the forest, and I heard the tremendously loud rhythm of a woodpecker. Against the bright wall of shrubs and high plants and below the dark wall of trees hung two white signs, one of them overgrown into illegibility; the other, framed by leaves and flowers, began: EAGLE MINING LLC. SYNERGY MINE 1 & 2. **WARNING:** EXPLOSIVES IN USE. The warning signal prior to each blast shall be from an air horn or siren and shall be three (3) short blasts of five (5) seconds duration . . . Leaf-edges sweetly cut the air, with ferns shining behind them.

On Monday at the bottom of this same Lindytown road I heard far up the mountain a noise like gravel sliding down a chute, echoing; while down where I was, below the sweet cumuli, a bit of thistledown spiced the air, and there were round purple clover blossoms and narrow bees, a meadow of daisy fleabane and buttercup-like yellow flowers, some Queen Anne's lace, and then the lovely dark green heights of the forest with rare flecks of silver-blue light halfway up and many green shadows, and far away, that falling sound, maybe like titanic engines starting and stopping or something massive falling over and over, after which the birdsongs around me hesitated a little. On Sunday it was just the same, except that there were no falling sounds.

After a steep mile or so (one vehicle-dependent Twilight man considered this road nearly unwalkable) there came a sharp skyward bend. Departing maples and beeches, one came into that wider blue sky of breeze and birds, everything so leaf-festooned, with ahead the stripe of reddish-beige horizon where the dead zone began. And so the forest gave way to first to grass, then to that trademark of mountaintop removal, the lip of raw dirt touched by clouds, and a few saplings just starting to grow. Then through the tall grass (which was perhaps comparable in its easy opportunism to the goldenrod and pampas grass of the Japanese red zones) I saw an ugly flat horizon, and just below that, a lower tier of pale sandy flatness. Thus the summit of Cook Mountain. Here it seemed appropriate to stand looking back down across the valley across into the deep green forest and across into the ugly terraces of what Mark Mooney had called *moonscape,* everything greyed down, a white machine on one of the terraces, and something like a plume of dust which was probably just light rising on the steep wall behind it.

At the roadside lay a rectangular settling pond, not significantly radioactive, with tall grasses growing out of it and reflecting themselves like their own roots. The water might have been turbid, but I did not understand what I perceived; other than the pancake frisker I had no tools of measurement; nor did I know what to measure for. It might have been a perfectly healthy pond, although I saw and heard no life in it.

If I were hiking back up the last quarter-mile of a mountain trail, with the parking lot coming into view ahead of me, I would feel both sorry and pleased to be reentering the grid after my exertions, with an electrically-chilled beer and a gas-grilled hamburger on the horizon. And the flattened top of Cook Mountain did look better than most parking lots, in part because so few vehicles crawled on it. Neither billboards nor litter afflicted it; it might well have been the best to which any parking lot can aspire. Moreover, the edges of that place had already been softened, like the tsunami wreckage of Fukushima, with tall grass. Later on that afternoon got hazy, so that the strip-mined horizon appeared almost natural. Stepping farther away from any edge I achieved denial's consolation: the grasses hid what I preferred not to see.

Just ahead and to my right on this place like a gravel parking lot ringed round with white roads was a gate and a sign for Cook Cemetery, accompanied by a sign that read NO TRESPASSING—BLASTING AREA. Someone had altered the latter to read BLASTED AREA. Sometimes in California I have seen old pioneer graveyards whose tombs were cleared away to speed progress or fight vandals; often bits of wood or bone remain. Cook Cemetery had been so thoroughly improved that without the sign I never would have known what it had been for. The most prominent monument was the ammonium nitrate tower, all metal, with its round platform reached by ladder, and the diamond warning sign 1.5 BLASTING AGENTS D1 . . .

Even now I cannot tell you who used to be buried there,* because not far down the road that circumnavigated that flattened mountain stood a small grove, right on the edge; at first I thought that someone must have neglected it

How they improved Cook Cemetery

---

* Perhaps the "two little girls who came down with some disease."

for some reason having to do with corporate costs and benefits. In this grove it was cool the way it must have been everywhere else up here; there were birds, mosquitoes and mossy tussocks half a foot thick and more. From its shade one could admire the so-called reclaimed meadows, the pretty good job that the boys had done. Here indeed lay the old slate markers in the moss, some graves sunken and desiccated, some swollen up with moss; and among those unknown graves (each delineated by two pointed slates, one at the head and one at the foot of each thick-mossed hummock) were two with names:

MARY C COOK
1820–1861

with another hand-carved stone that read DC (I guess for "deceased") SEPT. 1861; she lay next to her husband, Floyd Cook (January 9 1820 to August 20 1892); and I wish I could make you see the lovely old carvings on his grave— chiseled horizontals, and then the half-effaced verticals, sharp-angled like runes, and for the same reason: because carving straight is plenty tiring enough.

A year later, someone had placed pots of plastic flowers at each of those two graves, with each price tag still on. That was the year I heard not far away the rising and falling grinding of bulldozers, evidently grading the road. The driver had waved back at me as I walked up the road. Then another man rolled up in a dozer, wondering aloud what I was doing here. When I told him I was visiting the Cook graves he trusted me. He said that the mining was all finished up here except for the remediation. (A coal miner in Madison had looked around this spot; he disgustedly told me: "You won't believe it.") The dozer man smiled, waved and told me to be safe. I remember a sparrow or swallow flittering low over the graves, and then I discovered those plastic flowers. But on that summer day in 2013 there was no sign that anyone had been there for ages.

The coal miner's son who opposed the dismantling of Cook Mountain (he was Dustin White) "had family in" this cemetery. In the *moonscape* across the valley was Jerrell Cemetery, "which I have family in as well." I am just as happy not to know how *that* reclaim turned out.

If you have ever been backstage at a theater and gazed out at the brilliant place where the performance is playing, you will know how it was for me on that knoll with spruce saplings and tall lichened maples around me, trembling leaf-shadows on a reddish-brown puddle and pads of grey-green moss beside me in the dirt, as I listened to a woodpecker and a mockingbird; looking out through that screen of trees I could see only light.

Mary Cook's two headstones

"The tree-huggers say them graves got mined right up to the edge. You can see that's a lie. Of course they got deep mined underneath, but that was way under, a long long time ago," said the diabetic old security guard, whose name I omit for his own protection.* He got 11 dollars an hour, all the jobtime he wished on weekends (just now he had worked 38 hours without a break), not to mention the key to the gas pump, and I would not wish to say on the record that one job benefit might have been first dibs on any spilled high-concentrate ammonium nitrate so long as it was discreetly collected; this chemical, which had facilitated the work of the Oklahoma City Bomber, also did wonders for one's tomatoes.

Just past the truck in which the security guard had been dozing lay a rubble-pile, followed by another little cool grove whose sunken grave had a new cyclone fence around it. The inscription read:

---

* Both he and Mark Mooney were impressed by a college graduate. Mr. Mooney allowed that he couldn't read or spell too good. The security guard was very proud of his daughter, who had won a scholarship. He thought that world history was a waste of time (or perhaps worse). Only American history was worth studying.

The "moonscape" as seen from the "pretty good reclaim" of Cook Mountain

WILLIAM C. COOK
CO. I.
7 W VA. CAV.

His dates were December something of 1840 to August 18, 1914.

This was "Chap" Cook, the eldest son of Floyd and Mary, and the one the highway down below had been named after. How did they actually value him? I found it peculiar to stand by his grave and look out through the trees at the white rubblestones and the ocher rubblestones and a few blue mountains, some of which were not yet scarred. There were mosquitoes, crow's foot and strawberries. When I got tired of the mosquitoes, I strolled back out into the cool dry pleasant breeze of the seminude ridge.

The security guard told stories about the no-anesthetic surgery on his necrotic toe (done for free, during the doctor's 10-minute break, with a pair of scissors—the patient was grateful), and the time he helped carry a corpse up here to be buried, and the gravediggers were so lazy that the possums started eating it. He said: "There used to be good timber here. We'd come up hunting."

"Are you sad about those mountains?"

"I'm 77 years old. If I was a young man, I might be sad about it."

Then he said: "You ain't seen nothin'. If you like I'll give you a tour."

It was a long slow driving tour, all the way around the mountain. There was utter blasted rock. As the National Coal Association once aphorized: *Aesthetically poor reclamation fuels the fires of the abolitionist.*[*]

Having read that carbon sequestration "peaks in the second decade" of early-stage forest growth, I asked him when the forest would come back.

"This ain't never gonna come back. These pine trees, they won't hardly stay here in a flood. Good for nothing. They lay down just a tiny bit of dirt. The rest is all rock. They spray on the grass seed."

"How will it look in a hundred years?"

"Maybe in a hundred years, it'll look better."

He said: "In my opinion, nearly a hundred percent of the people in West Virginia would vote against mountaintop removal.[†] I'm not going to say it because I'm with the company."

---

[*] The essay in which I found this was entitled "Imaginative Plans Make Mined Land Better Than Ever."

[†] According to the *Coal Valley News,* that year an opinion poll found that "about two-thirds" of West Virginians were against it. The President of the West Virginia Coal Association expressed the matter as follows: "According to scientific polling of the people most impacted by mountaintop mining—those of southern West Virginia—the vast majority support the mining industry."—He did not say that they supported "mountaintop mining."

WHAT JIM JUSTICE LEFT US

And Stanley Sturgill said what *he* said—all praise to him!—precisely because he wasn't with the company. What could they do to him? He had black lung and a fatty liver; he was old and, as he said, hated; he was *free.* They could rearrest, threaten or maybe even crush him, but he'd already won the victory. If I lived to be old, I hoped to be as free and brave as Stanley.

He now said: "I want to take you to where I live at Lynch, and up the mountain to the top, Black Mountain, the highest peak in Kentucky, and just see what Jim Justice left us."

As it happened, a few months earlier I had read an item in the *Coal Valley News* about this titan of the *regulated community,* a regular hero pure and simple, whom Senator Manchin, having once more railed against "senseless EPA regulations," glorified as follows: "I thank Jim Justice for pursuing this wonderful new business venture, which continues to show his long record of bringing good jobs to our great state"—*yes!*—wonderful, good *and* great!—for Mr. Justice had just "repurchased" a Russian entity called Bluestone, with mines in McDowell and Wyoming Counties. If one of President Obama's men had done that, we'd be shouting Communist. But Jim Justice was one of *us.* As the great man deigned to express it: "You cannot tell the story of the United States without including coal, and West Virginia coal has been the engine of our nation for over a hundred years. The announcement shows that the coal industry and over 150 West Virginia coal miners will be going back to work . . ." How quickly that work would end, and whether it would take place in deep mines or stripper jobs the minuscule article did not explain; nor did it elucidate Mr. Justice's views, if any, on what the military would call "collateral damage"—which was precisely what Stanley Sturgill now wished to show me.

But first we arrived at another abandoned outpost of America's best friend; it was called North Side Winnifrede Mines, Carbon River Coal Corporation, about which Mr. Sturgill said: "That's an Arch Mineral lessee, or they may be part of Arch, I don't know . . . This was a U.S. Steel mine. The last I heard, Alpha bought all this . . ."—Please remember that the speaker had spent nearly all his working life in the coal industry and was describing a mine in his home town; even he couldn't keep track of all the fly-by-nighters. That was how it would also turn out to be with the many fracking entities of Weld County, Colorado.—I asked Mr. Sturgill whether we could take a look. "The only thing they're doing up there is fracking," he replied, but he was a good host and hadn't been up that way for awhile and maybe it was sweet for him to see it, for the sake of old times, so we turned up that hot sunny road of reddish gravel, with ferns and trees on each side. "You know what they built this road out of? Red dog. When you make a slate dump, and the slate heats up, it turns red. That's red dog. Real nice up here. Lots of animals. See where the bears pooped?" There was kudzu everywhere.

He had never seen bears underground, although he had heard of their presence; he did see raccoons from time to time.—"I was goin' one night and it was way underground, seven, eight thousand feet, and I was just ridin' a conveyor belt in and I saw these big eyes, and I thought, what in the world? I fixed my lamp on it, and it was like a huge cat!—And one of my guys, he was down on his hands and knees and he went through a mandoor, and there was a coonhound comin' out when he was goin' in, and they had quite a tangle; it was quite a sight. Sometimes snakes get brought in with timbers; I seen a big copperhead layin' across the track . . ."

I asked how working conditions varied from mine to mine. He said: "If the mine is ventilated well, it gets so cold you can't hardly stand it. But a hot mine, a lot of methane, there some guys work with their shirts off. And Scotia, now, in there, green, sticky crude would drip down onto your head . . . U.S. Steel was a good place to work, and they kept their houses up; they took good care of their people. And I think they knew what was coming, so they sold out. When they sold out to Arch, if you were Arch management, you did not live in the town of Lynch. U.S. Steel management had lived in town . . ."

"So you're saying that Arch was disliked?"

"You could say that. They knew how to talk the good talk, but they didn't do what they said."

We drove up several hundred feet, there on the Looney Ridge side, with cicada-music as steady as the heat itself.

Up on the wide flat road that appeared to curve all the way around the mountain he stopped the car, and there was a wide black stripe in the wall of rock that the road circumnavigated. "Now this bench, this is where the coal seam was. It was probably last mined in the early '90s. And there's a little seam of coal. And that one underneath that comes out of it, they call that a little bastard seam."

We strolled over to a frack pad; it was getting crude oil.*

---

* Reader, just in case you would like to see what that place looked like, I took a photograph of Mr. Sturgill standing by the nearest pad, with its tank of crude oil behind him. See the fracking chapter, p. 365.

The sign on the thing said:

> **Chesapeake Natural Gas Natural Advantages**
>
> P.O. Box 869, Gray, Kentucky 40734
> Phone 1-888-460-0003
> Lease: ArkLand Company
> Well No. 825425
> Permit 95817

He explained: "Those things are at different spots all around the bench."

We admired this object for awhile. Then he said: "Out of all this frackin', they won't sell it to us! Pipe it over to Virginia . . ."

"Why's that?"

"Buddy, you got me."

Then he said: "Jim Justice, he wants to cut a hole and come right down here and take everything."

Driving back down the mountain, we flashed through Lynch, where he pointed out a decrepit edifice and said: "This is the company store, and believe you me, you could get everything you wanted to. My wife got her first prom dress here. Now we have to go up into Cumberland to get groceries . . ."—A tree was growing out the window of a brick building which was once a satellite company store. Then Lynch fell behind us, and that highway kept wrapping up around Black Mountain, through kudzu and tall oak trees, dripping rocks and clouds; sometimes we glimpsed long drops down into the valley (not all industrially related; on one of them "they once had a ski run but the weather wouldn't hold out"); and here came a wide black stripe in the road cut. He pointed it out; that was the seam he used to mine. I asked whether anybody hereabouts still burned coal at home.—"I've got a neighbor up the street, and in the winter I can see him goin' to those coal seams and gettin' a bucket or two of coal."*—And we kept driving up.

Here was the place where his daughter's car had gone off the road one winter night, and two coal miners clambered down the mountain at some peril to themselves; they drove her to the hospital, and departed without giving their names.

"I used to drive this mountain going to work," he said. "I worked on both sides. We had to have Virginia mining papers as well as Kentucky ones."

---

* In one of his communications he wrote: "I can remember as a child living in our coal camp and buying it by the ton in order to keep warm throughout the winter. Many folks still do." However, that sort of coal use was definitely glimmering out. In 1990, American residential coal emissions remained measurable at 3.0 teragrams [3.3 million tons] of carbon dioxide equivalents. By 2008, and continuing up through 2012 (the latest date for this source), they "had almost flatlined at 0.05 teragrams."

At the side of the road by the summit (4,145 feet high) there was a sign:

POSTED—NO TRESPASSING
PENN VA COAL

and then:

WELCOME TO VIRGINIA—WISE COUNTY

I cannot say I liked what I saw of Virginia, because it looked like this:

Mr. Sturgill said: "I've got pictures of how it was before. We used to have two pretty good mountains here. I think they quit minin' it two-three years ago. I think it took about four years to do that. He's also started up there on the left, strippin' there. He wants to bring it out there through Virginia. They've started strippin' it out."

"I thought they were required to reclaim it better than that."

He chuckled. "Yeah, they are."

I should have known better, but it astounded me that these wicked people refuse to follow even the most watered-down laws. Why then would they help the rest of us in the matter of climate change? Reader from my sad future, please remember Stanley Sturgill as a good man who even when he had law and authority behind him could not enforce them against money.

In West Virginia they had just recently held public meetings (no doubt on terms congenial to the *regulated community*) regarding "proposed reclamation projects" for *the alleviation of health, safety and environmental problems created by coal mines abandoned prior to August 3, 1977.* So perhaps in Virginia a meeting or two to consider Jim Justice's time bomb might take place in about 2049. (Had I lived that long, I would have been 90 years old.)

Another car pulled up, and it was time for one of those Appalachian introductions. Like almost everyone hereabouts, Stanley Sturgill was distantly related to anyone who hailed from within a few miles. So they explored their consanguinities; I heard a woman explain: "She's the one her child got killed on the Harlan road."—Then we got in the car and drove down farther into Virginia.

The stripper job's first sign said:

> **9 MILE MINING INC.**
> Mine No. 2
> FED ID 44-07188

and below that smaller sign for A&G Coal 23, accompanied by the usual arrow for AMBULANCE ENTRY; by now I had seen quite a few of those.*

A little farther in from the highway was a sign on a dried-out post for the more perfect celebration of

> **MILL BRANCH COAL CORPORATION**
> 5703 CRUTCHFIELD DRIVE
> NORTON, VA 24273
> LOONEY CREEK TAGGART MINE
> PERMIT NO.: 1202109
> NPDESVA 0082109
> ANNIV. DATE 03-31-YY
> MSHA ID 44-07262
> DMME 14844AA

with a wall of leaves behind it.

---

* An Internet application which I shall call Gargle Maps represented that *Nine Mile Mining Inc. Mine #2* was the bare patch on the Kentucky side while *Looney Creek Taggart Mine* was the corresponding blot on the Virginia side. But then why should Nine Mile be proclaimed here in Virginia? Mr. Justice must have known the answer, for in the massive lawsuit filed against his entities by the EPA and various states, the environmental violations logged against Nine Mile bore Virginia code numbers.

Thanks to a private detective I managed to educate myself about these lovely members of the *regulated community*. Determining how they related to each other, and to Mr. Justice, proved slow and dreary; hence I shall relegate the details to a boxed item on page 29 of my already soporific source notes. In summary:

Mill Branch, the survivor of many peculiar New Year's Eve mergers, each of which took place with the *unanimous consent* of *sole shareholders* whose *Incs.* and *Corps.* all by happenstance occupied the same address (for instance, the 16th floor of a particular Bank of America Center), and whose manipulations involved stakes inexplicably comparable to those in a children's game of marbles (*Mountain Management has 100 shares of $1.00 par value common stock outstanding*), meanwhile kept digging coal right here in Wise, Virginia—and at some juncture fell into the orbit of (or for all I could tell had always been a satellite of) Don Blankenship's ungrateful heir, Alpha Resources. When Alpha declared bankruptcy in 2015, all its creatures followed. (Stalin: *Cut off the head, and the body dies.*) Thus it came out that Mill Branch's estimated assets ranged *between $1 and $5 million,* which hardly balanced the estimated liabilities: *More than $1 billion,* rendered all the more alarming by the estimated number of creditors: *Between 200 and 999,* including Don Blankenship himself, for a solid $3.5 million, which might have been part of his severance package when Alpha forced him out after the explosion at Upper Big Branch. Lower on the list, more alien creditors crept in, such as the U.S. Department of the Interior—for unpaid mining leases, perhaps—then the Environmental Protection Agency and the Mine Safety and Health Administration, which both awaited receipt of *undetermined* fines and penalties of a *contingent, disputed and unliquidated* nature. In other words, Mill Branch had cut corners, then kicked up a squawk; excuse me; I mean that Mill Branch was another innocent victim of the War on Coal. Compared to the great coal and nuclear disasters, Mill Branch's strayings must have been picayune . . .—for from 2007 to 2013 inclusive, the years of actual production— more than a million tons of coal!—Mill Branch paid out a mere $397,288 in fines.—What about those *undetermined, contingent, disputed and unliquidated* obligations? Ask Jim Justice.

Simpletons such as myself might imagine that $2 billion in bankruptcy debt would drag down a corporation, maybe even into coal-black oblivion, but in 2016, Alpha and all its underlings, including Mill Branch, magically "restructured" themselves into limited liability companies. Back to business!

Do you remember how Don Blankenship's crew used to outwit mine safety snoops back at Upper Big Branch? The folks at Mill Branch upheld that merry tradition—and "tradition" is the *mot juste,* for, after all, coal was a "heritage" fuel! Specifically, they employed underground text messaging devices to prevent G-men from disrupting their practices. In 2017 came the reckoning in a U.S. District Court:

> MILL BRANCH agrees to plead guilty to Count 1[:] . . . knowingly giving advance notice of . . . MSHA . . . inspections . . . MILL BRANCH agrees to pay the maximum fine of $10,000.000 . . . MILL BRANCH agrees to the imposition of restitution in the form of payment of $15,000.00 to the William M. Blankenship and Adam Justice Memorial Scholarship Fund . . .

With the extra $50 conviction fee, which added insult to injury, the punishment added up to $25,050. As it was, poor Mill Branch had already shelled out $235,488 for various "citations, orders, and proposed assessments."

Now for the consolation:

> MILL BRANCH, and its prior corporate parent, Alpha Natural Resources, Inc. ("Alpha"), have cooperated with the United States Attorney's Office[, which therefore] . . . does not intend to prosecute any of MILL BRANCH's corporate parents . . .

But what did Mill Branch have to do with Jim Justice? To answer that question I now lead you a few paces back to the first sign on that dirt road, the one for Nine Mile Mining, Inc.—which had been busily associated with several *operators;* indeed permit number 1102022 for Nine Mile Spur, LLC, referenced the *operator:* Mill Branch Coal Corporation.

So Nine Mile and Mill Branch were for at least one period connected, somehow. By now you may begin to perceive why it is that I spare you the pallid court filings, articles of restatement and puppet directories of the *regulated community.*

While Mill Branch was grubbing up its million tons of coal, Nine Mile had unearthed only 201,075 tons. But I was never an enthusiast of coal production anyhow. What did excite me was that the *current controller* (beginning in April 2010) for Nine Mile Mine No. 2 was our hero, James C. Justice II.

Since commencing operations in 2009, Nine Mile had racked up 18 pages' worth of fines! A typical page referenced around 25 violations. Most fines were only a few hundred dollars apiece, although several exceeded $1,000; a few had been pegged at exactly $4,000; one was $7,000; two were $25,500. I am happy to inform you that a certain $24,000 fine had even been *closed,* which hopefully means *paid,* but half of Jim Justice's fines were delinquent.

Current penalties, goodheartedly reduced from *proposed penalties,* came to $230,862, of which $85,262.87 had been *paid.*

The tally ended in 2012, when Nine Mile apparently got *abandoned.* But this company was as unkillable as Mill Branch—for in 2016, Nine Mile Mining, Inc., became one of the defendants in a U.S. district court lawsuit filed by several states and the EPA. *The Complaint alleges that the Defendants have violated the conditions and limitations of National Pollutant Discharge Elimination System ("NPDES") permits issued to them by the relevant State of Alabama, Commonwealth of Kentucky, State of Tennessee, Commonwealth of Virginia, and State of West Virginia . . .* The civil penalty was to be $900,000—half for the federal government, and the rest to the five states which the defendants had polluted.*

First on the list of defendants to be notified of any further legal shufflings was James C. Justice II, Executive Vice-President, Southern Coal Corporation.

Appendix F to this document lists approximately 90 violations against Nine Mile Mining specifically. And now let me resurrect the question of why Mill Branch, which escaped mention in the district court's decree, produced a million tons of coal, when Nine Mile, which certainly appears to have been a pseudopod of the same Mr. Justice, grinding away in the same neighborhood, scraped up barely 200,000 tons? Perhaps Nine Mile's purpose was to be a buffer of distraction, because whenever its violations got singled out—for impermissible quantities of manganese, iron, *settleable solids* and *total dissolved solids,* not to mention unacceptable pH and *flow*—the second column of Appendix F reminded us that Nine Mile existed in some realm where Justice is *operator only.* For all of Mr. Justice's other companies, page after page of them, this column read simply *N/A,* not applicable.

Fortunately, *the United States has determined that Defendants have limited financial ability to pay.* Was that the reason why, as Mr. Sturgill assured me, they didn't?

Passing a security guard whom he sweet-talked into letting us pass, we arrived at the stripper job. A little yellow grass had begun to grow on the older part of that grey gash.

---

* The last state, West Virginia, sweetly refused to be compensated, for a reason to be discussed on p. 107. Nothing stank about *that!*

I asked my standard question: "What do you think it will look like in a hundred years?"

"Probably like that."

Or, as the other side told it: *The restoration develops beautifully over time. Unfortunately, most never get to view these areas as mining areas still under bond are off limits for safety reasons.*

According to the West Virginia Coal Association, mountaintop removal should now be called "mountaintop mining." The Association defined it thus: *Surface mining technique which removes overburden at the top of the mountain in order to recover 100% of the mineral.*

And reclamation was *the restoration of land and environment after the coal is extracted . . . Reclamation is closely regulated by both state and federal law, and the coal industry's outstanding effort in this area has resulted in millions of acres of restored productive land throughout the country*—a most perfect example apparently being the FBI Center in Clarksburg.

Jim Justice's leavings constituted an outstanding effort in their own right. I saw rock dust and slate excavated from the mine. From somewhere came the smell of crude as the road began opening up into the ugliness of the supply yard.

As for the rest, seeing is believing. Let me show you Jim Justice's gift to the world:

Long before mountaintop removal a certain wise forester once wrote: *The farmer who clears the woods off a 75 per cent slope, turns his cows into the clearing, and dumps its rainfall, rocks, and soil into the community stream, is still (if otherwise decent) a respected member of society.* These words of quiet, patient bitterness had grown antiquated by the time I read them. It had grown too late for quiet patience. But that was all I had. So I exposed more frames of film, turned away from that double-entitled stripper job, and got back into Mr. Sturgill's car, wondering whether Jim Justice might be otherwise decent; no doubt he was a respected member of society—he must have been, for at the White Sulphur Springs resort (which he owned) he had just announced his candidacy for Governor of West Virginia! As he explained himself with statesmanlike eloquence: "We need someone to step forward who doesn't have a vested interest in trying to do something for themselves, and do something."

For awhile he couldn't be bothered to debate his opponents, and he saved himself trouble in other ways. One of his 97 firms, Justice Energy Company, had been sued for debt in 2013, missed four court appearances, and was

"By creating lakes, ponds and other permanent water bodies during the reclamation process, coal is the only American industry that is a net producer of vi-
tally important wetlands."—National Coal Association, 1993

finally held in contempt of court by a federal judge. Well, why should he care? The President of the United Mine Workers said: "Jim Justice is one of the good coal operators."—"Good" was an understatement; I'd call him a saint!—As of 2014 he owed not quite $2 million (or by another reckoning $2.1 million) in unpaid fines for breaches of mine safety—the third highest amount in eastern Kentucky. *The delinquent Justice mines committed more than 4,000 violations . . . 1,300 citations were classified . . . as reasonably likely to cause injury or illness . . . More than 500* of the citations flagged conditions *common in mine disasters, accidents and deaths.* A news analysis determined that *at mines with delinquent fines from before and after 2009, the injury rate was 71 percent higher than at mines where the companies paid health and safety fines.* Thus the definition of a *good coal operator.* In case you worry that those fines would have bankrupted Mr. Justice, let me inform you that they amounted to a fraction of what he gave to his pet charities. On the other hand, why *should* he pay? He deserved exemption—because as the newspaper pointed out, he was *the state's richest man.*

Now you understand how he ran his business.* And when it came to West Virginia, *Justice said he wants to run the state more like a business.*

Three months after I saw his handiwork on the Virginia border, the John E. Amos Power Plant (whose white coal-smoke you have seen on p. 45) bought three-quarters of a million tons of coal from Mr. Justice, *without fielding bids from other coal suppliers.* By the way, the offer was *unsolicited.*

In 2016, when he and the U.S. district court finally reached a "deal," his water pollution violations numbered 23,693, but only 837 of them occurred in the Mountain State, which graciously *declined to participate in the settlement,* so that Virginia, Tennessee, Kentucky and Alabama shared the payout with the EPA. Why did West Virginia walk away? Well, as this chapter has already proved, her citizens were so very, very prosperous that they didn't need Jim Justice's money. Moreover, as the DEP explained, *the environmental performance* of Mr. Justice's

---

* Another of his business dealings: In Wyoming County, Dynamic Energy, "a subsidiary of Bluestone Coal Corporation, a Jim Justice company," received an MTR mining permit. A Russian company bought the mine, then in due course sold it back to Mr. Justice. By then the residents of "a few dozen homes" along Cedar Creek Road had detected "changes in their water." Testing revealed high concentrations of strontium and other flavor enhancers. In 2014 these citizens filed suit, and a judge ruled that the company must bring in water for them. "But residents had to go back to court to get their replacement water when the company failed to supply it after Jim Justice repurchased the mine." They lost, of course. True to its inmost nature, the state Department of Environmental Protection "testified on behalf of the coal company." But because these allegations appeared in an environmental newsletter, any right-thinking patriot should reject them.

entity, Southern Coal,* *had improved, prompting state officials to believe it was "unnecessary to double down on them."*

In case you are wondering, why on earth would *the state's richest man* want to bother his head with a governorship? For one thing, it would be up to him to "implement" President Obama's carbon emission plan, assuming that the plan by some miracle had not been killed. Imagining his "implementations" was amusing. Better yet, the Governor got to appoint two of West Virginia's three Public Service Commissioners, whose task was the regulation of utilities. Reader, has your question been answered?

The gubernatorial contest was perfect: Tweedledum *versus* Tweedledee.†

"THE BILL COLE PLAN"

From the platform of Jim Justice's opponent

[who owned the Cole Automotive group and the Bill Cole Automall in Ashland, Kentucky]

- Control the out of control growth of state government.

· · ·

- Eliminate burdensome regulations.
- Harness our natural resources to put our energy resources back to work.

November 8th JOIN SENATOR SHELLEY MOORE CAPTIO [*sic**] IN VOTING BILL COLE FOR GOVERNOR.

---

* Mr. Cole, or his flunkeys and handlers, couldn't even spell his sponsor's name right.

Once the election race heated up (by which time he owned only 83 companies), Jim Justice actually condescended to debate his Republican adversary,‡ and

---

* You may remember Nine Mile Mining, whose signage stood proud at that unremediated site in Virginia to which Stanley Sturgill took me. It was the miners of Nine Mile who had sued Jim Justice for unlawful termination. One document filed in District Court during those proceedings was a "Positive Corporate Disclosure Statement . . . identifying Corporate Parent Southern Coal Corporation for Nine Mile Mining, Inc."

† The three other candidates, respectively representing the Libertarians, the Constitutionalists and the mildly leftwing Mountain Party, never had much of a chance.

‡ In 2017, perhaps on account of Trump's electoral victory, Mr. Justice became a Republican.

then, like a good boy, he even mused aloud about those unpaid fines and taxes: *He said he owns 102 companies, writes 7,750 checks daily, and in terms of paying back his debts, "may be a little late here and there, but I'll always be at the party."*— And what a party it turned out to be! For in October 2016, a few days after the Environmental Protection Agency *reached a $5 million deal* with him *to resolve hundreds of pollution violations,* Jim Justice became Governor-elect, and as I sat reading the paper I found every reason to expect that he would prove as brilliantly benevolent as his predecessor (who had endorsed him). For one thing, he possessed *a different way of thinking,* and it was exactly that he chose to employ *to take the Mountain State from worst to first.* One of his ambitions was to grub up West Virginia's woodlands as he had that Virginian coal seam whose remnants Stanley Sturgill showed us. I anticipated thrilling improvements in tree-related carbon sequestration. After all, this was still *the third-most forested state in the United States. Justice said he wanted to take advantage of federal timber subsidies to compensate logging companies with the goal of attracting furniture manufacturers* . . .

"BUT IT'S HOME TO ME"

Now let me tell you about another brave man, a self-employed carpenter whose name was Chad* Cordell. It was Mr. Barney Frazier, a retired lawyer whose back view was being encroached upon by one of those stripper jobs whose corporate ownership appeared to change whenever expedient, who first mentioned the meeting that would be held that very evening in Charleston. (You will come to Mr. and Mrs. Frazier's story in a few pages.†) Because Mr. Cordell and the Fraziers saw a common interest—for they had a common enemy, Keystone Development—it was Mr. Frazier, in spite of saying, "I'm not in that fight, since I've got enough of a fight over here," who actually led the way, driving sedately down the twisting roads from his hilltop mansion to the United Presbyterian Church, in whose window hung the following sign:

The presentation was marked more by sincerity than by any sort of procedural perfection. It convinced me.

---

* Or, in the legal filings, David C.
† Pages 163–69.

Mr. Cordell showed photos of giant white oak trees and American chestnuts the size of redwoods. He said that between 1880 and 1920 almost every tree in Appalachia was cut down.* The way he described mountaintop removal, first they clearcut, then they scraped off the topsoil, after which they brought in a blasting rig with diesel fuel and ammonium nitrate. Their explosions were "100 times the force of the Oklahoma City bombing." They detonated the rock apart, and dumped the rubble into valleys. He claimed that more than 1,000 miles of rivers in Appalachia had been covered over in that way.†

"Anytime we're confronting a deep-rooted and systemic wrong in our culture, it's gonna take a lot of work," he told us. "We want to stop this mine in our back yards . . . and we want to stop mountaintop removal everywhere. We actually have laws that have been changed because of the public's involvement over the years. There's nothing like this when you talk about mountaintop removal. Put the Governor on the spot. Make a stand."

The Kanawha State Forest was one of the few gems of West Virginia's capital city. If somebody could strip mine there, in plain view of one of the hiking trails, then no place was safe from mountaintop removal.—When I asked Chris Hamilton of the West Virginia Coal Association for a statement on this particular case, he called it unique; coal rarely appeared in deposits of such contested character: "You know, you're not going to find the large contiguous reserves that lend themselves to be mined through large-scale MTR methods, and that's why we don't mine in all the 55 counties. Presently the Kanawha Forest is the only example of where those two interests collide. Like I said, the top five or six counties do account for the top 70–80% of the coal we mine.

"Now I would point out in the defense of MTR," he continued, "when that was first raised with the level of concern, and that's been almost 20 years now, there wasn't a whole lot of planning and forethought that have been given to post-mine uses. We have a very active office of coal field development; the concept is that we'll talk about post-mine use as soon as possible. Years ago and even today if there's not a post-mine planned, we go in there and mine for about five or 10 years, and then the first thing we're required to do when we come out of there is remove all the buildings, the bathhouse, the mine office, we take every single road out of the area, then they require us to take out all sewer lines and power. What

---

* See "Carbon Ideologies Approached," I:79.

† But look on the sunny side, ladies and gentlemen! Let me quote from *Coal Facts,* which may as well be the Bible of mountaintop removal: "In West Virginia, the little hollows along which most people live often flood, wiping away lives and life's work in just minutes . . . The people . . . build their homes along these little hollows because there are no other good options . . . Former mine sites can be configured for residential development."

we started to do, while we're in there with our power, let's look then at what we need to do 10 years from now. There's truly good examples all over our state.

"I could take you to any industrial site—doesn't have to be a mine site—or putting in a new housing development or shopping center or a coal mine. It's not going to be the most pleasant thing, aesthetic thing while it's in progress, and I do make the comparison between them, but you come back in a year when the place is regraded and revegetated, and the highwall is removed, and it's going to look different. Now, if you're against coal, that's not going to make a difference."

Jim Justice's legacy in West Virginia did not strike me as one of those *truly good examples,* but at that time I had not yet seen it, so I contented myself with asking how negative the effects of mountaintop removal might be, to which Mr. Hamilton replied: "While living near or close to a mining operation causes some inconveniences, I think by and large those are relatively isolated to just a few areas and certainly not experienced on a statewide basis. Also, and let me just make this point, we're the second leading coal-producing state, only behind Wyoming, so we're known as a mining state. We also have a very, very robust tourism business. Generally speaking, there are just natural barriers between where you have a lot of mining activity and where you have a lot of tourism. I personally hold that we're a pretty good example of how you can have it both ways, how you can have a pretty strong energy industry that creates a lot of jobs and wealth for this state, and also a good tourism industry."

Letters to the *Charleston Gazette,* 2013

Eighth-grade boy from Chloe:

> Many people are all for taking the top off a mountain. Others would rather save the environment instead of create jobs for West Virginians. I support it all the way . . . There is no reason to put families in poverty just because you do not want a little dirt in your water. Dirt has never killed anyone when consumed. In conclusion, I believe mountaintop removal is a major necessity for West Virginia.

Eighth-grade girl from Millstone:

> I support mountaintop removal because West Virginia needs coal . . . It has killed people, but a lot of people would starve and die without the benefits they receive from coal. Yes, it destroys the mountain; but people destroy the land with skyscrapers and new buildings all the time.

I remember a dawn walk I took with my friend, a few days after the slide show at United Presbyterian. Entering the Kanawha Forest, we climbed up to the vantage point, which was only five miles away from the state capitol. From the spider-webbed green shade of birdsongs we looked down through fat and skinny beeches to a bright greyness of the opposing ridge where the mountaintop removal would be occurring, unless the activists won their struggle with the DEP. They lost, of course. The West Virginia Surface Mining Board *denied motions by the Keeper of the Mountains Foundation* and 10 individuals, among them Mr. Cordell, *to temporarily halt Development's new mountaintop removal coal mine on the eastern edge of the Kanawha State Forest and to delay a final hearing on whether the project is properly permitted.* In October 2014, a circuit court judge *wrote that the DEP violated state law by failing to give mine opponents an opportunity to be heard.* All the same, Keystone kept right on going. And so a year after that dawn walk I asked the young man who had given the slide show for a tour of the ongoing damage. The local paper had recently quoted him regarding "acid mine drainage due to inadequate sediment controls." He spoke of "a clear pattern of violations and a staunch unwillingness to comply with regulations and permit conditions meant to protect state waters from pollution and acid drainage." But what truly touched me was when he said: "I took a hike out there with the kids this winter and we parked down at the shooting range and we walked up the mountainside; it's super steep—you're almost on your hands and feet just crawling up it—and we got up there and we hiked along the ridge line for a ways and it was winter and the great thing about winter around here is that you get these long distance views that you don't get when all the leaves are on the trees, and so you could just see forever; I could see all across from the top there, all across to the neighboring ridgetops and I just kind of looked at my kids and said, you know what? Let's fight for this one."

He had been living hereabouts since he was three years old; now he was 36.

I will not say that we did or did not commit illegal trespass. When I was alive, the big boys of resource extraction (had they been inclined to read) would surely have characterized *Carbon Ideologies* as a work of fiction; they were correct, and I a fantastical slanderer. So let the following be fiction:

There were three of us. Walking up the former logging road that presently became a creek, the Kanawha Fork, which we waded in and out of, the water luke-warm as it entered our shoes, and cool shade consoling us for the extra 10 pounds of humid air we seemed to be carrying on our shoulders, we proceeded up the holler, with the mine unseen above us to our right. When we began to hear the beeping of a bulldozer in motion, Mr. Cordell advised us not to speak too loudly.

He was very singleminded. At first he expressed small interest in us, which I found appropriate, his focus being the endangered forest. Once we got to the steep and brushy part of the climb, which I, being 20 years older and 30 pounds overweight, found considerably more difficult than did my companions, he proved himself a prince of tactful patience, halting whenever I needed to; and when we gained the ridgetop I felt that outright camaraderie of hikers, lovers, soldiers and all their kin who press on together into lonely places.

As he went, he sometimes bent down to test the water for pH and conductivity. "For pH we typically see anywhere from 3.1 to about 7.0," he said.* "The standard acceptable pH is 6.0 to 9.0 by West Virginia legal standards."—As for conductivity, that indicated metal content. The previous year a district court judge in Huntington had finally ruled that high conductivity could of itself be damaging, and might signify the presence of a chloride, phosphate or nitrate. Anything above 500 was bad, said Mr. Cordell. He now measured one spring at 20; but some of his readings had been 600 or even 3,500 right here on Kanawha Fork. "Some areas you'll see in the stream, acid places, what you'll see is a metallic precipitate." The spring at our feet tested in at 20—a happy analogue to "normal background" radiation.

I remember tall whitish-barked sycamores along the stream, and maple and oak, and a few poplars thin and straight; there were many grand beech trees. "This area was probably logged several times," he said. To my ignorant perception the forest looked healthy—but here I should like to quote an old private detective who taught me a few things about Imperial County, California. He'd said something like this: *When I first moved here and looked at those fields, I didn't understand what I was looking at.* If only I were a chemist, a botanist, a hydrologist—and, of course, a climatologist . . . and a mining engineer with a special interest in coal mines! As for the medical aspect of the matter, let me quote a certain scientific abstract: *Rates were significantly higher in mountaintop [removal] mining areas for six of seven types of defects: circulatory/respiratory, central nervous system, musculoskeletal, gastrointestinal, urogenital, and "other" . . . Spatial correlation between mountaintop [removal] mining and birth defects was also present . . . Elevated birth defect rates are partly a function of socioeconomic disadvantage, but remain elevated after controlling for those risks.*

At any rate, Mr. Cordell understood much more than I of what we were seeing.

Here for instance clung another tree whose naked roots grappled down the steep bank and in spots groped in air, with dark hollows behind them.

---

* As the reader might know, this variable has been assigned a logarithmic scale from 1 (most acidic) to 14 (most alkaline). A pH of 7 is neutral. Six is 10 times more acidic than 7. Thus a polluted West Virginia creek with a pH of 3 is 10,000 times more acidic than distilled water. On the subject of mountaintop removal the National Coal Association advises: "It is impractical to attempt to reclaim spoil for farm uses such as hay, pasture, or row crops if the pH of the spoil material remains at 4.5 or below."—Everything is relative; the pH of vinegar is 2.4 to 3.4. But how would you like to drink vinegar every day and night of your life?

Mr. Cordell said that the stripper job above us had robbed the hill of topsoil that formerly held rain; accordingly, the side-gullies often flooded, increasing the speed and volume of washouts.

"Sometimes down here what we've seen is a greyish precipitate. Sometimes the water turns orange. And when you see these big cuts where a lot of water comes through, that's erosion caused by the mine above. Lately it's been very

Streambank eroded (according to Chad Cordell) by mountaintop removal. Note the exposed tree-roots.

turbid; now we're seeing more of that, that is especially after heavy rains; in these side streams it looks like mud coming down—which is incidentally a violation of West Virginia water law—which the DEP could enforce if it wanted to. The water runs off the top of the mountain just like the top of a parking lot."

But I am not denying that the forest looked pleasant, especially in the shadows as we strode up the red-mudded creek, admiring the jewelweed and spiderwebs, avoiding the stinging nettles and watching for poison ivy. I wished that Chris Hamilton were here with us to relate his point of view. Off a muddy bank, half unrooted, hung another sycamore, whose bark shone white. Down the hillside flowed what resembled pure pale water, dripping over into a dark fall and then eating away at muddy steps. We passed the stepped jumbles of a side-creek which was deeply cut in the ground; at first it seemed merely distinct, not wrong; ferns grew over its edge as it hissed and sparkled in the greenness, wetting logs, rock-knuckles and knife-edges of slate. It had eaten a narrow deep rut into one side of the old road, in a spot where according to Mr. Cordell the erosion had formerly rated "not bad."

How much of this had truly been caused by mountaintop removal? As I keep saying, had I only possessed some equivalent of the pancake frisker to measure coal's damage and danger, then I could have accomplished more in this volume of *Carbon Ideologies* than simply retailing my observations and the analyses of others.—But what pancake frisker could ever illuminate our unknown future as it used to light up the radioactive darkness of Tomioka? How many river-traits and coalqualities would go forever undiscovered?—Light-stripes were wilting across the dirt road as if an American flag had been laid over some unevenness of pebbles.

By now the noises of earth-moving equipment had become much louder; the mine lay right above us. So we began to bushwack up the opposite bank, finally forsaking this winding former road through which the river flowed (it looked mostly clear to me, although occasionally even I could see sediment). Sometimes the poison ivy was waist-deep. As soon as I got back to Charleston I would drop this set of clothes into a coal-powered washing machine, then shower, scrub hard and bask in the coolness of my coal-conditioned motel room. Meanwhile I huffed and puffed behind my companions, and we ascended into that sound of machinery. Mr. Cordell never got tired.

I think that the higher we went the more he liked me. Or it might have simply been that this was his true home, and he was happier the deeper into it he went. Dragging myself upward, shrugging off greenbriers and wiping my forehead on my dripping sleeve, I began to draw level with those noises of heavy equipment.

On the south-facing side about 200 feet up, on the west-facing slope above the river, we reached a terrace (Mr. Cordell's global positioning system read 37 0.002

degrees of latitude). We took a breather. Up higher where the hill began to flatten out again, decorated with boulders like stacks of great flat fists, and trees sprouting up between them, crowds of greenbriers, blueberries and sassafras awaited us: 480 feet above Kanawha Fork, absolute elevation 1,280 feet above sea level. By now the earthmovers had fallen silent; it was probably six o'clock.

Across the gap, other bluish-black terraces asserted themselves: the KD No. 2 Mine.

Mr. Cordell led us down boulders to a ledge blanketed with dead leaves; and then through damp-smelling rotten wood beneath a great mushroom-studded sandstone boulder, with wet tussocks of sphagnum moss in the blueberries and poison ivy, while a crow cawed far away. Gliding easily down from a boulder, he came to the edge of the mountain's shoulder and carefully peered out between the branches like the insurgent that he was. Through the thick growth came a glimpse of the mine:

. . . Through a hole in the leaves above my left shoulder I could see across that steep river valley to a partially silhouetted wall of pale-trunked trees slit evenly across, much as is a stretch of forest bisected by a freeway; and the edge of the hill (which, this being Appalachia, was called Middlelick Mountain) on which those trees stood curved gently up and down like an eyebrow. Beneath this the hillside had been sheared away. Within that wound, horizontal bands caught the light. They might have been roads and coal seams; Cook Mountain had offered a far nearer view of the grooves and terraces of mountaintop removal; from here it was difficult to tell, especially when I as usual could not understand what I was looking at. I asked Mr. Cordell to explain, and he said:

"From here you're seeing the main pit, which is also called the pavement, which is the lowest level, and the main haul road is also that contour there, and right above that we're seeing the highwall. They started by cutting the haul road out. I haven't seen up close and personal any other strip jobs, but this one, where it grows back up, they clearcut last year, and it's recovering as a forest tends to do. What we're seeing over there is where they're starting to blast to make spoil to cover what they've done."

There was another place in the leaves where I could see farther down, and that much-interrupted column of vantage revealed many near-white horizontal grooves which could have been terraces in some gigantic overgrown ruin. In this place, part of the brow beneath the trees on the crest seemed to retain shrubs or grasses a few paces down to where the cliff began.

We sat looking up between those dark and slender tree-trunks through a wealth of summer leafage into what I can only call a *handsome light,* grand and far-going, and to me who often finds so many things pleasingly feminine, somehow masculine in a way I cannot explain; this surely reveals more about me than what I was looking at, but (as Thoreau must have said) it may well be that the more we are in nature, the more we begin to see whatever lies within ourselves. Conversely, when we come into the manmade ruin of nature, we can describe it fairly exactly, in a way that other observers, even should they subscribe to an opposing carbon ideology, can recognize: The damage begins here and takes on such and such a shape—a human-formed shape. In the middle distance, lovely outspreadings of foliage caught the sun; then there was a gulf of brightness, and the far side of the valley a darker green beneath the humid white sky.

It was almost beautiful, like looking down into the Grand Canyon: reddish and purplish bands in the sunset. I admired the long grin of the main haul road.

(I had the pancake frisker, of course. The air dose measured 43 counts per minute, 0.12 micros an hour. From an alpha-beta-gamma standpoint this place was nearly as healthy as could be.)

Pointing at the sediment control ponds, Mr. Cordell said: "You've got all this acid-producing material from the blasting on the pavement that's producing runoff, this whole area that's picking up this acid, picking up heavy metals. They're using big tanks of caustic soda to bring the pH up. It increases conductivity. The DEP will tell you that high conductivity does not necessarily mean aquatic impairment," he said, popping a blueberry into his mouth. "I don't know that that's true."

I studied the cutaway forest above the forehead of the grinning cliff, and I decided that I did not like it.

Mr. Cordell said: "You've taken this

Mr. Chad Cordell

mountain, and you've taken this whole complex hydrological system that once gave you drinkable water and you've turned it on its head."

"What do you miss most?"

"From the aesthetic side of it, there's no match for me, this forest, with those incredible boulders and blueberries surrounded by Appalachian hardwood forest, and right here I'm in heaven. And when I look at that mountain over there, it's like a golf course. That's why I don't talk about the beauty. But it's home to me. My Mom is from Memphis, Tennessee, and I cannot imagine myself there, because this is where I was formed, and I see this as an assault. If you're living in your house and a company comes in and says, we want what's in your house, and they bring the bulldozers and bulldoze your home and then they scrape all this rubble back into a pile as high as it was and say, we've reclaimed it; there you go . . . When they finish mining, none of these systems work the way they used to. And they say: *There's your mountain.*"

"What's your opinion of coal? Even if you don't want to burn it, you must have seen the coal products tree, with those coal tar derivative paints and billiard balls . . ."

"Hmm," he laughed. "Coal is best left in the ground. I'll give you my opinion. *My* opinion is that we don't need this stuff! We don't need billiard balls and all that shit. When I think about energy, I see a civilization that's literally killing itself for a bunch of shit it doesn't need."

"Do you think it's too late?"

"I honestly have no idea. I don't talk a lot about climate change. Particularly around here, there's so much pushback, and it's not something I've studied. I talk about the things I can find agreement on, like childhood asthma rates, like mercury in the streams that makes them put out fishing advisories. So I tend to steer clear of climate change. But all I know is, I want to try . . ."

"To try what?"

"To do what I can. My friend who lives on the Coal River told me that the shellfish are gone and the fish are almost gone, the swimming holes gone . . ."

There was a fractured sandstone boulder with lichens on it, up which our third party proudly climbed for awhile; then we all sat gazing down into the evening, with a cawing crow and that blasted forest far over on the other side, very tiny, everything so far away on that cool green evening, the lovely purple-red grin of the stripper job going grey as the sun went down.

We descended into the dusk, with chanterelle mushrooms growing around us. Mr. Cordell said that if his legal fight against KD No. 2 Mine were defeated he would embark on civil disobedience. From time to time he knelt to test some pool or creeklet; and in the twilight, pallid toadstools gleamed at our feet.

THE *REGULATED COMMUNITY*

1: "No One Would Deny That We Have Shown Leadership"

One might suppose that myriad unwanted mountaintop removal mines, not to mention 100,000 coal miners' premature deaths, like the billions of fatalities which could result from global warming (reader, since I express my forebodings without certainty, let me again remind you that I lived in your carefree plenteous past), would have impelled our species to regulate its more callous members, but in my day, the faction enjoying preeminent rights consisted of what West Virginian business and political circles (much the same) liked to call the *regulated community,* by which was meant the coal lords, chemical cowboys and kindred entities whose profits ostensibly supported the whole state—never mind that the state took pride in its low corporate tax rates—and whose continuing environmental havoc constituted a necessity which only a traitor or Communist could have questioned; whose greed was noble Americanism and whose human casualties were unavoidable acts of God. When the Obama administration announced that thanks to its latest rules *drinking water for 117 million Americans will be protected* from pollution via streams, wetlands and the ubiquitous *Unnamed Tributaries* you will soon be reading about,* House Speaker John Boehner thundered that these rules would dispatch *landowners, small businesses, farmers, and manufacturers on the road to a regulatory and economic hell.*† If a river turned black or white, or the tapwater rotted away somebody's teeth, why should that be the *regulated community*'s concern? If the damage didn't show, it almost failed to exist. As for more conspicuous examples, perhaps the *regulated community* could offer them up as improvements: *Some geographers believe that if West Virginia's mountainous regions were flattened out,*

---

* See p. 154.

† In 2017, before Trump had been President for even two weeks, the Senate would vote, per corporate interest, to "reverse the Stream Protection Rule."—And because Exxon dreaded being boxed into "a competitive disadvantage," the Senate meanwhile "moved to reverse a separate rule requiring publicly traded oil, gas and mineral companies to disclose any payments to foreign governments for licenses or permits."

the area covered would extend well beyond the borders of the entire United States.
So why not engage in more mountaintop removal, for the sake of state aggran-
dizement?

Coming out of Pineville on Pinnacle Avenue, briefly trailing a chocolate
creek, then crossing a railroad track, one could drive along the southward windings of Highway 16, descending through the dank and mostly deciduous forests
of Micajah Ridge to Wolf Pen, Woosley and finally, suddenly, Welch; through
the windshield it all looked wild and good. Sometimes not a single other vehicle
showed itself on that rainy hill, whose best advertisement was this handmade
sign: DIRT WANTED. It was as difficult to believe in contamination hereabouts as
to feel the radiation in Fukushima. Joe Manchin (who remarked that *when the
industries see the D.E.P. coming onto their property he wanted them to feel comfortable*), Shelley Moore Capito and the other politicians tried to paint a similar
picture—the mining entities being not only powerful, but even popular: Coal
was America's best friend! (Stanley Sturgill, 2015: "I was down there at the post
office the other day and a lady hollered, *hey, you damn tree-hugger!* and I just said
yeah.")* But this was too perfect a state of affairs; bit by bit regulators crept in;
our labor barons, deserving only of hosannas, became the *regulated community.*
They fought back, and such was their pull that the rare do-gooding senator got
foiled . . .—but over time the United States swung leftward, not merely espousing
but even seeking to enact such godless doctrines as mine safety, community
health (the "community" here being not the poor *regulated community* but
whichever peons lived downstream and downwind of its effluvia), environmental
protection and the like; and so thanks to outrageous federal intrusions the *regulated community* grew regulated indeed—never enough for me, I admit, but far
too much for its resentful scions.—Future reader, possibly I overstate my case;
you may be getting along quite happily, checking the radiation weather reports,
pollinating your corn by hand, hiding underground in summer and drinking
the recycled urine of your neighbors. Perhaps regulation will save you. And if my
regulated community can find profit in regulating *you,* who can say what lovely
sulfurous sunsets you may enjoy? But when I was alive, the *regulated community* so fervently longed *not* to be regulated that it bridled at the mildest oversight. After all, who wouldn't act the same? (Gabriel Kolko, 1963: *Federal
economic regulation was generally designed by the regulated interest to meet its own*

---

* As I learned from a newsletter on the wonders of mountaintop removal, "the attacks are coming from
a combination of sources including leftwing radicals, professional protestors, a biased news media,
federal agencies and the Obama Administration."—Which of these was Mr. Sturgill, I wonder? I guess
one couldn't be more leftwing than a retired coal miner who had also been a mine safety inspector.

end, and not those of the public or commonweal.) That this was the case not only in West Virginia appears in the following tidbit (or tit-bit, as the British used to say) from the *Financial Times* of London: Upon the return to power of the Conservative Party leader David Cameron, *UK stocks enjoyed a brief rally, with the FTSE rising 2.3 per cent. It was led by energy and banking stocks, two sectors that would have faced increasing scrutiny under Labour.*—If escape from scrutiny raises a corporation's cash value, what does that say about the corporation's business practices, never mind the investors whose buying habits determine that cash value?*

The *regulated community* did learn to bend a little. When one of its projects caused a disaster, it kept quiet, expressed regrets or promised to learn from the experience and make itself even better than before. It pretended that it and its victims shared the same purpose, as they mostly did: who wouldn't want electric power on the cheap? When new restraints upon it were proposed, the *regulated community* protested. When outrage failed to meet the case, it delayed, bustled and prayed for rain. Pretty soon some terrorist atrocity or the birth of a princess would supervene. And then? *We have seen similar reports, written with the same good intent, gathering dust on the bookshelves of the national Mine Health and Safety Academy.*

From what I could tell, West Virginian politicians loved the *regulated community.* They did whatever they could to ease its business stresses:

A FIELD GUIDE TO WEST VIRGINIAN POLITICIANS,
ca. 2013–15

Each species herein remained by definition unique in form and lifestyle. Shelley Moore Capito, for instance, was an elegant, older sandy-blonde, sometimes observed with a string of pearls around her neck, which was no more wrinkled than mine. Josh Nelson was a veteran "and currently a pilot in the West Virginia Air National Guard's 130th Airlift Wing." Evan Jenkins was once discovered all suited up inside the Hunter Peerless Mine, with his hands folded at his crotch and two reflective stripes glinting across his chest. Nonetheless, all species featured in this table displayed an impressive homogeneity in regard to global warming, pollution and related matters. (It is hard to believe that they and I had a common ancestor.) When I was alive, they seemed nearly

---

* As Tim Bailey once remarked: "If you're a coal company and you feel the need to call everybody and look underground, clean stuff up" when the mining inspectors arrive, "you're probably breaking the law" in the first place.

invulnerable. I would expect that after our generation died out, their kind began to suffer habitat loss.

———

Ever since these mines had been started, the operators had controlled the local powers of government . . .

Upton Sinclair, *King Coal* (1917)*

———

Shelley Moore Capito, Republican, Representative

- Introduced the Affordable Reliable Energy Now Act in 2015. This bill read in part: *The following rules shall be of no force or effect, and shall be treated as though the rules had never been issued . . .*—meaning Environmental Protection Agency regulation of the greenhouse gas emissions of coal power plants.

- "Capito has repeatedly declined to respond to the Gazette when asked if she agrees with the scientific consensus on global warming."

Evan Jenkins, Republican, Third District Congressman

- "Democrat/Republican Evan Jenkins is the darling of coal operators. More than $2 million is being expended by groups allied with Massey Energy to unseat Third District Democrat Nick Rahall."

- Told the workers of the Hunter Peerless Mine "near Whitesville": "Don't let anyone fool you, coal is under attack by this current administration . . . It's incredible to me that they are using every tool to kill who we are[†] and what we do."

- When the U.S. Department of Interior introduced a "proposed stream protection rule," Jenkins said: "I am outraged . . . I've spent months talking with our West Virginia coal miners and know how worried they are that this rule will cost them their jobs." Was particularly infuriated that "among other provisions, the proposed regulations would require mining

---

[*] Sinclair remarked: "Such conditions are to be found as far apart as West Virginia, Alabama, Michigan, Minnesota, and Colorado."

[†] "Jenkins said his great-grandfather died in a coal mine and his wife has many miners in her family as well."

companies to collect water samples from nearby streams and rivers before and after mining and while operations are underway." Promised to sustain his counterattack against the Obama administration's "regulatory overreach and its war on coal."

Joe Manchin, Democrat, Senator [formerly Governor of West Virginia]

- Regarding that "proposed stream protection rule" [see **Jenkins**], Manchin said that it "was based on undisclosed data and was written without proper consultation with the states." He "urged his colleagues to pass the Supporting Transparent Regulatory and Environmental Actions in Mining (STREAM*) in order to rein in the newly proposed rule."

Patrick Morrisey, Republican, State Attorney General

- "Americans may be divided on what measures to take with the global issue of carbon dioxide emissions, but we should be united in our respect for the rule of law," and the EPA regulation of greenhouse gases was somehow "blatantly illegal."

- "Is surveying the state's 55 county commissions about whether they have prayers at their meetings."

Josh Nelson, Democrat[?], Representative, 23rd Delegate District, State House of Representatives. [All his punctuation faithfully reproduced by WTV.]

- Announced that "priority number one . . . was repealing West Virginia's Cap and Trade law that was passed under the Manchin administration, which he says cut the coal market by approximately 15 percent."

- "Cap and Trade is gone. I campaigned very heavy in getting rid of this liberal policy that has and would have continued to kill coal miner's jobs . . ."

- Warned that "the war on coal would decimate West Virginia's economy, raise energy prices, and send thousands of breadwinners to the unemployment lines."

---

* This acronym is a masterpiece of cynicism.

- "I recently did an interview with the national renown, Heartland Institute* where I spoke about the human impact of coal."

Nick Rahall, Democrat, Third District Congressman

- Denounced Romney because when the latter was governor of Massachusetts he said that *coal kills*.

- "On the grounds of the U.S. Capitol . . . rallied with hundreds of miners against the EPA's ideological war on coal."

- But did co-sponsor a miner protection bill, enhancing whistleblower protections and requiring mine operators to prove they were using limestone dust to reduce the risk of explosions.

Ron Stollings, Democrat, Seventh District Senator

- Helped repeal the Alternative and Renewable Energy Act, which required that "25 percent of all energy produced by electrical utilities in West Virginia be derived from alternative or renewable energy sources." *[Of course the state "gets 95 percent of its electricity from coal."]* "I rose in support of the repeal as I understand the value of coal in our district and state not only as a source of energy, but also a source of income."

Natalie Tennant, Democrat, West Virginia Secretary of State, wife of State Senator Erik Wells

- "Officially the U.S. Senate candidate of the United Mine Workers of America."

- Testified before Environmental Protection Agency hearing in Pittsburgh "opposing regulations on coal-fired power plants and challenging the President to invest in advanced coal technology instead."

- Assured an audience at West Virginia University that "coal is the foundation of our state and will be the cornerstone of our future."

---

* I would have liked to ask this organization about global warming. My fact checker tells the story: *Heartland: emailed on 2-23-16 to no reply*. They must have been too nationally renowned for me.

- "I am pro-coal. I am pro-coal miner . . .* I will stand up to the president. I will stand up to anyone who tries to hurt our coal jobs."

Earl Ray Tomblin, Democrat, Governor

- Said of the new EPA emissions rules (to reduce carbon emissions 30% by 2030): "These proposals appear to realize some of our worst fears. The only way to comply with these rules would be to use less West Virginia coal." What a shame *that* would have been.

- "The EPA's proposed rules would establish unreasonable limits on carbon dioxide emissions, devastating West Virginia and our region by eliminating jobs and unnecessarily increasing the cost of power across the country."

- "We must have reasonable electric rates to continue to be a world power."

- "Coal can, and should, power our country for years to come."

Through an intermediary I tried to ask these dignitaries the following questions. ("Your answers, if any, will be accurately quoted in the book. Refusals to comment will also be noted."):

1. *Do you believe in global warming?*
2. *If so, how much does it have to do with the use of fossil fuels?*
3. *If it has anything to do with fossil fuels, to what extent does coal affect this problem?*
4. *Is it true or false that some West Virginian drinking water supplies are dangerously acidified by coal mine runoff?*
5. *What is more of a threat to the next generation in West Virginia—the War on Coal, or continued use of coal?*
6. *In the hypothetical event that all coal miners in the state, employed and unemployed, could get jobs making solar collectors at their previous wages, would you be for or against it?*

The intermediary, who was famously polite, did what she could. Mr. Morrisey, she presently discovered, "had no available means of email communication advertised on his website." Mr. Rahall was "no longer in office." Messrs. Tomblin and Manchin she left in peace for some reason. As for the others, after half a year I received the anticipated manna: "So far, no response."

---

* It is possible that she actually was. During an election campaign she said of her rival: "While you and I were rallying in Pittsburgh on behalf of coal miners, do you know where Congresswoman Capito was? She was voting to cut and privatize Social Security and sneaking around with Patriot Coal executives who are trying to cut your retirement benefits."

Why did it have to be that way? Because, of course, there was no good alternative.

When we were alive it was an article of faith that elected officials somehow "represented" us. And they did:

> A key attribute of the period was that power did not reside in the hands of those who understood the climate system, but rather in political, economic, and social institutions that had a strong interest in maintaining the use of fossil fuels. Historians have labeled this system the *carbon-combustion complex . . .*

. . . And smiling, distinguished Nick Rahall rolled out the Regulatory Certainty Act, crowing: "This bill is a wrench in the gears of the EPA's machine." If he got his way, the government would find it much more difficult to stop dredging or filling in wetlands, rivers and streams. *Leading Democrats . . . praised Rahall as a "champion of coal miners and working people everywhere."* So did leading Republicans, I'll bet. *In addition, the third district congressman is at the forefront of legislation designed to strip the EPA of funding that would allow it to create rules to cap carbon emissions from . . . coal-fired power plants.**

In West Virginia they gave us what we asked for—just as they did in Japan.— Thus it was that while the tsunami at Fukushima did sadly succeed in vibrating into our hearts, and the resulting nuclear explosions transformed an emergency into an international crisis, three, four and five years later the vibration was dimming down, my American neighbors and even many Japanese turning, as we have seen, forgetful or indifferent after the fashion of humankind—for we must confess that there was "no immediate danger." The proprietress of a dried food store in Kesennuma said: "Memories are quickly fading away. Even some of us, victims of the disaster, have started believing tsunamis will no longer come." And if they wouldn't come, then all those reactors along the coast would be safe after all, in which case why hold back the poor *regulated community*?

An *independent Diet probe* concluded that Japan's *disaster was "profoundly man-made" owing to management lapses and collusion with government regulators.* I have read that for a half decade ending in the year of the great horror at Fukushima, a certain Mr. Tanaka Satoru took in 600,000 yen from the not coincidentally named Tepco Memorial Foundation; that after the accident, when the Democratic Party of Japan established the Nuclear Regulation Authority, it

---

* But he wasn't perfect. "Republican groups have also blasted Rahall for receiving support from a left-leaning super PAC, which has donors who are carbon-capping advocates."

forbade the appointment of anyone who had been paid 500,000 yen or more *from a single nuclear plant operator over the preceding three years*—a fairly lenient rule, I should say—and that the Liberal Democratic Party which then took power *has signaled it will ignore this rule,* and accordingly nominated Mr. Tanaka—perhaps out of compassion for the *regulated community.* It might have been a coincidence that right about then the Japanese were preparing to restart one reactor—not yet in Fukushima, of course—and if that proved acceptable, why, then they might bring more of them back on line. You may recall what that information technology student in Iwaki* said to me: "In the beginning, people were talking about it. Nowadays there's nothing said. It seems that no one is worried"—a mere 150,000 nuclear refugees probably excepted; and maybe even they weren't worried anymore.

(On August 11, 2015, Kyushu Electric Power Company's Reactor No. 1 recommenced fissioning at the Sendai Power Plant,[†] whose *second reactor is scheduled to be brought back online later this year.* And so two out of 43 Japanese operable reactors were once more up and running! By July 2017, five nuclear plants would be fulfilling demand! Reader, wasn't that better than burning more coal?)

Parking lot wall in Williamson

---

* See above, I:341

[†] Despite the name, this complex was not in the Sendai that you and I visited in the nuclear chapter (I:266ff.), but in the south.

From the extremely practical standpoint of the *regulated community,* there was "no immediate danger" anywhere! So why shouldn't the regulators be gentle on them?—Stanley Sturgill, 2015: "A lot of those coal operators still won't pay the fines I wrote up—and I retired in 2009." The *regulated community,* of course, pretended that not it but any regulator was the outlaw. Sensitive to such feelings, kindhearted regulators left the *regulated community* in peace.

Reader, when I was alive nobody thought much of the fact that the new director of West Virginia's Division of Air Quality had *worked in a high-tech segment of the petroleum industry,* whose burnoffs brightened our nights in ever so many sectors of this warming planet. I suspected that he'd pose "no immediate danger"—at least not to the *regulated community.* Perhaps I was unfair, and he became a good watchdog. Some regulators were more aggressive than others. (Example: Lori Ann Burd, who was environmental health director at the Center for Biological Diversity, alleged in a lawsuit that in a single three-month period in 2014, West Virginia *failed to conduct 171 required inspections* of coal mine sites.)* The regulators meant well, no doubt, but why did so many of the same lethal coal mining practices continue after Upper Big Branch, while coal slurry dams menaced various localities of Appalachia even as Karen Elkins over on Buffalo Creek could assure me that there were no more coal slurry dams? (Stanley Sturgill, pointing past butterflies in Lynch, Kentucky: "Anyhow, that road went back up into Machine Shop Holler. That's how they hauled that sludge up there. That slurry is probably within a thousand feet of here. Now that was a big mine.") Thoreau again: *On the whole, it was not so impressive a scene as I might have expected.* And if the social vibration caused by even so great a catastrophe as Fukushima could so easily weaken, how much more so the sustained humming chord of climate change!

Once upon a time when I was alive the newspaper warned: *May 2015 has been unseasonably warm,* but it went on to say: *across the northeastern quarter of the nation.* So the other three-quarters might have been fine. *Over all* [*sic*], it said, *average temperatures this month have been running anywhere from four to eight degrees above normal,* which sounded ominous, but then it undid the effect: *A cold front will bring an end to the warm weather Sunday and Monday.* In that same month, a glossy magazine showed a picture of a skinny half-naked man squatting *on his parched land* in India, *as a brutal heat wave continued to blister the country, claiming at least 1,100 lives and causing roads to melt in New Delhi.* Well, maybe there too a cold front would come.—How foolish and useless to judge a permanent global problem from ephemeral local data! But we humans

---

* Many or most of these appeared to be mountaintop removals.

learn best from our own experiences. If the weather blows hot and cold, and the chord hums only now and then, what can we do but turn away?

I learned to trust my pancake frisker, and so I never got any radiation burns. I tried to make sense of the weather reports, and then I turned on the air conditioner.

In 2014 I had asked Chris Hamilton, the Vice-President of the West Virginia Coal Association, what he thought about global warming. *Hamilton, in pinstripes and tasselled loafers, is a constant presence in the State Capitol,* although I actually first met him at the Coal Festival in Madison. He had no reason to give me an hour of his time, and yet he did, although he might have repented when I admitted to being a Californian. He asked me to tell him straight up if I were an environmentalist. I said that I wasn't, not really. How could I have called myself one, and lived the way that I did? Mr. Hamilton remarked that it made no difference anyway; he would say the same things regardless. "And I really enjoy speaking to high school and college groups and environmental groups." On the subject of global warming he now told me: "I am not technically competent in that field to permit me to make independent judgments, so I have to read, and I have to rely on my own devices. I personally think climate change is occurring. I have reservations as to man's contribution. I'm of the belief that climate change has always occurred and will always occur. I'm intrigued and even find it a little humorous when I hear people say that the winters today are not what they were when I was a child. I personally point to the winters of 1976 and 1977. I have not seen a winter as bad as those. My children point to the winter of '93. My parents point to the winter of 1950–51. I have a ten-year-old grandson who says he's never seen a winter as bad as last year.*

"Now here's what I do know," he continued, and most of his subsequent words could have equally been uttered by the United Mine Workers.† "Again, the United States is not a planet, and West Virginia is certainly not a planet. To the extent, yes, it is occurring, I believe it is occurring, I believe man's contribution to it is substantial; you have to reasonably conclude, it is a global problem and you have to have a global solution, and that's where I have a disconnect with the President and the EPA. I do not buy the statement that the U.S. must demonstrate leadership. I think *we have been providing leadership throughout my entire adult life*! We have

---

* To these anecdotes I would reply with a quote from *Dracula:* "Remember, my friend, that knowledge is stronger than memory, and we should not trust the weaker."

† Gearing up for an anti-EPA rally in Pittsburgh a month later, the UMW's President insisted that climate change "demands a global solution, not one that punishes American coal miners and their families after they have provided the means to power our economy for 150 years." He claimed that the new EPA rule on coal plant emissions "will perhaps cut about 1 percent" of greenhouse gases "worldwide by 2030." Wouldn't 1% be better than nothing?

been making major upgrades at a cost of hundreds of millions of dollars. We have ratcheted down all airborne contaminates, 90%! We have reduced sulfur dioxide, we have enacted major legislation and all the oxides of nitrogen and ozone, particulate matter, mercury through major mercury mac* legislation, and we have made tremendous improvements. Now maybe we're not where some would have us be, but no one would deny that we have shown leadership. Every day, other countries are going on line with coal-fired plants that have not followed our lead with scrubbers, first after-treatment devices to take care of human health issues. Why would we be so naive as to believe that they would shut down their economy for the sake of CO_2, which will not have any immediate human health issues?"

He was surely correct. Who would shut down an economy even for immediate health issues? Not West Virginia! *This shipwreck had not produced a visible vibration in the fabric of society.* Hence this letter to the editor from 2014: *The world's aflame with global terrorism but President Obama's major effort is climate change. He's weakening us with regulations . . . especially [for] coal . . . He may be gambling with the security of us all.*

In brief, *there is no immediate danger,* a lullaby whose American equivalent runs: *there are no immediate human health issues.* That must have been the crux of it. The cardboard partitions I saw in Big Palette, marking off futons, bottles of water, worn cushions, neatly draped towels and blouses, blankets and plastic bags of snacks, were tokens of homelessness. The sorrow and anxiety of those nuclear refugees could not be ascribed to anything but the accident. But I never saw anybody cowering away from greenhouse gases. (From the Intergovernmental Panel on Climate Change: *For average annual N[orthern] H[emisphere] temperatures, the period 1983–2012 was very likely the warmest 30-year period of the last 800 years (high confidence) and likely the warmest 30-year period of the last 1400 years (medium confidence).*) Radiation sickness proclaims itself, but what about carbon sickness? Oh, what an unconscionable omission!

2: "No Vision for Your Future"

Comprehending coal's effects, much less evaluating them, proves slippery in any number of ways—in part because *West Virginia tied with Nevada as the least transparent* state *in America.* Once upon a time, Nicholas "Corky" DeMarco complained in a letter to the editor of the *Coal Valley News* (are these perchance a coal miner's words?—Oh, now I get it: an issue from the following year prints

---

* I did not understand this word. My colleagues at Viking propose: "Mac" may have been "MACT," aka "maximum achievable control technology."

his photo—business-suited, rightwardly smiling, almost entirely bald—captioned **Executive Director, West Virginia Oil and Natural Gas Association**): *The government's "social cost of carbon" calculations—done in a bureaucrat's back room* . . . and hidden deep inside a technical report—are a hammer intended not to build but to knock down our energy resource industries.* For all I know, these calculations might indeed have been "unfair" to Big Coal—but Mr. DeMarco never offered to make them himself, and if he did, I would be no less skeptical of his numbers and methods than of anyone else's—for although we would be negligent not to tot them up, the lamentable fact remains that all social costs simply resist quantification! The benefits of coal remained as murky as power plant smoke; the liabilities must in part be ascribed to such noncarboniferous matters as rural poverty, corporate capitalism and globalization.

I have already told you that coal country newspapers sometimes reproduced a "coal tree" diagram, depicting with branches and sub-branches the many commodities which we derived from coal: cosmetics, toothpaste, batteries, laughing gas, perfumes, billiard balls, baking powder, photographic chemicals, artist's pigments, floorwax, etcetera. In this respect, nuclear power fell short.[†]

And from the standpoint of short term economic gain, coal miners might win out over nuclear engineers and their teams of construction workers. Constructing a reactor was a one-time process, while exhuming coal from a seam could employ myriads of locals for years. (Never mind that ever fewer were needed, thanks to mechanization.) That Charleston lawyer Tim Bailey, who often defended miners in accident claims,[‡] told me: "I think anyone would have to say about the extractive industries, *why don't we have enough factories here?* and they tell you, when you really talk to the insiders, for all the smoke and mirrors that we hear, it all comes down to one thing, topography. So what have we historically been able to do here in West Virginia? We've got these extractive industries, like the coal industry. The shorter the distance you have to move it, the better off you are. So we had steel factories, glass plants, lumber mills. Anyone who would try to say that coal has been devastating, horrible to West Virginia, that would be an overstatement. It's been the core of our development for hundreds of years. The bad part is that we've been so dependent that we have not

---

* To me this accusation better fits the *no comment* of the *regulated community*. Mr. DeMarco might have known a few things about back rooms, for as his obituary explains (he died in 2016): He "didn't begin his career in the oil and natural gas industry, previously working as the director of operations for the state of West Virginia under the late Gov. Cecil Underwood. However, that certainly didn't appear to stifle his passion for the industry."

† But let me celebrate medical isotopes and the irradiation of food.

‡ One of his firm's newspaper advertisements began: **"MASSEY LINED ITS POCKETS WHILE COAL DUST LINES YOUR LUNGS."**

adequately prepared for the move away from carbon-based fuels. I don't want to see anybody without a job. But the reality is that there's going to be less coal. The leaders and the industries in this state have been shortsighted. There are a lot more things you can do with coal than burn it. Now we're sort of behind the curve, and transitioning away from coal, and all the sudden, we're not willing to transition. That coal tree, there's a lot of things we can be doing. We hear about the power plants and the War on Coal, but it has a whole lot less to do with the EPA than with the cheapest energy we've had for tens and tens of years, which is this natural gas boom we're in now.* It's just so much cheaper."

If we set aside direct income and loss to producers, and considered only "consumers" and the environment, then nuclear power's advantages and disadvantages were both more definite than coal's. I remember Japan's complacent

"Friends of Coal Miners" at the West Virginia
Coal Festival. In the upper left we see mention
of one of FOC's sponsors: Bucci Bailey & Javins
LC. Bailey was Tim Bailey.

---

* The next section of this book will concern itself with natural gas, which I like about as well as nuclear power and coal.

prosperity in the days before March 11, 2011. In those earlier visits I occasionally used to raise the subject of carbon ideologies, only to find my coal-burning nation served with resentful disdain. Nuclear energy was clean, the Japanese most often said; it was progressive; its power plants even paid subsidies to the neighbors! Never mind that coal produced a higher proportion of their global warming than ours:

COAL COMBUSTION'S PERCENTAGE OF ENERGY-RELATED CARBON DIOXIDE RELEASES FOR SELECTED COUNTRIES, 2010,

in multiples of the Swiss figure

	1	
Switzerland		1.4%
	25.79	
United States		36.1
	26.57	
Japan		37.2
	28.71	
Germany		40.2
	54.79	
Estonia		76.7

Source: International Energy Agency, 2012, with calculations by WTV.

No, never mind the facts; once upon a time, I too might have wished to dwell in an atomic paradise.—In 2014 a certain labor activist in Iwaki, leonine, with silver hair but a smooth young face (he was 40 although he looked 50), told me that when his family moved to Fukushima "on account of my father's business," "I was in elementary school. Around that time,* No. 1 was built . . . *and we*

---

* He was then 10 years old, he said; hence the year would have been 1984. By then Japan already produced 10% of the world's nuclear power, which supplied 30% of her electricity.

thought this was just a power generation plant!" he said, clenching his fist. "I still remember that we visited Tepco on our school excursion to Okuma. We saw beautiful ladies welcoming us, and explaining how nuclear energy is efficient and safe! *It's a myth of safety.* A huge earthquake wouldn't come, they said . . ."— When it did come, the myth lost credibility, and the culprit was undeniable.

But, as I said, to what extent could West Virginia's myriad difficulties be fairly blamed on coal mining alone? And what deserves to be called a difficulty? In the town of Iaeger, just before turning onto the Vietnam Veterans' Highway, I stopped to look down onto the articulating spine of a long freight train, and saw vines growing on the telephone wires just as in good old radioactive Tomioka. A man was slowly trimming kudzu by the front of City Hall, whose backside from across the river had nearly lost itself in greenery. But that could almost have been planned, or at least tolerated; I myself lazily permit ivy to grow on one side of a building I own in California, and please don't tell me my property is decrepit! Downtown Iaeger appeared almost pleasant, especially from a distance. Some of its two- and three-storey brick edifices were vine-grown even over their windows, and one building sported a terrace whose entire surface had been pretty well creeper-strangled. To be sure, on either side of the town's decaying center it was by no means difficult to discover stout brick houses in good repair, with clean vehicles parked before their immaculate lawns, not to mention shuttered two-storey homes whose clean porches looked down on weeds. Investigating that icon of democracy, The Little Fellow's Bank, I found that its parking lot, trash can and shaded sidewalk were all empty and its doors shut at the height of business hours. At least there were no weeds growing up through the asphalt as in so many parts of Tomioka. I lacked a carbon dioxide meter, but anyhow not much seemed to be burning. Sumacs strained at the light, rising high from the brownish-green river, and jungle-like forest guarded the town, ascending the hills behind it, ready to entomb everything as at Tikkal or Angkor Wat. Iaeger had lost population, for a fact, and no doubt the decline of coal mining had exacerbated or even caused that decline—but if the burning of coal turned out to be *at this moment* the worst energy generation choice for our planet, of which (thanks to Fukushima) my slow mind had not yet been convinced, wouldn't the end of coal mining be necessarily a blessing?

Well, it would hardly be good for everyone in West Virginia; apparently it had not done wonders for Iaeger. In the bitter words of Cecil H. Underwood, who was Governor in 1999: "I have received criticism from some coal critics who look forward to the 'post-coal' era in West Virginia. They have no vision for your future except welfare or relocation. That is not acceptable to this Governor, and it never will be."

This speech must have been well received, given its audience: a Boost Coal

Downtown Iaeger

Luncheon. What lovely sport to blame our defeats on others! If *they* blame coal mining, *we'll* blame coal critics!—Indeed, as coal demand declined in West Virginia, with few other industries coming in, welfare and relocation did loom larger. In that same year 1999, nearly 14% of West Virginia's families lived below poverty level.—But according to the West Virginia Coal Association, the "record production year" was 1997! So where was all the coal money going?

There were five occasions in that decade* when West Virginia qualified as the poorest of all our 50 states. A particularly pathetic statistic: In 2013 and again in 2014, West Virginia was in such trouble that even its indigent burial program had to cease making payments. Meanwhile, sheriff's sales in collection of delinquent property taxes told their own sad stories:

- Ticket 4048: Yvonne Oliver's half-acre going for auction at the courthouse door *between the hours of ten in the morning and four in the afternoon,* because she owed $1.98, which had magically increased to $72.13;

---

* More exactly, in 1991–2003.

- Ticket 2932: Sandra Lee Vint's strip of 125.37 × 15 × 127.34 × 15.13 feet going under the hammer so that her debt of $3.80, now $74.09, could be laid to rest;

- Ticket 1924: James Mitchell's lot of 50 by 60 feet at Pond Fork being sold away from him on account of delinquent taxes amounting to a full 12 cents—total now due: $79.80.*

Incidentally, I would now like to recall to mind that dated aphorism of the National Coal Association: *Each percentage increase in Gross Domestic Product... has resulted in just over a 1 percent rise in the demand for electricity.*[†] Perhaps it didn't work the other way—for West Virginia's per capita energy consumption[‡] was 389 million BTUs, a good 31% higher than the national average. But somehow West Virginia did not strike me as 31% better off than the rest of us. (As it happened, the lion's share of her energy use went for a category called "industrial." I suspect this meant coal mining.)

In 2015, West Virginia's population was in more rapid per capita decline than any other state in the Union's, and its rate of fatal drug overdoses—more than twice the national average—the highest in the nation. Shall we point our fingers at job loss? In 2014, West Virginia and Mississippi had tied for the highest proportion of adult obesity (35.1%, which *marked the first time that a state has surpassed 35 percent*.) The closures of restaurants and food markets in the Appalachian coal country surely had something to do with that. But how much of the whole mess could be blamed on *coal critics*?—Quite a lot, I suppose, if *coal critics* got us into *this natural gas boom we're in now*. I was astonished to see that the U.S. Department of Labor's *Occupational Outlook Handbook* for 2012 (the only year available at the Madison library) lacked any entry specifically on coal mining, although the bulletin consoled me that material moving machine operators (entry-level education: less than high school) might sometimes *work for underground and surface mining companies. They help to dig or expose the mine, remove the earth and rock, and extract the ore and other mined minerals.* Everybody agreed that coal mining was going away, in West Virginia at least.—"Oh, yes," said old Mr. Arvel Wyatt when I interviewed him at the Flat Iron Drugstore in Welch. "Coal ran this county[§] for many years. My Dad went

---

* All these taxes were only a year past due. Thus the miracle of penalties, fees and compound interest.

† See I:51.

‡ 2012 data.

§ McDowell. By some measures, as you will see, this occasionally qualifies as the poorest county in America. Welch is the county seat.

in when he was 15 or 16 years old, but he died early of a heart attack, at 47. He enjoyed the work. He was an electrician, working on the continuous. My whole family was NR & P: at the New River, down at Pocahontas. I lived down the holler from the mine. They did away with the mine, so that's one of the hollers they picked to unload their trash. It's still one of our main sources, but I just read in the paper today there's two more places shutting down coal and putting in natural gas. It ran the county for many years. It's still a big business here in the county. Still a few punch mines. The big dimplers are all gone. No more orders for coal; no demand for it anymore. The punch miners don't hire near as many as before. That's where the unemployment came in. It started here in the '60s,[*] but in the '80s it got worse."

Active Coal Mining Sites in West Virginia and Total U.S. Carbon Dioxide Emissions for Coal-Generated Electricity

2012	184 sites and	1,511 million metric tons* CO_2
2013	152	1,571
2014	96	1,569

Between 1973 and 2015, the highest emission level was 1,987 MMT in 2007; the lowest was 812 in 1973. The 2015 figure was 1,353.

"We dealt for several years with the fact you couldn't get federal permits; now they've taken on the power plants and put out those rules on greenhouse gases . . ."
—Bill Raney, President, West Virginia Coal Association

Sources: Coal Valley News, 2014; U.S. Energy Information Administration, 2016.

---

*1 metric ton = 1.1023 U.S. tons.

As it happened, between 1960 and 1990, American coal consumption nearly doubled, from 1.014×10^{16} to 1.912×10^{16} BTUs—in other words, from 10 to 19

---

* He was misinformed. As the 1960s ended, Americans were digging out 10–15 quads of coal every year. One quad is a quadrillion BTUs—"roughly the energy contained in 200 million barrels of oil." And they had plenty of orders. From 1973 through 1981, world coal production rose from 65.4 quads to 79.7 quads, while the Western Hemisphere's coal-harvest swelled by 41% "in spite of the slight drop in production reported by the United States in 1978 and 1981." In 1980 "the highest ranking coal exporters were the United States (29 percent of world coal exports), followed by Australia at more than 15 percent and Poland at 12 percent." (In light of her later stellar nuclear record, it is interesting to learn that "Japan, the largest single national importer, purchased 2.1 quadrillion BTU—25 percent of the world total" imports.)

A sign of the times in Madison

quads!—From the *International Directory of Company Histories,* 1991: By 1989, Peabody Holding Company, Ltd., incorporated in 1890 as Peabody Coal Company, current sales 1.78 billion, employees 10,400, was "the largest U.S. producer" of low-sulfur coal, much of which came from West Virginia. *In the early 1990s more than half of the electricity used in the United States was generated by burning coal, and 10% of that coal was supplied by Peabody . . . The electric companies were counting on Peabody; as the nation's largest coal producer, it seemed likely to deliver.*

Coal use as a percentage of total fuel consumption: U.S.A., 1870, and World, 1991

U.S., 1870	26.82%
World, 1991	26.48%

Source: Encyclopedia of Chemical Technology, 1991, with calculations by WTV.

To be sure, eastern America's bituminous wares had long since, in quest of some un-American chimera called clean air, run afoul of the Environmental Protection Agency, and so, as one journalist put it, *power companies turned to the far less efficient but cleaner coal out west, where very large-scale strip mines became coal's new cash crop,* and this must have been one cause for the longstanding anti-government anger of the *regulated community.* But that journalist was writing about Ohio and Pennsylvania, not about West Virginia. As I said, the region where Mr. Wyatt lived used to be rich in *low*-sulfur coal, which once the

Environmental Protection Agency came out against sulfur in the 1970s became correspondingly more valuable to coal-burning utilities whose executives declined to invest in smokestack scrubbers. In 1992, Appalachia still dug out nearly half of the nation's coal—and more than half of America's electricity was generated through the burning of coal. (I take it that the Clean Air Act was not quite as apocalyptic as pictured.) Coal production in central Appalachia* actually did not summit until 1997; but once the nation's scrubbers were in place, cheaper, *more* sulfurous coal such as Wyoming's could be burned, so that by 2013, "every few hours" coal trains sped through the Crow Agency in Montana, *some on their way north to Canadian ports for shipment to Japan and South Korea.* Cloud Peak Energy, the mining company, slavered for "many more trains" to head for "proposed export terminals in Washington State," while Richard Morse, managing director at "an energy consultancy" called SuperCritical Capital explained to us all: *The future of the U.S. coal industry is at stake.*—Meanwhile, southern West Virginia was getting mined out, leaving as one expert put it only "either thick and dirty or thin and clean" slabs of coal.† In short: "No more orders for coal; no demand for it anymore."

Regional Coal Prices, *per ton,*

April 15, 2015

Northern Appalachian	$61.15
Central Appalachian	**$53.06**
Illinois Basin	$40.32
Uinta Basin	$38.13
Powder River Basin	$11.55

Sources: Energy Information Agency; National Mining Institute.

---

* Central Appalachia "includes the southern portion of West Virginia, Kentucky and Virginia" and "has the lowest sulfur coal in the country." Northern Appalachia, "which includes Pennsylvania, Ohio, Maryland and the northern portion of West Virginia," still had enough thick coal slabs left unmined to justify longwall mining.

† But according to the West Virginia Coal Association, "the coal seams of Appalachia (West Virginia) are not in significant decline. In fact, underground productivity is at near record highs. Our overall productivity, however, has declined due to the reduced use of highly productive surface mining. Falling coal prices are directly attributable to the war on coal ..."

It was not the *coal critics,* but that capitalist deity "the market" (somewhat influenced, to be sure, by the anti-sulfur rule), that had no vision for West Virginia's future.

3: "You Could Hardly Push Your Way Down the Street"

What was that future? I will tell you; I was there; it was business failures, dying towns, depopulation—or, in the words of Kevin Crutchfield, CEO of Alpha Natural Resources, *transformational opportunity.*

The slowness of life in the coal counties as I knew them was epitomized by the old man who came up to me in the fast food restaurant in Bradshaw, telling me how he took a wrong turn and ended up at somebody else's family reunion by mistake; he advised me where to go if I cared to see the prettiest, the deepest or the best of everything, then finished his sandwich, got into his truck, pulled halfway out of the gas station, opened the door with the engine running, hung his leg out and chatted for another 10 minutes, calling me buddy; it was he who said: "I worked my ass off in the coal mines! We broke a bunch of laws to stay in business . . ." His mother had run a restaurant in Iaeger for 55 years, and closed it down in 1998 for lack of customers. He said: "You should have seen Iaeger 30 years ago. You could hardly push your way down the street, there were so many people." As he sat there in the gas station parking lot, not a single other vehicle went by.—But does that fact signify post-coal economic stagnation or should it be credited to the unhurried pace of Appalachian culture?

The man in Bradshaw

And that drive-in in Welch—the only restaurant in town that remained open—what about its slowness? There might be one waitress, or even two. The tables on the checkerboard floor were mostly empty. So were the places where cars used to park. It must have been busier once. Although Ross Macdonald's novel *The Doomsters,* published in 1958, was set in southern

California, parts of it memorialized a midcentury Anywhere, U.S.A., including the Red Barn (whose counterpart in Welch was Sterling's Drive-In), an establishment

> accentuated by neon tubing along the eaves and corners. Inside this brilliant red cage, a tall-hatted short-order cook kept several waitresses running between his counter and the cars on the lot. The waitresses wore red uniforms and little red caps which made them look like bellhops in skirts. The blended odors of gasoline and frying grease changed in my nostrils to a foolish old hot-rod sorrow, nostalgia for other drive-ins along roads I knew in prewar places before people started dying on me.

At Sterling's, where two wide old men were having eggs, biscuits, ham and Coca-Cola for breakfast, chatting about a friend of theirs who'd just had a heart attack, the waitresses' uniforms were orange. There was frying grease, but no gasoline smell, and the coffee had that baking soda taste of treated mine runoff.* Once or twice I ate lunch at Sterling's; I took dinner there, too, since restaurants that stayed open past midafternoon were pretty hard to find in Welch; and I remember how the yellow light-tube glowed with increasing apparent brightness against the awning's corrugated silvery belly while color drained out of the bluish-grey sky, and at a far table, a middle-aged ponytailed man and three young fellows with shoulder-length brown hair sat hunched over their burgers, murmuring something about some girl, while the two waitresses were slowly and brightly chatting in their drawls. There were photos of cars on the wall of that huge, mostly empty, extraordinarily clean room with the black-and-white-tiled floor and the white tables with orange and white chairs. Looking out the grand windows at the orange awning of the drive-in and then at the road and the hill of not yet budding trees combing the ridge-clouds, I tried to imagine how Sterling's must have been in the years before coal jobs went away. I could tell once again that the coffee had a least a little of the tapwater in it, or maybe it was only that the mug had been washed in that water, for the taste did not come on until after I had drunk it down, then lingered on my tongue, not quite stinging it.

What was cause, and what effect? One early afternoon just after a rain in Welch, with a lovely breeze cooling down the world to 75°, I was looking from the parking lot of the women's shelter down across the steep bright vine-grown

---

* See below, p. 152.

Sterling's Drive-In

grass-hill to the triple railroad track from which the smell of creosote rose up as if in humid breaths (the scene's most active component being a dump truck slowly advancing and reversing over the ballast, a close second being the ballast-heaps themselves: clean gravel, white and almost bright against the wall of greenery behind), I waited for something to happen, and something did: A string of open grainers came slowly creaking in, loaded not quite full with more gravel, then stopped just as it began to rain again. Should anything else have occurred? Rain was sparkling on ivy and honeysuckle; rain darkened the walls of wet reddish rock and the hills of Welch, which had been sliced rawly open. In the grass stood dandelions out of flower, with rain-pearls on their leaves. It was summer, the slanting wall of rainy trees making freshness and stretching over us all.

Since carbon sickness, if it even exists, takes ever so many discreet forms (thus it was, at least, back in the years when I was alive), I found myself hard pressed on my journeys to distinguish the decrepit from the simply rural. There were Appalachian stars affixed to certain houses as one rolled out of Welch on Highway 52 North; then came another Methodist church and Bulldog Hollow Road, soon after which a motorist would find himself definitively out of town, dipping down into the winding forest and passing another closed United Mine Workers office,* the Ark of Safety proposing to save his soul, not his body. If these were the people who as their politicians said *kept the lights on in America,* they sure as hell weren't wasting much power on themselves, and that was why so much of West Virginia was "scenic": Welch through Indian Gap to Premier, Roderfield, Clear Fork Junction, Wilmore, Sandy Huff and then Iaeger, which lies near the northern edge of McDowell County.—Although West Virginia boasted the third highest rate of car accident fatalities, and the second greatest number of roads "in poor condition" (only Connecticut was worse), I found the latter statistic less surprising than the former, because those winding back roads were so often *empty*! Instead of glowing convenience stores, wide highways of petrol-burning traffic and other monuments to electrochemical work, I found a turquoise creek by the kudzu-overgrown railroad track; Soldier Drive, Marine Way, a long string of boxcars and grainers flashing through the trees across a concrete trestle bridge; yes, the river was now foul-colored and filmy with a strange

Retail store, McDowell County

---

* Tim Bailey told me: "The local union halls, those locals are run by the miners themselves, and there might still be union miners there, but they're not working, and if they are, they're not working in a union mine. So the halls are closed. They're so job scared. When the coal companies decided to bust the unions, they did a smart thing: *If I could just save the amount of the money that I have to pay into the union pension! This union, and all the money they're costing us, we can't afford it and I'll have to shut down. I'll pay you the same money and the same health insurance.*"

vinegary smell but the mountains were so green around Iaeger; I loved it here . . .—It certainly must have been different back in the days when one had to push one's way down the crowded street.—What had changed?—Coal was going away.—But *why*?

No demand for coal, Mr. Wyatt had told me, *no more orders.* Would you still prefer to blame the *coal critics*? But it may be worth quoting a few lines from a handbook for coal mine operators, dated 1978: *A good place to turn for higher profits in a mature surface mine is productivity—digging more coal with the present work force or using fewer persons to dig the same amount.* This notion attracted those whose God was higher profits. (They have no vision for your future except welfare or relocation.) And God smiled on them, if not on the toilers underground: *Direct coal employment dropped from 70,000 miners in 1985 to 35,600 in 1997, while production peaked.* Near that century's end the *Encyclopedia of Chemical Technology* reaffirmed the utopia of productivity: *Coal mining has changed from a labor intensive activity to one that has become highly mechanized. In 1988 the average output per person per day in underground mines in the United States was 17.7 metric tons. For surface mines the output was 42 metric tons per person per day.* Imagine that! Forty-six U.S. tons of coal disinterred in one day by a single human being! So the *regulated community* kept digging more coal while using fewer persons, and mountaintop removal slimmed down their mining crews still further.

In short, by the late 20th century coal's absolute contribution to West Virginia's economic prosperity, like that prosperity itself, might have been exaggerated. But in relative terms, well, what else was there?—A woman down in Keystone (her name was Jean Battlo) once said to me: "I'm an environmentalist, so I'm very concerned about the damage, and as a matter of fact I used to be all on that side and then I went to a meeting and heard a man say to me he was tryin' to raise his three kids and coal was his only hope." *His only hope!* How much could that man have worried about climate change?—Or let's suppose that he did worry. Remember Stanley Sturgill: "As far as the people around here, coal companies has beat it into 'em: If you want your job, you better not talk about climate change."

4: Fourteen More Reasons to Love Coal

Yes, it was convenient for the *regulated community* to blame *coal critics* for its own profit-driven layoffs—and even better to blame the hated regulators.

I remember the weary pride in Chris Hamilton's voice when he said: "I think we have been providing leadership throughout my entire adult life! We have been making major upgrades at a cost of hundreds of millions of dollars"—and, as you recall, he listed them. I for one respectfully salute the coal industry for the many reductions in air pollution already accomplished. To him it must have seemed that the demands on the *regulated community* were never ending. And so the West Virginia Coal Association dug in its heels.

But as much as the *regulated community* had done, it was not enough. With three issues already outstanding—coal slurry dam failures, deep mine methane explosions and black lung—let us now extend the list a trifle, from acid mine drainage to aluminum, then anhydrous ammonia, *used to treat mining runoff in waste water ponds* (and also to make methamphetamine*); arsenic, carbon dioxide, coal ash, which was the heavy-metal-rich residue of the coal we burned back when I was alive; coal dust (not only explosive, but also a *questionable carcinogen with experimental tumorigenic data[†]*), coal tar (*confirmed human carcinogen . . . mutation data reported*—very useful in cosmetics, however), DT-50-D, methane again, which not only incinerated miners in explosions but also warmed the atmosphere with extra facility; 4-methylcyclohexane methanol, a licorice-reeking contaminant in drinking water (from now on let's nickname it MCHM, for it will soon play a star performance), and never mind radionuclides, selenium and sulfur.

Some of these evils might not have been intrinsic to coal itself; they could have resulted from the way coal happened to be mined, refined, shipped or burned. For all I know, several might have been trivial. As I have said, no matter what *Scientific American* said, my scintillation measurements of coal always came out innocuous. So the radioactivity scarcely compared to Okuma's or even Tomioka's . . .—and possibly many other items on that list of coal pollutants could be rendered outright salubrious![†] For instance, in 1983 a certain pro-nuclear scientist who wanted to make other forms of energy generation look bad alleged that coal burning caused about 10,000 annual American deaths. But what if coal-smoke could be better filtered?

All the same, compiling that list made me hope that the *regulated community* would never be left to itself.

---

* The practical allure of that process explains why it was that in 2008, two men tried to steal some anhydrous from a coal mine—and mistakenly let it loose into the air. They were sentenced to prison in 2013.

[†] "Variable toxicity depending on [silicon dioxide] content."

[†] Fracking waste could be quite radioactive. See p. 321.

> **From a local newspaper, 2014**
>
> **"Safe enough to eat":**
>
> A lawsuit filed on behalf of 77 people against American Electric Power and related companies claims they were exposed to dangerous chemicals in coal waste from the Gavin Landfill in ... Gallia County, Ohio.
>
> The suit alleges employees were told that coal waste was "safe enough to eat ..."
>
> Six employees had died by the time suit was filed in August, and two more have died since ...

5: "Just Everything Falling Apart One Stage at a Time"

Let contaminated water be a case in point. The first time I headed for McDowell County, wishing to learn why the tapwater down there was supposed to be dangerous, I left Bluefield for Bramwell, where a roadside cross spelled out, hand lettered: **Jesus Wants You**. Leafless trees, reddish fallen leaves and hills of dirt accompanied me to Coaldale (unincorporated); and beyond the Coaldale Freewill Baptist Church, rolling gorges of leafless trees and brown trees took me right into McDowell County, which some have called the poorest county in the poorest state. In 1935, McDowell's miners generated 19.3% of the state's economic value of coal products, and a respectable 5% of the American total. One columnist claimed that McDowell *has produced more coal than any county in the country*.* In 2014, conveniently for this book, it not only contained the highest proportion of the state's food stamp recipients (an impressive 34%) but was also declared the unhealthiest county in West Virginia.†—But was it really the poorest? Maybe not in May of the following year, when jobless rates increased in all but 10 of West Virginia's 45 counties, of which the worst off was Pocahontas County at 12.3% unemployed, followed by Mingo, Wetzel, Webster, Marion and Logan, after which we pass below the double digits. In 2008, Clay County had

---

* In that case, shouldn't it have been rich?

† The state's five unhealthiest counties, McDowell, Wyoming, Mingo, Logan and Boone, were all in the coal fields. McDowell's ranking came in part from its great number of lethal drug overdoses. How blighted its people were in relation to Harlan County, Kentucky, I cannot tell you, but in 2012 Stanley Sturgill claimed: "Gallup polls have shown that we are the sickest district in the United States," for which he blamed mountaintop removal.

registered a significantly lower per capita income than McDowell's, although Clay's median household income had been higher. One way or another, McDowell was not what I would call prosperous. Old white houses stood guard along the river. Mayberry's homes were silent—could so many folks be at work in this underemployed place, or were some homes abandoned?—and a black man in a jacket, his hands in his pockets, paced back and forth on the highway. Following chilly hills, riding this nearly untrafficked road beneath the cloud-hematoma'd sky, I came into Elkhorn (unincorporated) and looked down on the spread-out houses and the railroad tracks; I passed through Upland (unincorporated), as a truck of logs rolled by, then Powhatan, and Kyle, where brown vines grew on a power line. A Burlington Northern & Santa Fe unit was pulling a long string of freight cars through Northfork (unincorporated), whose librarian, a pleasant soul named Mary, said that the town had been (my italics) *five years without water:* "It's no good for eating and drinking. On account of population decline, they don't fix the pipes. It might never come back."

She said there was one supermarket eight miles away where she got her bottled water.

Downtown Welch

Down the street lay Shupester's The Dinner Bucket Miner's Family Restaurant. That was the only place to eat in Northfork, so I ordered the special, which was barbeque meagerly served on a thin and soggy bun, with corn on the cob and cherry cobbler beside it; then I asked the waitress how long the water had been bad.

"Since July."

"It was okay before that?"

"Well, I wouldn't have drunk it," she replied a little shortly.

"What caused the trouble?"

"Just everything falling apart one stage at a time."

That was a pretty good answer, I guess, for it served southern West Virginia well. Consider the town of Prenter (population, maybe 60 souls) whose single well ran dry in December 2015. *Even under normal circumstances, the water in Prenter is not safe to consume,* explained the newspaper, and the activist Chad Cordell had worse things to say.* *However, residents use the water for sanitary purposes.* One them, a lady named Jennifer Massey, testified that *in a neighborhood of 10 houses* there had been half a dozen brain cancers, four of which proved fatal. Her neighbor Maria Lambert also came forward, because "when I leave this earth, at my funeral I want them to be able to say, 'Maria done what she could.'" Accordingly, like Stanley Sturgill, who'd called himself a "hated person," she became "unapproachable" to many other Prenterites: "You don't want to get into too much of a conversation with her because she has joined forces with them." (They were "afraid the coal company was going to turn the power off to their water supply.") Here is one of many gruesome details Ms. Lambert provided: "My children lost some of their teeth from decaying before [coming] through the skin. I mean, it's just like it was coming through, breaking through gums, their teeth was decaying and they had to have them pulled . . ."—And why wasn't the water safe? Ask a Boone County administrator: "It's a coal mining area. Coal companies owned most of the property, and they owned the well." According to a coal miner's son (he was also an organizer at the Ohio Valley Environmental Coalition)[†]: *It was Massey (now Alpha) and Patriot that caused the contamination in Prenter. Both were conducting slurry injection. They were putting coal slurry into underground mines that was seeping into the groundwater and into residents' private wells.* And indeed, in 2009, *Massey and four other coal companies* paid a whopping settlement of $45,000 to the citizens of Prenter and the neighboring town of Seth, but *the companies admitted no guilt,* so it must not have been their fault—and as usual, on the *regulated community*'s side stood the good old DEP, whose director laid all alarms to rest: "We studied specifically the possibility the slurry injection had migrated into the water, and there's not a geologic connection between where it was stored and where their problem is."—I'll bet his children weren't losing any teeth! Well, then, what *did* cause all those cancers, not to

---

* See footnote, p. 154.

† Dustin White, re-introduced below, p. 213.

mention the iron and sulfates in Prenter's water? I propose to blame the copper thieves who stole wire from the substation that pumped the well; why irritate the thin-skinned *regulated community?*—In short, things in Prenter were just falling apart one stage at a time.

The waitress at Shupester's said that after the East Mine closure in about 1985 the town had gone downhill.—By all means, let's blame the *coal critics*! Those villains preferred coal from the Far West, which had less sulfur in it . . .

Looking out through the backward **OPEN** sign whose oval-enclosed neon pulsed feebly blue and red, I got a fine view of the railroad tracks and of the tall narrow houses set into the hill behind them; they were white, yellow or sometimes pale green and resembled truncated pyramids; mostly they were two or three storeys high (one of them was burned); then behind them rose leafless trees and brick erosion wall. Once in awhile, someone in a hooded sweatshirt or a parka would walk by, hands in pocket, while a pickup truck rode slowly up and down the track. Most attractive was the track itself: trains hooting by, the red and white striped crossing signal going up and down.

Funeral home, McDowell County

Continuing north, I passed a few shops at the north end of town, then the House of God, and almost immediately found myself in Keystone, where the top of a brick building was smashed in not unlike the awning of the funeral home down the street. This town, I was told, "collapsed after a bank embezzlement." Should one blame coal or high finance for that? Rusty spikes lay alongside the track, and lumps of coal were scattered in ballast as the sun passed in and out of the clouds over Keystone, where a single dogwood stood gloriously white-flowered. The houses, most of which were also white, were all quiet. The white steeple of a grand brick church pierced up through the hill's leafless trees. Up the other hill, on some quiet mining track, two black men sat on a fence waving.

Following the windings of Elkhorn Creek down into Elkhorn Canyon, where truckloads of coal thudded by, I reached Landgraf (unincorporated), then, failing to notice Vivian and Bottom Creek, rolled into Kimball, where I remember from just before Carswell Hollow Road a huge banner: **TAKE AMERICA BACK!**—as they indeed would do come the 2016 elections.—In Kimball they said there was no problem with the water.

There came a working mine with its coal heaps and elevators, and after Big Fork, Superior and Maitland came Welch, the county seat: laid out along the steep and winding valley, Elkhorn Creek rustling, a frog loudly croaking under the bridge.

Coal-smoke can be visible at long distances. Charles Darwin saw the bituminous fumes of steamships 70 miles away and more. I spied no smoke in Welch— because the place was poor.

"Welch has the worst tasting water. It's not even fit to make coffee out of. Your coffee looks green," said a lady who worked at the newspaper office.

"I have been drinking the Keystone water," I told her.

"Ours is not that bad."

The hotel proprietress said: "Well, I just drink it." Her tapwater carried a taste not unlike baking soda; it stayed on the tongue, not quite burning but persisting.

Welch panorama

What caused the trouble?—Something that faded away long ago.—"In West Virginia it all ties into mining," aphorized Dr. Tom Jones of the Department of Integrated Science and Technology at Marshall University.* Since trees gather in toxins through their roots, long after which they may sometimes fossilize, coal can contain any number of heavy metals. Dr. Jones explained that *iron is mixed in with coal,* in the form of *iron pyrite usually (fool's gold),* which *when exposed to water and air produces sulfuric acid,* and, *once coal seams are exposed to air and water, makes acidity, which dissolves metals and compounds of base materials.* In case you are wondering how coal seams get exposed to air and water, feel free to admire that old photograph of 11 miners posing in the mouth of the 18.6-mile-long drainage tunnel at the Bishop Mine, whose white-lipped rectangular mouth gapes blackly behind them: *The tunnel drained 12,000 acres of coal. Each of the workers involved in the massive project received a medallion from the Pocahontas Fuel Company.* Another tunnel which came out *near the town of Amonate* achieved a comparable victory, for it *drained 12,000 acres of Pocahontas Number 3 seam coal into the headwaters of the Dry Fork of the Big Sandy River,* and through the toothlike tines of this wide black arch, water comes foaming and leaping out into summer forest. *The project was started in the fall of 1931 and completed in 1936, opening up unmined coal reserves of 400 million tons.* Then what? A pollution text supplies the following equation: *Pyrite + oxygen + water* $\longrightarrow$ *"yellow-boy" + sulfuric acid.* Yellowboy is "iron and aluminum compounds that stain streambeds."

An unclaimed mine, said Jones, *is generating acidic water that lowers the pH of the water into being more acidic than stomachs can take,[†] and our stomach acid is a filter for bacteria. Moreover, the acidic water dissolves everything else, so all these heavy metals concentrated in water (aluminum the nastiest) looks whitish in the water, like white sediment in the bottom—extremely toxic and causes fish gills to hemorrhage. Water with low pH will remove all the bugs except for very hearty things like crayfish because they have lots of calcium in their shells which acts like a buffer.*

In Welch, the creek that runs through town looks very pretty. You can smell the water, and when you touch it it feels slippery. It is very toxic and will kill plants. Even the treated water is not great. The way you treat mine water is to catch it in a

---

* This interview was conducted, partially on my behalf, by Prof. Laura Michele Diener, who kindly allowed me to quote from her unpublished notes. I was not present.

† This might be a slight exaggeration, since the pH of human gastric contents ranges from 1.0 to 3.0, the latter being a typical pH for mine-polluted West Virginia tapwater. All the same, habitually ingesting pH 3 water cannot be easy on the esophagus.

pond, and use some substance like limestone to treat it. In south West Virginia the preferred route is sodium hydroxide, around [pH] ten (extremely basic)... You turn a spigot and let it drip into the pond to make the water less acidic. The problem is that the ponds are small and get overwhelmed when it rains, so mine water runs over into the creek... Rain spikes affect people's drinking water: every hill in Welch has either old or new mines, which fill up with water... so mountains are literally filled with acidic water... This kind of acidic water can cause mouth ulcers and loose teeth, can also corrode lead pipes, which some outdated systems still employed, *so they are also freeing lead into the drinking water...—a new kind of toxin.*

("Tapwater, I drink it all the time," said Arvel Wyatt. "It's not bad if you're used to it. Twenty years ago it had the same taste.")

Homes and small towns have tried to adjust by buying filters, but metals and acids break down filters, toxic water can overwhelm filters and make them last for shorter times than indicated, and poor people/areas are not replacing their filters often enough. Besides, filtered *water can still have toxins without the taste and the smell.*

The baking soda smell: comes from calcium carbonate, used for treatment in mines and in filters, so calcium carbonate is basically trying to raise [pH] from 3 to 7 to make it neutral. Sometimes this chemical likewise took out the bad taste while leaving in toxins.

So was this all coal's fault?—Jones reminded us that *there is no regular trash pick-up. So people burn it and then throw the debris and unburnable stuff off the hill into the creek, even cars;* and Jones's own *family did that repeatedly, repeated a thousand times over the generations, so people start perceiving the water as trash and keep treating it as trash.* Perhaps the mining companies felt the same.— "People need to realize," said that Charleston lawyer Tim Bailey, "that if you look at the '20s and the '30s and the '40s and the '50s, all the way up to the '70s and the '80s, and how aware we are of the chemicals used all around us, nowadays in the supermarket people turn around laundry detergent to see what is in it, but you go back into the post–World War II industrial boom, people didn't go into those things. This idea of longterm monitoring, what happens to the stuff after you've used, it wasn't considered. From time to time, we all read about cases where there's some sort of allegation in the community about contamination, and sure enough, sometimes you find some things that today, if you had that product and you had to get rid of it, you sure as heck wouldn't have done what you did back in 1960."—Fair enough, perhaps, but what if they *were* still doing what they did in 1960?

6: Unnamed Tributaries

When I think of coal mining, I think of unacknowledged substances spewing into unnamed tributaries. On occasion, the rivers might run black or white. In 2013, Governor Earl Tomblin told the United Mine Workers of America: "I grew up in Logan and I can remember when the river ran black. That's not the case anymore." (Rather surprisingly, he then said of the Environmental Protection Agency: "We want to work with them.") Almost as he uttered these words, some residents of Van who were taking their children to school happened to perceive the accuracy of his narrative arcs—for, indeed, the creek wasn't black today; instead it was "white as snow or looked like milk"—for which we may thank a 2,000-gallon leak of a certain coal dust suppressant called DT-50-D, courtesy of Eastern Associated Coal, "a Patriot subsidiary." Ms. Kathy Cosco, the spokeswoman for the West Virginia Department of Environmental Protection (and for one six-month period simultaneously "Communications Director" for Senator Joe Manchin's "Political Campaign"), *wasn't sure what the affects [sic], if any, the contaminated water could have on any well water . . .*—and if she wasn't, I'd bet the citizens of Van weren't, either. *The incident has affected about five miles of the stream, but at this point there hasn't [sic] been any fish found dead, Cosco added.* Her advice: Wash with soap and water. (Which pollutants were in *her* washing water?)*

In October of that year, Kathy Cosco moved on to *represent Frontier Communications before the WV legislature and state agencies;* in other words, to *communicate the company's interest . . .* I trusted in her continued success; after all, her explication of the Van incident had certainly not impeded the interest of Eastern Associated Coal. Besides, in her photo she had a beautiful smile.

When she and I were alive, nearly every issue of the *Coal Valley News* carried four or five legal advertisements of the following sort:

> Notice is hereby given that Coyote Coal LLC, 500 Lee Street East, Suite
> 900, Charleston [*in the adjoining advertisement, Wildcat LLC presents the*

---

* The anti-MTR activist Chad Cordell told me: "These chemical spills happen all the time and generally don't make it into the newspapers." He told another interviewer: "I was talking to a friend, actually a guy I just recently met from Prenter [see above, p. 149], which is about 20 miles south of Charleston. Prenter has had poisoned water for years and years and years because of underground coal slurry injections, which poisoned all the groundwater. They were all on wells. This guy is in his early 20s, grew up drinking this water. His father has cancer, his mother has kidney disease, he has kidney disease, which causes him to regularly urinate [blood] and throw up blood. His doctors estimate that his kidneys will give out by the time he's 30 years old. And you go down that holler, and you hear the same story over and over and over."

same address, although on the next page, Independence Coal Company, Inc.,
is on 782 Robinson Creek in Madison, while Elk Run Coal Company's ad-
dress is a P.O. box in Whitesville] has submitted an application to the Department of Environmental Protection (DEP), Division of Mining and
Reclamation (DMR), located at 254 Industrial Drive, Oak Hill, WV
25901 for an Article 3 permit for the surface disturbance of approximately 107.23 acres in order to surface mine in the Lower 5 Block . . .

The proposed operation is discharging into Unnamed Tributaries of/
and Moccasin Hollow of Left Fork of White Oak Creek; Unnamed
Tributaries of/and Laurel Fork of Coal Fork, *etcetera, etcetera.*

Strange to say, Coyote Coal LLC (which according to the DEP had racked
up only three violations, two "inactive" and one "reclamation";* *it appears all fines*
are paid) remained exempt from stating just *what* its mine would be discharging
into those unnamed tributaries. While perhaps not even the helpful Kathy
Cosco could have provided a complete list, here is one likely ingredient:

In March 2013, Governor Tomblin, who evidently liked to add value wherever he could,† issued an emergency decree to enrich the waterways of the Mountain State with *thirteenfold and forty-sixfold increases in aluminum concentrations,*
depending on hardness levels. Aluminum pollution derives, of course, from coal
mining. Do you remember those Tom Jones notes? *Aluminum the nastiest . . .*
looks whitish in the water . . . extremely toxic and causes fish gills to hemorrhage.
Environmentalists and other leftists might have expected objections from the
Department of Environmental Protection, but I'm thrilled to report that har-
mony prevailed. *According to DEP, the emergency rule is justified to protect "the*
regulated community" from "unnecessary treatment costs."—The long-suffering
regulated community must have been grateful, and perhaps the rivers only grew,
say, 12 times and 45 times more poisoned than before, because *when hardness is*
greater, DEP said, higher levels of aluminum are not as toxic.

A year later, on the far side of the Pacific, other carbon ideologues (the

---

* However, the DEP also listed 15 violations of various sorts between 2009 and 2013 on permit no.
S502799, associated with Coyote and with Hobet Mining LLC; three trivial-sounding violations between 1998 and 2016 on permit no. U303693, linked to Coyote and to Catenary Coal Co.; and 11
violations from 1994 through 2016 (mostly for "effluent limits") on permit no. P057400, also associated with Coyote and Catenary.

† Some members of the *regulated community* might well have loved him, for consider this newspaper
headline: **Men charged in mine kickback scheme are big Tomblin donors.** But don't
think bad thoughts; their contributions added up to "only a tiny portion" of Tomblin's war chest.

Japanese Ministry of the Environment, which at least gave out fancy pamphlets from conveniently accessible urban offices staffed with people who knew something) explained the *Future Direction of Rivers and Lakes in Overall Image.* Reader, do you remember a certain minor inconvenience that occurred at Fukushima, leaving souvenirs of, for instance, cesium-137? Good news about that! *More than 99% of radiation is shielded at the depth of 1 m[eter].* Hence the multicolored arrows indicating a green dam and blue lakes and rivers: *Decontamination will not be implemented because of [the] water-shielding effect that prevent[s] radioactive impacts to the surrounding environment.*—Whether this was the same vicious, selfish thinking that must have informed Governor Tomblin's emergency decree, or simply the most practical method of addressing the irremediable, only a gaggle of civil engineers could have told you; either way, I have faith that it was *justified to protect "the regulated community" from "unnecessary treatment costs."*

In the same month as that emergency decree, West Virginia had already found another way to make her streams more precious. Let Representative Joshua Nelson (Republican, Boone County) bear the good news: "This week was a huge win for coal at the house. We passed House 2593 which re-vamps the States [*sic*] Selenium standards, which are currently based on false 'science . . .' Matter of fact, Selenium is sold as a vitamin for humans to consume!"

About this vitamin the *Hazardous Chemicals Desk Reference* explained: *Poison by inhalation and intravenous routes. Questionable carcinogen with experimental tumorigenic and teratogenic data . . . Chronic ingestion of 5 mg of selenium per day resulted in 49% morbidity in 5 Chinese villages.*

Well, but we weren't Chinese, were we? And didn't our *regulated community* deserve eternal help? In the words of Josh Nelson, who qualified as a genuine Demosthenes: "This bill brings the standards into reason, so that our waters are protected, and we have a reasonable environment to get our people back to work . . . We must continue to fight for legislation that puts the EPA in check. Our WV DEP is very well capable of taking care of us. After all, they are mountaineers too!"*

---

* The DEP's capacity for taking care of its citizens remains on record in another such advertisement, this one from a certain Wildcat LLC. After the usual announcement of discharges came this paragraph: *The Department of Environmental Protection is seeking information on private surface water intakes for human consumption located in the above listed receiving streams . . . Please provide your name . . . and the physical location of the intake. The information needs to be submitted to the address above.* In other words, the DEP lacked that information, and was expecting people to read the fine print in this small ad and then do the work of alerting state officials that certain families might be on the verge of drinking poison.

7: "What Makes My Coffee So Bitter?"

The town of Bim, which a resident said had been named after a cartoon character, was not the largest community ever. In fact one could easily have driven through Bim and missed it. Its attractions included the sealed-up Lightfoot mine.—"I don't know where all the people here come from," an old woman told me. "Now, this is not what you call a coal community. Now, Barrett, all *those* houses, they were built by the coal company to draw people in there. A lot of 'em come from Kentucky and Ohio and them places."

On the front porch of a brick house in Bim sat a long retired coal miner named Bill White, who by one neighbor's account might have been the oldest resident, and around him many colorful birds were sipping from hanging feeders. He said: "I was born right up there about 500 feet, or at least that's what they told me. Then we lived at Mount Cazy,* about three miles up the road by the Baptist church, because my Daddy worked at the sawmill; I remember living there for awhile. So for 80 years I've lived within a four-mile radius. I don't guess I'd trade it for anything, even though there's nothing here. I don't want to be anywhere but the mountains."

"Where do you go for your supplies?"

"Three or four times a week we go grocery shopping at Walmart or Sam's Club in Madison. We go to Sam's Club and buy as high as 12 cans of water at a time."

Since I expected that the Whites would not pay for bottled water without a reason, I asked: "How did that chemical spill affect you?"—for that year an outfit called Freedom Industries had managed to pollute the tapwater of 300,000 West Virginians with that coal cleaning chemical MCHM. (Freedom will receive its own tribute on page 170.)

"I just turned my icemaker on yesterday," Mr. White answered.[†] "Was afraid of it. We had them hard rains and so it spilled again. When I was real young we used to get black licorice. That's what it smelled like."

"Have you had other problems with your water?"

"It goes way back. I had a well here I had drilled back in 1970. And then my daughter and I started a restaurant. Those two fellers stopped there all the time; they had a core drilling rig; they drank coffee all the time. They was all the time putting sugar in it. I asked: *What makes my coffee bitter?* They said: *I believe it's*

---

* Local spelling.

[†] This was June 23—a good four and a half months after the spill.

*your water.** So in '83 I had the man in that drilled the first well: 51 feet to start with. He went on to 156 feet. He said: *There's mines under us two or three times! There's seams of coal right under us.* I don't remember the year, but on the 18th of December, after four in the morning, they had a pillar[†] collapse, and the whole house shook. I thought somebody had run into the house with a truck! And after that the well wasn't worth squat. When the feller first drilled the well 51 feet he said to me, he said, *I'm afraid I'm gonna hit a coal seam, so I'm gonna stop here.* I think we've been undermined more than once. I hear things sometimes at night. There's a popping. There used to be four of us on my well. They decided they had to have a swimming pool; they brought it up out of that well, and it was like swamp water. So we had no choice but to go on city water. That was after the second well had been dug. I don't know if there's anybody now without city water; the coal mine ruined 'em all."

"So what did that make you think about the coal company?"

"Not good. This big pillar thing, it messed up the well so we couldn't use it. Can't see no water in it now. I took a mirror and tried to look down in it the other day; I could see a wet place but it was pretty far down.

"Never have drunk city water," he said then. "Still don't trust it."

"Is there plenty of coal still in Bim?"

"The coal industry here, it's gone. It's just a matter of time. In the '70s, I think, our immediate foreman, he said, *we got 60 or 75 years up here.* I worked 38 and a half years up here—about 12 years underground. I bid out an outside job, and I got it because I had a lot of seniority. It's pretty good work, especially if you have a good crew, but it's dangerous work . . ."[‡]

(His younger brother Harry had said something similar. That couple were situated on their shady upstairs porch overlooking the Baptist church, and when I asked Harry White how it had been when he was a coal miner he said: "It was a job. We had longwall mining and we had longshear and we had a plow. The plow just had a bit that would go back and forth, scratch the coal off. I worked

---

* Do you remember Maria Lambert from Prenter? One of the first things that happened with her family's water was that "Mommy started noticing that . . . her water didn't seem right, and she really hated using it. She had to clean her coffee pots really often, and she would have to go and buy new coffee pots often because they'd kind of quit working." In due course the mother got lung cancer, the father got thyroid cancer, and Ms. Lambert, her husband and her sister all got "major stomach problems." Mr. White was wise to be cautious with his water.

† In their room-and-pillar coal mine.

‡ "It's like, just a tunnel," he said. "They call the walls ribs. You know, all the years I worked on the mine, I never saw anybody get mashed up." In regard to the latest accident, which had killed two miners, he mentioned a dead man's name and said: "I knew his Daddy pretty good. Hard to believe he's dying of cancer."

on an old plow, but I didn't like it. I was the youngest, so I got stuck on it. It was more dangerous. You had no protection from the rocks falling behind you. I had 19 and three-quarters years in when my back went out on me. I needed four more months to keep the medical."

(His wife said: "Well, the good Lord took care of us. So we didn't need it."

(She had lost a breast to cancer, and suffered a stroke along with several other ailments.)

I asked Bill White: "How harmful is coal to the air?"

"Four thousand airplanes a day, just about anytime, and they don't say anything about *them* pollutin'! But burnin' coal, it's bound to have some pollution, but . . ."

"By the way, what's that smell of gas?"*

"It's a gas line, goes across there somewhere, and when you get up 12 miles south you smell it again. They drill wells and got a thing on them to let some gas off. They put something in there to make it smell . . ."

"What's your opinion on global warming?"

"It's not true to the extent that they stress. There was a story one time in the *National Enquirer,* and I wish I could tell you how many years ago—30 or 40 years ago—he said that the U.S. Air Force had acres and acres of radio antennas in Alaska shooting radio signals into the ionosphere and it would eventually cause the weather to get worse and now you can't stop it. One time I talked to a man from FEMA and I asked him, *does it seem to you these floods are getting more frequent?* and he said yes. He did agree that the storms are more frequent, more furious."

8: The Lucky Places

But let's take comfort. When I inquired how safe the water might be in McDowell County, Tim Bailey, who was a lion when it came to coal mine accidents and a lamb in reference to pollution, reassured me: "You know, so much of that's dependent upon how close that is with all those creeks and hollers and where the coal refuse is, and abandoned mines that are leaking, and old chemical dumps around the shops. I think there's some places that you'd find there's nothing there."

---

* Thanks to Stanley Sturgill, I now know that I should have asked: "What's that smell of crude oil?"

"WE DON'T COMMENT PUBLICLY"

1: How Many More Buffalo Creeks?

By now the *regulated community,* wearily fighting the good fight against any false science which might lower profits, had grown adept at that self-defensive miracle, the *no comment.*

In one of the introductory volumes to the *Great Books of the Western World,* an idealist insisted: *The spirit of Western civilization is the spirit of inquiry. Its dominant element is the* Logos. *Nothing is to remain undiscussed. Everybody is to speak his mind. No proposition is to be left unexamined.*—That is the spirit in which I wrote *Carbon Ideologies.* I sincerely wished to hear each and every one of coal's self-defenses.

But it was always smoother to deploy what Hitler used to call the *big lie* (the more preposterous the fiction, the more easily the masses will swallow it), and so the *regulated community* rolled out the whoppers—as when it called global warming a leftwing fiction invented on purpose to harass the innocent coal industry. A kindred trick was to give a selfish action a good name, as when a certain "legislative initiative from the West Virginia Coal Association" called itself the Coal Jobs and Safety Act.—Here were some of its provisions:

The panel that reviewed concentrations of underground diesel fumes would be eliminated. (Let 'em choke!)

Until now, metal sideboards on shuttle cars had been illegal, because they occluded the driver's view. In 2010 a section electrician named Wilbert Starcher got run over by the No. 1 Joy shuttle car as he was ascending the No. 2 entry inside the Pocahontas Deep Mine. According to the "victim information" sheet, Mr. Starcher had accrued 35 years of TOTAL MINING EXPERIENCE, including that of foreman, but this had been his first day in Section No. 2. Apparently he was never trained to make himself known to the drivers of moving vehicles. But the first-listed *root cause* derived from this: *The driver's vision was obscured by pieces of metal that had been welded onto the vehicle,* evidently to protect him from coal spillage. This improvement was evidently ad hoc, for *none of the three shuttle cars had sideboards of the same design.* And so the No. 1 shuttle

caught Starcher by the left foot. They found him face down in the mud, and it took them a quarter-hour or longer to cut him loose. *Vital signs were checked, but none were ever detected.* The relevant company, White Buck Coal (which from the citation must have been a subsidiary of Massey Energy), faced prosecution under the old law. *This violates a provision of the West Virginia Code, is of a serious nature and involves a fatality. In addition,* it *creates an imminent danger.**
Well, the *regulated community* couldn't have that! Sideboards would now be re-legalized. At least they must bear cameras . . .

Next improvement: When heavy equipment was moved here and there within the mines, it used to be mandatory for any miners in the way *to be kept in safe areas.* No more.

Until now, escape routes had been 500 feet from wherever the miners were working, but in the gloriously innovative spirit of Upper Big Branch, the Coal Jobs and Safety Act would allow them to be 1,500 feet away. After all, anyone who couldn't outrun a methane fireball didn't deserve to stay on the payroll!

Furthermore, this benevolent measure (House HB 2566 and Senate SB 357) *would give coal operators a shield against citizen lawsuits over Clean Water Act violations and a long-sought change in the state's pollution limit for aluminum.*—I wonder whether the aluminum limit would go up or down? Any guesses, fellow citizens?

Senator Art Kirkendoll from Logan County summed it up best: "I think the content of this bill is as perfect as it could be to give people a chance to get back to work."

For some unfathomable reason the Sierra Club and the West Virginia Environmental Council opposed it, while the President of the United Mine Workers wrote bitterly:

> A new majority in the state Legislature is advancing legislation . . . designed to remove decades of laws and regulations that they believe have made West Virginia "uncompetitive" . . . The [coal] industry's supporters say that to fix that problem, we need to roll back safety regulations . . . Will making it impossible for grieving family members to sue a coal company that is responsible for the deaths of loved ones be

---

* Among the other violations in this marvelously well-run mine were five that recalled to mind Upper Big Branch. Do you remember the miners' complaints of "not enough air"? Here inside the Pocahontas Mine, faces number 3 and 4 showed not much over half the minimum legally required quantity of air. Face number 7 approached the minimum, but on faces 5 and 6, airflow was so inadequate that the anemometer's blades would not turn for the investigator.

enough? . . . Maybe the companies can once again charge the miners for the tools and equipment they use.

To this Chris Hamilton (now Senior Vice-President of the West Virginia Coal Association) replied:

> Yes, we also have worsening geology and lots of lower-cost gas, but rest assured, as everyone knows who is remotely close to the coal industry, its misfortunes today principally have been brought about by our president's* attacks (which the UMW repeatedly have embraced) . . . We have tried the United Mine Workers' way for the past 83 years, and it failed. It is time for a change to make our mines competitive and save West Virginia jobs.

He added that the state legislature *is taking a close look at the body of law . . . to ensure they adequately address miner safety and streamline important safety procedures.* I suspected that the close lookers, advised and abetted by the *regulated community,* would find miner safety more than adequately addressed but insufficiently streamlined—a shortcoming they would be pleased to fix.

But sometimes the *regulated community* lacked the appropriate *big lie.* Then it was opportune to roll out the *no comment.*

In 2011 the U.S. Office of Surface Mining tested the density of various coal slurry dams. The crushed rock employed to hold back the liquid mining waste had been adequately compacted in only 16 out of the 73 trials. This indicated that these dams might be at risk for "internal erosion," which was part of the reason for the 1972 Buffalo Creek accident.—Do you remember what Karen Elkins told me? "I think they quit putting them up here." She probably slept better for believing that.

One dam was very likely the proud accomplishment of Energy Marketing Co., Inc., *associated with the non-producing Century 101 mine,* which *the Mine Safety and Health Administration has labeled . . . "high hazard," meaning a failure would likely cause fatalities.* The company's owner, Mr. Dominick LaRosa of Potomac, Maryland, declined to reply to a Department of Labor complaint against the dam. Well, that settled that.

Another masterpiece of the *regulated community,* namely, Massey's Brushy Fork impoundment out by Whitesville, right by the Upper Big Branch memorial, received federal permission to increase its volume *to a height taller than the*

---

* President Obama's.

Hoover Dam ... The engineer long responsible for the impoundment was involved in illegal ventilation plans at Upper Big Branch, where those 29 miners had died for the sake of Massey Energy's bottom line (excuse me, I mean, they died to keep our lights on). *Public records show a failure at Brushy Fork could create a 100-foot wave that would hit Sherman High School in 17 minutes.**

In 2013, when citizen activist groups obtained a one-page summary of the dam tests, the OSM *refused to release the complete data.* As for the state Department of Environmental Protection, its functionaries claimed that their own testing, carried out in 2012, *found no violations or safety concerns. But like the OSM, DEP refused to release data from its testing.*

Why these two governmental organizations would decline to inform the public of issues which might kill multitudes can, I think, only be explained either by their arrogant incompetence or by their being outright pawns of the *regulated community.*[†]

2: "We Don't Comment Publicly"

Next story: Mr. Barney Frazier was a lawyer from Charleston who chose to retire out by Rush Creek. He liked pretty views. The *regulated community* could not care less. (And here again it struck me that *how coal was mined* was among the most odious elements of this particular carbon ideology. Supposing that "scenic values," the safety of slurry dams, mine ventilation and the non-introduction of aluminum into unnamed tributaries all had to be respected, would coal no longer "pay"? And if it stopped "paying," would it stop being mined?) The day came when a certain Keystone Industries decided to bring down 375 acres of mountaintop in plain sight of the Fraziers' back deck.—This was the same KD No. 2 Mine that Chad Cordell's coalition was fighting. (I have seen sweet photographs of them standing before the state capitol, with their handmade signs, each with a red heart over a green **KSF**,[‡] and above the heart, **BIKERS** or **LITTLE OLD LADIES** or **KIDS & CRITTERS** or **SCOUTS**.)—As might be expected, the DEP

---

[*] But why worry? Massey had been taken over by Alpha Natural Resources! Do you remember their famous "Running Right" safety program? (See p. 63.) In 2014 a federal judge determined that Alpha's water discharges at Brushy Fork had illegally high selenium levels. All we could do was laugh about little things like that, back when we were alive.

[†] I myself was refused interviews by, among other persons and entities, the Friends of Coal Ladies Auxiliary (who had originally agreed to meet me, but then changed their mind, either because of my filthy carbon ideology or because a picture-book of me dressed up as a woman had just soiled the market), the U.S. Attorney Booth Goodwin, who had prosecuted several cases relevant to *Carbon Ideologies*—he might have had no time for me simply because he had no time for me—and Mr. Bill Bissett, the President of the Kentucky Coal Association.

[‡] Kanawha State Forest.

approved the plan *(after all, they are mountaineers too!)*, so Mr. Frazier and most of his neighbors appealed to the Surface Mine Board; and once Keystone Industries consented to leaving a buffer zone, they all lived happily ever after, at least until a "new owner," Revelation Energy LLC, replaced Keystone. In point of fact, why should Revelation be bound by any of Keystone's namby-pamby and now irrelevant acquiescences? So it was that Revelation requested an amendment from the DEP, leaving the buffer zone in place but proposing more than 80 acres of "contour mining" in a previously unmined sector.—*Guy Branham, who lives across from Frazier, said he worries what other future plans Revelation Energy may have . . . "How many amendments are they allowed? There's no end to it."*—To this concern the CEO of Revelation, a Mr. Jeff Hoops, replied: "We don't comment publicly."

3: "They'll Just Go Through the Holler and Take It All"

Wishing to hear Mr. Frazier tell the tale, I gained permission to meet him at home. His wife Jackie had prepared some hors d'oeuvres, and he prayed over them as we sat around the dining room table. They were a likeable couple. He was a slightly plump, roundfaced, balding man, bespectacled.

Barney and Jackie Frazier. The MTR mine is visible at Mr. Frazier's right shoulder.

"Mr. Frazier, how do you feel about coal?"

"Well, it's one of the energy-producing alternatives that we have. My personal opinion is that the use of coal is not all bad and not all good. There are tradeoffs as with any."

"Do you see a relationship between coal and global warming?"

"I'm not convinced. I'm somewhat convinced of global warming, but skeptical that it is caused by human activity.* If we use cap and trade, and follow the new rules to make the atmosphere cleaner, it will make very little difference. If you watch public television and you see any pictures of eastern Europe, China and some of the developing world, you will realize that very little of the pollution comes from the U.S. of A."

"How do you feel about mountaintop removal?"

"I would be much more opposed to mountaintop removal as a general rule. See, I was born and raised in the coal fields. Auger mining, deep mining, underground mining, those things, provided the law is obeyed, I have much less problem with, provided the law was obeyed. I was a heavy equipment operator in a deep mine. I reported every day to an underground mine. I loved it. They would send rock out of the mine that they would haul away. When I graduated from Marshall, the first thing I did, I was an elementary school teacher in Wayne County, then in Putnam County for four years each, then I started my mining career for four years. I did not consider it to be extremely dangerous. But mountaintop removal, if I was king, I would . . . I do not think that I can envision a time where I would approve of mountaintop removal. Now, you need to understand something, Bill. I generally, myself, do not oppose mountaintop removal all over West Virginia and Kentucky and Ohio and so forth. But what I am opposed to is *mountaintop removal behind my house.* In Logan County where Jackie and I were born and raised, you might not be able to find a person who's opposed to it. I generally don't try to interject my beliefs into other places."

"Could you explain what happened to you here?"

"Jackie and I bought this piece of property, and we moved in in about February of '04. In July of '04, after five happy months, we came home and there was a notice in our front door saying you are within so many feet of a blasting area and you are entitled to a home inspection if you so choose. So we checked into it and we realized that the notice we received was because the mining company had applied for a permit. The mining company had been out here since about 2001, but there are at least four or five permits they are mining under. I would see the aura of their night lights for about four years. Now, the DEP has a process that

---

* His wife did believe in human-caused global warming.

you go through if you are a citizen who objects. So my neighbors and I, we got organized, just about everyone, 20 families, and we also contacted everyone from here down to about Whispering Woods, and we had several meetings in our home, and I hired a pilot who circled and took pictures with our camera of the mining going on, and I put up a letter and sent it out to everyone, and then I and my neighbors and Jackie started going to the city council meeting at Charleston and the Kanawha County Commission meeting, and I went to see Representative Capito, wrote a letter at that time, and we did everything we could to bring attention to this permit that was pending out here. So we went through the process; it's a very long, convoluted process; we went to a final hearing, where the coal company brought all of its workers and all of its men, and we brought all of our neighbors; the DEP recorded it, then closed the record: I think that's how they kill it, Bill. They decided to grant the permit. Well, my neighbors and I, we had another meeting and we decided we would appeal that. We secured a lawyer down in Lewisburg, and he appealed, and what happened in the end was, the coal company said: We'll settle it with you; if you agree to drop your appeal, we'll amend our permit to not devastate everything, and our attorney thought that was a good deal. The coal company at that time was Keystone Industries, and had always been operating under that name.

"Now, since then, we've put up with smoke, dust, blasting, shots, vibrations, alarms, pictures shaken on the wall and things of that sort, even though we reached a settlement. If you look out there to the left, you can see a ridge where the trees are cut. And what the company agreed to do was they would mine their coal there from the backside of the area, and not disturb the trees and all that.

"Sometime around the end of 2012, early 2013, a new company took over, called Revelation Energy, and it was taken over by Jeff Hoops.* He and his family deserve a lot of praise for what they have done for the care of children down at Marshall"—and, indeed, as I sit here writing up Mr. Frazier's words I have before me a clipping from the *Herald-Dispatch* with a color photograph of Mr. Hoops, suited and smiling, with a red necktie and thinning hair, with his beautiful blonde high school sweetheart of a wife smiling beside him: **CITIZEN AWARD: CITIZEN OF THE YEAR. Hoops embodies spirit of giving.** *Every morning when Jeff Hoops wakes up, he says a small prayer. "I get up*

---

* The owner of Keystone had been a Mr. Tom Scholl. According to Mr. Frazier, Scholl's name remained on Revelation's permit. Mr. Frazier said that "according to the Kanawha county assessors and sheriff's department, they [Keystone] were the biggest tax delinquent in the county. They owed about $25,000 in back taxes." He thought that this was why Revelation had taken over the mine. Two years later, when this story finally ended (see p. 208), the permit holder remained Keystone Industries, although its domicile was now listed as Jacksonville, Florida.

every day, and really the thing that enters my mind, the first prayer is, 'God, let me make a difference today.'" (Well, he certainly made a difference to the Fraziers.) Fifty thousand Bibles for the Russians, an orphanage in India, a hotel in the Dominican Republic to house young missionaries, a residence hall at Appalachian Bible College, not to mention $3 million to the Hoops Family Children's Hospital in Huntington! That was how the coal barons played the game. Don Blankenship, who presided over the Upper Big Branch catastrophe, distributed Thanksgiving turkeys in his home town of Matewan, where he also built a pro-coal museum, a Little League baseball field, etcetera. Jim Justice, that future Governor and scofflaw remover of mountaintops, donated $5 million to Marshall University, $10 million to the Cleveland Clinic and $25 million to the Jim Justice National Boy Scout Camp. That was how those individuals warmed our atmosphere, polluted our land and still got glowing press!—"So he comes in and takes over these permits," Mr. Frazier continued. "Mr. Hoops and his Revelation folks in about March or April 2013 sought an amendment to the permit, to where they'll come up through the head of the holler, they'll go all the way up the left side of the holler and then within 300 feet of the closest house. That's our thank you for settling seven or eight years ago. They'll just go through the holler and take it all.

"We went through the same process, DEP process, regarding that, organized again. All the same thing again, right up to Representative Capito again, for their permit amendment again. Everyone was up here July of last year, Mr. Hoops and his superintendent, and Mr. Tom Hill, who was the DEP officer at Oak Hill. Sometime around September, October of 2013, the pit permit amendment closed on that, and now we're just waiting to see what the DEP is going to do about granting or denying the amendment."

"Are any of the same people in Keystone and Revelation?"

"I cannot speak with authority. The superintendent for Keystone is not there, I am told, and the foreman is not there, but I am told the miners are there."

"How was the meeting at your house? Was it tense?"

"Me, myself, I don't think it was tense at all. Not everyone is like me. Just at my life at my age now, I try to separate the people from the plan.* If I was going off the hill today and one of these miners had driven in the ditch, I would do everything I could do to help them. I don't associate the mining with the people. I am adamantly opposed to the mining itself, and I will fight that as long as I can."

"Did Hoops ever contact you?"

---

* A noble policy, which I have tried and sometimes failed to follow in *Carbon Ideologies*.

"He did call me once to ask me something, Bill: Could we meet? I said, yes, the answer is, we can meet, but so many are involved, and they had said, Barney, if there's meetings, keep us in the loop."

"Do you feel the DEP is objectively poised between the two sides?"

"Well, you know, the DEP in West Virginia, I believe, has been given laws they have to work under, and rules they have to work under, and as long as the mining company complies with those rules and laws and are willing to amend their mining plan and mining permit, it appears that DEP is always inclined to grant the permit. I was told there was only one permit that was flat denied for mountaintop removal. But you should do your own research, Bill. In the end, the coal company is going to mine their coal. Maybe not all of it, maybe not all in the way they would like to.* But I am very concerned for my neighbors, that if the permit amendment is granted, if they start up the holler and go this way, that their wells would go dry. In one of the conversations here at my table, Mr. Hoops said, what is it you and your neighbors really want? What do you object to? I said, we object to everything."

"What did he say?"

---

* Asked to comment on the paragraph up to here, a DEP "Environmental Advocate" replied: "I believe that is a pretty accurate view of how permitting works in any of the DEP divisions. If the law allows for mining coal and the application meets the requirements of the law, then a permit is granted. *I am not familiar with the 'flat denied' comment or know a statistic on that* [italics by WTV, who had hoped and imagined that someone of her job description *would* know]. I do know that final permits often look very different when issued versus when the application is filed. The review process is a back and forth exchange and improvements and minimizations are made to the final permit. It is frustrating to the public to learn that there is no magic regulation that will prevent mountaintop removal mining. The laws, as written, allow for [it] . . . For that to change, there needs to be a change to the laws and regulations on the federal and state level."—Offering a slightly different spin, the DEP's communications director pointed out that there were "permit provisions that the applicant must meet before the permit is approved . . . While *we try to ensure the permit review process goes as quickly and efficiently as possible* [italics by WTV] we cannot approve permits until state and federal law compliance can be met," and therefore "depending on the point of view . . . you could also say a great deal of them are initially denied." Since those initial denials were mere *delays,* her reasoning failed to impress me; but never mind; she did helpfully report that "someone in our Mining program was able to determine there were five surface mining operation applications denied since 1996."—Five permits denied in 20 years! She went on: "There were a total of 89 withdrawn or terminated permits in that time period."—These ladies deserve my thanks for their trouble on behalf of *Carbon Ideologies.* The "Environmental Advocate" even appears to have been liked by environmental organizations. But no good deed goes unpunished, and a few days after the above communications, their punishment came. You see, Jim Justice was now Governor. Pursuant to his ideals, he installed at the head of the DEP a certain Austin Caperton, who "previously worked for A.T. Massey Coal and for his family's company, Slab Fork Coal." No conflict of interest there! "He's worked as an energy industry consultant since 1989, but a list of his clients has not been made public." In his newspaper portrait, Mr. Caperton looked young and personable, with a wide awake smile. He had been in charge of the DEP for less than two weeks he when fired both women. The chairwoman of the West Virginia Highlands Conservancy said: "I'm almost speechless. If they wanted to alienate the citizens and the environmental community, this is the way to do it." In a glorious blaze of *no comment,* "Caperton did not return phone calls, and, on orders from the Governor's Office, has rejected interview requests . . ."

"I don't think he said anything. Mr. Hoops, he seems to be like a good man. I separate who he is, from what he and his company aim to do out here."

I nodded, although to tell you the truth, the way that Mr. Hoops seemed to Mr. Frazier was not the same as the way he seemed to me, because when a local citizen whom his operations might be hurting expressed a worry about his future intentions, he had replied: *We don't comment publicly.* I never saw much goodness in a man like that.

And Mr. Frazier was now saying, in much the same words as had his neighbor Mr. Branham: "If this amendment is granted, and they come around and do all this, and that's Amendment 1, what's to prevent Amendment 2, or 3 or 4 or 5? There was no response like, well, Barney, this will be the only amendment, and then we're finished. In West Virginia, a property owner like Jackie and I who owns the mineral rights is a real rarity. The worst is that they would mine their mineral under your house, and the mining area begins to subside, and your house then has cracks in it. The fissures in the rock hit your well water, and that finishes it. The pure dollars of it, we spent a small fortune on this house, and I don't ever plan on leaving here. That's a pretty good burden on how your property can be ruined. I'm not a very good quitter."

I wished him luck; I wanted him to win. I didn't think he would. But in 2015 I heard that Revelation had ceased mining behind his house, evidently for financial reasons. In the Kanawha Forest, Revelation kept forging ahead with that stripper job against which Chad Cordell fought his continuing battle. As for the Fraziers' backyard view, perhaps another company would resume stripping there in 2017 or 2022 . . .

4: Exempt from Inspections and Permitting

That horrendous chemical spill in Charleston, which contaminated the tapwater of 300,000 West Virginians in 2013—by the way, it *involved a chemical used in coal processing. But it didn't involve a coal mine—and that's a point state officials are trying to convey to the public**—exemplified the same spirit of corporate helpfulness.

The culprit was an entity with the apt name of Freedom Industries; oh, yes, *that* company (CEO: Mr. Gary Southern) apparently considered itself as free as

---

* This was true in its way. Setting aside my ingenuous wonderment as to why state officials might have been so anxious to convey that particular point, I now quote our old acquaintance, the miners' lawyer Tim Bailey: "Knowing how they ran that place, and the reason that chemical was used in that plant, well, the guys that ran Freedom, that could have just as well have been wood preservative, and then it would have been the timber industry's fault."—As it happens, it *was* the coal industry's fault.

a bird. As the newspaper explained, *Freedom Industries is exempt from DEP inspections and permitting since it stores chemicals, and doesn't produce them.*[*] Reader, wouldn't it be grand if you and I were also exempt from inspections and permitting?

As for the pollutant, that was 4-methylcyclohexane methanol, familiarly known as MCHM (miners called it "hex-meth"), and *used,* as you have just read, *in cleaning coal to prepare it for processing.* If you prefer, please call it a "foaming agent." In the state's hundred-odd "prep plants," mining technicians employed MCHM to *separate and float particles of coal and clay,* after which "much of" the MCHM "and other chemicals" *ended up in the slurry that it piped to huge waste impoundments.* (It was one of these impoundments that had killed those 125 people at Buffalo Creek.)[†] What next? Then the MCHM might *bind to the solids in the slurry and stay in the impoundments. But others described that as a largely untested theory.* MCHM appeared to be leaching into groundwater here and there. By a remarkable coincidence, *the waste coming from the prep plants is not monitored or regulated for the chemicals.*[†] When the *Coal Valley News* thought to ask the regulators about it, *a promised interview on the subject with a DEP official never took place.* One had to laugh about things like that, back when we were alive.

At any rate, why was a 35,000-gallon tank (which might have actually been 40,000) of MCHM so brilliantly situated at the edge of the Elk River, just upstream of a drinking water intake? Whose coal needed to be cleaned? Well, 7,500 gallons of MCHM leaked out. And in due time, Freedom Industries pulled the same trick that Tepco had long since mastered with its leaks of radioactive water at Fukushima: *Freedom Industries has revised its estimates to approximately 10,000 gallons as the amount of Crude [sic] MCHM/PPH blend that leaked . . . on Jan. 9.*—A scientific investigator later wondered: *Was this including the 11.5 G[allons] P[er] M[inute] discharge for 24 hours before the spill was discovered?* (Just to entertain myself, I did the arithmetic: $11.5 \times 60 \times 24 = 16{,}560$ gallons.) The scientist also inquired: *Didn't liquid escape the tank even after the spill was detected?* For the sake of jollity I hereby declare that it didn't.

---

[*] Possibly Freedom Industries did more than store and less than produce, because the company identified itself as "a full service producer of specialty chemicals for the mining, steel and cement industries."

[†] See above, p. 53. According to the West Virginia Coal Association, "it is not unusual for a plant to handle over 1,000 tons per hour of coal refuse."

[†] Laissez-faire "regulations" facilitated all the carbon ideologies. Consider this gem from *The Japan Times,* September 2015: Tepco's Plant No. 1 "has released rainwater tainted with radioactive substances into the Pacific Ocean at least seven times since April . . . [A] prefectural expert on atomic power . . . wants the N[uclear] R[egulation] A[uthority] to place radiation limits on rainwater immediately. However, the NRA's position is that there are no laws that regulate radiation-tainted rainwater and therefore it cannot set numerical limits."

WHAT DID COAL, NUCLEAR AND OIL HAVE IN COMMON?

An Anecdote from a Children's Book on the Deepwater Gulf Oil Spill of 2010

"In order to respond to the spill, people needed to know how much oil was leaking. B[ritish] P[etroleum] originally said it was 1,000 barrels per day. Then government scientists said it had to be at least 5,000 barrels per day. In the end, scientists estimated that was it was 53,000 to 62,000 barrels per day."

Total oil released: 210 million gallons
[= 5 million barrels]

A year later in North Dakota, "accidental releases" from other oil operations allegedly "turned out to be far larger than initially thought, totaling millions of gallons."

News from January 30: *Now comes this warning for hundreds of thousands of West Virginians: They may be inhaling formaldehyde while showering in the tainted water, which was declared safe for human consumption . . .*—A few residents were still detecting the trademark licorice odor of the stuff as late as March 20, but perhaps they were too sensitive: MCHM may be discernible to some human noses at 0.15 parts per billion. Shortly after the spill, concentrations of that chemical crested at 1,787 times higher than that; for that matter, *levels inside residences were greater than the highest . . . level detected in the water distribution system on some days, and no one conducted indoor testing.*

It was a horribly unexpected fluke accident. How could the pitiful *regulated community* have predicted any such occurrence? A legal brief eventually revealed that Eastman Chemical Co., the stuff's manufacturer, *knew* that MCHM could eat through Freedom Industry's tanks. But had I harped on that, I might have hurt the *regulated community*'s feelings.

Tim Bailey recalled: "I was in Saint Albans,* and I do remember my wife calling me on my cell phone and asking me where I was. My son was working out on the trainer, and he says, apparently you need to buy water; we're all gonna be needing it; and I went to Kroger, and it was surreal, watching the parking lot fill up, and there pretty much wasn't any water, so I went to CVS, and then I went

---

* About 15 miles northwest of Charleston.

back to Kroger, and the only thing left on the rack then was Fiji water because it was too expensive for most people, and out of the back of Kroger, a big lift truck comes out and drops a pallet of water, and it was like a magnet, so I, being a big tall guy, was sort of able to push my way to the center there, and I figured out the way to handle it was I picked up a case of water and handed it into my buggy and picked up another case, and I handed it back to someone else. So I got my buggy, loaded it with ten cases of water, and that was the last case they had, and the checkout line was all the way to the back of the store, and it really for the first time made me realize, it's one thing to have your power out, but here, what if something did happen here so my water's gonna be ruined and we're gonna be in some valley where we can't live anymore? What if there'd been some attack? Oh my gosh, did the value of my house just get cut in half? What just happened here? And then, after you check out, people in the parking lot are kinda sitting there looking at you, and you think, tonight it might be okay, and if this goes on three or four days without water, people might get violent here. We ended up shocked and amazed at how little everyone knew about Freedom and how much they should have known about Freedom. And I was a little surprised about how little was known about not just this particular chemical but also about everything else that's being stored around here. I believe that everyone in this valley would have to tell you they feel a little less confident about their world. My law firm filed one of the first cases when I realized what this would do to businesses, so we represent quite a number of the businesses. Today I just drafted discovery for the clients.

"I can still smell that licorice smell," he said. "That stuff's everywhere up here. And it doesn't go away."

Back when I was alive, and Mr. Bailey flourished in Charleston, how could he or I have thought to draw any parallel between his situation and that of a nuclear evacuee from Tomioka? In those days (you from the future will find this hilarious), he and I supposed that our children would always be able to drink right out from the tap!

West Virginians telephoned the Poison Control Center. I am pleased to report that a good *36% of the total number of calls . . . could be answered due to call volume challenges.* What about the other 64%? Reader, let's pray they learned the golden lesson that the early bird catches the worm!

In the fifth month after the spill, one week before old Bill White in Bim dared to resume the use of his icemaker, I lunched in a Charleston restaurant whose table information contained the following: *You may take comfort in the fact that we will be using "safe" and bottled water . . . for the foreseeable future. We feel this is not a choice but a necessity . . .* As for the *regulated community,* it offered nothing; it saw no necessity.

Although the *regulated community*'s arch enemy, President Obama, had seen fit to issue an emergency federal disaster relief decree affecting Boone, Cabell, Clay, Jackson, Kanawha, Lincoln, Logan, Putnam and Roane Counties, and although the front page of the Sunday paper regaled us with photographs of Army National Guardsmen in camouflage uniforms preparing to lower pallets of bottled water into care of the Culloden Volunteer Fire Department, *Freedom Industries has said it doesn't believe the product to be harmful.* ("Federal health officials" were not quite so sure; they declared the stuff safe to ingest in concentrations below 1 part per million.) The superintendent and principal of Scott High School in Madison *both said there was "plenty of water to last us for a long time" in storage at local schools. Neither said when they might expect students to drink the WV-A* water again.*

A Culloden resident named Carol Sheets told a reporter: "The first two days we took showers anyway. And then I noticed I started having a breakout on my face so I thought we'd better not do that anymore . . . We just heated up water and took sponge baths." Culloden was the place where the MCHM could still be smelled in March. A month after that, slightly more than one-third of Kanawha County's residents dared to drink their tapwater.[†] The rest must have all been un-American complainers.

Since Freedom Industries remained so conveniently off the hook, one might have hoped for some explanation, warning or clarification from West Virginia American Water, the equally free-spirited company which controlled the water supply in question. But Cabell County's Office of Emergency Services never could get WV-AW to return their phone call. That brings the art of *no comment* into the realm of genius.

In the first month of the crisis, Mr. Jeff McIntyre, the President of West Virginia American Water (and all-around Demosthenes), flapped his oratorical tongue: "We don't know that the water's not safe. But I can't say that it is safe." *He holds several professional certifications in water and wastewater operations and engineering.* Perhaps these should have been revoked.

In the second month, at a hearing in Charleston, Representative Shelley Moore Capito exceeded my expectations of her by inquiring whether the water was safe. Mr. McIntyre, in the newspaper's words, *dodged.*[†] *Water companies do*

---

* West Virginia American Water. See paragraph after next.

† Among them were the Fraziers. Mr. Frazier remembered: "We made it through, probably three or four days it was; we had water to drink but you couldn't make coffee, couldn't take a shower or flush the commode, and then it was obvious we were going to have to do something. After that I would get bottled water from the distribution station. I would have five-gallon cans here, just in case. Whatever they said, we started taking showers in it, but we never did drink it until they said."

† In 2016 he would refuse to turn over call records about the spill to a scientific investigator.

not set safety standards, McIntyre said. *They just follow them, and West Virginia American is "in compliance with all the standards."* If the state of West Virginia set them, no worries!

Dr. Letitia Tierney, commissioner of the state Bureau for Public Health, answered that same question as follows: "That's in a way a difficult thing to say, because everybody has a different definition of safe. As I've used the example before [*sic*], some people think it's safe to jump off the bridge on Bridge Day."

No wonder it was so difficult for her to say! Four months after the hearing, Dr. Tierney's entity, *which,* by the way, *regulates public water systems,* finally *confirmed that the agency had not taken any of its own water samples, and was relying on the water company's data.* How lovely for the *regulated community*!

In a shining example of civic responsibility, West Virginia American Water presently sent out a "decommitment letter," announcing that the company "would no longer financially support public/private investments," and might or might not do other things, "but may serve new projects . . . using West Virginia American Water wholesale tariff rates." What was actually meant by these words will presumably be hidden from us until we all burn up, but I do know that the Public Services Commission compelled the company to keep fulfilling its responsibilities in the various districts, *including the Regional Development Authority of Charleston/Kanawha Valley*—which seems to imply that West Virginia American Water aspired to abandon the people whose water it had contaminated. Am I wrong? I once telephoned the company in hopes of a clarification, but—what do you know?—was instantly advised that *we don't comment publicly.*

MCHM was not listed in the latest edition of the *Hazardous Chemicals Desk Reference* available at the Charleston Public Library.* At their own expense, a professor from Alabama and his graduate students sampled water in people's homes and concluded that *crude MCHM is more toxic than previously reported by Eastman Chemical Manufacturing*—eight times more so, in fact. Who would have guessed? Some undreamed of calculus may yet prove that MCHM is safer than sugar, in which case my heart will contritely bleed for the poor old *regulated community.* After all, the Alabamans test-poisoned water fleas, not people.†

---

* The closest match was methylcyclohexane no. 4, which is to say C_7H_{14}. "Moderately toxic by ingestion . . . even at the level of 500 ppm, exhibits only a very faint odor . . . Believed to be about three times as toxic as hexane, and has caused death by tetanic spasm in animals."

† Jennifer Sass, "senior scientist with the Natural Resources Defense Council," remarked that this experiment *shows the problem with relying on limited studies by chemical makers when deciding what is safe for the public.* "But often, we have nothing else available," Sass said. "*Government funding for scientific studies is in rapid decline, and, unfortunately, government over-reliance on industry-sponsored data has become a bad habit.*"

As it happened, there was also a "second chemical" in the spill. Freedom Industries *immediately knew* of this and *told its workers in an email,* but *did not let state officials know . . . right away.* What was it? It was called "stripped PPH." Don't worry; the ever trustworthy Gary Southern pronounced it *less toxic.* I hope he poured it on his morning cereal.

In fact, the spill contained two different isomers of MCHM with divergent properties, not to mention 1,4-cyclohexanedimethanol, 2-ethyl-1-hexanol, etcetera, etcetera, not to mention *other unidentified compounds.*

After the accident came new *multiple chemical spills at the Freedom Industries site,* some of which *prompted WVAW to shutdown [sic] its intake.* These incidents seem to have been ignored in the official Chemical Safety Board Report.

In time there were federal lawsuits filed—against the airport for digging new runways and therefore increasing rainwater runoff down into the tank farm, and against the landfill *that accepted 228 tons of Freedom Industries wastewater mixed with sawdust from the Jan. 9 spill cleanup.* The landfill's guardians called for a dismissal of all charges, noting that the city and county complaint *fail to allege that the Crude [sic] MCHM constitutes a hazardous waste* under the federal government's Resource Conservation and Recovery Act—and if it were not hazardous, then why on earth should the tired *regulated community* be bothered?—Indeed, the landfill people brandished a letter from a Mr. Armstead at the Environmental Protection Agency, assuring the world that MCHM is not *a hazardous waste.* To this the city of Hurricane and the county of Putnam, West Virginia, replied: "We know nothing about whom Mr. Armstead is—aside from the fact that he has access to EPA letterhead." There's just no pleasing some people.

As for Mr. McIntyre, he did eventually tell the public: "We recognize that the presence of the odor is not acceptable to our customers"—as if it were some minor aesthetic question.*

The West Virginia Coal Association took a similar line.—"Well, I think everybody experienced inconvenience," Chris Hamilton told me. "I think the anticoal activists tried to tie that into the coal industry. I view that as an accident. I think there's enough blame to be spread around. I think our government, our water company, our legislature worked very close together to come up with a meaningful fix."

What fix was that? Well, in a doorway in Charleston I found this leaflet:

---

* But maybe it truly didn't upset him; who am I to say that he was not one of those stoic he-men who could have calmly drunk his own piss? After all, "his additional experience includes positions with . . . the Royal Canadian Navy."

TURN UP THE TIPS!
SUPPORT LOCAL BUSINESS
BEAT THE SPILL!
MAKE AN IMPACT TODAY!
VISIT YOUR FAVORITE LOCAL RESTAURANT &
TIP YOUR SERVER A FEW EXTRA BUCKS.
TURN UP TIPS / PEOPLE FIRST

—as if *we* were obligated to step in for the *regulated community.*

Six months after the spill, U.S. Chemical Safety Board Supervisory Investigator Johnnie Banks stood in his suit and tie by the podium in Charleston. The Chairman of the board had already sadly admitted that "we find self-policing is not preventing accidents." So there was Mr. Banks, questioning the air for our benefit: "How do you get a situation where you have a chemical plant this close to the intake of a water system that treats water for 300,000 people? We hope to learn from that . . . ," he said ingenuously. Do you from the future suppose that he ever did learn? If so, his eureka moment never got reported in the newspapers.[*]

But like all pollution stories, this one had a happy ending, because on a Saturday in the middle of June, *Freedom Industries proposed . . . to double its runoff pumping capacity and post contractors around the clock at its Elk River chemical tank farm.* This was, the newspaper explained, *in response to demands from state regulators and in the face of strong criticism from West Virginia American Water following two spills at the site in as many days*—for a runoff trench had discharged another *small but undetermined* volume of overflow back into the burdened Elk River on Thursday and Friday. Well, why worry? Mr. Mark Welch, Freedom's "chief restructing officer," whatever that might have been, now explained to us all that *officials from the bankrupt company "understand the importance of these issues" and are "taking action to address them."*[†]

---

[*] In September 2016, a civil engineer from Purdue University who together with colleagues and students had investigated the accident on site wrote in an open letter that he was "highly disturbed" about something: "CSB Team Leader Johnnie Banks spoke with *WV Metro News* . . . and claimed 'the findings of these two independent tests suggests that there is no threat to humans' . . . Mr. Banks [*sic*] statement and similar statements by the CSB in the report are simply untrue . . . [It] must be retracted and revised in the interest of public health." The report also failed to acknowledge "three other states were affected by the chemical leak: Ohio, Kentucky and Indiana . . . The incident affected nearly 1 million people up to 700 miles downstream, not just 300,000 people near Charleston."—Well, but wasn't it wonderful that Mr. Banks played down the negative? We needed positive thinkers when I was alive.

[†] They were certainly taking action, all right. In the first six weeks after they declared bankruptcy, although they "made little progress" in identifying the spill's hordes of potential claimants, their seven consultants, lawyers and advisers billed the estate for $1.8 million, much of which the bankruptcy judge

So did the beneficent legislators of West Virginia, who on February of the following year would pass House Bill 2547 and Senate Bill 423, communitarian measures (at least from the standpoint of the *regulated community*) whose provisions *would largely nullify the regulations imposed on above-ground chemical storage tanks in the aftermath of the spill.*

From West Virginia American Water's brochure, *ca.* 2015

We are invested in West Virginia. $365 MILLION IN UPGRADES TO YOUR WATER SYSTEM. And yet, tap water is still about a penny a gallon—a real value!

. . . To comply with the new requirements of Senate Bill 363 [*sic;* it was 423] related to above ground storage tanks . . . , West Virginia American Water . . .:

• Commissioned a $200,000 project to create an advanced, dynamic, automated source water protection tool and contaminated information database . . .

Had I lived anywhere in that state and had I cared the slightest bit about myself and others, I would not have liked those two new bills. At a minimum, I would have demanded maps and diagrams of tank farms in their regulated and unregulated states. But the *regulated community* had now invented an even more preposterous way of saying that *we don't comment publicly.* In July 2015, West Virginia American began arguing to the court that *making public the information in question, including drawings and descriptions of water system equipment, would make that equipment vulnerable to terrorist attack*—this claim being concurred with not only by the Division of Homeland Security, which loved to make everything secret, but also by West Virginia's Bureau for Public Health, which surely should have been less interested in hypothetical terrorist threats than in open discussion on the dangers of situating a water intake downstream of chemical tanks.

---

characterized as "nebulous things" accompanied by cost overruns. Three months later this judge "expressed concerns over how much Freedom is spending to secure and protect documents it fears could be seized by authorities." Soon, it appeared, "the organization could consume all available cash," with none left to clean up the pollution. In 2016 (see next page), a district judge fined Freedom the maximum of $900,000, adding, "it's all symbolic, anyway," because the penalty would go unpaid. In 2017, Freedom's owner, the "coal industry giant" J. Clifford Forrest, contributed $1 million to Donald Trump's inaugural.

Legalizing the *no comment:*
From the text of Senate Bill No. 423:

A list of the potential sources of significant contamination contained within the zone of critical concern or zone of peripheral concern . . . may only be disclosed to the extent consistent with the protection of trade secrets, confidential business information and information designated by the Division of Homeland Security . . .

Any person who makes any unauthorized disclosure of such confidential information . . . may be fined not more than $1,000 or confined in a regional jail facility for not more than twenty days, or both.

For a comparable bill for the greater good of Colorado frackers, see p.358.

In 2016 the DEP proposed a new way of calculating carcinogenic water pollution: the so-called "harmonic mean." That way, poison concentrations would register as more dilute than they would have before. At the public hearing, every citizen who stood up to speak expressed opposition to this trickery. The DEP consoled them that *changing to harmonic mean . . . does not automatically allow a tripling or higher of the amount of carcinogens being discharged.* Then again, maybe that *tripling or higher* would be allowed *non-automatically;* who was I to say?

Back to 2014: The tank farm was scheduled for demolition in July. The DEP, presumably not wishing to further harass our poor *regulated community,* explained that since *"no screening value . . . for safe inhalation levels of MCHM"* existed, nobody at all would monitor chemicals in the air once the tanks came down.

From the *Charleston Gazette,* Friday, June 27, 2014: *Rejecting the recommendation of his own experts, Gov. Earl Ray Tomblin said Thursday he has no immediate plans to conduct additional testing of home tap water . . .*

The tank farm was a 10-minute drive from downtown Charleston. When I rolled out there in June (in case you wish to visit, the address is 1015 Barlow Drive), the Elk River lay thick and still and brown down between its lushly vegetated banks, and then, just past the industrial park, cylindrical white vessels towered behind the fence, although the open trench remained invisible from the road. A redbearded fellow, doubtless one of those round-the-clock contractors whom I had just been reading about, was sitting just inside an opened automatic gate. What a busy bee he was! The instant I stepped out of the car, he caused the gate to close, then hid himself away. The stench reminded me of synthetic licorice.

LEGAL CONSEQUENCES OF THE MCHM SPILL, 2014–16

Defendant: Gary Southern, President of Freedom Industries
Crime: Negligent pollution by Freedom Industries of 300,000 people's drinking water with MCHM, January–June 2014
Possible sentence ["for the three pollution crimes he admitted to"]: Three years in prison and a fine of $300,000. This excludes 12 felony counts that were ultimately not filed.
Verdict: Pled guilty to three misdemeanors.
Actual sentence: 30 days ("well below the range of 18 to 24 months recommended by the advisory guidelines"), a fine of $20,000 and six months' probation. *Judge Thomas Johnston: "Like the others, this defendant is hardly a criminal."* Apparently not, for the government "promised to return" to him "more than $7 million and a Bentley luxury car."

The company's co-owner, Dennis Farrell, pled to two counts and also got 30 days; four other officials squeaked by with probation. Jeff McIntyre was never charged with anything. The company was fined "the maximum of $900,000 . . . but . . . the bankrupt company is unlikely to ever pay."

Meanwhile, a new chemical company, Lexycon LLC, had already been registered on Marco Island, Florida, where Gary Southern just happened to have a house. When Lexycon registered in West Virginia, with Dennis Farrell as the "sales contact," its address and phone number turned out to be the same as that of Poca Blending, "a longtime affiliate . . . that merged with Freedom" a week prior to the Elk River spill. Lexycon issued a statement insisting that it had no connection to Southern or Farrell.

5: "We Can't Give You a Personal Answer"

In the *regulated community*'s defense, I must admit that *no comment* proved quotidian; for compounding my narrative burden in these West Virginia sections was the guardedness of some local people, not merely the *regulated community* itself, one representative member of which was the Hobart mine just north of Madison, welcoming the world with a **NO TRESPASSING** sign. It was midday and only 92° on that day which smelled like mildew and sounded like silence as I looked across the creek bridge as if across No Man's Land, puzzling over the smoke and light-strings on the facility's strange bridges, and as soon as I had photographed the sign that warned against parking or turning, the security guard began to drive toward me on the bridge. Why on earth did I feel he might not open up his heart to me?—Chris Hamilton had assured me that the near impossibility of my touring a coal mine derived from the financial loss my presence would cause, since up to four people must be pulled off other duties to escort and protect me, and it might be necessary to idle whichever longwall or other zone I gawked at.* I believed Mr. Hamilton, but I also suspected that even if I could recompense the company for whatever productivity my presence cost them, my observations might still be considered, like a mining inspector's, no gain at best, and at worst an outright obstruction. So I never got to go down in a coal mine, not in West Virginia or in Kentucky, either—which disappointed me because the mines sometimes sounded magical as well as sinister. Old Bill White from Bim, the man whose coffee had gone bitter, told me: "In the mines I saw some things that are hard to . . . I just don't have the words. A tree with . . . perfectly lined up in the rock formation, *trees,* like, straight, like *breasts* if I can use that expression, like one big log, perfectly in line, like one after another, but it was just rock. Oh, I've saw some funny-looking things! One of 'em, we used to call it kettle bottom. Used to be, at old Wharton No. 1, or Eastern Associates Coal as they called it then,† several men got killed because it would fall out. It would be round, and as it went up, it would taper a little bit, and the sides was as slick as could be. The policy was to shoot it out."‡—But the mining operators had no time for me.—And there were towns in West Virginia whose citizens perhaps took a similar attitude to outsiders. Japan was never like that, although I met my share of rudely scornful people, fearful evacuees in a hurry, bought men and, more commonly than any of these, individuals whose mouths remained unauthorized to open on a given subject.

Chris Hamilton knew how to speak to me as one American to another, in that folksy, personal way that Americans were suckers for, back in the days when Americans were alive. As you may remember, when I asked him about global warming, he brought up his parents, children and grandson. I suppose we

---

* When I asked to interview just one coal miner, he referred me to Roger Horton of Citizens for Coal, "a non-profit organization dedicated to helping maintain the vitality and productivity of the coal industry in West Virginia." Since Mr. Horton's booth at the Coal Festival was busy with leaflets, I left him alone, suspecting that he might not be a typical coal miner. In 2015 I was surprised to read that even he now had issues with Big Coal. "I am a retiree from Patriot and like many of you, I am looking at the loss of my pension and my health care . . . I am angry. I am sad . . . We have to get up and start taking the fight to those who are hurting us"—although perhaps by that he did not mean Patriot Coal but rather those politicians and institutions who were trying to protect us against climate change. For his bitter "apology" for keeping our lights on, see p. 217.

† In Wharton (unincorporated) there was now a Patriot Coal establishment with lights on its towers and bridges, with smoke coming out. It was also known as Wells Plant Eastern Coal.

‡ That sad coal miner at the Upper Big Branch memorial in Whitesville also used to discover them. He remarked that in the mine he used to find fossils all the time, fossil leaves from trees just like the trees right here; they were difficult to cut out; but he had a bunch of them and gave them away.

Americans had a naive tendency to believe that when a man said anything about his grandson, he was being honest at that instant. When interviewing those public relations officials from Tepco I began to hunger for such personal details. Mr. Hitosugi Yoshimi, section manager of the Corporate Communications Department, was explaining to me: "It will take 30 or 40 years to decommission the reactor. So it's important to make the working environment better within the premises. Lowering the radiation level is also important. We would like to make an area where you won't need any full facemasks, just a half mask. We would like to extend that area. We are trying to make some place where the workers can take a rest and eat something warm." This of course led me to inquire what was the maximum dose these workers could receive, and Mr. Hitosugi earnestly replied: "There is a law. A hundred millisieverts for five years, or 50 millisieverts per year. But in actuality, the workers' exposure is much lower, and it's getting lower and lower! You can check it on our website."—As soon as he told me to check the website, I could not help but distrust his answer a little, words on a website being so ephemeral and deniable. So I asked a more pointed question: "How many millis per year would you consider acceptable for your family?"—I was answered by Mr. Togawa Satoshi, deputy manager, International Public Relations Group, Corporate Communications Department: "We can't give you a personal answer, but the radiation, there is a law about that."—As would many another American, I then felt (quite wrongly in this case) excluded and evaded. What kind of person wouldn't say what was good for his family? Had he said that he would be perfectly satisfied if his children got 100 millis a week, I would have been horrified but I would have thought, well, this fellow certainly stands for something—whereas Mr. Togawa might have considered his family none of my business, or else perhaps had been explicitly constrained by company rules to answer only those questions which I had submitted in writing and in advance—or (and if this was the case I was right to find his answer unsatisfactory) feared to assert any particular figure, in case it might get him in trouble with the company.—But although I frequently ran into situations of this nature, I never encountered reserved or unfriendly *localities,* except for that time in 2011 when everyone was fleeing Kawauchi in a panic—whereas in, for instance, Madison, West Virginia, our car neared an open stretch of curb that fronted a certain yellow house, an old house which hunkered just below State Street, and I thought this a plausible place to park for the 2014 Coal Festival, but although the car actually pulled in several lengths away from this house, the three slender crewcut young men sitting on the porch kept staring over their shoulders at me, quite unsmilingly, so I got out and called down to them: "Would you mind if we park here?"—"You know, we kinda would."—Then there was my ancient landlady at the Madison Hotel, who, although I had now given her my business for two years' worth of Coal Festivals, when I inquired as to whether she could tell me any stories about this town in which she had probably lived most or all of her long life replied steadily: "No, I don't know anything. Your best bet would be to check at the library." When I think about it, her answer reminds me of Mr. Toga-wa's. Down behind the Madison Hotel lay a mining rig. I asked a teenaged couple who were walking past it if it were inactive. The girl said nothing. The boy said he didn't know. Then they walked away. I suppose they had probably passed it most days since they had been born. As for the lady at the Coal Museum, which was dedicated to mining, when I asked her how I might meet a coal miner, she coolly proposed that I try on the street; there would surely be plenty of them at the Coal Festival. Thus her version of *no comment*.*

6: The Bolter's Story

None of those last four encounters could be definitely labeled unfriendly, not even the first, but to me they came to appear typical of Madison, where even such unsatisfactory meetings as these might not come easy, as on that June evening on the front porch of the Madison Hotel (the same place I had stayed the year before en route to Cook Mountain) when for an hour and more, hoping to know whether a room might be available, I had awaited the above-mentioned landlady, who turned out to be at her sister's for Sunday dinner, according to the only lodger I could find, a man upstairs who was lying on his bed with the door open, watching television—a coal miner, as it happened, and when I asked him how coal mining was, he said, "Well, nothing to brag about." Hearing that I hailed from California, he immediately expressed a wish to try "the herb," but unfortunately I had brought no illegal substances. So I went back out to the front porch, where the air was 97° Fahrenheit and humid, with State Street essentially empty, excepting birds, chattering radios and two teenaged couples kissing on the roadside. Across the street rose a long flight of steep cement or concrete steps, the weeds pressing in on them from either side, and then a glorious great tree which hid most of the house and porch. As I sat watching this house, feeling as limp as the flag in front of the Terrell & Hill Insurance Agency, a white Madison police cruiser rolled past, and

---

* I wish I could blame the *no comment* exclusively on the four wicked fuels without which I would not be sitting in my air-conditioned palace, writing these words by night. In fact, anybody who got ahead in life, and "upgraded" his business into a fully sharktoothed member of the *regulated community* became enamoured of the *no comment*. Reader, wouldn't you consider it harmless to ask a manufacturer of photovoltaic cells about their efficiency in generating electricity?—"You know, that's not a service that we offer," explained a California solar equipment distributor. "And we wouldn't be comfortable sharing our customer and support base with you." I told her that that was just perfect, and she told me to have a nice day.

then a crewcut man pushing a baby in a stroller; he nodded at me and called me *bud*. That reminded me that even in Madison people could be friendly enough. So I waited and waited for the landlady, sinking into that gentle humidity in which others seemed to swim; it was not as bad as Louisiana or the Congo, but all the same I saw small reason to change my shirt, since in 20 minutes the new one would smell like the old. Now it was five o'clock, the pancake frisker reading a moderate 75 counts per minute* near the brick fireplace in the lobby, after which I returned to the front porch in case the landlady might be arriving, and the perfume of old cigar smoke came out, every now and then mingled with the smell of honeysuckle.

The landlady did come, and one room was available, so I became a guest, and saw that miner again.

He had been a bolter for 20 years, supporting the ceilings of mine tunnels. A journalist once described his specialty as *the worst job of all . . . drilling six-foot rods up into the top of the mine as the coal got dug out and the earth complained, . . . collapsing more or less on a regular basis.* Half of the miners who perished in Upper Big Branch had been personally known to him.—And just this year two men had died in an accident at the Brody Mine.[†]

"I found 'em," he said.

"I hope they didn't suffer," I said.

"Some things ain't meant to be talked about." Later that evening he said: "I can still hear their screams." It had been two months.

A piece of coal came loose, he said. He had never seen anything like it.

As the newspaper told the tale, MSHA investigators *issued three citations that allege Patriot's Brody Mining LLC did not protect miners from hazardous conditions, did not tell the government about a similar incident that happened three days before the deaths, and allowed the destruction of evidence about that earlier incident.* As befitted its estimable rank in the put-upon *regulated community,* Patriot promised to *vigorously challenge the citations and orders that were issued.* And a federal judge voided those citations, so it all worked out quite happily.

The roof bolter was estranged from his wife, and asked for a ride so that he could pick up his clothes. When I knocked on his door the next morning, it opened by itself. His belongings were there, but he was gone. The old landlady entered the room and peered around. He had disassembled the lock and left the pieces strewn about. What does this personal detail reveal?—Nothing, I guess. Perhaps he was running away from his memories; I think he would have told me

---

* 0.24 micros an hour, not far above the national target air dose for Japan.

† Tim Bailey described Brody thus: "It's one of those mines that's non-union, but because of the contract that they used to have, they have to call these guys off the union panel."

more if I had found him . . . but he was one of the very few people who said anything fundamental to me in Madison. Like the sad miner's words at the Upper
Big Branch memorial, the bolter's story haunted me in its incompleteness. My
aspiration in *Carbon Ideologies*—to see the lives engendered by our four fuels—
would never be adequately realized.

7: How the River Got Away from Me

Nor could I blame *no comment* on my own failures in various specific investigations. After I had done my feeble mite to hear local stories in connection with the
Dan River coal ash spill in North Carolina,* a young lady from Greensboro advised me: *I wonder if perhaps something was lost in the translation of Southerners?
You know we tend to regard Californians with suspicion! . . . especially those journalist types carrying notebooks. Sometimes, just [by] hanging around for hours and
listening and waiting in large spaces of silence, they'll tend to reveal more, little by
little, and their stories will meander like a crooked river in the telling . . .*

No doubt my ignorance dragged down the result. But the ignorance was not
only mine.

That first morning in the Greensboro shopping mall I had asked the girl who
made my coffee for news of the coal ash spill, and she replied: "The what?" But
after all, the Dan River Power Station lay a good hour and more northward on
the rainy freeway, past the Hungry Church and Moving & Storage; and probably
the television news had let go of the matter six or seven weeks ago. Anyhow, in
Greensboro there would have been "no immediate danger." *Although any creatures that lived on the bottom of the river were immediately killed, the poisons in
fishes, turtles, birds, and animals in or near the river accumulate over time and may*

---

* "On February 2 downstream from the City of Eden an estimated 39,000 tons of coal ash stored in a
lagoon beside the now shut down Dan River Power Station poured into the Dan River when a metal
stormwater pipe collapsed . . . Wake Forest University Professor Dennis Lemly, an expert on coal ash,
estimated the long-term economic cost of the Dan River spill to be up to $700,000,000." The culpable
entity, "the nation's largest electricity company," was called Duke Energy. The governor of the state had
been an employee of Duke Energy until his election campaign. No doubt he was very, very impartial.
How much of that $700 million do you suppose Duke Energy would have to pay? A year after the ac-
cident, with federal charges finally in motion, it published its fourth-quarter earnings estimate, which
mentioned a debit of "approximately $100 million, or 14 cents per share, related to the company's as-
sessment of probable financial exposure." The *regulated community* expressed contrition by way of
Lynn J. Good, Duke's president and CEO: "We are accountable for what happened at Dan River and
have learned from this event."—Presently Duke Energy pled guilty and promised to pay $102 mil-
lion.—You may be interested to know that Duke Energy also owned the Oconee Nuclear Station in
South Carolina; that plant lay downstream and across a long scorpion-shaped lake from the Jocassee
Dam. Five days after the tsunami disabled the reactors in Fukushima, an official of the Nuclear Regu-
latory Commission remarked that "'the results of it sure look like what I would expect to happen to
Oconee.' Except, he added, that Oconee would receive more water."

not be apparent for months or years. Scientists are gathering fish from above and below the spill . . . No doubt Fukushima was far worse.

It had now come time to select Exit 153: Reidsville/Eden/Yanceyville, and a billboard warned: WILDFIRE EMBERS CAN TRAVEL MORE THAN ONE MILE, the Haw River meanwhile flowing like dirty fog around slender leafless trees. Crossing the Dan, glimpsing the smokestacks of what must have been the power station, I approached Eden's dogwoods pink and white. Another billboard said GOD BLESS AMERICA; I could have been in West Virginia. The riverside's semi-industrial sprawl struck my intuition as interview-poor. Following Washington Street all the way to Leakville (established 1797), I found the Leakville Moravian Church closed, while Kathy's Mini Mart gave me a rude brushoff. The Jesus effigy was captioned: **HE DIED FOR YOU.** I wrote that down. Back to the sprawl, then; it was lunchtime: pancakes for the driver and a sandwich for me. Having thus purchased the right to ask questions, I inquired of the diner waitress who should be blamed for the spill—or had it merely been (using what I supposed might be a local expression) an act of God?—"I don't worry about any of that," she answered. "All I know is that we're safe, because it happened below our filtration system." Thus human nature.—On the road, a truck full of long wet skinny bark-covered tree-trunks groaned back across the Dan River. The rain had moderated, so I pulled my cap down tight, descending through the mucky grass from the strip mall toward the Harry Davis Bridge, while ahead and above a billboard invited me to continue on to WHERE CURIOSITY BLOSSOMS, as indeed I did, trying not to step into the standing water between the ivied trees; now I had attained a curious view, even a sort of *blossoming* view right here upstream of the bridge and the power station; let's call this the uncontaminated part of the river, although *Eden's Own Journal,* alas, had just admitted that the town had spewed so many "excessive Sanitary Sewer Outflows," which it defined as *when sewage flows out of manholes or broken pipes,* that by the beginning of 2012 the Environmental Protection Agency's killjoys had placed Eden under an administrative order to do better. From 1987 through 2012 it had to be said that *Eden averages 2,028,819 gallons in SSOs on an annual basis.* An environmentalist or kindred fool might wish that Eden's citizens would feel some responsibility toward the river they fouled, and for all I know they did, but the newspaper's headline expressed only the usual dreary American concern for dollars and cents: **Administrative Order forces city to increase projected utility costs.** And in this book about competing carbon ideologies (among them that most wondrous solution, nuclear power, which has proved so safe and clean at Fukushima), I shall now teach you success's magic rule: *Get someone else to pay expenses.* A sanitary sewer overflow, like a coal ash spill, is nothing to be feared if it hurts only the neighbors downstream. *I don't worry about any of that. All I know is that we're safe, because it happened below our filtration system.* What I saw was a taffy-colored blob impaled on a snag, with many more pallid clots worming down upon the brown-green water, the trees just coming into leaf as swallows chittered unseen, and this foulness kept spreading out as it went under the bridge, so that on the far side it almost resembled pale ducks and white ponds sliding along on the brown water. There was a vague chemical smell, but was it coal ash or something even more American? The broken pipe could not be seen. Motoring downstream along Town Creek Road past a caramel-colored creek adorned with colonial houses, horses and donkeys, the inaccessible power station now long past, I reached Draper Landing where the water was muddy and eddying, meeting a weeping willow, a plastic Easter bunny and a sign: **PREPARE TO MEET GOD!**—In Draper some people were just about to enter their little bungalow house, so I pounced.

"Well, she went down there just the other day," said the old man, gesturing to a woman.

"What did the river look like?"

"It looked yellow."

"What do folks say about it?"

"Well, we're not too happy about it. We all drink bottled water."

"Even though they tell you it's safe?"

"Well, we just don't know. We don't know if it's coming up through the ground."

Thanking them, I continued to parallel the Dan as well as the highway permitted, never seeing yellow water and no longer smelling that chemical smell, although farther down river, I think it was Chumney, I learned: **ONLY ONE WAY TO HEAVEN . . . JESUS.** Then came the Lincoln Elementary School, Oregon Hill, Victory Baptist Church, where **GOD NEVER LETS GO!**—and turned left on Business 29, parallel to the railroad track, hoping to get back to the Dan River, crossing a lovely green rolling country by the railroad, where a white picket fence and lovely pink dogwoods were my best interviewees. Before I knew it, I was all the way in Danville, Virginia, where four ducks in a row adorned the Dan River, whose water looked perfect. (Among the heavy metals in coal ash are arsenic, lead, mercury, cadmium, chromium and selenium. And by the way, the newspaper informs me that *coal-fired power plants generate more than 130 million tons of various ash wastes every year*.) At the water treatment plant, several hard-hatted workers puzzled over a flowing channel. At a coffee shop I asked if anyone drank Dan River water and they said: "No problem there."—Let me quote from *The Grapes of Wrath:* "When a bunch a folks, nice quiet folks, don't know nothin' about nothin'—somepin' goin' on."—Following my usual policy (herewith recommended to global warming alarmists), I left the

spill behind me and breezed into Virginia's Entrepreneurial Region, antici-
pating the outlined ridge-trees like the fringe of some textile as one approached
West Virginia; and this is all I ever learned about it.

8: Why I Love West Virginians

*Sometimes, just hanging around for hours and listening and waiting in large spaces
of silence, they'll tend to reveal more;* perhaps it was like that—but in West Virginia I frequently did wait and wait, sometimes for nothing. I could tell you
about the Boone County Republican Executive Committee whose secretary, a
resident of Van, did not return my phone call (his answering machine might have
been defective), and whose President declined to let my accomplice communicate
with him on his Facebook page even though she, a Democrat, agreed to electronically "like" the Republican credo for awhile (perhaps he was too busy to
notice her); I could mention the District Attorney in Charleston whose Secretary, having taken down my name one evening, called first thing next morning
to assure me that her boss would find time for me neither this week nor next.

But I could also tell you about the regular customers of a Charleston bank
who, turning round from teasing the tellers by pretending to be bank robbers,
greeted me and called me buddy, the middle-aged waitress in Pineville who called
me baby, the teenaged waitress at the drive-in down in Welch who did the same,
and, most happily of all, the contrast between the Transportation Security Agency
in San Francisco and in Charleston. At the former airport, when I, having passed
through the cylindrical inspection chamber where each passenger must raise his
hands like a surrendering bank robber, politely asked a TSA officer whether I
might turn on the pancake frisker and send it through the X-ray, he glared, then
instantly, venomously said: *"No."*—At Yeager Airport in Charleston, when I
asked to run the frisker through, the officer said: "Don't make no difference to
me!" Afterward two TSA agents in full uniform strolled all the way to my depar-
ture gate and sat down with me for a quarter hour, talking about radiation, coal
mines and what I thought about West Virginia; one of them was the son of a coal
products salesman who had been inside a mine and told me about it, and the other
was from Alabama. Both of them believed in global warming, although in their
opinion it might have no human cause. They shook my hand and introduced
themselves by their first names, and I, who have always wanted my government to
like me, was very, very happy. I told them, as was true, how kind West Virginians
are, and how fond I have become of them; and they were truly pleased. When they
left they shook my hand repeatedly. One of them remembered me two years later
and shook my hand again. I am still happy when I think of this.

No, I would never call West Virginia an unfriendly place. But coal had its secrets, and secrets live underground.

9: Regulated No More

Then came Donald Trump, and the *no comment* achieved an all-American perfection to which the following headline gives homage:

Staff Tells of Rampant Secrecy at EPA: Mission to Weaken an Agency Once Known for Transparency

"HOW COULD MANMADE EQUIPMENT PUT UP ENOUGH SMOKE TO MAKE A DIFFERENCE?"

1: "I Hope This Doesn't Hurt Coal"

And because these secrets were so discursive—because coal can run through so many dark and crooked seams—the effects of coal were nothing like some nuclear accident that can be probed by going deeper and deeper into red zones and learning how to measure wisely. When I was alive, West Virginia's desperate attachment to coal, reinforced by habit, economics and religion itself ("God put it on earth for us to use") was abused by corporations with shameless ingenuity, the mountains leveled and the hollers below them poisoned, the streams acidified and worse, beauty made ugly, the land stripped above and below and then the coal towns left to hollow out. How could I represent this malignant chaos of effects as anything but a bundle of broken stories?—But I had at least persuaded myself of the banefulness of large scale coal mining, and of the astounding resignation of its victims—except of course for such isolated gadflies as Stanley Sturgill: "We had two mine blowouts back in March of this year, and it absolutely opened up Highway 160 right by my house. They come up and did a little temporary job on it, and they left it that way. It went on, March, April, May, June, July, and I got tired of it." So he wrote a letter to the editor.—Was it a cause or an effect of Appalachian people's isolated poverty, that they met calamities with such stoic, forbearing and sometimes even self-blaming patience?

In Twilight, Mark Mooney said to me: "We're worried about Patriot. Right now I'm paying a hundred sixty, a hundred eighty a month, for insurance; when all them retired people have to go on Medicare, all we'll be working for is our insurance. I'm scared to death. If they get by with this, then those big other companies, they're gonna pawn it off, too. There could be a war on the first of July when their insurance cuts off. Then the old union guys will cross the picket line."—But they didn't.

When I asked Arvel Wyatt whether people in McDowell County felt that miners had received their fair share, he replied: "I don't think they have any ill will toward the coal companies for shutting down. Business is business."

Ms. Bonnie Wireman, 81 years old and resident of Dry Branch, West Virginia, achieved her moment of fame in the Huntington *Herald-Dispatch*. *The widow of a coal miner, Wireman was angered about the chemical spill that's deprived 300,000 West Virginians of clean tap water for four days, but doesn't blame the coal or chemical industries. "I hope this doesn't hurt coal," said Wireman, who lives in an area known as Chemical Valley.*

Barney Frazier had this to say about the spill: "When I was a little boy till I was nine or 10 years old, in the coal camp where we lived was a well. Dropped a bucket down there and we pulled it back up. Now I was used to carry water when I was a little boy. So I really didn't think it was much of an inconvenience. I'd just go down to the distribution center with Jackie and do it once a week. In all that, I cannot say I ever smelled that licorice smell. I was already gonna be off the hill for some other reason. That was gonna be one of my stops."

I asked whether he blamed anyone for what had happened, and he replied: "It's easy to say in hindsight that the state should have had an inspection system for tanks. There are releases of gases and chemicals here in the Kanawha Valley quite often. It's hard for me to be very critical of West Virginia America Water. They thought their carbon filters would catch and hold that. You have to realize there are hospitals and some places that you just can't turn their water off, at least not without notice."

"And how dangerous is that MCHM?"

"Well, I would think that someone would have to put the stamp of approval on a chemical before Dow could sell it to the general public."

That might or might not have been so. It does seem to have been so that no one had to put the stamp of approval on a chemical before it was stored directly upstream from drinking water.

In 2014 I asked another former coal miner now living in the southwest part of the state to explain what happened at Upper Big Branch. He began by remarking, "I really couldn't say about that," and then, once I had promised to refrain from identifying him in this book, continued: "We were living here during that disaster, and a lot of these guys, we were told that even though they were trained and had things provided for them by the company like the respirators, if you were working for like what they call a longwall cutter, and it was cutting up and down 18 inches deep, and they would go about a thousand feet in a day, now, those guys should never have worked on that without a respirator. Those guys, if they got word inspectors were in the mines, they would wear them. Other times, they

Americana at the West Virginia Coal Festival

would say it's inconvenient, uncomfortable; and they would take their chances."*—In other words, he seemed to intimate, how could we blame management for cutting the same corners as the workers themselves?

One of this forgiving Christian resignation's motives was economic. If a person had the prospect or even merely the hope of financial gain, it might make sense to look the other way when miners were killed and mountains destroyed.— "If you don't mine coal, you're gonna be working at the Walmart," said Tim Bailey. "If I can get to be an electrician or topnotch continuous miner, the number's gonna be 30, 35 dollars an hour. It's nothing for a good worker, a guy at a mine that can really produce the coal; 75, a hundred thousand a year, that's nothing. Foremen, superintendents can get into six figures, 120, 130 [thousand], and when you've left them no alternative, where else can they go? Working men and women, they don't vote the way you expect them to vote."

Meanwhile, *notice is hereby given that Pine Ridge Coal Company . . . is*

---

* The official report implies that in fact the miners' bosses were the ones compelling them to "take their chances." Gary Wayne Quarles to a former co-worker: "Man, I'm just scared to go back to work . . . Man, they got us up there mining and we ain't got no air. You can't see nothing. Every day, I just thank God when I get out of that coal mine that I ain't got to be there no more." He was killed in the blast. Testimony of Dean Jones's wife: *At one point her husband told her he shut down the section for lack of air, and "Chris Blanchard called the dispatcher and told him to tell Dean if he didn't get the section running in so many minutes he would be fired."*

discharging into Lavinia Fork, Frozen Hollow, and Unnamed Tributaries of Lost Branch of Little Jarrells Creek and Big Scott Hollow of Hopkins Fork of Laurel Creek of Big Coal River of Kanawha River . . .

And because they had swallowed every other coal lie, they could be expected to give coal the benefit of the doubt on climate change. Coal was America's best friend.

Doorway in Williamson

2: Why Fukushima Was to Blame for the Oysters

And anyway who was I, or anybody, to pronounce a certain doom? Hurricanes Sandy and Katrina had been killers, to be sure, but climate change's involvement in those specific events could not be quantified. In future floods, droughts and tornadoes, the immediate cause of death would remain those floods, droughts and tornadoes. The planet warmed a smidgeon more, while we dragged out debate for profit's sake.

A teacher explained: "In West Virginia you have to be careful. You can't come out and say something's bad if their Daddies are in the industry."

The science coordinator of West Virginia's Department of Education confessed that "lessons on climate change have been embedded in the state's curriculum for several years," but fortunately "it isn't as though we're telling students what to think. Teachers should show both sides of the argument"—which reminds me of how Marx summarized Proudhon's views on slavery: Encourage the good side of this economic category and discourage the bad. "So," the science coordinator continued, "we might say that coal is done as well as it can be done while still protecting the environment or [we might] talk about the best technologies to do it safely for miners and the environment." Hard-hitting stuff indeed.

No, we would never wish to tell students what to think! Every idea was correct, because we lived in a democracy.

In late July of 2014 I read in *The New York Times* that ocean acidification was proceeding especially alarmingly off the coast of Washington State. Oysters and starfish were declining rapidly. When I happened up that way in August, I started asking people what they made of it. In a certain oyster farm on San Juan Island they did not like any such questions, since I might be hinting that their shellfish might be contaminated. A woman allowed that some ranger or other would "train" her in this matter pretty soon; if I came back next year she might have something to tell me. In Roche Harbor the oysters on my po boy sandwich were tiny. In the clam shack by the Pike Street fish market in Seattle I ordered a platter of fried oysters. There were fewer oysters than I remembered, and more french fries. I asked the server, a certain Javier, whether he had noticed any decline in shellfish and starfish, and he said he hadn't heard about it. I asked about the acidification of seawater, and he said it came and went in cycles. I asked whether global warming could be causing it, and he said: "You know what I think? It's that radiation from Fukushima Daiichi." He was proud of knowing this name.

3: "The Earth Is So Large"

"Do you believe in global warming?" I asked Pastor Blevins, whose life story began this coal chapter.

"You know," he said, "I believe in climate change. Anybody that can't believe in that is really not looking around. I don't understand enough to know that our global warming is the result of our poor management of emission. Not long ago we flew to Atlanta, and from Atlanta to Africa, and we stayed six weeks, and we flew back, and I noticed how much more land mass of timber and trees there is in North America and in Africa than what's been developed. Here you do see the smokestacks and you know that they do put off the smoke and everything, but it seems to me that the earth is so large and there are so many trees and everything that how could manmade equipment put up enough smoke to make a difference? I personally think that the biggest thing you need to do is update technology. I would really like to make a visit over to an operation near Wise, Virginia, near a worked-out coal mine. They built this power plant, and they don't have smoke-stacks. And the smoke goes through a tunnel, and is channeled back through a worked-out coal mine. And that seems to me to be the very best thing. There's one mine in this area, and you go about thirteen hundred feet, and you have tunnels radiating about five miles. Now, if they would pipe the emission into that, and seal that off, that's the type of technology they need to have.* I think for wind power and solar power, they're real good, but again, how long will it take to get this? Duke Power† had purchased a lot of land near Bluefield; they were going to put a wind farm on that land. The county, they're governed by supervisors, I think they had five, and during that process they elected three new ones and they ruled out the use of that land, and now they're on the second year, going into court. They said it would ruin that beautiful mountaintop between Clinch Valley and another just before Shenandoah Valley. This would have produced enough power just about for the whole county. It seemed to me such a win-win. Now I could understand if you have a subdivision here with a hundred homes and these windmills are going to encircle you, you might not like it. But a lot of good could come out of it. It seems to me like more of everything and not cutting back on anything is how to go."

"Where do you come down on mountaintop removal?"

"Well, I favor mountaintop removal, and one of the reasons for that, I'm very familiar with that in eastern Kentucky; they moved a mountain that was not of a great deal of value other than the beauty of it, trees and so forth, and the top of that mountain was restored; they graze cattle on it, and have a large shopping complex up there, a Holiday Inn, and all types of commerce. And they're using what was not usable! I know one of the big things is the streams; you do have a problem with them, but there again, I think that some of those, they could use drainage. And the majority of the mountaintop removals, they want to keep the

---

* A month after this interview, the world's largest carbon capture venture began in Texas. The Petra Nova Project, underwritten in part by Japan, would compress 90% of one coal-burning plant's carbon dioxide, pipe it into an underground oil field, and with its pressure drive out the oil, then seal the gas where the oil had been. The notion of eternal sequestration sounded as ludicrous as the idea of isolating nuclear waste forever. But maybe the stuff wouldn't leak until I was dead.—In 2012 a "review of current literature" pertaining to coal-generated electricity had concluded that "the energy penalty of CO_2 capture, defined as the percentage decrease in electricity output per unit of fuel input, ranges from 12% to 48% . . . Therefore, . . . this approach cannot be treated as a single safe solution for industrial decarbonization."—But what did those reviewers know? They were only scientists!—In 2016, the Kemper power plant in De Kalb, Mississippi, whose carbon capture technology was supposed to be "a central piece of the Obama administration's climate plan," stood unfinished, years behind schedule, billions over budget, and tainted by a whistleblower's extremely credible allegations of dishonesty. However, in that same year Chevron announced one project in Barrow Island, Australia, in which "approximately 120 million tons of carbon dioxide is expected to be safely injected," and another in Alberta.

† Duke Energy.

dirt, and move it out, get the coal out, then level it off, put the dirt back, so they will have more land up there to be used."

"Do you support nuclear power?"

"Well, I favor nuclear power, very much so. I think it's a good idea. Well, in Africa, we lived in a small city that had a uranium mine about 40, 50 mile out, and employed about 3,000 people; the ore was called yellowcake and placed in vinyl containers and shipped by rail to the ocean, and they did have a lot of it that was sent out by plane in a specially designed 747, in Namibia. The second place in Africa where we lived, they mined copper, lead, arsenic; it all came out of a pipe and went into a spiral about a mile deep. This copper mine was owned by stockholders in Europe and Australia. The American ambassador took a little ride in the mine, and she told me, she said, if you don't have to go down there, don't go. High humidity and you mine with combustion engines down there . . ."

4: "When You Guys Get It Right"

"There is global climate change occurring and it is a problem," said Chris Hamilton. "I have been disturbed over the degree of environmental degradation that I have personally experienced in China and Japan. It reminds me of Pittsburgh and Steubenville, Ohio. In Wheeling, West Virginia, I remember being sent home from school during a snowstorm, because they wanted us to play on white snow as kids. I remember standing in downtown Beijing and Shanghai and not being able to see across the street at 4:00 in the afternoon. I found the air quality in China and Japan a lot worse than I had anticipated. Now, to their credit, talking to some of the China officials about the pollution, I inquired about the environmental quality and was told that outside of your major commerce hubs, China is still a major Third World country. What was said was, we're still trying to get a light bulb in our provinces, and until we get a light, we're not going to be all that concerned about the air emissions. The other thing that was said, by one of the ambassadors in China was, they are watching us, because cap and trade legislation was very much active. Their country is watching us and monitoring our trial and error period. The conclusion was, when you guys get it right . . ."

Encouraged thus far, I asked: "How do you feel about cap and trade?"

"I've been involved in the politics of that, and I've found it very concerning. I've been conditioned to be so against cap and trade," he laughed, "that I probably can't give you a real objective response to that."

He was tactful with me, because I was from California; he knew my kind. In a page or two you will hear how he sang to the choir.

5: Yes, but Not Tomorrow

When I asked Tim Bailey about the contribution of coal to global warming, he said: "I don't think that anybody incredibly honestly can say that science doesn't support that any fossil fuel, coal, gasoline, anything you burn, *the science is there,* and anybody who wants to say it won't happen is just sticking their head in the sand. Our country needs to look at coal, and look at the entire spectrum of energy sources available, and you have to have a 20-year, a 50-, a 100-year plan for this country. The people who want you to stop burning coal tomorrow are foolish, and so are the people who want to never stop burning coal. Let's look at what your bill would be, what your products would cost you. Come on, guys, we're gonna need coal for the next *x* number of years. Embrace it, give them tax credits, I think it'll work.* In the meantime, I think you can plan, by that time you have invested in the other things you can do with coal, and some of the other power sources, like solar and wind, whose unit of power is still very expensive. It just seems like we're locked in this either/or battle here in West Virginia."

---

* Here he sounded not unlike Chris Hamilton, who told me: "There will be a natural transition period that will be absolutely mandated on us as a society, and in all probability that's going to be within the next one hundred years, because coal really is a finite resource, and when you think about that and how much coal we consume within our energy mix, we're not prepared as a society to make it through that transition today. When you think of a hundred years, that is not very long. That is three generations. It doesn't allow for a long period of time to accelerate and expand the other energy sources. And I'm not sure we have before us the next generation of energy sources. Maybe it's a refinement; maybe it's the next generation of nuclear. The fact is, we're going to be out of coal. And we're just so unprepared to fill that void. Rightfully or wrongly, we're so dependent on electricity, and there's only a handful of things used to make it."

SOLDIERS OF COAL

1: "Paying for Those Freedoms"

have mentioned the several monuments to victims of Upper Big Branch. *Also located in Whitesville is The Big Coal River Veteran's Memorial. The memorial commemorates the sacrifices of the men and women of the Big Coal River area who served in the armed forces of the United States.* Indeed, I often found coal miners conceptually associated with soldiers—an effective trope in the state with the highest per capita number of veterans in the nation. On a brick wall in Williamson, courtesy of C & G Electrical, hung the following sign: **WE SUPPORT THE VETERANS. KEEP THE COAL BURNING.** At the opening of the 2014 Coal Festival in Madison (an event not quite as busy and bustling as the previous year's), two coal-resin statuettes were presented to a pair of ancient D-Day veterans who sat in the hot evening sun facing the courthouse steps, and after a colonel uttered a rather touching homily to them, Chris Hamilton, mopping his glowing face, told the people (most of whom were sitting or standing under the lawn's edge shade trees): "After all, this is where it all started about 250 years ago!" (He also said that closing down America's coal plants would make a difference in ocean rise of less than three sheets of paper. Maybe that was true.) He mentioned the founding of West Virginia, 151 years ago now, and said: "We've been mining coal in Madison all of those years, and a hundred years before that in Virginia." Then he began to speak of the miners themselves, saying: "It's because of what they do every day that affords us all the liberties and privileges that we enjoy . . . *low cost industrial power*."—All the liberties! All the privileges!—These were sweeping words.

The chairman of the state's Democratic Party rather grandly explained that *West Virginia . . . won two World Wars. West Virginia turned mountains of coal into rivers of steel for ships, tanks, and rifles. And her servicemen fought those wars . . . Over 100,000 Americans have died mining coal, and 100,000 more died from black lung. No one has counted all the costs paid by West Virginians to bring coal up from underground.* He was another preacher to the choir.

A work of local history similarly asserted that *the story of McDowell County,*

Veterans and kin at the West Virginia Coal Festival

West Virginia, is the story of American freedom, although people who live in the economically challenged and job-starved southern West Virginia coalfields have borne the brunt of paying for those freedoms so that others might reap the benefits. Now, *that's* what I call inspiring. A visit to practically any of the scarce and wretched grocery stores in the coal counties, where the main product for sale was diabetes, showed what war prizes these soldiers had won. Upton Sinclair, 1917: Coal *would go to the ends of the earth, to places the miners never heard of, turning the wheels of industry whose products they would never see.*

Once I met a waitress named Robin who had lived and worked in McDowell for awhile. She had moved to Charleston, doubtless in order to be one of those others who reap the benefits. She was missing teeth and toiled six days a week in a diner. Talk about liberties and privileges! About McDowell she said: "You're either a coal miner, you're a prison guard or you're in prison."*

---

* Between 1994 and 2014 the yearly average inmate population in the prisons and jails of West Virginia's Department of Corrections nearly tripled, increasing from 2,392 to 6,763. Robin's statement bore some truth—for in 2014 the McDowell County Correctional Center held 444 of these unfortunates "on a per diem contract basis." Now for touch of local color: It "participated in several inmate baptisms on two separate occasions." Best of all, "the facility continues to see improvements and expansions ... Razor wire was extended and installed on exterior fences ..." But MCDO (as we endearingly called it) was not the state's largest establishment. The Mount Olive Correctional Complex in Fayette County (an area whose fracking accomplishments will earn my commendation on p. 321) could swallow 1,030 people! Hence the DOC boasted that summing up the employee and prisoner populations "makes MOCC larger than many communities in West Virginia." No wonder that

And if you were a coal miner, how much better was that than the other two alternatives?—I remember a tall, wide, overalled Twilight man whom I met in a Van gas station, just after those two men were killed in the Brody Mine. One of the dead men had been his hunting buddy.—"Yeah, it's a hard life," he said, resigned. Unlike Upton Sinclair's characters, he blamed nobody.

Not only were the miners soldiers who fought and sometimes died so that we could keep our lights on, they were also defenders of a people, region and tradition now under siege. Tim Bailey said to me: "When bad things happen to coal miners, the price of coal fluctuates, and the economy fluctuates, and for the margins for coal to be profitable, well, sometimes they are truly fighting to stay afloat, and I will tell you that with this feeling in West Virginia that the whole world is against us, the environment, the EPA, President Obama, it's this entire . . ."—His voice trailed off.*

Tom Jones, who explained the mine-induced acidification of tapwater to us earlier in this chapter, once told a colleague: "I usually say I'm a biologist, not an environmental scientist, because immediately people say, *tree hugger*[†]—and can be aggressive, because *that's my job and my family's job*—and can be, well, really upset, because unfortunately too many of them have bought into what the coal companies say through their view of themselves or through their talking heads that talk through them, that this environmental stuff is shutting down the mines."—He was prudent around those soldiers of coal.

When I tried to interview Mr. Jack Caffrey down in Welch, I made the mistake of asking (because I didn't yet know) whether coal was a good or a bad thing. Bristling, he replied: "Well, it's a good thing, okay? I'm a coal miner by birth and education. My Dad worked in a coal mine. I was born in a coal camp. We grew up in the coal mines. They were a part of our life . . . I dunno if you're for us or against us." He was a fighter, all right. I must have been against him. So the interview ended.

---

complex got its own ZIP code! (By the way, only 5.60% of the West Virginians who were locked up in fiscal year 2015 possessed "post high school education.")

* Had I asked Chris Hamilton what would happen to Americans in the event of a complete prohibition of the mining and use of coal, I suspect that he would have painted for me a frightening picture, while the equivalent question, posed in that windowless conference room of Tepco officials, elicited from Mr. Togawa a pointed refusal to consider that hypothetical case for atomic electricity generation: "There is an energy basic plan. In April the cabinet office made a decision that in this plan, nuclear power is considered one of the important baseloads, and safety should be the primary factor. So we would like to do our best to explain to the residents who live in areas surrounding the power plants that this method is safe. We do not foresee or assume any prohibition." Perhaps it was precisely because he trusted that Japanese nuclear power would return to life that he eschewed speechifying as if he were Leonidas at Thermopylae. Perhaps doom-saying was simply not in Tepco's culture.

[†] You may recall (p. 121) that Stanley Sturgill's neighbors sometimes greeted him with this term.

As a woman whose father once "wore coal proudly" wrote in to the newspaper: *Coal is falling around us, and there is not much left and that saddens me. People think of coal and see dollar signs, filth, global warming, and much more. We see our past, present, and declining future but refuse to go down without a fight.*

And so Congressman Evan Jenkins sent out his latest leaflet: **Fighting For Our Future.** (That he certainly was; you who have been condemned to exist in that period must enjoy worshipping at his statue.) Smiling likeably out at us, hale and handsome, with the American flag behind his right shoulder, he explained (italics his): "*I believe a good job solves a lot of problems.* As your representative, I'm fighting each and every day to stop the war on coal . . ."

In case that failed to strike you, here comes his most armor-piercing bullet point:

- I am fighting the EPA every step of the way by using the power of the purse to cut funding for the EPA's war on coal.

What necessitated his fight? Please don't say *climate change*!—When I asked Chris Hamilton, "What do you think is the basis behind the war on coal?," he replied: "I think it boils down to economics and politics, not necessarily in that order. Now, we can explore those individually and jointly. Now, I believe I see the war on coal being more applicable to a regional application. I don't see it in a universal sense. By that I mean, the central Appalachian region has been by design singled out more so than any of the other mining regions throughout the country. I mean the Illinois basin, the western Powder River basin, the Texas lignite, or even the northern Appalachian region. We have experienced since the day this President* took office the objection, objection after objection to any newly proposed mining activity within a four- or five-state region. We saw what we felt were was an unnecessary refusal to issue new mining permits. In fact, the period in question coined the phrase a *permitorium,* because there was a moratorium on permits that were previously approved by multiple federal agencies."

"Why Appalachia?"

"I think there was some consideration to certain political constituencies. I think there has been strong opposition to certain surface mining activities and mountaintop removal specifically, for perhaps the previous seven or eight years

---

* Obama. Back in 1999, Governor Underwood had of course blamed the Clinton-Gore administration for "the assault on coal."

since this President took office, and I think that the opposition were supportive of this President. There was also the imposition of unprecedented requirements, generally imposed through the permitting process, unprecedented standards, layers of requirements. We also saw unprecedented activity to withhold or revoke mining permits that were previously issued. The President's home state, that region of mining is doing very well today. If you want to expand the economic stream of coal mining in southern Illinois, you're going to have to ratchet it back somewhere else. A lot of our miners in southern West Virginia are moving to southern Illinois."*

"And it's the same kind of mining?"

"Exactly."

In his Coal Festival speech, he expressed a higher level of embattlement: "Don't let anyone tell you that we don't have the best miners and the best mining engineers! . . . There are those in Washington, and there is a President, and he is *blank*-bent on weaning this country from fossil fuels. We've lost 50 million tons of coal production right here in Boone County over the last three years. You can shut all 12 hundred coal-fired power plants down in this country today and you will effectively reduce less than 4% of CO_2 emissions. This is all about less than 1% of CO_2 emissions, and most are in China and India and they are never going to follow such a dumb, ill-conceived plan."

When it came to carbon dioxide reduction, even the United Mine Workers, who to defend their constituents against the Don Blankenships[†] of the *regulated community* must stand across the divide from the West Virginia Coal Association, agreed with Hamilton, calling the EPA emissions rule *a neutron bomb aimed directly at the heart of West Virginia's economy.*[‡]

The Attorney General had already gone all out: "West Virginia is in a battle for its life . . . the ability of the Mountain State and others to mine and use coal has been steadily eroded by the Obama administration and its Environmental Protection Agency . . . That is why our Office has focused like a laser beam on everything EPA-related coming out of Washington, D.C."

How refreshing to blame Washington! When Ms. Samantha Davison, a "corporate communications spokeswoman" for Alpha Natural Resources, the huge

---

* In August 2014 a certain Nick Carter, the big cheese of Natural Resource Partners LP, explained that coal mining hereabouts "is expensive." "He said many operators in Central Appalachia are moving to the Illinois Basin and will not be coming back."

† When Blankenship was indicted for Upper Big Branch, Tim Bailey said: "This . . . confirms information that we have heard time and time again that the practice of advance notification and the 'War on Safety' at Massey was . . . condoned at the highest level."

‡ Hamilton called it "a runaway train, wrecking everything in its path, to include our state's fragile economy."

corporation that had bought out Massey after the Upper Big Branch disaster, announced that "due to tough coal market conditions, we're having to adjust our operations"—meaning that Alpha was giving pink slips to 160 Boone County miners—Boone County Commission President Mickey Brown performed the exegesis: "We just keep getting battered by the current market situation in the coal industry and the stricter EPA standards." When Coal River Mining and Patriot Coal laid off miners, they made their own similar, separate accusations. Those EPA standards were needless, hence cruel; they had nothing to do with improving the health of Appalachian communities, or helping Mr. Brown's grandchildren to endure climate change.—From a leftwing periodical: *Even conservative estimations state that in order to keep the average global temperature rise below 2 [degrees] C—let alone below [the] 1.5 [degrees] . . . being called for by more than a hundred countries . . . 89% of Europe's coal reserves have to be kept in the ground.* Meanwhile Mr. Tom Harris, the executive director of the International Climate Science Coalition in Ottawa, wrote a special column for the *Coal Valley News,* comparing *the highly debatable premise that we can control Earth's climate merely by adjusting our CO2 [no subscript] emissions* to an asteroid-deflection laser boondoggle aimed at an asteroid unlikely to strike us. *It would only be by demonstrating that the impact threat was false that anti-[Bureau of Asteroid Deflection] advocates have any chance of winning the debate, but they dare not bring it up.* Harris castigated the UMWA for not speaking out against *the climate scare.*

At the Heritage and Cultural Center in Madison, his comrade in arms, Senator Shelley Moore Capito, brandished all her characteristic eloquence to warn us that President Obama's "clean power plan" was "his new plan to cut carbon emissions which, if it's even possible to further devastate our region, it's going to further devastate our region." Across the state line, another mining company went Mickey Brown one better, and blamed closures on the killjoys who dared to punish polluters: *Amid litigation over thousands of water quality violations in Ky., Frasure confirmed last October that it had stopped mining in that state. The company told the court that their precarious financial situation made it impossible to pay the amount demanded by environmental groups for violations and penalties.* (Those "environmental groups" must have been the government.) In faraway Colorado, another town lost its coal mine, at which a county commissioner remarked: "Many of us feel like there's a target on our backs and the federal government keeps aiming for us."

AMERICAN GREENHOUSE GAS EMISSIONS FROM FOSSIL FUELS, 1990 AND 2012,

in multiples of the 1990 figure for natural gas

All levels expressed in [teragrams of carbon dioxide equivalents]. *1 Tg = 1 trillion grams, or 2.2 billion pounds.*

These numbers compare the *absolute emissions* of American fossil fuels for these two years. They do not compare those fuels' pound for pound carbon releases. (See I:200–201.)

1990

Natural gas [1,000.3].	1
Coal [1,718.4].	1.72
Petroleum [2,025.9].	2.03
Total [4,745.1].	4.74

2012

Natural gas [1,351.2].	1.35
Coal [1,593.0]. *The only absolute decrease.*	1.59
Petroleum [2,127.6].	2.13
Total [5,072.3]. *An increase of slightly under 7%.*	5.07

Source: U.S. Environmental Protection Agency (2014), with calculations by WTV.

You from the future may be sufficiently goodhearted to set aside your own woes long enough to ask me, "Well, what were those West Virginia politicians supposed to do, but back coal? It was practically their only support of revenue."— And it probably was. We just heard from Mickey Brown in 2013. In 2014 Alpha Resources warned of over a thousand potential layoffs come autumn; two of the mines concerned were in Boone. Mr. Brown said: "We try to make life good for people in Boone County. But it's getting harder and harder." The County Commission had already reduced the budget from $20 million to $6 million on account of falling revenues from the coal excise tax, which went for water systems and other necessary infrastructure. Were Alpha to relocate mining equipment from Black Castle and Institute, the two coal sites where layoffs impended, "the loss in property tax would cripple the county government." He also said, and I could not help but pity his situation: "Without coal, it's going to be hard to survive. Because we don't have much diversity. We don't have any flat land."*

It would seem, then, that since it polluted land, air and water, and since its productivity increases and declining market share allowed for ever fewer West Virginian jobs, Big Coal would provide a benefit in only one category: taxes.

In 1987 the state established a business franchise tax at *0.55 percent of a company's capital accumulation and stock value.* The taxable percentage changed as follows: 1989, 0.75%; 2009, perhaps due to pushback from the *regulated community,* 0.48%; 2014, a mere 0.1%. *At one point the tax brought in more than 100 million annually.* The obvious way forward: Repeal it! *That* would *make life good for people,* all right! In a guest editorial in the *Coal Valley News,* Governor Tomblin, always looking out for those people, explained: *Coupled with the reduction in the Corporate Net Income Tax and the dramatic decrease in workers' compensation, these changes have helped our state secure additional investments and will continue to pay dividends now and for years to come.* What can I say to this? Could any mind not stupefied or prostituted by the *regulated community* possibly conclude that because injured coal miners would now get smaller benefit payments, and the already impoverished state would take in even less revenue from the corporations that mined it, somehow, somewhere, *dividends* would be paid? To whom? And what precisely were these wonderful *additional investments?* How would they make up for the new hole in the budget?

Coal revenues kept falling, and there came a time when the *regulated community* enlarged its enemies list. When Murray Energy laid off about 532 coal

---

* You may recall Mr. Yoshikawa Aki's telling me just this regarding Fukushima's dearth of rice-growing areas (I:450): "Because the land is very narrow here. It's hard to live . . . They need some industry, like nuclear power."

miners in the state, *the company laid the blame on President Barack Obama's administration, the surging use of natural gas for electricity and West Virginia's tax on unearthing coal.* So there it was: West Virginia, already desperately and defiantly open for business, and habituated to blaming leftists in the federal government for closing down the companies that paid the coal severance taxes that county commissioners such as Mickey Brown relied on to pay for local services, was now, so it turned out, another one of those anti-coal villains! Who would have thought it?

Ignoring such treacheries, West Virginia fought on.—Who was the enemy?—It could never be the *regulated community*.

Near the end of 2015 we learned that the municipality of Flint, Michigan, with due American regard for the bottom line, had switched sources of drinking water, the new source proving sufficiently acidic to corrode the insides of pipes and thereby insure that pregnant women and children were ingesting enough lead to vandalize growing brains. It was a heartwarming West Virginian kind of story, with disadvantaged people suffering from the incompetence, indifference or greed of others. Since the MCHM spill had happened less than two years before, I would have hoped that residents of the Mountain State could figure out who to blame for such mishaps. One of them wrote a Solomonic letter to the editor:

WHO WAS TO BLAME FOR FLINT, MICHIGAN?

One Man Explains, 2015

"This unbelievable story begs the question, 'Why did the EPA do nothing?' ... If the EPA cannot intervene in situations like the Flint tragedy, while at the same time systematically savaging the American coal industry, American families and the American dream, then it is time to close its doors, de-fund it, send the President's sycophants scurrying and send the money we save to help provide health services for our under-served veterans."

I would have thought that if the EPA had really done nothing, then that organization should have been shaken up, strengthened, invigorated, not liquidated for the convenience of foxes in the *regulated community*.—But what did I know? I needed brain surgery!—And so when the *regulated community* kept busily warming up our atmosphere, why, that must have been the fault of the regulators.

Coal's soldiers had their generals, drum majors and sky pilots—such as Roger Horton of Citizens for Coal. *We are also asking you to please come to the Charleston Civic Center . . . and attend the . . . hearing on the new Stream Buffer Zone Rule. This threatens to kill what surface mining remains in West Virginia and even threatens our underground operations. We need you.* Sometimes they wore real uniforms, as did for instance Josh Nelson, my favorite soldier of coal, who even while serving as a West Virginia delegate continued to be a pilot in the West Virginia Air National Guard's 130th Airlift Wing; he stood up in the House of Delegates to assure his fellow veterans: "Our weapons may have been different but the fight in the warrior is not. My oath of office . . . the one where we swore an oath to protect the Constitution against all enemies, foreign and domestic, has no time limit."

And even though the War on Coal kept going against them, those valiant defenders of the *regulated community* still won battles:

From a local newspaper, 2015

House votes 95–4 for repeal of alternative energy law:

The vote followed more than an hour of floor debate, with proponents calling the bill a way to preserve coal mining jobs and keep electricity rates low . . .

The tritium spilling into the ocean at Fukushima, the cesium-137 washing away here and concentrating there, few within *their* reach dared to call them acceptable—until they normalized it. So why wouldn't all carbon ideologies be normalized in West Virginia?—What was global warming to them but some propagandistic absurdity perpetrated by the socialist tyrant in the White House? Even if it were real, it remained something far away.—"Nuclear, I like it," said old Arvel Wyatt in his drugstore in Welch. "It's going to be controllable. Fracking, I don't have no opinion on that." That was three years after the Japanese accident.— Meanwhile, coal mining effluents discharged into the streams of Kentucky and West Virginia were problematic only due to federal regulation and those nasty "environmental groups." Even Upper Big Branch could be blamed on meddlesome regulators. (From the scathing report by the Governor's Independent Investigation Panel: *When deviant mining practices led to the terrible tragedy of April 5, 2010, . . . the message was direct: MSHA* made us change the mine's ventilation*

---

* The Mine Safety and Health Administration.

*system in ways that were dangerous.** As you know, MSHA actually remained so weak that it could not even subpoena witnesses—not after a mine accident of any size.) Meanwhile the carbon dioxide and methane from coal mining rose up as tastefully invisible as radiation itself.—Heading up north from Williamson toward Charleston I saw another long coal train: black hoppers and then silver ones, two black Norfolk Southern engines pulling everything up along Highway 119, beneath the sinuous cliff of summer trees, no cars on the road, 23 miles yet to Logan, one side of the highway in the light, the other now shadowed, both thickly forested and sometimes blackly rock-striped, and the hoppers passed into shadow. And so Senator Manchin enthused on West Virginia's 150th birthday, even as the newspaper noted a heat wave with "all-time highs" in parts of Alaska (which, like "the hottest summer on record" in Japan one year later, could have been a fluke): "The abundant natural resources of our state and the hard work and sacrifice of our people have made America stronger and safer. We mined the coal that fueled the Industrial Revolution, powered the railroads across the North American continent, and still today produce electricity for cities all across this country." The man who made my sandwich at the fast food restaurant down in Bradshaw said it even better: "Well, the coal's about gone, so we're about gone. They keep takin', takin' from this beautiful land, and there's nothin' left to take."

He was wrong about that. And Chris Hamilton proved quite right when he attributed coal's defeats to *economics and politics, not necessarily in that order.* You see, there was still something left to take. What could it be? It contained hydrogen atoms, bonded to something called *carbon.*

In 2016 the defeats continued—and one of coal's high-ranking generals took a bullet. Don Blankenship began to serve his year of prison time. In a presorted, 67-page glossy mass mailing (the copy I scored had been directed to the P.O. box of some West Virginian "Or Current Resident"), he informed the world: *The prosecutors, the President of the United States, Senator Joe Manchin, and the judges have used their positions of influence and decision making [sic] to deprive me of my rights as an American citizen . . . They imprisoned me for political reasons. I am in fact an American Political Prisoner.*

Then another bullet struck home! Do you remember the activist Chad Cordell? I was astonished to read his communication: **WE WON! The KD#2 Mine is Permanently Shut Down.**

The Kanawha Forest Coalition had saved about 300 of the 413 acres of

---

* To this Tim Bailey said: "At the end of the day you can't blame someone driving 95 miles an hour on the fact that the cop didn't pull them over."

Middlelick Mountain—although the *Gazette-Mail* called it *a bittersweet victory for the citizens [sic] group;* for as Mr. Cordell pointed out: "We now have perpetual pollution, including acid mine drainage, entering tributaries of Davis Creek," thanks to the great work which Keystone Industries had already accomplished. The next step: reclamation, however deficient.

When I asked him how he felt about his struggle, he wrote back:

> It seems a small and insignificant victory in the face of global habitat loss and mass extinction. It feels frustrating to know that people can take a stand and win, as we did, yet so many people continue to passively accept the myth of their own powerlessness. It makes me angry that we had to spend years of our lives working to protect a tiny spot of land from the greed and insanity that threatens it, and that we could spend our *entire lives* doing the work of protecting the land yet still mourn daily for the many places we continue to lose to this sick culture to which nothing is sacred except exploitation. It feels good, really good, to stand up to powers that seem so invincible and systems so broken and corrupt . . . and beat them . . . It's inspiring to find and work with those few people who do care, and are willing to take action.

Up in Rush Creek, Barney Frazier sent out a mass e-mailing that concluded: *It is certainly great news for all of us who fought so hard to achieve this result,* and it pleased me to think that he and his wife finally had their tranquil retirement back. Meanwhile Mr. Cordell reminded the public:

> The DEP continues to try to weaken water quality standards, continues to permit massive strip mines in other communities, and continues to put our mountains, streams and health at risk while our legislators do nothing to address these issues.

That must have consoled the *regulated community* and all those exhausted foot soldiers—and never mind his closing phrase: *For the Mountains and People!*—because, as the author of *Carbon Ideologies* knows all too well, words are only words.

Concurrent with that consolation came hope, and then an outright miracle—for in that same year of twists and turns, a Presidential candidate named Donald Trump took the stage in Pittsburgh, promising to *end the war on coal and the war on miners.* An environmental economist at Harvard responded bluntly: "The primary cause of the tremendous fall in coal employment is low natural gas prices, due to increased supplies of natural gas from hydraulic fracturing. If the Trump administration wanted to help coal, it could ban fracking. But he can't have it both ways"—although Trump certainly meant to: Even as he stood up for Big Coal, he invoked the moneyworthiness of *the shale energy revolution.** Oh, heavenly hydrocarbons, let us praise your abundance forever and ever! And if rival *regulated communities* ever turned upon one another, why not blame the regulators again? That played so well in Appalachia!

"If You Don't Like Coal Don't Use Electricity" (Harlan County, Kentucky). The license plate reads "Friends of Coal" and "Coal Keeps the Lights On!"

Mr. Trump would never take the prize, of course. The Democrats all said so. Even on Election Day it took them a long while to admit the impossible. Presently Chad Cordell wrote me: *I am bluntly reminded, once again, with this election, that people will defend only what they identify with. When people become dependent on, and identify with, systems of exploitation, they will defend those systems of exploitation.*

(He was a sore loser, it seems. Wouldn't it have been more courteous to speak of *systems of interlocking friendship*? Let me mention one of those. Four months after Trump's swearing-in, the EPA *dismissed at least five members of a*

---

* Regarding fracking, the West Virginians I met in 2013 and 2014 had barely mentioned it. Scarcely the faintest whispers of employment came hissing out of the ground. In 2015, the year after "the hottest summer on record," local periodicals began carrying items about gas pipelines of potential relevance, although none of these had yet engaged Mr. Cordell's resistance; nor did any soldiers of coal I met raise up pro-methane banners.

major scientific review board. Who would replace them? Why, *representatives from industries,* of course! The EPA's spokesman explained: "We should have people on this board who understand the impact of regulations on the regulated community.")

In late June 2015, while the *regulated community* went about its God-appointed business, five days of *unusually high temperatures* in southern Pakistan (up to 113° Fahrenheit), killed more than 1,000 people. By the beginning of July, *countries from Spain to England* began *setting record high temperatures.* Being prey to false ideas back when I was alive, I wondered if fossil fuel emissions contributed to these situations, but a newspaper set me reassuringly straight: *The heat is the result of an unusual bulge in the jet stream, which is allowing a massive ridge of high pressure to surge over the continent, bringing in extremely hot air from North Africa.*

And we kept burning carbon of one form or another—no need to single out coal!—while coal's soldiers crouched grimly in their trenches, still unsuspecting that President Trump would be anointed to save them.

2: "But Everything in Our Country Is Short Term"

Why so many of coal's soldiers had fallen in the tunnels of battle was a question with several answers. Mining is perilous in and of itself. When Pastor Blevins and I were discussing the accident at Upper Big Branch, he remarked: "I do know that if you've got gas and you've got dust, I do know that a lot of the electrical equipment in that mine, no matter how safe you try to make it, you get arc. I know how easy it is to set off gas. We have a member of our church here, who retired recently, after 60-odd years in the mine, and he shared some things that I was somewhat familiar with, but he was talking about the air flow; he wore long underwear, a mining safety suit with stripes on the side, and an overcoat, and then he kept a raincoat, and he said he still, because of the force of air, sometimes he would have frostover, pointing to his forehead. He would have to wipe it off with his raincoat. He had a little machine that he went from place to place with measuring the air. He took the height and width and this machine would tell him how many cubic feet of air they needed to keep the gas moving. If it gets in a pocket, and that gas settles there, a spark can set it off.* The company he worked for, over in Wyoming County, they would complain, you're freezing us to death, and the bosses would say, well, you need to just dress up.

---

* This is what happened at Upper Big Branch.

"I didn't work in a gassy mine," he added. "You need to just force enough air in to breathe. Most of the gassy mines are in the shaft. I went in, almost a half a mile."

For him, I thought, life had often been dark and difficult. The bosses might do their best, but "no matter how safe you try to make it," miners were going to get killed now and then.

Tim Bailey had less sympathy for the bosses. He said: "If you really wanna make people do what you are telling them to, give them something to fear. So everybody has a real nice safety program on paper, everybody be careful today, and in the meantime, the underground supervisor, you hear from him: *Guys, you got to cut 300 foot today. If we can't maintain* x *number of feet, we're gonna shut down.* If I tell you one time to be safe and 10 times to produce, what do you think happens? And those poor men, and in this I include these underground supervisors, and they are being ridden by rented supervisors, so, corners get cut."

"Do the coal executives you've had dealings with lack concern for the miners who work for them?"

"I've not sat down with anyone that I thought was indifferent to human life. But they don't realize that the constant emphasis on budget and production has effects; they're not getting it. The production demands and fear and terror about the ability to keep this job are . . ."

"How would you characterize Don Blankenship? Some people seem to hold him partially responsible for Upper Big Branch."

"He's one of those men raised to work on close margins, like Henry Ford. Henry Ford would figure that if I use four bolts instead of six, but I won't drop the price of the car, I'll save fifty cents. When you have that sort of mentality, what's gonna happen? If you ever get to see, and I have seen, the way Blankenship ran the company, he was the ultra micro-manager. It's this top down pressure to stay on budget. Here's what I would be willing to say about Blankenship. Based on the investigation and the public documents on cases such as Upper Big Branch, there were such an emphasis on managing costs and staying on budget, and that message simply drowned out whatever safety message they were trying to send at the same time. I don't look at this as people intended for anyone to be unsafe."

"How often do miners come to you for redress in cases where it's the miner's fault?"

"Pretty rare. Even if the miner had been actively involved in cutting the safety corner, it was a practice that had been condoned for a long time. And the horrible thing is, they come back and say, he should have known better, he'd been mining for the last 31 years—but this is the way they'd been doing it for the last 20."

"Do you believe that coal mining is more or less safe than it used to be?"

"One thing made miners safer, and that's the one thing that is going away, and that's the union, and the union is going away. My grandfather was a hand loader. He ran the first coal cutting machine that Allen County ever bought. Every coal miner that I talk to today when there's an unsafe condition, the story's always the same: If I open my mouth, I get fired. You don't want to be the guy rocking the boat. The 800 numbers that MSHA says to call, they're a joke, because there's always retribution, When you had union mines it was better. I represent miners that work in union mines and don't work in union mines. The latter will literally take on the company.

"I think coal miners need to realize the value of a union. There is value to a union, and one day they're gonna wake up and realize it. They shouldn't have walked away from the union. And we need to develop an energy plan for this country that shows an orderly transition from burning coal to doing other things with coal so that we can still mine it, but this is a plan that has to be so long term, but everything in our country is short term, and every time a President changes, or a Speaker changes, it's just like . . . I think at that point you should probably give some stability to the coal industry."

3: War Aims

So the soldiers of coal kept fighting to keep our lights on, but the generals called ever fewer of them to that quotidian toil on the mountaintops and underneath the ground. They never stopped dying on the job.

In two of the 12 stories left behind by that 26-year-old West Virginian genius-suicide Breece Pancake (died 1979), the ghosts of coal casualties haunt their sons. A miner meets a small boy gathering fossils at a place the miner knows quite well, *the rotting tipple where his father was crushed only ten days before they shut it down, leaving the miners to scab-work and DPA.* And a lonely mechanic drinks in the sympathy of a boardinghouse hooker when she listens to him speak of *his dad, who sucked so much mine gas, they had to bury him close-coffin because he was as blue as jeans.* (According to the MSHA, *West Virginia has repeatedly had the highest coal mine fatality and accident totals in the country.*) Amidst all the other hard sad people in Pancake's stories, the two dead fathers stand out exactly as they should, as no more or less important than anyone else. Perhaps they never existed in "real life," so why should we power-consumers be concerned? As for the 100,000 Americans who died from black lung in the 20th century, I didn't meet them, either.—Thank goodness we solved *that* problem:

From a local newspaper, 2014

Black lung's worst form roars back:

Progressive massive fibrosis 10 times more prevalent than 15 years ago.*

---

* And here I would like to single out for a special remembrance by you in our future Senator Mitch McConnell, Kentucky's truehearted hero in the defense of coal, who hated the President's affordable medical "Obamacare" program so stoutly (perhaps just because it was the President's) that he sought to repeal every part of it, even including entitlements for black lung.

. . . And even the horrendous leavings of Jim Justice in the 21st century might prove through the working of some bleak socioeconomic calculus to be a price worth paying for keeping our lights on. After all, I who wrote this book never had to pay it. So wasn't it the right price?

In 2014 a young Boone County man named Dustin White, who has already twice appeared in this book,* became *one of several people on a hunger strike against mountaintop removal last week,* saying: "I never really wanted to take this stance, but when I had my eyes open and realized that I was outliving people I had played with as a child, I knew that I had to do something." He had called Cook Mountain his "ancestral home." That spring, coincidentally or not, his coal miner father had died of cancer. But what if mountaintop removal were the most exalted expression of Western civilization?—When I was alive, I read that in northern China people died, on the average, five and a half years sooner than in the south, "because they breathed dirtier air" due to coal burning; in the year 2000, almost 7,000 Chinese coal miners perished in accidents; perhaps those deaths were somehow likewise worth it (if not to me). Maybe coal truly was America's best friend, and China's faithfulest lover:—In 2015 the utility Appalachian Power increased residential electric rates by 11.8%, in part because of the closure of coal-fired plants. No doubt poor people suffered as a result; maybe some died that winter who could otherwise have stayed warm. How many died

---

* See pp. 84, 93, and 149.

that way, and how many from drinking coal-contaminated water? How many lost mining jobs cancelled out coal's cancer deaths?

All this had grown almost trivial—for the acts perpetrated upon our atmosphere fell into a new category. And if this chapter has shown anything, it must have indicated a pattern of careless or callous irresponsibility, of business for the sake of the few and poisons for the many, of a *regulated community* prone to shrug off the few feeble restraints that could, however inadequately, have been counted toward the protection of people, land, water and air. The more I studied the carbon ideologies, including nuclear power, the more strongly I felt that their corporate proponents could not be trusted to respect our longterm benefit. Why should the entities that brought us Buffalo Creek and left more coal slurry dams ready to fail all around Appalachia, that caused Upper Big Branch (after which *state regulators . . . failed to enforce tougher rock dust standards or put into place the stricter methane standards contained in the law*), that fouled the Dan and Elk Rivers and then left most of the cleanup to others, refrain from cooking us to death?

In other words—Tim Bailey's words: "I've not sat down with anyone that I thought was indifferent to human life. But they don't realize that the constant emphasis on budget and production has effects."

4: Last Battles

Not far south of Welch (where I liked watching the grainers and boxcars slowly rasping at the base of the hill which was just beginning to bud; they kept creeping out from behind the brickfront and crossing behind the parking lot through the window of Raymond's, and once in a great while they crawled accompanied by open cars of coal) was a high school that somebody touted; it had been built upon the leavings of a mountaintop removal mine, so they called it Mountain View High School. Driving up the mountain, one gained a decent impression of near-naturalness, thanks to the tall trees, and up where the crest used to be it was flat, to be sure, but very grassy; what if it had always been a mesa?

At the administration office I asked when the mining had been carried out, and the woman didn't know, so she sent me to a teacher named Daniel Phillips who had been here a long time—a nice man, she said, who wouldn't mind someone interrupting his class.

Mr. Phillips supposed that the mountain had been decapitated "probably in the '70s and '80s."—"I had an uncle who owned that property near there, farther down the mountain. The state, if you don't pay your taxes they take it. I think maybe something like that happened when they was getting ready to blast out there. It was just a wooded area out here. Plenty of deer."

He said: "Mountaintop removal is a problem. You don't wanna see Blair Mountain removed, because that was the one place where the U.S. government bombed their own people.* Now they're trying to remove that."

Mr. Daniel Phillips

"Any time you take the material that's been settled, locked into its source, and you remove it, you make a problem. You never finish it," he said.

"It's not one sin," he said. "It's the sin against nature in *all* aspects. In the old days, there was greed. For reclamation they throwed a foot of topsoil and it looked pretty from the highway, until there was erosion. Here, they did a pretty good job. Over on that side, though, there's more flooding than before."

"So the reclamation's not effective?"

"Why did DuPont get out of hairspray? It's the money. They got out 'cause of class action lawsuits. Then who gets involved in cleanup? People who wanna make money. They clean it up only to look pretty. That's human behavior. Russell, don't shake your head at me. Get back to your seat. All these things is a world problem. Now fracking's taking over. Iaeger used to have good water. No more. All that water underneath there, they left transmission fluid in here, as the mountain collapsed. You get a release of strychnine out here. So we've got all kinds of uncommon cancers out here."

"I think the water in Welch has a whole lot of salts in it," he added. "You release the salts when you disturb the ground."

* This event loomed large in the historical memories of West Virginians, although it was not necessarily taught in school: "We never were told how outsiders cheated ignorant hill folk out of their mineral rights or about the 1921 Battle of Blair Mountain, pitting thirteen thousand striking coal miners against police and National Guardsmen armed with tanks, machine guns, and biplanes."

"Do most people drink out of the tap here?"

"No," said all of his students at once and instantly.

"You boil it, but you might be concentrating the material," explained their teacher.

One of his students, who was accomplished at bagging squirrels, rabbits and deer, volunteered that it was getting more difficult to hunt around here now.

Mr. Phillips nodded and said: "When the mining companies go in the timber, they get the best trees out, the nut trees. Then the squirrels go away . . ."

Like most West Virginians, he was resigned to almost everything. About stripper jobs he said: "It got too expensive to get coal out. So then they took up mountaintop removal. If you take it out in order and then put it back in order, you still disrupt the heavy metals just like fracking, but you don't have so much acid rain."

He read me a poem that he had written. I copied it down just as he spelled it:

Black Lung

I spent 8 days in the Hospital talking with Dieing Men
I saw their dieing Stories
And I saw their dying End

They had all mined coal, of course. Mr. Phillips spoke their epitaph: "All the money they made, it was keeping them alive just to pay for their dying."

"So how do you feel about coal?"

"I'm not against coal. But let me tell you something. Hereabouts there's only one book in the house, the Bible. You only want to read one thing in the paper: *Who got locked up?* You got children reading to the parents because the parents can't read. It's all from greed. People from out of state makin' the money."

Then he said: "It's cultural devastation to lose families in the coal mines and it's cultural devastation to have families break up when men can't feed their loved ones. When you make a product, and you base it on the labor of men's backs, and then you take it away, you turn us into a Third World country."

When you from the future judge us for our trespasses, perhaps Mr. Phillips's words will touch you. I don't doubt your judgment will be the same as before. But I myself, who knowingly and unknowingly devoured electricity with the rest of us, left that classroom feeling still more sympathy for the soldiers of coal, who fought for very little, and what they fought for was poisoned, but once that little was taken away, how would they live?

> ### *From* A COAL MINER'S APOLOGY TO TODAY'S AMERICA
>
> #### by Roger Horton, 2014
>
> "I am sorry for mining the coal that provided heat and light for 773 homes for the past 30 years. Apparently some would rather sit in the dark and cold, so I offer them my humblest apologies for making their lives more comfortable. I am sorry for mining the coal that defeated the Nazis and ended the Holocaust . . . I am sorry for the freedom of speech, of religion and of thought that my work and that of so many others purchased."*
>
> ----------
>
> * The "apology" gets more truculent as it continues.

Mr. Horton was right to be bitter. For even as America turned away from coal, leaving West Virginia poorer and more hopeless than ever, other parties kept right on digging it.

In 1980, world coal production totaled 71.2 quadrillion BTUs. By 2011 it had more than doubled, to 152.5 quads.*

It had to, to accomplish all the thermodynamic work we commanded. After all, given (as I keep repeating) that our power plants needed to burn three pounds of it in order to utilize one pound's energy,† coal didn't stretch terribly far. As the following table shows, keeping the lights on *for a single hour* in the reading room of one of those West Virginia libraries that I frequented while writing *Carbon Ideologies* might take 20 pounds of that good old brownish-black stuff—which meanwhile released 49 pounds of carbon dioxide . . .

----

* In 2011, global energy production of all kinds totaled 518.3 quads—of which coal thus comprised more than 29%.

† See I: 152–53.

COMPARATIVE STANDARD ELECTRICAL LOADS FOR SPORT, COMMERCIAL AND INDUSTRIAL LIGHTING, U.S.A., 2002,

in multiples of the load for an outdoor skating rink lit with high pressure sodium lamps,

showing pounds of coal burned and CO_2 emitted while lighting each facility for 1 hour.

All loads expressed in [watts per square foot]. All headers over 10 rounded to nearest whole digit. 1 watt = 3. 413 BTUs per hour.

"Wattages shown are for white light with incandescent filament lamps. Where color is to be used, wattages should be doubled."

A plausible average size for such a space might be 3,000 square feet. [My studio, a former restaurant, is 3,300 square feet.] Making this assumption, I have calculated the *<pounds of average-HHV West Virginia coal needed to illuminate the given facility for 1 hour>*.

Multiplying this by the typical emissions figure for bituminous coal, I give the **{pounds of carbon dioxide released in order to illuminate the facility for 1 hour}**.

1

Outdoor skating rink, lit with high pressure sodium lamps [0.10 watts/ft²]. *<0.33 lbs coal/hr>*   **{0.82 lbs CO_2/hr}**

1.6

The same, lit with metal halide lamps [0.16]. *<0.53>*   **{1.31}**

2

Swimming pools, lit with high pressure sodium overhead floodlights [0.20]. *<0.66>*   **{1.63}**

3.2

The same, lit with metal halide floodlights [0.32]. *Underwater metal halide floodlights should be 250 to 400 watts per area.* *<1.056>*   **{2.60}**

20

"Coal breaking, washing, screening." Also: dance halls [2 watts/ft^2].
<6.6>   {16.28}

30

"Machine shops: rough bench- and machine-work" [3].
<9.9>   {24.41}

40

Art galleries, bakeries [4]. *For paintings in art galleries: "50 watts per running foot of usable wall area."**
<13.2>   {32.55}

45

Cutting, inspecting, sewing light-colored cloth. Also: grading flour [4.5].
<14.85>   {36.62}

50

Church pulpits, operating rooms [5]. *"Operating tables or chairs: Major surgeries—3,000 watts per area."*
<16.5>   {40.69}

60

Cutting, inspecting, sewing dark-colored cloth. Also: library reading rooms [6].
<19.8>   {48.83}

70

Drafting rooms; offices requiring "close work" [7].
<23.1>   {56.97}

Source: American Electricians' Handbook, 2002, with calculations by WTV.

---

* Compare to "Large cities: Brightly lighted district—700 watts per running foot of glass." No watts-per-square-foot value provided.

So coal became, if not necessarily America's, then many other countries' best friend. For one example (this one taken from that ungodly, coal-hating source, *The New York Times*), in 2016 *China is scrambling to mine and burn more coal . . . Mines are reopening . . . Chinese coal is the world's largest single source of carbon emissions from human activities.*

"No more orders," Arvel Wyatt had said. Not so—except in West Virginia.

Or had there begun some fundamental American change? A Dutch "Science for Policy Report" now related the strange news that *CO_2 emissions in the United States decreased considerably, in 2015, [by] 2.6%, which was mainly caused by a drop of 13% in coal consumption, representing the largest relative decrease in any fossil fuel in the United States over the past five decades.*

Other nations also began to waver. It came out that between 2013 and 2016, Chinese coal production slipped by 7.9%. World production was down 6.2%, while British production *and* consumption declined to *levels last seen almost 200 years ago . . . with the UK power sector recording its first ever coal-free day in April of this year.*

West Virginia, it seemed, had nearly lost the War on Coal. But on November 8, 2016, a slim majority of the electorate, or in some versions of the tale a minority, with a few more string-pullers on his side in the Electoral College, opened a new frontline: They chose Donald Trump for President. About the Environmental Protection Agency this orangehaired warrior said: "We are going to get rid of it in almost every form."

WHEN GOD STEPS IN

For six days he had offered many kilowatts of prayer, but static kept him from being heard On High.

Roger Zelazny, 1967

I began this book by asserting* that had we all received adequate information about the relative costs, risks and goods of various modes of energy generation, we would have improved the odds for our species and our planet—but even if such knowledge had been ubiquitously disseminated, it could never have sufficed. Short term selfishness was the primary enemy; but one must not forget the stubbornly irrational component in human affairs. I myself would have loved to forget about carbon ideologies, and had some prophet of Good News persuaded me that my generation could keep polluting and squandering without repercussions, how gratefully I'd have shrugged off the dreary labor of these pages, together with all their apprehensions! Indeed, just these enticing reassurances do offer themselves to interested parties. In that reliable West Virginia standby, the *Coal Valley News,* one used to find a religious column entitled "For a time such as this," from which I extract the following advice on "guarding our minds":

> Our minds can be weakened by entertaining non-essential philosophy that comes from the world's arena, not being cohesive with the Word of God. We should reject all thoughts and notions that hint at such things as evolution, other gods, and the ability of man to solve his problems within himself, making himself to be his own god, which are the tenets of modern day humanism.†

That such instructions appeared in a pro-coal newspaper may or may not be coincidental, but it was certainly convenient. If even *the ability of man to solve his*

---

* See I:11.

† This last word was defined in another column: "Perhaps the most modern form of idolatry is present day humanism. This ugly doctrine raised its head in 1933 with the adoption of The Humanist Manifesto. Humanism is the elevation of the self as being its own god . . . and aims to obliterate the One True God from society."

problems within himself was construed as a mental weakening, then what could we do but leave the solutions of our problems up to spiritual authorities—who, by the way, never seemed to consult the Environmental Protection Agency?* (The unfailingness with which the Word of God supported and perhaps even facilitated coal mining was doubtless another coincidence.)

Correcting the West Virginia Board of Education's science education standards, 2014

Sixth-graders should learn to . . .

Previous language: . . . ask questions to clarify evidence of the factors that have caused the rise in global temperatures over the past century.

New language ("made at the request of school board member Wade Linger"): . . . ask questions to clarify evidence of the factors that have caused the rise **and fall** in global temperatures over the past century.

EXPLANATIONS OF BOARD MEMBERS

Wade Linger: "Doesn't believe human-influenced climate change is a 'foregone conclusion.'"

Gayle Manchin, President: "Said the fossil fuels industry hasn't influenced her view on the changes. Doesn't believe human-influenced climate change is a 'foregone conclusion.'"

---

* Organized religion necessarily contains an anti-intellectual streak, firstly, because every creed, once consistently expressed, must crystallize into dogma—which inherently resents investigation; and secondly, because religion seeks to comfort us in the face of what cannot be helped: poverty, sickness, unhappiness, death—and this inner consolation deemphasizes the material outer world.—And why not? I admit that in and of itself, learning cannot make us better. Science, which teaches how to purify water against disease, also teaches how to poison wells. For that matter, the scientists of my acquaintance appeared no more at peace than anyone else (less so, perhaps, if they understood global warming). Hence Buddha preached: "Learning is a good thing, but availeth not. True wisdom can be acquired by practice only." More pointedly, Thomas à Kempis wrote, *circa* 1413: "Some do not live sincerely in My sight, but, moved by curiosity and conceit, wish to know My secrets and fathom the high mysteries of God, while neglecting the salvation of their own souls. When I refuse them, such men often fall into great temptations and sins through pride and curiosity." It was just such thinking that led to the humiliation of Galileo, and the burning alive of Giordano Bruno, precisely because they asserted what we now accept to be true. Four centuries later, blasphemously prideful climatologists sinned comparably against coal and the good Lord.

"Manchin's husband, Sen. Joe Manchin, D-Wa., has said he has 'never denied the human impact on our climate' . . . and, on Tuesday, introduced the new Senate's first bill, which would build the Keystone XL oil pipeline." "Kelly Goes, state director for Senator Joe Manchin[,] said President Obama's recent speech encouraging climate action is a shock of 'cold water' thrown on the industry. "'We are not going to let this war on coal be the defeat of our state,' Goes said, adding that Manchin believes Obama's climate agenda is a 'pie in the sky plan' that 'ignores reality.'"

As to the relevance of carbon emissions to global warming, Ms. Manchin said: "I cannot begin to answer. I cannot even begin to tell you how much gas mileage my car gets, but I appreciate the people who are out there doing that research." *[Goethe, 1829: "Nothing is more frightful than to see ignorance in action."]*

Michael Green, Vice-President: "Referred questions . . . to [Ms.] Manchin . . . When asked if he believes . . . emissions are the predominant reason for a global rise in temperatures, he declined comment."

Tina Combs: "Referred a reporter's questions to Manchin."

William White: "Declined to discuss the issue with the Gazette . . . , declining to give specifics before saying he was in the middle of something and hanging up the phone."

Now, what *is* the Word of God? The Jehovah's Witnesses knew it as well as anyone, and they informed me that God promised exactly what I was longing for:

- Contrary to what doomsday prophets say, God assures us that our planet Earth will never be destroyed.—Psalm 104:5, Ecclesiastes 1:4.

- Humans will be in the process of ruining the earth when God steps in and stops them.—Psalm 92:7, Revelation 11:18.

I could not disprove this. Had I even tried, I would have become one of those who seeks "to solve his problems within himself."

When I was alive I used to love viewing the many churches of West Virginia, which sometimes bore beautiful names (my favorite being the Zion Temple of the Heavenly Sunlight Church, with the Living Water Church a close second) and whose inspirational marquees never hesitated to be pithy. On the road to Beckley, the Deliverance Temple announced that **GOD IS AT THE END OF**

Another Kentucky stripper job, not stopped by God

YOUR ROPE. Not far from Iaeger, Calvary Baptist Church, whose establishment might have been a former dentist's office, urged me to **FIGHT TRUTH DECAY—STUDY THE BIBLE DAILY**. The brick wall behind an old swingset in McDowell County had been painted with the likeness of a first-prize trophy cup and the legend: **BE ON GOD'S TEAM!** Meanwhile, **GOD BLESS AMERICA**, read the corner marquee in a closed brick building in Williamson. The same request appeared on the cinderblock wall (0.12 microsieverts an hour) of Doten's One Stop in Barrett; Mr. Doten appeared to have long since departed the scene. On the wall of Shupester's restaurant in Northfork, amidst framed snapshots and below an American flag, hung the following "Coal Miner's Prayer":

> Each day I descend into the hole
> to earn my living digging coal . . .
> I pray to my Father in heaven above
> That I may return to those I love . . .
> if somehow death I should meet
> I want to wake at Jesus' feet
> I want my loved ones to be sure that
> in His Arms I am Secure!*

---

* Full text. Ellipses and other punctuation as in original.

While I admired the writer's cheerful fatalism, I thought how convenient that sentiment must be for this world's Don Blankenships, for the sake of whose cost-cutting measures all too many coal miners encountered Jesus early.

Coal was not the only carbon ideology over which God held His benevolent hand. When I interviewed the retired oil magnate Archie Dunham, I found him unworried about climate change. I asked him whether our planet might have a carrying capacity, and, if so, whether we were reaching it. He replied: "Maybe God takes care of that, with plagues and so forth."*

And one midafternoon in Sharjah City, United Arab Emirates, with the souks just opening after Friday prayers, I once became the passenger of a very sweet Pakistani taxi driver whom I asked: "What do you think about petroleum?" (He didn't recognize the word "oil.")

"Oh, very good, sir, *hamdullya!*"

"They say it makes the air warm. They say the weather is changing."

He had not heard about that; he didn't know; he wasn't worried: "That's up to God."

So we had often told ourselves, even back in our polytheist days. In 438 B.C., Euripides's play "Alcestis" ended with Death losing a wrestling-match for a

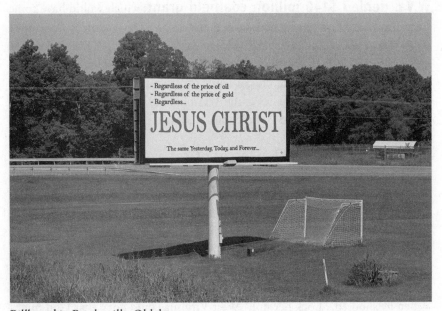

Billboard in Bartlesville, Oklahoma

* See p. 456.

woman's life! And the chorus proclaimed: *What men look for is not brought to pass, but a god finds a way to achieve the unexpected.*

Often the churches did good work. Pastor Blevins helped his mountain community get tapwater. Interfaith Worker Justice, Religious Leaders for Coalfield Justice and the Homeland Ministries of the United Church of Christ all stood up against Peabody Coal when the company declared bankruptcy and tried to defund its pension plan. About Patriot the Catholic Committee of Appalachia said forthrightly: "Through manipulation of the laws, some people exercise power over the workers."

Shortly before Christmas of 2015, the wife and teenaged daughter of a newly unemployed miner entered a church. One of their consolers recalled: *I have to tell you that we were broken-hearted for this little family. So many in our area have experienced the same thing. We all, in some way, have been affected by the loss of coal.* And she hugged them. *One of my favorite verses is Joshua 1:9 and [I] keep it written on little cards. I gave each lady one, encouraging them to trust in the LORD, telling them that the best stories in the Bible had started out the scariest.*

I hope that she also gave them food or money, because a message on a little card will go only so far. Wouldn't it have been nice if the federal government had spread some dollars around? But a month later came the following headline: **W.Va. denied $140 million coalfield grant**, which would have come from the U.S. Department of Housing and Urban Development *to help the struggling southern coalfields.* HUD explained that it had only $1 billion to give and received $7 billion worth of applications, which is understandable and plausible; all the same, West Virginia got sent away from the table, by a government which kept determinedly fighting the War on Coal, so who did those people have left to turn to but church folks handing out little cards? Who would step in but God?

Another headline: **Coal job losses offset everywhere: Gas, wind, solar industries make employment gains, but not in W.Va.**

What were West Virginians supposed to do? When Big Coal went away, and nothing else came to feed them, were they supposed to like it?

Some of my California friends were shocked by the election of President Trump. They shallowly supposed that pillaged Appalachia had voted against its own interest. But Obama, so it plausibly appeared, had taken with one hand and not given with the other. Trump was rather godless, but so was his opponent. When he shouted, "Lock her up!," West Virginia could hope that God might finally smite the hated Democrats.

I happily remember smells of leaves and grass by a wrecked Gospel Lighthouse on Highway 80 South, and then the Little Wonder Church of God as one descended the cliff-shaded road; more ruined houses, then Bradshaw by the steps of Red Rock; the Red Springs Baptist Church; and after a right turn on 83 up onto the top of Bradshaw Mountain where Pastor Blevins posed for me beside his Ten Commandments sign, there came other sweet churches, prominent among them the rocket-shaped Bee Branch Primitive Baptist Church, founded in 1873; and a cool breeze rose up from green hollers, oak leaves rustling, tall grasses bobbing, the yarrow-flowers beginning to twitch. The white cleanliness of the Bim Free Will* Baptist Church, and the high steeple and long low seven-windowed barnlike whiteness of the Garland Church of the Living God, reflected in a still summer river down below the highway; and in Wyoming County the way the Church of God in Jesus Name† aimed like a gunbarrel at the forest from the top of the hill at the highway's edge, are likewise counted in my memory-wealth.

When God stepped in, West Virginians often won peace. Sharon L. Hill of Chapmanville once wrote a letter to the editor of the *Coal Valley News* in order *to assuage the fear produced by the written and televised news. Personally, if I didn't believe that God has a plan, I'd have jumped off a cliff long ago. There's the ebola*

---

* These two words were more commonly joined together, as witnessed by the Quinland, Bim and Martha Freewill Baptist churches.

† The apostrophe was frequently omitted in this particular possessive.

plague, ISIS's beheadings, sexual perversions, etcetera, and right at the bottom of her list, *weather disaster.* Maybe weather would never climb higher on her list. *What helps me is that I don't "feed" the fear; I feed my spirit with the good news of the gospel.* So I guess she didn't have to worry about any "weather disaster," which I myself sometimes used to ascribe to the very worrisome quantity called climate change, until God turned me around.

A Bank of Oklahoma Vice-President* once told me: "Churches are on a scale. There gonna to be some churches that are extremely all the way on the fundamental scale, and that means Earth's only 6,000 years old, and they have logical arguments, and carbon dating is based on the idea that it's uniform, and we know that there are times when it wasn't uniform. You're gonna have that. You're gonna have people that say that God said that we're in charge of the Earth, and the Earth's not in charge of us. You can see the movement is, if we evolved that it makes logical sense that we can't kill any animals, because we should all become vegetarians; I don't know. They have logical underpinnings, it's called theology. I don't think that everybody realizes they made a leap of faith somewhere... I'd say, they're not in the business, churches, generally. They're in the business of saving souls, not getting into social issues. There's a whole group of churches that are salvation-oriented. Some friends of mine would say, all they want is your scalp. Then there are some that are more like the Sierra Club than a church, it's all Mother Earth, and God only gets mentioned once in awhile. I go to a Baptist church, and there's no way I'm gonna hear anything about global warming from the pulpit. Definitely not liberal, but not Free Will."

Meanwhile, God's reassuring shadow fell over a coal rally in Pittsburgh against the Environmental Protection Agency. Our old friends were there: Governor Tomblin, Attorney General Morrisey (whom I remember for presenting the Fourth Circuit Court of Appeals with an amicus brief "favoring prayers at meetings"), West Virginia Coal Association Vice-President Chris Hamilton, etcetera. *Prayer was led by Joel Watts of the WV Coal Forum. The event concluded with the singing of "God Bless America."*

---

* Mr. Sam Hewes, interviewed in 2016. Most of his remarks appear beginning on p. 486.

A MESSAGE FROM AN ENVIRONMENTALIST

n 2016 Chad Cordell concluded his letter to me:

> You . . . framed the question in terms of what I would want people reading the book in a few hundred years to know about our work. I would want them to know that we are working within systems so entrenched, so corrupt, and so violently exploit[at]ive that all our work feels too tiny and insignificant to stop it. We are doing what we can, but it isn't enough. I hope they still have clean rivers and streams in a few hundred years. I hope they still share this earth with . . . millions of other species. I hope that this violent industrial civilization can be stopped or redirected before it destroys everything beautiful and magical and sacred in this world. The odds are slimmer by the day.

A MESSAGE FROM AN
ANTI-NUCLEAR ACTIVIST

O f course we had to keep on destroying. There was no good alternative. What else should Mr. Cordell have expected? Absent some new magic, so long as we continued to increase demand, our "solutions" to global warming and the so-called "energy crisis" could only be shell games.

In 2017 I asked Mr. Yamasaki Hisataka, whose bleak remarks about Tepco's faux pas at Fukushima have already appeared in *Carbon Ideologies,* * how we should address global warming if nuclear energy were removed from consideration.

Solar, he replied, his e-mail conducting itself into the interpreter's computer thanks to God knows what form of energy generation, then passing into her printer in order to be represented by means of electrostatically charged ink-dots on two sheets of paper which winged oil-power then conveyed from Japan to California, at an unknown cost in total BTUs,

> can generate power only in daytime, but wind can do it . . . all the time [when] . . . wind is available. But in residential Japanese areas, the low frequency noise is an issue, and in the national parks, bird strike [that is, birds being killed by striking the turbines] is an issue. The most likely area for wind turbines is currently far from the consumption area, so transfer cost is also an issue. Continental countries might benefit more from wind power [than Japan].

This was not sounding promising. Mr. Yamasaki continued:

> Marine wind power has been developed a lot in Japan but this cannot replace fossil fuels so soon . . .

Why should this have made me sad? As a result of horrid experience he was against nuclear power, and he was practical; therefore, he stood up for fossil fuels,

---

* See I:327 and I:338.

> ...so we have to keep to relying on L[iquid] N[atural] G[as]. The highest expectation is that LNG and thermal will reach 90% efficiency by using cogeneration which means using two power sources. If [Tepco] can achieve [80% or more through cogenerated thermal plants], they may be able to reduce LNG to 2/3 of current volume.

Forgive me for saying so, but that reduction might not have sufficed. So let us hasten on to America's best friend—and why not Japan's?

> Coal thermal power plants also aim at higher efficiency ... [Coal gasification plants] are being built to extract the fuel portion of coal in a gas form to burn. The most advanced equipment has 50% heat efficiency ... Cothermal power generation used to emit a lot of CO_2, but by using the newest method they say the exhaust can be reduced to the level of oil. Still, it's more [CO_2] than natural gas. In China and India they have a daunting task of doing something about their coal generation [emissions]. That's why they are shifting to nuclear. Rather than going in that direction, they should achieve higher efficiency of coal generation. Just by increasing the coal generation [efficiency] from 30% to 50% they can reduce the coal to be burned by half, and they can also reduce the air pollution substantially. If they have money to invest in nuclear power, they should invest in this.

In other words, if not nuclear, then fossil fuels! No good alternative, my friends!

To be sure, *the next stage is to use renewable energy in a high priority way.*

Mr. Yamasaki then admitted: *Renewables will rapidly increase for certain, but we may not have not enough time to stop global warming. That still is a daunting issue.*

A MESSAGE FROM A PROUD WEST VIRGINIAN

And so coal's godly soldiers kept resolutely defending their way of life, and although some colonels and generals might have begun worrying whether the war was lost—in those faintly "progressive" years before the election of Donald Trump (the years, that is, of Upper Big Branch and of streams running strange-colored), the chairman of the West Virginia Democratic Party even began to craft the most favorable terms of some hypothetical surrender, calling for carbon regulation to *be accompanied by fair compensation and support for the cities and towns that will . . . bear the burden for what the country as a whole has chosen. Like the miners of Patriot Coal, West Virginia has earned the right to that*—most other members of the *regulated community* refused to consider the unbearable end, not of our planet as we knew it but rather of their declining but still perfectly practical short term profits; and on their behalf Representative Shelley Moore Capito uttered these brave words: "As a proud West Virginian, I have a message for President Obama and Senate Majority Leader Harry Reid who put their liberal base ahead of American jobs. We won't let you turn off the lights on West Virginia."

Postscript: West Virginia Sees the Light

Fortunately, thanks to natural gas, although West Virginia showed *the smallest job growth of any state* since 2010, it was somehow becoming *one of the fastest growing economies in the United States.* By the beginning of 2016, six pipelines, *much of the work in West Virginia,* were waiting for federal approval . . .*

And then, as it magically happened, West Virginia didn't need to betray American's best friend for the sake of natural gas. With Trump in the White House, she could have them both! And in one of her mass e-mails to "friends," Shelley Moore Capito gushed (above a photo of her standing beside an American flag, while Trump, who appeared unusually smooth, smiled with his right arm around her shoulder and his left thumb victoriously up): *Enjoyed seeing @POTUS & @VP during retreat in #Philly today. Glad he discussed putting #coalminers back to work.*

On October 9, 2017, Mr. Scott Pruitt, one of our most carbonaceous ideologues and therefore glowingly suited to be the EPA's new commander, touched down in Appalachia and declared victory: "The war on coal is over."—I had a sore throat just then. The sky was hazy with smoke from the 15 conflagrations *stretching almost the entire length of the state,* in what the newspaper called *one of the most destructive fire emergencies in the state's history.* Thank goodness I lived in Sacramento, not in Napa or Sonoma or some other ecosystem somewhere! Scanning down the illustrated column, I found Governor Brown's explanation for our situation: "The heat, the lack of humidity and the winds are all driving a very dangerous situation and making it worse." No mention of climate change, thank goodness! You see, *an unusually wet winter produced ample brush, and the state's hottest summer on record dried it to tinder.* (With luck, the summer of 2018 would be hotter, and then the leftists would have to stop calling the summer of 2017 a record-breaker.) Now back to Mr. Pruitt: "Tomorrow in Washington,

---

* Coal lovers did not despair. According to the Australian government, "India is projected to emerge as the world's largest coal importer, as soon as 2025, driven by the competiveness and quality of imports in some coastal locations." I would have gone there for *Carbon Ideologies,* but the Indians denied me a visa.

D.C., I will be signing a proposed rule to roll back the Clean Power Plan. No better place to make that announcement than Hazard, Ky."

IN RECORD HEAT, A PERFECT TIME TO ROW

Enjoying the unusually warm weather, Cindy Westbrook of Carmichael glides across the calm water Monday in her scull at Lake Natoma. Temperatures in the Sacramento region approached or topped the record of 76° set for the same date back in 1977, but forecasters say by midweek, the heat will dissipate, rain will douse the Valley, and a foot or two of snow will cover the Sierra Nevada at all times.

The Sacramento Bee, February 2016

3

"Today or Tomorrow It Will Have to Come Out"

May 2015

In view of the dwindling nature of gas availability in Bangladesh, coal is being considered as the major option for power generation . . . As such the use of coal will be increased manifold within a short span of time . . .

Bangladeshi Government Report, 2012

The panel noted a clear trend to move "dirty" resource and labor intensive operations to less developed regions of the world.

Journal of Cleaner Production, 2005

"WE ALL SHOULD BENEFIT"

The labor union president pleaded with me to tell his workers' story; all they wished for was a decent wage. I remember the man whose pallid face was blotched with pigment; he wore a laminated identification card around his neck, and the stripes on his shirt swirled like wood grain as he fixed his patient eyes on me, with the union president's face replicated on the wall behind him over and over. He stared at me, seeming skinny, wary, helpless and hurt—and the president, whose picture was everywhere on the walls on those newsprint-quality posters behind the line of workers, now took over the discussion in order to tell me: "It is unfortunate that there is no coal policy in Bangladesh."

The president's name was Rabiul Islam Rabi. He said: "This is black gold money and we all should benefit. And we're all suffering! Please get the word out. I spent 26 days in jail, because I was trying to get us higher wages . . ."

"Is there a minimum wage here?"

"We don't have it," he replied, but Nakeeb the fixer interpolated that one day's eight-hour shift here in Barapukuria paid 297 taka.*

There were three grades of labor: foreman, skilled, and ordinary. Trying to fill any one of them would have killed me.

Mr. Rabi was certainly the suavest, best groomed and most confident. While the other men sat back against the wall, with their hands on their knees, he leaned forward to explain their problems, being as active and agile as an otter. The instant he appeared upon the scene, they went silent in mid-sentence. I remember one who was slender and white-bearded; his eyes were round as he gazed at me, and rootlike tendons worked in his long neck. Beside him sat a younger, stronger man who had crossed his legs and now studied me as if he expected no good; and on his right was a dark small man in whose lined and tilted face I seemed to see some kind of dreamy resignation. Their throats shone with sweat. And the president leaned alertly forward, resting his elbows on his widespread

---

* At the 2015 exchange rate of 80 taka to the dollar, this would have been about $3.71—but Nakeeb's figure may not be so exact. See the footnote on p. 243.

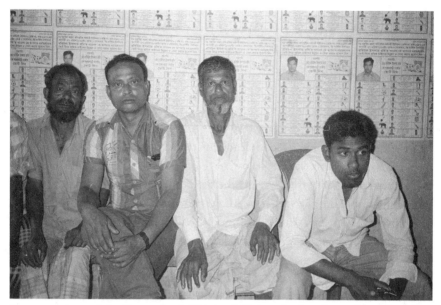

Some members of the Barapukuria Workers' Union. In the photo above, President Rabiul Islam Rabi is at lower right; his likeness appears on all the posters on the wall.

knees. He made sure that he appeared in all my group photographs. The pigmented man, who was as white as an albino but did not have red eyes, sometimes was in them and sometimes out. There were many others who had no room to sit against that wall; they came into the portraits when they could; of them I most remember one dark short man whose sad eyes barely reached me from behind another worker's shoulder; there was also a balding man in a sweat-stained shirt who stared at me as if he expected the camera to kill him, and a cool fellow whose calm was neither friendly nor unfriendly; one young laborer's right eye drooped as he stood there with his head swaying... and in the doorway stood ever so many more whose stories I would never hear; one worker smiled at me, and another upraised his arm from behind two others; some of them craned their heads over the shoulders of closer men; one stood handsome, sweaty and grim.

They had opened their grating especially for us. They had been lined up and waiting in that sweltering concrete cell of their organization, the Barapuku-ria Workers' Union; and so I will try to tell a portion of their story as Mr. Rabi asked me to do. The fan was shaking on the wall, sweat ran down our faces, and from those identical posters Mr. Rabi's likeness watched us on the dingy turquoise walls.

As happened so sadly often when I performed such researches in poverty-riddled places, the sheer preposterous unlikelihoods of my presence and desire to listen gave rise to something that could have been hope—my reified improbability, in short, causing an impossibility whose void might have been temporarily and perhaps significantly satisfied had I given them all the money on me. It was not my business (nor, given my small powers and the conflicting pressures on the three people fixing, driving and interpreting for me, my capability) to discover for you how their families managed, how unpowered except I suppose by wood or charcoal must have been their hot dark homes, how disease-infused their drinking water, much less who they were. The longish autobiography of Pastor Bob Blevins at the beginning of the Appalachia chapter conveys a trifle of the man's hardworking competence, his quietly worldly wit, his empirical reasoning process and of course his faith, all of which together add context and (at least for me) even a sense of inevitability to his views on climate change. The Tohoku people whom I interviewed for the Fukushima chapters, although more removed from me by interpretative gulfs both linguistic and cultural, not to mention the time limits imposed by radiation exposure, frequently bore witness such that I could not only feel but even partly comprehend their differences from me; meanwhile, those time limits, whose raisons d'être affected all of us, brought us closer together even though they were victims and I a mere visitor. The bitter and contemptuous bravery of Mr. Shigihara Yoshitomo,* who led me maskless to his contaminated home in Iitate, expressed itself and its reasons over the few hours we were together, until I knew I would remember him for as long as I *could* remember. The uniqueness of each of our species' sufferings haunted me through him, with his isolation compounded by his dialect, and his anger crystallized around talk of mutant butterflies. But these Bangladeshi men whom I met only there in Barapukuria and for less than hour must remain in my mind and telling mere ambulatory causes (and imagined future sufferers) of climate change. That they would indeed suffer, assuming that they lived another 20 or 30 years, seemed likely enough. But *how* that suffering would fall upon them, and how each one would go on or not, was not for me to say. *And indeed,* writes Trotsky, *one cannot demand of a prognosis that it indicate not only the fundamental tendencies of development* (or in this case of disintegration), *but also accidental conjunctions,* such as Mrs. Shigihara's steadfastly neutral face when, as I told you, *I offered to frisk anything in their abandoned place that she might wish to have with her, but she calmly replied that the neighbors would be too frightened if it came out that she had brought anything back from* there. Enough. Trotsky again: *Consciousness is nevertheless determined by conditions.* And what were those conditions? *What moves things* in a steam engine *is not the piston or the box, but the steam.* For him that steam was revolutionary energy. When I was alive, it was carbon's combustion products. Steam turned our turbines; carbon dioxide and methane rose up. If I cannot know these men, I can at least reveal what is to *Carbon Ideologies* the most relevant manifestation of their conditions.

In 2014, each American's share of carbon dioxide pollution from fuel combustion was 35,754 pounds. The corresponding Bangladeshi figure was 862 pounds—in other wordvs, 42 times less than mine. To what extent were our diverging consciousnesses determined by these facts? I must leave that to you from the future, who presumably detest all of us: I for consuming so much more fuel than you could hope to do, and these union men for doing their own mite to increase per capita fuel combustion.

Without exception they worked in connection with coal, although none were presently miners. One man had been employed on the conveyor, but the heat there was so fierce that he applied for a transfer to the signals department. I remember his stained blue shirt and striped *lungi.*† He would have told me more, but that was when Mr. Rabi arrived.

Before the mine began operation, they had all been fieldworkers. A man in

---

*See I:429.

†The skirtlike garment worn by many Bangladeshi men.

a yellow shirt explained: "Here it's like government. You work from nine to five and you get a salary. This is better than agriculture: If you work more, you get more."

Perhaps you recall a certain nuclear evacuee, formerly of Tomioka; his name was Mr. Endo Kazuhiro, and he said to me: "When we were young, when the Tepco plant was built here, we were very thankful to finally be able to work locally. In this area, when you say, I got a good job, that means, I'm a Tepco employee or a civil servant."* You may remember the West Virginian coal miner's widow Mrs. Patricia Wheeler: "The coal made a very good living for so many people, because what would we have done without it?"†—Like these Bangladeshis, we all aspired to get more if we worked more. That certainly sounded like paradise, and the tunnels of the mine must have been empyrean. According to that government report from 2012: *Coal is friable and prone to spontaneous combustion . . . High temperature (39° C‡) and humidity (100%) ma[k]e the working condition[s] difficult.* In 1998 the mine was *inundated . . . due to in rush [sic] of water. Consequently, underground development . . . was suspended for about 30 (thirty) months. Dewatering of the mine required high capacity submersible pumps which had to be imported from China due to non availability of such pumps in Bangladesh.*

Another former miner remarked: "They were supposed to take Bengalis to China for training, but to cut costs they just made us watch videos."

A one-legged man, Mr. Idu Sarker, now told his story. Bearded and moustached, not yet old, he sat in a chair beside me; in spite of his infirmity he had put on no weight. (None of them had; that tall white-bearded owl-eyed man was as skinny as a rickshaw driver; the union president looked the sleekest, although not even he came close to being fleshy. Would this have anything to do with minuscule per capita CO_2 emissions?)—The accident occurred in 2004, when Mr. Sarker was earning 2,700 taka for six hours, seven days a week:§

"On that day," Nakeeb interpreted, "he went to underground at 7:00 p.m. and his shift was until 6:00 p.m.¶ He was drilling 115 times underground and then his China boss told him to dig and throw coal onto the transport, which is like a train. The transport was fully loaded, and it fell on his leg. The company

---

* See I:377.

† See p. 71.

‡ 102.2° Fahrenheit. At 100% humidity, the "humiture" or apparent temperature would exceed 130°, and my weather table notes: "Extreme Danger."

§ This would have been about $4.80 a day. Mr. Rabi said that the daily wage was 3–400 taka ($3.75–$5). The union men agreed that coal miners got 10–12,000 taka per month. Mr. Sarker's monthly wage would have been 10,800 taka, or about $135.

¶ It was not clear to me whether his shift had been switched or whether he was working extra hours. Nakeeb might have meant 7:00 a.m.

came and gave him 50,000 money.* He cannot work anymore. After five years, they gave him another one lakh."†

I asked the man: "Do you feel satisfied with this compensation?"

"No, sir."

Nakeeb said: "Actually there is some job in factory that he can do. He just want a job . . ."

Mr. Sarker showed me a scar on his forehead and another on his remaining leg.

"Is coal good or bad for people?" I asked the president.

He replied: "There's no substitute for coal in Bangladesh, and we have really good bituminous coal."

"What's your feeling about global warming?"

"Maybe this topic is difficult," said the other fixer, Pushpita, but I insisted, so she spoke with them all for awhile, then reported: "They're not familiar with the concept."

---

* About $625 U.S. in 2015.

† A lakh [see I:589] of taka was about $ 1, 250. Hence, Mr. Sarkar's total compensation would have been $ 1, 875.

THE PRINCIPLE

How do you feel about the coal situation in Phulbari?" I inquired.

Mr. Rabi answered at once: "There's no alternative. Today or tomorrow it will have to come out. And it has to be open pit; they have no choice."

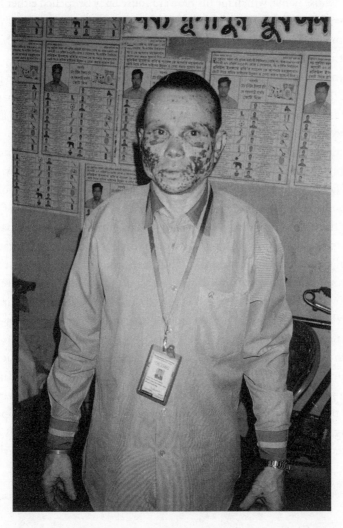

AUGUST 28, 2006

1: The Monument

I n the northwest reach of Bangladesh, in the district of Dinajpur, which almost touches Indian West Bengal, any reasonably large map will show a certain town whose name, which means Flower House, may be spelled either Phulbari or Fulbari. Employing a widely used electronic search engine, Pushpita had found images of an open pit coal mine in Phulbari, and so we drove there, only to find that the mine did not exist.

My interview in the Barapukuria mine had been assured until the previous day, but then all six managers mysteriously got called to Dhaka for a board meeting just as I departed Dhaka to come see them. So we continued on to Phulbari, which lay only 15 or 20 kilometers from Barapukuria.

The deposit in Barapukuria (seven seams) was 6.68 square kilometers. In Phulbari (11 seams) it ran 24 square kilometers, and the company involved called it *possibly the largest coal reserve in Bangladesh.* The mineral in both cases was *good quality low sulphur bituminous coal,* of the Gondwana type. And from one place to the other, above the mineral beds of Phulbari and all around Barapukuria, stretched lakes of greenish-golden rice with chocolate-colored water in it, ducks in the water, and islands of bright-robed women bending over their harvest. Around the lakes it was mostly pastureland; I remember a massive cube of grass moving across a ricefield upon a man's slender brown legs. According to one source, *Phulbari is a very densely populated area and has recently emerged as the bread basket for Bangladesh.*

From Phulbari I remember women with covered heads and bodies,[*] their faces shining with sweat; they were walking slowly past loudspeakers mounted on corrugated metal walls. Three slender men were speeding past on a motorcycle, and then I saw a rickshaw driver in a stained green T-shirt and a striped *lungi* standing on one skinny leg with the other one up on the floor of his chariot, straightening, then looking down the street for business; here came a girl in a red

---

[*] A *borkha* is what Bangladeshis called the full covering; it can be a hijab of one color and a high-collared, long-sleeved, ankle-length skirt of another. Hence it is far less conservative than the face-obscuring, monochrome *burqa* of Saudi Arabia, Afghanistan, etcetera.

skirt and sandals and a covering striped green and gold which was tied in a diamond shape down the back of her head. Pushpita, who was thirtyish, remarked over in her lifetime she saw more and more covered women. "That's how we were, actually. You know, you were a Bengali before you were a Hindu or a Muslim or a Christian. Our weddings are very, very Hindu, but nowadays people are saying, you can't have that because it's not Muslim." The majority of the men wore white prayer caps. Not infrequently one would see a man who had henna'd his hair and beard, because the Prophet was supposed to have done so.*

As for Nakeeb, that filmmaker and activist was 28 years old. He said: "We are so kindhearted. Did you know that Bengali is so much enriched language? And our literary history is rich."

He said: "I'm not a writer. I'm a common man. In Phulbari, there is some critical situation. On 2006 on 4 August, government tried to go there and open some coal mine. People suspect foreigners, outsiders, because all of them would be victimized, because they would have to leave their land. All of the companies told them they will pay, and they do not pay."

He explained that on account of his "discipline" he would never lie to me even though there might be things he would not tell me.

He always wore white because "I am a dead living body in a shroud."—"My profession is same as Issa, Musa,† Muhamed, Buddha: to help the people. Bill, are you an artist? Bill, I am a work of art."

When we stopped at a gas station, he shoved his fist toward me, opened it, let loose a yellow butterfly. He believed his soul would live a thousand years, because he was an optimist. He said: "In five years I will sing and it will be cinema. In ten years I will go out."

Entering Phulbari on a narrow asphalt road, with a trailerload of grass almost blocking our way, we found people bending and reaping the ricefields on either side. An old woman kept chewing as she dragged a skinny white cow down the road. Then came three women in saris of many colors, carrying silvery cooking vessels. An ancient woman was staggering down a berm between fields, bearing cornshocks, while ripe ricefields sweated in the sun and other ricefields darkly glittered with water. My shirt stuck to my belly, and hot sweat popped out from my face. At 4:30 came the loudspeaker's call to prayer, succeeded by sounds of hammering, a multicolored bus honking and nearly scraping our car, Pushpita

---

* I remember my host's neighbor, whose beard was henna-dyed yellow and who kept shaking the youngest daughter's hand and reinforcing her in the Muslim greeting: "Salaam alaykum."—"Walaykum asalam." When I told him I would see him again, insh'Allah (God willing), he snappishly informed me that that was what one said in Afghanistan, not here.

† Jesus and Moses.

meanwhile receiving calls on her cell phone at very short intervals. We reached downtown: truckloads of immense plump sacks, vast watermelons along the side of the highway, then a river bridge where Nakeeb told the driver to stop. Taking my hand, he led me across a rice-threshing yard, through a gate, and down a slippery path to the river's edge. An old man and a boy came with us. That was how I saw the monument to the three *shaheeds*.[*]

Monument to the three *shaheeds*. Nakeeb is on the right.

Said the old man: "There are many good men. Life is nothing. They gave their lives for the country."

"Did you know them?"

"All the men here are friends. This is my motherland."

Nakeeb said: "In 2006, 28 August, the people of Phulbari, there were at least 5 to 10 *lakh*[†] people protesting. This was the spot. Three were killed by the police and many wounded."

The next day, when I asked the youngest victim's mother to describe the son she had lost, "he was very fair.[‡] He was 14," she said weeping. "He was a carpenter for a year. He was an apprentice in a shop, four or five kilometers from here; he used to go by bicycle . . ."

---

[*] A *shaheed* (more commonly transliterated "shahid") is a martyr. All Bangla transliterations in this chapter come courtesy of Pushpita.

[†] Half a million to a million.

[‡] She might have meant "light-skinned," but I am guessing that he was simply her beautiful child.

"Was he a part of that movement or was it just an accident that he happened to be there on that sad day?"

She began to sob, and then covered her face with her robe.—"He went as part of the community," she finally said.

His name was Al Amin.* The mother was Rehena Begom. Her home was very poor.

I had hoped for a photograph of the boy, or for some story to help me better know his memory, but that brown face ancient with grief (for all I know she might have been only 40) sufficed to wind up my journalistic schemes. I asked my last question: "How do you feel about coal?"

She just sobbed; I declined to torture her anymore; I never would have come had Nakeeb not agreed that this would be a good idea. As she sat there shuddering, the golden glass bangles clittered together on her arm.

"He was shot in the head," the neighbors explained.

2: Pushpita's Maxim

Our host in Phulbari was Mr. Mohammed Aminul Islam Bablu,† who was chairman of the National Committee to Protect Oil, Gas, Mineral Resources, Power and Ports. Of the exact scope and nature of this committee I am not entirely certain. In Nakeeb's words: "Committee is a consortium of different organizations." When I hounded Pushpita to define its politics, she finally wrote: *Left and progressive.*

That evening Nakeeb and I were sitting side by side on a charpoy‡ in the concrete parlor with the fan going, while Mr. Bablu sat in a chair facing us, and beside him a boyish activist whom Nakeeb called "a very dynamic young comrade."

Mr. Mohammed Aminul Islam Bablu

---

* The other two *shaheeds* were called Tariqul and Salekin.

† Another interpreter told me he was Mahmud Hasan Babu.

‡ A hollow-rectangled bed strung with ropes, hence bearing some of the advantages of a hammock. I have reclined on them in Bangladesh, India, Pakistan and Afghanistan.

"Phulbari is not yet a mine," began Mr. Bablu. "But we know there is coal. There are just ricefields."

He pointed down at the floor.—"We are sitting on coal right now," he said. "That's 300 to 1,200 feet. The coal seam occupies 135 square kilometers. That would have displaced 200,000 people—one million people actually, taking land, making railroads and so forth . . ."*

His complexion was smooth and chocolate-red. He wore spectacles. He was bushy-browed and slightly sweaty, and his hair was still black.

"So in the 1960s when we were still part of Pakistan, there was some kind of government research here, and they realized there was coal but it was not profitable yet. And then in 2004 or 2005 Asia Energy came, and I went to some community meetings."

The younger activist put in: "And before that was BHP, Broken Hill Properties. They were based in the U.K.[†] From 1998 BHP were doing surveys, and then they handed over the contract to Asia Energy."

"Where does Asia Energy come from?"

"They are Australian, but their headquarters is in England and their business license is in Honduras. The gentleman who came was Garry Enlight,[‡] a British, and with an Australian wife I think. Then came some Bangladeshi officers, and they contracted out a Malaysia firm to survey, to dig, and then there was a Bangladeshi firm. They did some research on animal life and human life, how they would be affected."

So already it was the usual story of coal: One company gave way to another, so that nobody could say who did what, where the money travelled and what liability or responsibility any entity might have.

"The Bangladeshi representative said if there is a coal mine here this area will benefit. He said it will be an open cut mine[§] and it will be on top of the ground and people will have to be resettled. They will get compensated for their land, a lot of money for their land, and all the infrastructure that has to be destroyed will be compensated many times over."

---

* He kept an archive of documents in his house. Many of his figures derived from that 2012 government report, which Nakeeb photocopied for me.

† *Circa* 1991, Broken Hill Proprietary Company Ltd., headquarted in Melbourne, was "Australia's largest commercial organization" with nearly 51,000 employees and sales of U.S. $7.72 million. BHP's first prospectus, which involved silver ore in New South Wales, dated back to 1885. The company's "three core activities" were steelmaking, exploration and production of iron, coal, copper and manganese, and petroleum exploration and refining.

‡ He spelled it thus, but I falsely guessed that the name was "Enright." Another source called him Gary Lye, which was actually his name.

§ We would call it an open pit.

The young activist interjected: "In 2005 the Phulbari *thana**** was producing most of the rice, number one of any *thana* in Bangladesh. If you take away the land, you put a huge dent in food production. In 2012 it was number three..."

And here I would like to quote a few extracts from that government report:

> Irreparable loss to the social and ecological spheres could be far more than the apparent benefit from the increased out put [*sic*] of coal...

> Geological features... suggest that there are large-scale tectonic forces in the area of the Phulbari coal deposit... When the major faults penetrating through the coal beds are activated, [they] will cause [a] devastating flow of water with high pressure resulting [in] land slides and other calamities...

> Overburden[†] dump would be a time bomb... During heavy monsoon season these materials will spread beyond the mining area... and clog the river system up to the Bay of Bengal...

> High risk of social unrest and conflict is involved in relocation of about 1 million people.

> The loss in agricultural products due to [open pit] coal mining... could be such that the prospect of adequate food grain production in the country may be at stake.

But why concern ourselves with insignificant considerations? The economy had to keep growing at 6% per annum, and that was all there was to it:

> Considering the dwindling resources of natural gas, Bangladesh will have to put great emphasis on proving new coal basins and on developing the existing ones to diversify the sources of primary energy.

Mr. Bablu continued: "I just asked him [the Bangladeshi representative] who would be resettling people and compensating them. He said the government would resettle and compensate people, but this was all wishy-washy, and the

---

* Subdistrict.

† The waste earth, rock and vegetation removed from a mine. It was estimated that 25 metric tons of overburden must be removed for each metric ton of coal.

company changed their words often. I was at three or four of these forums, and they would give out these documents, and they would say that maybe eleven hundred people would get permanent jobs, and what they would say was very different from what was printed in these brochures. When I told them as much, they would respond that how many we employ directly at the mine is not important because there will be other jobs—but this was not happening at the other mine."

Evidently it still was not happening at the other mine even yet, or if it was, those other jobs left something to be desired; otherwise why would those sad men in the Barapukuria Workers' Union be begging for a decent wage?—But I was not surprised that "what they would say" varied from "what was printed." For here was a company so determined to dig that mine that they had even faked up a picture of it on the Internet.*

"Now, when did all this happen?"

"This was still in 2005. I did not become chairman of the National Committee until 2009. At that time I was district chairman of the Peasant Organization.† So initially they said they needed 10,000 hectares‡ of land, and then when protests started they said 6,000 and then eventually just 2,500. Initially we were actually positive about having a mine here, but then we started seeing a lot of discrepancy. In their report there was no mention of what Bangladesh would get from the extraction. But we had a secret link in the research department, and so we learned that Bangladesh would get only 6% and we could buy another 14% and the remaining 80% would be exported, which seemed to us very exploitative.

"Then we became concerned with the water situation, because this is a very agro-dependent region. For the mine you'd have to dry out the water table. They claimed it would only affect four or five miles around, but a Bangladeshi research organization said it would affect 40 kilometers around!—The unique thing about the water in Phulbari is that it's completely arsenic free, and you know arsenic is a big problem in Bangladesh."

From the government report:

> Both upper and lower seams of the Phulbari basin lie beneath an unconfined aquifer . . .
>
> Depths [of] groundwater range seasonally from 2 to 9 metres . . .

---

* Pushpita, who had arranged this trip, fully expected that we would soon be riding down into humid darkness; I wish you could have seen her astonishment to find no hole in the ground!

† He sometimes referred to that group as the UP, which must have stood for Union of Peasants.

‡ One hectare = 10,000 square meters. Hence, 107, 580 square feet.

> Dewatering of the UDT* aquifer to ensure safe mining conditions will require extraction of groundwater over the whole of mine life . . .
>
> The extent and magnitude of dewatering effects are uncertain . . . Grave consequences may be arisen [*sic*] due to the tremendous dewatering activities of the project.
>
> Chance of arsenic (As) removal in ground is also a risky and uncertain issue . . . If "Oxidation theory" proves true, ground water [*sic*] As contamination will . . . emerge as an [e]nvironmental disaster.

In other words, whether or not the aquifer was poisoned with arsenic would depend on untested chemical hypotheses. Would the reduction theory or the oxidation theory be valid?—What a pleasant gamble!—But this was how we conducted business back when we were alive, hoping for the best, and leaving you to face any possible worst.

On my arrival these two men had interrogated me, not without suspicion. Although Nakeeb, who was obviously a comrade of theirs, emphasized my reputed benignity, the Phulbari people had learned to be on the lookout for tricky foreigners, and so I had proven my bonafides with facts and photos from West Virginia. They were especially interested in the MCHM spill in the Elk River, I think because water contamination was such a worry to them. I told them whatever I could, and when they asked what the West Virginians had done to fight back, I had to describe the trusting submission and resignation of those people, the almost endless acquiescence to acidified drinking water and aluminized rivers, to unknown entities spewing whatever they liked into unnamed tributaries while old ladies said that coal had been placed there by God for the nation to use. And then I asked: "Why are you so much more organized than West Virginians?"

Mr. Bablu replied: "On account of that inside man, the Bangladeshi researcher who worked for us. And we had those layers of discussion, which revealed the issues. And everybody here knows something about the land."

When I think back on this distinction, it seems that I failed to penetrate to its cause. The West Virginians certainly *knew something about the land.* I remember Mark Mooney in Twilight, whose mother still lived in a log cabin on six acres up there in the holler; and Bill White in Bim, who loved the mountain where he was born. What had Mr. Mooney said? "My folks, when the coal companies came in, they offered 'em a little bit of nothing, and they didn't know better so

---

* Upper Dupi Tila, the name of the principal aquifer.

they took it."* Why had they taken it? Hadn't there been those *layers of discussion, which revealed the issues*? Had there been no *inside man*? How tight-knit were they before the coal companies came in? I might never know. Now there was only fear of losing whatever work they hung on to.

"What happened next?"

"The company said, after drying out the water, once we finish mining we're going to inject water to fix it, and we said, where have you done it?—and they said, it will be an experiment on Bangladesh! We didn't like that.

"We wanted 100% of the coal to remain in Bangladesh. We have such few sources here! If that coal is exported and we have to buy it back at a much higher price, well, people understood this idea of ownership. And we were able to make the public understand the effect on our environment. They tried to bribe people, drug addicts and whatnot, to protest against the protesters, like hired thugs. They intimidated the meetings."

Outside, the call to prayer echoed through the loudspeakers. It was getting dark and a little cooler. I asked how and when the protests had begun.

"In 2005, April, 16 April, they started with a peasant assembly, the National Peasant Agriculture Day Laborer Collective Organization."

"Were they socialist, communist?"

"They were not a political organization, but politically conscious, and supported by the left," said Nakeeb, sitting beside me on the charpoy.

"What happened in the first protest?"

"We were initially waiting for an extraction method that would not harm the environment. In the beginning there were only three to five thousand people, and then some other agricultural issues were brought up in this same assembly. The company's response was that they were backed by the government and what is backed by the government will happen. And the people responded, that the events of 1950 and 1971, these were all people's movements."†

"And when did the protests meet with violence?"

"On 26 August 2006."

"And what exactly took place?"

"Once people in different layers of society, such as teachers and lawyers, became aware of this issue, then the mayor started the Phulbari Protection Committee on 18 June 2005, and all the political parties joined together. We were

---

* Above, p. 87.

† I do not know what took place in 1950. Indian and Pakistan were partitioned in 1947; Pakistan's first constitution was enacted in 1956. "The bloody birth of Bangladesh" occurred in 1971.

about to celebrate the one-year anniversary of this committee, and then there was a clash between different segments of the committee, with one faction already using the local bureaucracy to creating a model town where we would get relocated from the mine, so we needed to do something more urgent.

"The National Committee to Protect Oil, Gas, Mineral Resources, Power and Ports existed before. So the activists in the Phulbari Protection Committee reactivated their chapter. The Phulbari Protection Committee had been co-opted and dismantled and no longer accomplished anything. So all people united under the banner of the National Committee. And when the Phulbari Protection Committee was dismantled, we gave this program the plan; we will surround the Asian Energy office on a certain day."*

Now it was time to sleep, and I had the interesting experience of lying side by side with Nakeeb in that narrow charpoy. My bedmate always wore white, because he considered himself a living corpse. For sleeping he wrapped himself up in a clean white cotton shroud, and he even had an extra one for me. Since it was sweltering, I took a cold shower, pulled on my underwear and wrapped the shroud around me. We slept together like brothers. Nakeeb lay perfectly still. I hope that I did not snore and keep him awake; at dawn he was on his back with his eyes open, trying not to disturb me. He was a very noble and sincere young man, and I felt fond of him.

When I awoke it was a cool cloudy dawn, birds already singing, Nakeeb rising and meditating.

"What did you dream?" I asked him.

"Secret," he replied.

Then my host's two little girls came in: "Bill uncle, you are beautiful."

Mr. Bablu and the younger cadre entered our room. Sitting cross-legged on the charpoy, Mr. Bablu resumed:

"So we reactivated the National Committee. In the beginning, we had hunger strikes, forums, blockades, sit-ins and so on—all completely nonviolent. But these had no effect; the government responded that even if the contract with Asia Energy was no good, Bangladesh could not get out of it! The people felt very desperate, and Asia Energy was already starting their work. The Asia Energy office is about a kilometer from here. We gathered in a school, far away to show our nonviolence—even indigenous people were there!—and then . . ."

According to one typescript that Nakeeb later gave me, more than 80,000 people had now joined in this *gherao*. Perhaps the authorities panicked.—I

---

* For this action Mr. Bablu used a formal Bangla term: *gherao*.

watched Mr. Bablu's face. In a low voice he went on: "When suddenly the police starting launching tear gas and the BGR* fired . . ."

"Why?"

"I feel that the BGR were bought off by the coal company."

Wishing not to interfere with his story, I kept quiet, and presently he resumed: "After the tear gas I was escorting some of the National Committee delegates away to save them. I heard firing and could not make it back to the bridge. We had some vehicles; the others who were critically wounded we carried away . . . Basically the leadership were giving speeches at this time. That was all we were doing. The police later said that tear gas would have been enough. The BGR shot to kill; there were 200 people wounded."

Nakeeb's typescript claimed that 100 were injured, 50 by bullets, and the rest by a baton charge.

The younger activist in the dark shirt was Mr. Shahriar Sunny.—"On 26 August 2006 at the time of two," he now said, stretching his arms,[†] "there was a gathering of almost 70 to 80 thousand people, from four *thana,* including all of us and people from all of Bangladesh. Some intellectuals, some university students and teachers and political leaders and cultural activists attended on that day. At 4:00 p.m., almost 5:00 p.m., when we thought it was a quite good day and we had completed our gatherings, the Border Guard Regiment fired at us, and we don't know why."

"How many BGR were there?"

"There were army, police, BGR, around 1,000 or more. Maybe 2,000. The previous day the army invaded that location and that day they also raided with arms. It was Law 144: Only five people can gather."

"And what were the police doing when the BGR surrounded you?"

"They shooted with autos. It was prethinking, because the police shooted, and when they shooted, they shooted very close, I think on purpose to kill. After two or three hours, the polices went around until 7:00 p.m. at night. At night the polices and BGR they went to some of the activists' houses and tried to find them; their intention must have been to torture them, because it was night. After that night, next morning, at first, some little groups of women came to the street

---

* Bangladeshi Guards Rifles, or, as the younger activist interpreted the acronym, the Border Guard Regiment. According to a secret cable by U.S. Ambassador James F. Moriarty, the BGR were "a paramilitary force charged with securing Bangladesh's borders." Also, "the BGR is responsible for border security and control of smuggling."

[†] The English sounds different here because Pushpita interpreted for the other man and Nakeeb performed that office for this one.

with sticks and stones and they showed their protest, and after that, little by little, 10:00 a.m. to 11:00 a.m., there was a gathering again.* Muslim, when someone is dead, we pray before we bury him. *Galbi janajah* is the prayer for unknown
person. Because some of them were missing, so we thought that some of them
were dead. And there were lots of people; there was a huge gathering; and the
police went away. After that, many times they tried in different names, in different times, but they were not succeeded. All the time we defeated them. The company, to Global Coal Management they changed their name, but we protested
them."

"When you saw the police shooting, how were you feeling?"

"Anger, fear."

"How do all of you feel about coal?"

"Evil," said Nakeeb.

"It is harmful for our home. It is harmful for our environment," said
Mr. Bablu.

Mr. Sunny said: "If you compare water and coal, just, if you avoid all other
things, there is a good quality coal and there is a best quality water, in that situation you have to choose the water."

To Mr. Bablu I said: "So do you feel that the people won a victory?"

"I feel that what the company did was to put fear in people. *We do not protest
anymore.* From the 26th to the 29th, this place was not under control of the local
government. The people set fire to the houses of those who owned the company;
we broke the drilling stock, pulled up train lines, broke up culverts. The mayor
from nearby Rachai did not feel safe coming here. So then we sat together—four
thana. The authorities were scared about Phulbari, so they signed an agreement
with us:†

1. Asia Energy was to be kicked out.‡
2. Coal mining would occur only with permission of the locals and
   only in an environmentally friendly way.
3. Those responsible for inciting the shootings would be brought to
   justice.

---

* In the words of Nakeeb's typescript, "Women took the lead from 27 August and made a historic
uprising." After this, the National Committee called a strike all across Bangladesh.

† Officially called the Phulbari Agreement. Nakeeb's typescript called it a "people's verdict written by
blood."

‡ All these numbered clauses were written down by me as they were spoken. Mr. Shahriar Sunny later
summarized the first clause thus: "In Bangladesh, there won't be an open pit coal mine."

4. They would pay compensation for the injured and killed.
5. The government would build memorials.
6. All charges must be dismissed against the people who had de-stroyed the books which Asia Energy had given to children.

"They paid some compensation. They investigated the shooting but did not disclose their findings. They sent 4 lakhs* for a memorial, but then the government changed. Asia Energy still has not been fully kicked out of Bangladesh.

"A few days after the contract was signed, the current Prime Minister, then the opposition leader, brought five lakhs of people from 16 districts, and she called them *a burning people,* the people of Phulbari, salute, salute; we will honor them when we come to power . . .—but I said we would wait and see."

"And the pit mine has not been dug."

"Correct."

"So was it a victory, then?"

"Yes, it's good, but for the last 10 years we've had to be on our guard. The Asia Energy people keep trying to come back in. The ones not directly affected have been pretty much bought off. They have given huge amounts of money to public relations firms, which say that so many people will be given jobs . . ."

Up until then I had thought that I understood the man's position: He was against coal extraction in Phulbari due to the environmental and economic ruin it would cause. Mr. Sunny had put the case pretty clearly. And Mr. Bablu had said, "We were initially waiting for an extraction method that would not harm the environment," so that I had presumed that he must have realized that while deep mines might be less egregious than open pit mines, there was no harmless method.—I now asked him: "What is your opinion of coal?"

"In our country," he said, "we need an energy policy. We don't have that. Ownership is a huge issue. I'm not against coal itself; we need the coal. We have very limited environmental resources. We have three crore† unemployed here, and one crore employed. Our labor is cheap, and if this could make energy cheap . . .—But they might give us a lot of money; we might move away, but that wouldn't benefit the country."

"What do you think about global warming?"

"We're not responsible for it, but we're going to have to pay for it!" he replied bitterly.

---

* Of taka, I presume. Equivalent to slightly more than $5,100 U.S.
† Thirty million.

"If you mine your own coal and use it, you'll be responsible also."

"Of course a lot of carbon will be released, but it won't be as much as the U.S.'s and China's."

I thought that ingenuous (if not perhaps disingenuous). But as Pushpita said over breakfast (pumpkin-mustard mush, very pungent like Chinese mustard; a few bony hunks of goat, tiny bitter melon pieces, dal and rice; the quantities meager, because the family was poor; everything eaten together with the hands): "It's only the activists who really understand about the water and other environmental issues. The others cannot understand until it happens to them."

3: Déjà Vu

And what *would* happen to them?

Once again I choose to quote the Intergovernmental Panel on Climate Change. (If somebody were to ask why I believe that organization instead of one that made opposing claims, I suppose I would say: About scientific matters a scientist is more credible than a non-scientist. A large panel of peer-reviewed scientists, expressing a common judgment, with the caveats and qualifications denoting honesty, increases this credibility from the beginning, while I start by distrusting a lobbyist who was paid to say a certain thing. Scientists may be as corruptible as anybody else, but why was it that the *regulated community,* with all the money at its disposal, found so few individuals in lab coats who would oppose the climate change Cassandras?—To which a true believer could always say: "I don't care about that, Bill. I rest easy. You'll see how wonderful it will be once God steps in.") So:

> A large fraction of anthropogenic climate change resulting from CO_2 emissions is irreversible on a multi-century to millennial time scale . . . Surface temperatures will remain . . . at elevated levels for many centuries after a complete cessation of net anthropogenic CO_2 emissions . . . Ocean warming will continue for centuries. Depending on the scenario, about 15 to 40% of emitted CO_2 will remain in the atmosphere longer than 1,000 years . . .

In other words, Mr. Bablu was correct in remarking: "We're not responsible for it, but we're going to have to pay for it!" Without or without coal mining, Bangladeshis would pay. To console us, the panel envisioned delightfully "*low scenarios* in which atmospheric GHG concentrations peak and decline and do not exceed values that are equivalent to 500 ppm CO_2."

> The frequency of the most intense storms will *more likely than not* increase in some basins. More extreme precipitation near the centers of tropical cyclones making landfall is projected in . . . South and Southeast Asia . . . (medium *confidence*).

This sounds vaguely unpleasant. It fails to inspire appropriate horror and fear. As long as Death keeps himself out of sight in our hot dark future, we need not face facts. We can prioritize unemployment, and even say: "I'm not against coal itself; we need the coal."—But once Death strides forward, pulling a mask over his featurelessness so that we apprehend *specifically* how we must die, then the horror begins. In the case of Bangladesh, Death dons a drowned man's face.

It is virtually certain *that global mean sea level rise will continue beyond 2100 . . .*

Even in these [low] scenarios sea level continues to rise up to the year 2500 . . .
And Bangladesh was a low-lying place.*

4: Déjà Vu

The others cannot understand until it happens to them. Isn't that the sad truth of all carbon ideologies?

5: "The Best Way Forward"

But certain others understood all too well.

On July 29, 2009, Ambassador Moriarty sent one of many cables classified CONFIDENTIAL and obligingly released to the public by the organization called WikiLeaks. The cable began:

---

* Extracts from the 1976 *Britannica:* "The inundation of most of the fields during the rainy season makes it necessary to build homes on higher ground . . . Storms of very high intensity . . . sometimes create winds with speeds over 100 miles per hour, piling up the waters of the Bay of Bengal to create crests as high as 20 feet that crash with tremendous force onto the coastal areas . . ." And we then read about the cyclone-tsunami of November 1970, "killing at least 250,000 people and rendering millions destitute."—And this was before global warming became conspicuous.—In 1997 the United Nations estimated that a one-meter sea level rise would cover 17% of Bangladesh. In 2016, when I was finishing *Carbon Ideologies,* it appeared that the rise might be much more than one meter.

1. (C) The Ambassador recently urged the Prime Minister's Energy Adviser, Tawfiq Elahi Chowdhury, to resolve several pressing issues, including awarding offshore blocks for natural gas exploration and authorizing coal mining.

Why were they "pressing"?—Because money talks.

The Adviser indicated that Conoco Phillips would likely be awarded two of the uncontested blocks and that Chevron would likely soon receive permission to go ahead with the first of three compressors necessary to improve flow in Bangladesh's main gas pipeline. With respect to coal, Chowdhury requested technical assistance for evaluating the technical and environmental problems associated with different types of coal mining...

4. (C) The Ambassador noted that Bangladesh's coal reserves were vast and of the highest quality; coal appeared to provide a potential way to at least partially resolve the country's energy crisis. Chowdhury replied that because of global warming concerns about green house [*sic*] gas emissions and air pollution, multilateral financial institutions had become reluctant to finance coal mining projects . . . He asked for technical assistance from the U.S. to determine what type of mining would work best in Bangladesh and how the environmental impact could be mitigated . . . The Ambassador promised that the USG would look for ways to help the G[overnment] O[f] B[angladesh] and added that open pit mining seemed the best way forward, if the rehabilitation of land could be done properly.

5. (C) The Adviser remarked that the proposed coal mine in Phulbari was politically sensitive, in light of the impoverished, historically oppressed tribal community residing on the land . . . He said the government would seek to ensure the rights of the local community and build support for the parliamentary process. (Note: Asia Energy, the company behind the Phulbari project, has sixty percent U.S. investment. Asia Energy officials told the Ambassador on July 29 they were cautiously optimistic that the project would win government approval . . .)

A 60% U.S. investment! No wonder the Ambassador was involved. I hoped that the open pit mine would never be approved.

6: "Only as Much Coal as Is Necessary for the State"

In the courtyard underneath a slanting corrugated roof sat Mr. Sayed Saiful Islam Juuel, whose title was "National Committee, Convener, Phulbari Committee." He must have been an active sort, for the previous year he had presided over a 12-hour *gherao,* explaining that "people were spontaneously participating in the blockade. We will not allow open-pit coal mine in Phulbari and elsewhere in the country." Two vehicles belonging to Asia Energy or Global Coal Management or whatever the company now called itself had been vandalized, so that was progress.—Another undated typescript presented to me by Nakeeb had Mr. Juuel say: "We were already dead in Aug 26, 2006 during the Kansat incident. Now, if we live we only live to protect Phulbari from Asia Energy."

Mr. Juuel brought out the roaring green G.F.C. fan, at which mosquitoes pelted away, then settled back in our sweaty hair and on our sweaty arms and hands. It was warm, all right. As the Bengali author Wasi Ahmed remarked in one of his short stories: *The hot season had begun by the beginning of March. There was hardly any summer season anymore; the hot season seemed to last the whole year . . . Climate change . . . lay in wait to ambush and lay waste to mankind . . .* Well, what did he know? He was merely a fiction writer.—Laundry hung on three clotheslines; tiny green mangoes sat ripening on a concrete windowsill. Mr. Juuel went and put on a shirt. He said: "No foreign companies, no export! The ownership of the coal has to be Bangladeshi. Then comes the question: If the owners are the people of Bangladesh, how can we take this coal out?"

"If the coal is taken out, will your water supply be damaged?"

"So that is the crisis: land, water and social loss. Open pit will cause them. We want to only extract as much coal as is necessary for the state."—(What a dreamer, I thought.) "We do not want to harm the environment. For example, we hear about coal gasification.* We're waiting for something like that . . ."

"Are you familiar with global warming?"

"Open pit has noise pollution and a big impact on greenhouse. That's why we don't want open pit to happen. These are the more technical issues, but we are more concerned with the immediate effects on us . . ."

(In other words: *The others cannot understand until it happens to them.*)

They brought out bananas, cookies and homemade sweets in the form of brown or tan balls. My clothes were sodden, the sun a huge and reddish-orange

---

* See p. 231 and I:557 (entry on IGCC).

disk three-quarters down the hazy sky. And Mr. Juuel said: "We can't let some foreign company take our resources."

"What are the advantages and disadvantages of coal?"

"Coal is an outdated form of fuel. Renewable is better. Taking coal out makes a huge impact. New innovations will be better."

He was very agreeable, talking out of both sides of his mouth. Perhaps he wanted to please me, or for all I know he saw no contradiction.

"Lose life and trees and water and what do you have?" he asked himself. "Coal!"

Then he said, in reference to the Asia Energy gentleman, "He sold the sand and soil and water that needed to come out before he ever did the deal. So if he could do that, why not someone else?"—meaning, I think, him and his neighbors.—In a way I felt relieved that Mr. Juuel was no nobler than I.

For him as for the labor union president, it seemed that the coal could not be left alone. *Today or tomorrow it will have to come out.*

THE MINE

1: "It Need Develop"

Unlike the phantom open pit in Phulbari, the longwall mine established by Barapukuria Coal Mine Company Ltd. ("a company of Petrobangla")—just out of town and through the forest and then past rectangular mud-ponds by the ricefields, where a skinny man pulled a cartload of great fat sacks by a bamboo-palisaded lagoon—was no figment. Hence those union men I had interviewed; hence their president who had said: "This is black gold money and we all should benefit."

On the outskirts of Barapukuria a roadside sign announced:

> **Shu-Mon Traders**
> Boulder Coal,
> Thin Coal and
> Dust Coal

—and beside it, chickens and goats waded in a pool of coaly water. That coal cost 1,800 taka* per kilo—too expensive for the neighbors when they prepared rice, which to facilitate dehusking at the mill needed to be pre-roasted, so they made do with burning hay instead. If only more coal came out of the ground, the price might go down.

The Barapukuria deposit, which had first been located in 1985, consisted of *weakly-caking bituminous coal with an average sulphur content of about 0.52% making it an ideal fuel for power generation, which is the primary identified market for the coal.* (In short, it would hardly help the rice farmers any.) Its high heating value was claimed to be 11,040 BTUs per pound—barely above average for American lignite and nearly 25% below the calorific output of several West Virginia

---

* U.S. $22.50, or $10.20 per pound.

264

bituminous coals.* More than 200 million tons had *been calculated for the entire area;* but these entrepreneurs hoped merely to get out 63 million tons in the "first phase." The principal target bed, Seam VI, was about 36 meters thick. To access it, two shafts bored straight down for about 320 meters. One stretch of ricefield above its galleries had already subsided by seven or eight feet, which nobody had expected to happen so soon, although, as usual with environmental damage, the event was not all that surprising, not only because life tends to deviate from easy progress, but also, as one basic reference book points out, in longwall mining *all coal is recovered from the mined panels; hence subsidence occurs.*—The sinkhole in question had long since filled with water; and had a man not explained what it was, I would have mistaken it for an ordinary pond. According to one of Nakeeb's typescripts, *mine operations at Barapukuria have destroyed roughly 300 acres of land, with devastating impacts on at least 2,500 people in seven villages so far . . . People in 15 villages have lost their access to water, as huge quantities of water pumped out for the Barapukuria mine caused a rapid drop in water levels.*—I failed to prove or disprove this.

Now I saw the mine conveyor, like a freeway overpass, shading some men who sat sitting and chewing on flatbeds mounted to bicycles, while grey pigeons went about crossing the road, and goats picked their way through retail establishments. In West Virginia that windowed corrugated corridor would have been off limits; and a security guard would have faced the world from within an air-conditioned vehicle. In Bangladesh it was simply a concrete-pillared part of life, offering puddles, bicycles, firewood, shoes, metal-roofed vegetable stands.—How could it be otherwise? This crowded nation could not afford the luxury of segregating her industrial production zones. I remember on the highway to Phulbari seeing dark brown bodies bathing in green-brown pools that reeked of brickyard smoke; and very likely people were dwelling in the corners of Dhaka's horrendous garment factories (permission to see those had been denied me†).

Beneath that conveyor a man who was weaving fishing nets said: "It's not so bad when they don't explode something underground. The dust is bad . . ." —Beyond it, the mine announced itself by means of a white wall with outward-leaning barbed wire.

From what the company claimed, Barapukuria coal must have been a

---

* See I:211–13, table of Calorific Efficiencies, multiplication factors **120, 158, 160, 161**. Factor **39**, however, is a 1974 international average for lignite of 3,600 BTUs/lb.

† I could, they said, photograph the machines on the shop floor, but only if no workers were present. The authorities' strictness on this score resulted from the Rana Plaza collapse of April 24, 2013, when more than 1,100 garment factory workers lost their lives. According to an Ontario lawsuit, "the subcontractors often operated sub-standard and unsafe factories which put the garment workers at significant risk of severe personal injury or death."

Business under the mine conveyor

money-maker: a per ton production cost of 2,076.15 taka, and a price of 3,420 taka*—making for a good 40% profit margin.

The contractor was China National Machinery Import & Export Company, familiarly called CMC. It appeared that the Chinese brought superior skill and experience to the job, for as a designedly nameless Bangladeshi administrator (in a yellow shirt, bespectacled, well-groomed and wearing a silvery watch) explained to me, 300 Chinese, "25 of them expert," and 1,000 Bangladeshis worked the mine together, proceeding as follows: "They have to do like this: two passages to let oxygen in." (Pushpita inserted: "He has not been in there, so this is only what he's heard.")— "The original design is high-risk, so the Bangladeshis are not prepared for that."

The administrator had reluctantly come out to have tea with us in this open hut which was surrounded by men peering over my shoulder at my scribblings while the proprietor slammed down half-empty glasses of sweet milk tea with tealeaves swollen in the bottom like algae, while flies crawled across the top of the refrigerator.

The company brochure said that the coal was 0.53% sulfur. The administrator said that it was actually only 0.46% sulfur—and very black, very hard. He informed me that the mine had already grown to seven square kilometers. They were presently extracting a million tons per year. The company's owners were "all

---

* U.S. $25.95 and $42.75, respectively.

public, Bangladeshi, no foreign owners." About CMC he said: "They contract out the Chinese workers. Whatever they get paid is their policy. So the Bangladeshis are involved only in the extraction. Whatever happens on the surface is our business, but the Chinese are underneath."

He went on: "The experts get their own houses with their families. Workers get dormitory style housing. Our workers are trained by Chinese. They can speak Chinese since they have trained for 10 to 12 years. Local Bangladeshi engineers have to learn Chinese."

I asked to meet a Bangladeshi engineer, but none was available.

But in the Chinese barracks of the mine, Mr. Zhang Wen happened to be sitting on his bunk, with the mosquito netting cast aside, in a narrow dark air-conditioned room. He spoke English. Three years ago he had come from "near Nanjing." The administrator had said that all the Chinese were on 72-month contracts, in which case Mr. Zhang had another three years yet to go.—I asked why he had chosen Barapukuria.

"The first reason was that I wanted to go abroad, see outside, get some experience. The second reason was for some income. The pay is not high but medium. But I was younger then. I got my high school teaching certificate; then I trained in the National University. Now I am in materials supply. Maybe two years ago I married. This afternoon again I go underground, just to have a look at the materials and at our staff. It will be only myself and my men. I put on clothes and cap and only then you go down, once you are complete."

"What was it like the first time?"

"It was a little amazing, and I was very a little afraid. Very black! I think it's good, because if you prepare in the safety, no need to worry . . ."

"Is coal a good energy source?"

"From my point, what I saw at home three years ago,* the coal was very important in China. *It needs the coal!* The coal is very important. China should develop another coal more, for productivity—but only one, since there are many environmental harms. So only one coal more will not do so much harm."

"What do you think about global warming?"

"Now the government is making so many kind of rule to control so much industry, since global warming get warmer and warmer. Now China realize this point and this to solve. But I think in Bangladesh conditions are a bit different because they develop not fast. And so many huge population—and average family income is low. Rich man very rich and poor man live some bad life condition. It need develop."

---

* I have emended this muddled clause.

2: "A Tough Question Here"

In Barapukuria, as usual, I was not allowed to go underground, so all I can tell you is what I saw on the surface. Just inside the entrance gate was a waist-high concrete guard station, where two sentries in blue uniforms sat in folding chairs (these were "generals," said Nakeeb); immediately past them stood a pair of rifle-bearing soldiers. Then came a long wide brick-paved avenue between whitewashed walls and trees white-painted up to waist height. A Mr. Islam was one of our conductors. He explained: "There's no alternative to coal, since we've run out of gas here."

The premises were vast and almost parklike. The Chinese barracks lay on the left; we walked

Mr. Zhang Wen

rightward, toward the central offices, passing many trees and a canteen. On the auxiliary shaft tower, two great wheels slowly turned in opposite directions.

I had been prohibited from photographing or interviewing anyone en route, but Mr. Islam was kind, so I stopped a "worker for underground ventilation."

"We bring the oxygen," he proudly said.

"What does it look like down there?"

He smiled. "It's dark. It's concrete. It looks like a concrete drain, and then there are angles. We're not scared. The main walkway into the tunnel is paved, and then there are metal walkways . . ."

"When was the last fatal accident?"

"We crack heads, simply crack heads, but a long time ago there was a fatal accident . . ."*

"Is global warming true or not true?"

---

* The administrator had said: "We're really careful about security. This is the most risky type of mine but we have the lowest accident rate. Since we're at the foot of the Himalayas, when we open up some underground chamber the water sometimes rises in there so quickly it could drown people, and then

Top of the second shaft

"That's a tough question here," said Pushpita.

"Ask anyway."

Everyone fell silent. Cross-ankled, he leaned against the seat of his bicycle, clutching his keys, and Pushpita presently reported: "He's not familiar with global warming at all."

3: "We Are Not in a Position to Answer That Kind of Question"

Reader from our hot dark future, what do you think about this? Are you, as I first was, saddened by the man's ignorance? Are you astonished?

Yes, at first I thought to have discovered something special in this answer. And then I remembered interviewing those three Japanese nuclear executives in Tokyo—conscientious men who had received my questions in advance, and written out replies which they read aloud, sitting in a row on the far side of the Tepco conference table (also present was a woman who had no business card, "because it's actually my first day," and said absolutely nothing during the interview except

---

we must close that place right away. Also there is a very high risk of carbon monoxide . . ."—The company report had detailed a couple of fatalities, one of them being a European who died of heatstroke.

at the end, when in English as good as my own she coolly inquired: "Do you represent the American coal industry?"): They had touted nuclear power, of course, because, in the words of Mr. Togawa Satoshi, deputy manager, International Public Relations Group, Corporate Communications Department, "unlike oil, nuclear fuel is available over much of the world, as in Australia and Canada, where the political situation is stable. Moreover, nuclear power generation does not emit CO_2, so it should contribute to the prevention of global warming. Also, the cost of electricity is comparatively low, and the fuel is long lasting, so you can generate energy at a stable price. For the future, you can expect nuclear to be a stable power generation method for a long time. A mix of nuclear and hydro is the best way for Japan."

"So you believe in global warming?" I had asked, and Mr. Togawa replied:

"The average temperature is going up. That is for sure; that is a fact, because of the numbers, because of the measurements. So, yes, we believe. Also, according to the report of the IPCC,* since 1950 it's very probable that the temperature increase observed since 1950 is due to human activity. However, regarding the irregular climate or some extreme weather, we don't think that all those strange events are due to global warming. It's a rough theory. In Tokyo, for example, there's a heat island phenomenon, and people sometimes confuse this phenomenon with global warming. This issue of global warming, we must always deal with it in a calm way."

"What would be the worst case for Tokyo regarding global warming?"

Mr. Hitosugi Yoshimi, section manager, Corporate Communications Department, took over. He said to me: "We are not in a position to answer that kind of question."

I thought about that then and I am thinking about it now. From a practical point of view, his response and that of the "worker for underground ventilation" were the same.

4: Planting Trees and Whatnot

When we were in that teahouse by the great conveyor I had asked the unnamed mine administrator for his views on coal, and he had replied: "As sources of energy, oil and gas are better and cleaner and their prices are lower. But here in Bangladesh we don't have oil or gas, so obviously coal is better."

"What about global warming?"

He flexed his thumbs.

---

* International Panel on Climate Change.

Extracted coal on the premises

"So there are problems with coal dust, with dust and smoke being generated, but the government is taking steps, planting trees and whatnot . . ."

He might as well have said: "We are not in a position to answer that kind of question."

5: "Environmentally Friendly"

In the central office sat two officials. The one who answered my questions gave me a business card which read:

MUHAMMAD ANISHUR RAHMAN

B. Sc. (Hons), M. Sc in Forestry,
Deputy Manager (Personnel),

BARAPUKURIA COAL MINING COMPANY LIMITED,
a company of Bangladesh Oil, Gas & Mineral Corporation (Petrobangla)

---

while Mohammad Sana Ullah, B. Sc (Hon's), M. Sc (Math), Manager (Admin), kept silent. They heartily wished us out of there.

"How efficient is your coal?"

Mr. Rahman looked at a piece of paper.—"Eleven thousand BTUs per pound":—which agreed with the company brochure.

If the West Virginia Coal Association were to be believed, that figure was not so enviable, the average being 12,500 BTUs. Some low-volatile bituminous coal from West Virginia burned at 14,780 BTUs. But as we shall see, it was not so terrible, either.

I requested their views on climate change.

"Basically," said Mr. Rahman, "we can say that our coal has a lower sulfur content than others. It is environmentally friendly."

That was awfully happy news. Moreover, he informed me, their coal contained no selenium. As for climate change, perhaps he was another person who lacked any notion of what it was.

MARVELOUS SOLIDARITY

1: A Man Who Cared

And after that we were escorted off the premises, and then the labor union president had pleaded with me to tell his workers' story; which now I have, what little I know of it. Those skinny, sweaty men crowded into the union office, staring at me as if I might somehow be able to improve their lives, speaking one at a time until their president took over, didn't they deserve as much as you or I? In the words of the Bengali writer Syed Shamsul Huq: *A life that is like a story arouses expectations of a lovely life, a life where a wish-fulfilling wind blows through it . . . A life where there is no sorrow, misery, wounds, or insults. But do we know of such a life? Can it possibly be true?* Who wouldn't want that? I had been so habituated to electric power that my lovely life would have been impaired without it. As for them, since in fact they possessed so much less than I that toiling on the conveyor or even down in the ghastly hot humidity of those longwall galleries was something to hope for, if only their wages would increase—they never said how much but I suspect that they would have been grateful for almost anything—how could my heart not go out to them?

And what would *their* work be for? Every year two million Bangladeshi children perished from waterborne disease (diarrhea generally carried them off). The newspaper quoted the man in Old Dhaka whose family tap produced "dirty-colored water" for six hours and more each day: "It not only causes diarrhoea but it is also harmful to our skin and increases hair-fall." If the electricity generated by burning Barapukuria coal could sterilize more water, cool more fevers, light more hospitals and produce more antibiotics, who was I to oppose it for the sake of you from the future who are probably doomed anyhow?

As you know, the activists in Phulbari were leftists. The Barapukuria Workers' Union defined itself as apolitical. All the same, those two groups were almost neighbors, and, so I would have thought, comrades in arms against coal-related exploitation. (I remember from Phulbari a woman in a vermilion hijab and a green robe riding a motorcycle behind her brother or husband, shadows growing on the red and yellow tanks of compressed gas, a woman in a short *borkha* and loose yellow trousers walking beside a more covered lady. I remember two white-bearded men in white prayer caps, one in white, one in pale grey; they stood facing the front wall of a hardware store which displayed pictures, not photos, of fans, light bulbs and drills; and straight across these ran the legend: **SECURITY SECURITY SECURITY**. And I remember from Barapukuria people whose difference from the Phulbari people I could not distinguish; weren't they by some criterion brothers and sisters?) So when I asked the legless man what he thought about the three *shaheeds* of Phulbari, I expected at least some sympathetic pieties.

"Actually," interpreted Nakeeb, "he's not conscious enough to care anything."

When we were alone I expressed my sorrowful surprise, at which he elaborated, or perhaps corrected himself: "Actually there is insecurity and inferiority complex. The issue of jobs is totally sensitive. The Phulbari issue they are not interested in talking openly. If someone talks and a spy hears, he could lose his job."

That was understandable, if not admirable—but on second thought not so understandable after all. Wouldn't the legless man's already communicated grievances imperil him? If Mr. Islam Rabi the union president had already gone to jail for the cause, weren't they all blacklisted anyway?

I can never be certain, but Nakeeb's first interpretation rings more true to me. The legless man's answer was brief, and his affect indifferent.

"Not conscious enough to care anything." This could have been a local corollary of Pushpita's maxim: *The others cannot understand until it happens to them.*

Meanwhile, that unnamed administrator who came out of the Barapukuria mine for this reluctant cup of tea had assured me that there were still *plans to develop coal in Phulbari.*

2: "Environmentally Friendly"

And how could there not be, with Ambassador Moriarty and his ilk on the case? In 2008 that fellow cabled (confidentially, of course):

> 5. (C) [NOTE: . . . Asia Energy's project in Phulbari in northwestern Bangladesh, which would use U.S. financing and equipment, has been stalled in part due to the lack of coal policy. END NOTE.]

In the same cable he wrote bitterly:

> 8. (C) Comment: Bangladesh's power and energy sector has failed to develop adequately in part due to special interest groups that have opposed numerous projects on social and environmental grounds . . . The

G[overnment] O[f] B[angladesh] is developing new sources of energy at a glacial pace.

He seemed to see Bangladesh as a playground for resource extraction companies. The following might be my favorite paragraph from his confidential little cable:

2. (C) Malaysia's PowerTek . . . which planned to use Siemens equipment, was the sole bidder for the development of a power generation plant near the Bibiyana natural gas fields in northeastern Bangladesh . . . The Bibiyana power plant will be located near Chevron's Bibiyana gas field . . .

In West Virginia the *regulated community* operated with shocking impunity. Here in Bangladesh the *regulated community* consisted of foreign-based multinational corporations, with the American Ambassador exerting his influence on behalf of at least one of them. And since he was pushing for Asia Energy, why not perform the same offices for Chevron? Meanwhile, should or should not one suspect the Malaysian government of operating comparably on behalf of PowerTek?

In short, I gave up on Phulbari. The corporations must have fixed their claws in it long since. To be sure, when I was finishing *Carbon Indeologies,* their open pit mine remained the merest slavering hope. (By the way, *the mine would involve the eviction of more than 50 villages.*) My certainty that it would be dug derives from the following: *According to the Ministry of Power, only around 60% of people in Bangladesh have access to electricity and demand is growing 7% a year.* Who could argue against demand? Even the activist Mr. Bablu had said: "I'm not against coal itself; we need the coal."

As for me, I went home to my air conditioner. A year later, *The Guardian* announced that *Bangladeshi villagers* staged further protests . . . after police opened fire and killed at least four people demonstrating against the planned construction of two large Chinese-financed coal-fired power stations.* So now there were four more *shaheeds* in a new place—or, if you like, another four blasphemers against demand.

You in the future won't care; all of Bangladesh must be under the ocean.

---

* "In the coastal town of Gandamara near Chittagong."

THE BRICK KILNS

1: Moloch's Altars

The unnamed administrator at the Barapukuria mine had remarked that their highest allocation priority was electricity generation. And indeed, every day, the government entity called Bangladesh Development Power turned 2,500 tons of Barapukuria coal into smoke.*—As a West Virginian might say, that kept the lights on—not to mention the coolness in the quiet vastness of my five-star hotel in Dhaka with its copper-potted palm trees. A literary festival had put me up there. (Shall I confess my discomfort in the sweltering family home of Mr. Bablu, where I stayed for just one night, tingling and itching all over from bites from possibly malarial mosquitoes, the webs between my fingers sticky, my bluejeans sodden, sticking to my knees and beginning to stink? You from the future must be better adjusted to that than I.) In the lobby a worker carefully lifted out from the elevated planter potted plants that had wilted and replaced them with fresh ones from a trolley, packing pebbles around them; the old ones went into a box. Just outside, the blue-uniformed guards stood beside (a) the metal detector which was probably turned off because it never beeped, and (b) the X-ray belt which passed through the window into the lobby; that latter device did function, as my pancake frisker proved, so perhaps my ostensible security was courtesy of Barapukuria coal.

Whatever was left over from electricity generation went to brick manufacture.[†] The unnamed administrator supposed that a large brickyard required about 10,000 tons of coal per year.

Anywhere from Phulbari to Dhaka one might see walls of bricks along the side of the highway. To be sure, they were of less visual interest than the smoldering garbage spread along the roadside, the bamboo-pillared vending sheds, the bamboo poles laid out to season. Here came another family in the back of an open pickup, whose teenaged member was chatting on her cell phone, looking very pretty in a pink-and-black floral hijab; I liked her better than any number

---

* This would have generated 5,492,666.27 kWh, or 18.3 billion BTUs.

† What the miners extracted in three months took another two to "develop," which must have meant pulverizing and washing. The administrator said that the stores of saleable coal often ran low; hence only 5 lakh, or 50,000, tons of Barapukuria coal went "outside," to the brickmakers.

of bricks! We crept south toward Dhaka, back across the very long bridge, waves bright and pallid, and the stacks of bricks grew more numerous; at last those smoking kilns appeared ahead, towering over banana plantations and brickyards. By the Turag River just north of Dhaka were multitudes of them, warming our atmosphere in black or white.

Someone could have said (as was the case with so many other topics in *Carbon Ideologies*) that this particular subspecies of manufacturing hardly deserved to be singled out. The annual energy spent on Earth to produce bricks might have been one-quarter of that used to make glass, which as you may remember* was one-thirtieth of what we used to produce steel. And it may have been true that brickmaking kilns of Bangladesh took no conspicuous place in our global energy budget. But (speaking strictly as an uneducated wanderer) it did seem that they gave off a lot of smoke.

We pulled in at a promising-looking establishment, whose name, I eventually learned, was MESSGRS SYC BRICK. Nakeeb took my hand to help me down the crumbling slope below the highway, while in the interests of appropriate gender relations Pushpita clambered alone. The closer we drew, the more it resembled some Old Testament vista of slavery and human sacrifice. A hundred and fifty men and women toiled on a wide clay plain, some of them carrying the raw stuff up onto what could have been an elevated altar to Moloch, others creeping and digging, feeding coal into the furnace, stacking finished bricks 10 and 15 feet high, loading them into trucks for sale. (One truck could carry 3,000

---

* See "About Manufacturing," I:132.

tons.)—What did those laborers look like from here? I cannot do better than quote the Bengali writer Hasan Azizul Huq (no matter that he was depicting a different species of proletarians): *And to tell the truth, they were dark like burnt pieces of wood, skinny and ill-shaped, or somewhat hunched; in other words, they were all ghostly, much like shadowy spirits.* And these burnt-wood figures directly and indirectly concerned themselves with the burning of coal.

Messgrs SYC brick kiln

The smokestack was far higher than it had seemed from the road. As we descended we could see other brickyards all drearily alike, each narrow stack rising high into the humid air from a wide clay plinth approached sometimes over filthy little creeks and sometimes through narrow and temporary lanes between rising stacks of bricks; to ascend the plinth one needed local advice, because these convenient-appearing clay stairs, or that steep dirt path over there, might not bear an adult's weight anymore ("very risky," said the proprietor of MESSGRS SYC BRICK); and the plinth itself was equally precarious, being riddled with wells five to seven feet deep in which the bricks gestated, baked by subterranean coalfire whose water-filtered smoke rose white or black from the stack. The heat around each altar rendered trivial the broiling humidity of the ricefields.

The proprietor was named Mohammed Maksud Hossain. He sat in his shack of an office, doing accounts. He was very obliging, but I took care not to occupy more than a few minutes.

He explained he did not use Barapukuria coal since that mine sold only small

amounts, and those only irregularly—"and we have to pay in advance, and the coal does not come on time." So he got his coal from Indonesia. "We don't distinguish between types of coal, but we look at it to see if there is rock, which is bad, or oil, which is good."—He bought 500 tons of it every five months. In that time he made 60 lakhs of bricks,* "depending on strikes and blockades." This works out to six bricks produced per pound of coal burned. His best bricks sold for seven taka each.†

About global warming he opined: "Not good. We have changed the way we burn. Government regulations make us mix the smoke with water."

I must remark how strangely it struck me that in this country where most of my interviewees had never heard of climate change, Mr. Hossain, who must have operated on a tight margin, which those government regulations would have further tightened, could call global warming "not good," while so many West Virginians, who, however poor they were, could not have been *this* poor (or in

---

* Six million.

† Less than nine cents. But according to Nakeeb, one brick cost 10 to 15 taka [12.5 to 18.75 cents]; perhaps that range reflected the retail price, while Mr. Hossain must have been a wholesaler.

some sense could they?), denied it and railed against regulations. I suppose the reason was that rather than being simply neglected, they were aggressively mis-educated by their politicians and the *regulated community*. But perhaps you will draw better inferences.

By phone I chatted with another brickyard owner, a friend of Pushpita's who did not wish to be named because "he's paranoid and doesn't know what will happen" with new environmental regulations. I will call him by his initials: R.C.

"We get coal from a lot of different places," he said. "Barapukuria coal is the best we have. Unfortunately it's a very seasonal thing, so we never know if it will be available. We always have a standing order."

Indonesian was the best at a calorific value of 6,400 joules per kilogram,* and that was what R.C.'s establishment usually burned. Next came South African at 6,100, and finally Indian at 5,700.† This last "has extra sulfur," he said; the smoke was extra whitish. Although India contained the fifth largest coal reserves in the

---

* He must have meant kilojoules [kJ], in which case the respective fuels would have come in at 2,751.5, 2,622.5 and 2,450.6 BTUs/lb.

† Indian coal was by no means the most calorifically inefficient in the world. According to Robert Finkelman at the U.S. Geological Survey, "lignite ... has the lowest Btu or hhv," and the Indian coal was almost certainly lignite. "From my experience," he continued, "the lowest rank coal (and therefore the lowest Btu/hhv) was in Eastern Europe—Poland and the former Yugoslavia. But this is not from a comprehensive survey." Perhaps other lignites offered even less energy in exchange for the carbon they released.

Workers at Messgrs SYC

world, it was still a net importer of coal and busily burned it,* so that the sulfur-high, poor-quality home-dug stuff found its inevitable way to Bangladesh. "We suffered using the Indian coal," R.C. said. "When we first started our business, we tried to save money by using it, but it actually cost us more and was worse for the environment."

"We recycle the smoke," he said. (Perhaps that meant that he, too, water-filtered it.) "Then we can see the yellow smoke."

To make 100,000 bricks cost 12.5 tons of Indonesian coal, 13.5 tons of South African coal or 20 tons of Indian coal. I calculated as follows:

COMPARATIVE CALORIFIC EFFICIENCIES, 2015,

in multiples of the number of bricks produced by 1 pound of Indian coal

(from Mr. R.C. and Mr. Mohammed Maksud Hossain)

The average weight of one of Mr. R.C.'s bricks was 1.6 kg [3.52 lbs].* I have assumed the same weight for one of Mr. Hossain's. I have also assumed that the original ton-nage figures were metric.

All levels expressed in [bricks per pound of coal combusted].

1
Indian coal, burned by R.C. [2.3 bricks per lb].

1.5
South African coal, burned by R.C. [3.4].

1.6
Indonesian coal, burned by R.C. [3.6].

2.4
Indonesian coal, burned by Mohammed Maksud Hossain [5.4].

---

* Therefore 1 pound of his three respective coals produced respectively 14.08, 13.02 and 8.8 pounds of brick.

---

* In the glowing words of the International Energy Agency (2012), "the People's Republic of China, the Russian Federation and India constitute three of the four countries that produced the most CO_2 emissions in absolute terms. Their shares are likely to rise further in coming years if the ongoing strong economic performance currently enjoyed by most of these countries continues."

A tabulation of high heating values told a peculiarly different story:

COMPARATIVE CALORIFIC EFFICIENCIES, 2015,

in multiples of the thermal energy of Indian coal

(from Mr. R.C., Mr. Mohammed Maksud Hossain and
various published sources)

All levels expressed in [HHVs where known; otherwise in unspecified calorific efficiencies].

1

Indian coal, burned by R.C. [stated by him at 2,450 BTUs/lb]. According to the Australian government, the average efficiency of Indian coal at this time was 6,296 BTUs per pound, although the lowest quality stuff came in at 5,600. But the "useful heat value" of that weakest coal the Australians rated at 2,338 BTUs per pound, which is within 5% of R.C.'s figure.

1.07

South African export coal, burned by R.C. [Stated by him at 2,623].

1.12

Indonesian export coal, burned by R.C. [Stated by him at 2,752].

2.4

Indonesian coal, burned by Mohammed Maksud Hossain [no stated HHV, but multiplying his brickmaking efficiency of 2.4 times R.C.'s Indian coal's gives $(2,450 \times 2.4) = 5,904$].

2.63–3.86

South African bituminous steam coal, for domestic consumption [reported by University of Cape Town at 6,449–9,458].

3.19–4.28

Indonesian lignite [advertised by China Energy Holdings at 7,808–10,477].

3.69–5.09

Indonesian export coal [reported by World Energy Council at 9,028–12,468].

4.51

Barapukuria bituminous coal ("high volatile") [advertised at 11,040].

4.73

South African bituminous export coal [reported by University of Cape Town at 11,608].

4.97–5.29

Subbituminous Indonesian coal [advertised by China Energy Holdings at 12,167–12,971].

5.10

"Average coal," according to the West Virginia Coal Association [12,500].

The heating values of all the imported coals combusted by that pair of brickmakers were preposterously low. Perhaps they could not accurately rate their own feedstock—and the inconsistent rankings of South African and Indonesian coals in the two tables above may suggest that—or else, as the calorific efficiency of the Indian coal in the table just above would indicate, the high heating values hovered irrelevantly high above the dreary "useful heat values" with which these brickmakers contended: Inhabiting as they did one of the poorest nations on earth, they surely had to burn the worst available, pebbles and all.

And so it struck me that not only would Phulbari's coal someday almost surely come out, since nearly all parties desired that, but also that the global environmental consequences of it *not* coming out might be even worse than extracting it, because otherwise these brickyards would have to keep burning, in decreasing order of thermal efficiency, Indonesian or South African coal, whose transportation alone increased global warming—or, worse yet, that Indian coal, which did indeed lie closer at hand, but whose thermal efficiency was so inferior and whose added benefit was sulfur pollution.

2: Carbon Arithmetic

Better still from this standpoint would be to import West Virginian coal for the burning:

Barapukuria coal ranked against major American coals, *various years*

[Calorific efficiencies in BTUs per pound]

Northern Appalachian	12,997.5
W.V. Coal Ass'n average	12,500
Central Appalachian	12,487.5
Illinois Basin	11,763.7
Uinta Basin	11,701.2
Barapukuria	11,040
Powder River Basin	8,750

Climate change has its own arithmetic, whose results come out similarly.

You may recall that lignite emits about as much carbon dioxide per BTU volume as the most prevalent Bangladeshi fuel of all—firewood—which tops the scale of carbon fuels, being 112 times worse than the apparent best of them, natural gas.[*]

While West Virginian bituminous coal emits somewhat more carbon dioxide per unit volume than Indian lignite, its calorific efficiency is far higher. By my calculation, if R.C. could somehow use what the West Virginia Coal Association would have been ever so happy to supply, his contribution to global warming would be almost three times less:[†]

COMPARATIVE CARBON DIOXIDE EMISSIONS TO MAKE ONE BRICK,

in multiples of the average emissions of West Virginian bituminous coal

All emissions figures in [pounds of carbon dioxide per brick].

All efficiency figures in {bricks produced per pound of combusted coal}.

1

West Virginian bituminous [0.211 lbs CO_2/brick].
{11.74 bricks/lb}

---

[*] See the table "Carbon Dioxide Emissions of Common Fuels in multiples of natural gas's," I:199. I call natural gas the merely "apparent" best on account, of course, of methane (I. 216, header **259**).

[†] Of course this ignores the energy that would be consumed in shipping the coal from the U.S.

1.13
Barapukuria bituminous [0.238].
{10.36}

2.88
R.C.'s Indian lignite [0.607].
{2.3}

Sources: R.C. interview; table of Calorific Efficiencies; table of Carbon Dioxide Emissions of Common Fuels, with calculations by WTV.

Anyhow, these were but smoky fantasies of mine—for R.C. could not afford the stuff from West Virginia, and the Barapukuria coal remained mostly unavailable. In a better world, he would have burned better fuel.

3: The Best of All Possible Worlds

According to a group called Waterkeepers Bangladesh, *from the environment perspective, using coal energy to generate power in Bangladesh is not feasible as severe water pollution occurs at every phase of using coal from extraction to final usage . . .*

So wouldn't it be better for Bangladesh if the pollution stayed in West Virginia?

That would have pleased Chris Hamilton of the West Virginia Coal Association, for he told me: "World use of coal is on the increase. It's going to become the world's number one fuel source. Without question it will help our state's economy. We just happen to have some of the more expansive coal deposits, actually in McDowell and Wyoming Counties, some of the best coking coal for steel production that you'll find throughout the world. We have been a global economy forever, because of our exports and our metallurgical coal reserves. We find ourselves competing for markets in China and Europe with countries like Colombia, Indonesia, Australia, and they beat us on raw mining cost, the transportation cost, and the big part of that part of that cost is that they don't have the same level of protection of human safety, human rights, environmental quality as we do when mining. And in some of those countries, they use child labor still!," he said, smiling and shaking his head. "I'm not suggesting wholesale regulatory reform, but to the contrary! Our government is working us every way they can. The world use of coal is increasing. We have more than anybody. But we don't have the right marketing model."

Mr. Hamilton might have been too modest: In 2011 West Virginia had been America's primary coal exporting state: 35 million short tons sent abroad! *Analyses suggest U.S. exports* of coal *could be reducing by half or wiping out completely the pollution savings in the U.S. from switching plants from coal to natural gas. The nexus of the challenge can be found in and around Norfolk, Virginia, which exports more coal than any other place in the U.S. and is already experiencing one of the country's fastest rates of sea level rise.*

Brick kiln workers at Messgrs SYC

"EXCESSIVE EASE AND PROSPERITY"

Once upon a time the famous dissident Aleksandr Solzhenitsyn, raging against the capitalist West for ignoring the suppression of his writings, ultimately dismissed that zone as follows: *Excessive ease and prosperity have weakened their will and their reason.* When I first began researching *Carbon Ideologies* that sentence clutched at me with talons of truth. Of course the lazily affluent hordes to which I belonged preferred to keep the carbon burning; that was so much nicer than sweating like Bangladeshis, or struggling to rein in the *regulated community*! In Japan, where nuclear ideology continued to resuscitate itself from the contretemps at Fukushima, certain citizens protested, but the vending machines glowed on, and electricity continued giving birth to all customary magical things. In West Virginia, perhaps, blaming coal-worship on *excessive ease and prosperity* was not quite fair. And in Bangladesh that phrase rang grotesque.—"There's no substitute for coal in Bangladesh," the union president had insisted. What kind of complacent monster would I be, to accuse this man who had gone to jail for the sake of the workers, of embodying excessive ease, let alone weak will? And when Mr. Bablu, who had fought for the people of Phulbari, and seen the three *shaheeds* shot down, insisted, "I'm not against coal itself; we need the coal," was I to convict him of my own middle-class American softness? As for Mr. Juuel, that organizer of blockades, when he expressed the issue as "how can we take this coal out?," how could I pretend that he had not resisted the status quo, determinedly and perhaps even desperately?

But I repeat: *The coal did not need to come out.* Bangladeshis could have achieved some un-excessive measure of electrical ease and prosperity through solar power (the sun was breathtakingly strong there), not to mention sensible limitations on consumption—for instance, if people had grown up without air conditioning, and probably could not afford it, was it wise to addict them to it? (So I theorized, hoping to keep air conditioning for myself. At that time only 59.6% of Bangladeshis "had access to electricity."*) As for coal's supposed local economic benefit, if West Virginia were any indication, a great deal of the extraction would be performed first by foreign workers such as Mr. Zhang Wen, and

---

* The equivalent worldwide figure was 84.6%.

288

secondly by labor-saving machines. Meanwhile the farmers of Phulbari would become refugees. Based on the conclusions of the government report, far fewer Bangladeshis would gain livelihoods then lose them.*

Where was the ease and prosperity?—Not here.—In Bangladesh as in Appalachia, business interests offered the only game in town to people whose longing for a "decent wage" could be satisfied by pathetically modest compensation—and these people, of course, did not understand climate change.

As *Carbon Ideologies* sometimes likes to ask: What was the work for?

Do you remember Hasan Azizul Huq's portrayal of the Bangladeshi underclass? *And to tell the truth, they were dark like burnt pieces of wood . . .* The work was not for them. Reader from the future, I fear that Mr. Huq's description may fit you best, now that Bangladesh lies under the acid ocean and you begin to consider the possibility of extinction. You know full well the work was not for you!

The beneficent Mr. Moriarty had advised *the prime minister's energy adviser* that *open-pit mining seemed the best way forward.* It must have been, then. And we remember Moriarty's aside: *Asia Energy . . . has sixty percent US investment.* That was whom the work would be for.

In 2015, the company that wished to mine Phulbari called itself GCM Resources. The Chief Operating Officer was still Gary Lye. He was also Chief Executive Officer of GCM's subsidiary, which was still named Asia Energy Corporation. *Gary . . . has been with the Phulbari Coal Project since 2004. Gary led the exploration programme and Feasibility Study,* gushed the home page for GCM. *He is a qualified geologist and geotechnical engineer . . .* In that case, he must have comprehended quite well what his open pit mine would do to Phulbari.

According to a leftwing source, *capitalism in Bangladesh is characterised by corporate land grabbing and dispossession of people.* That might well have been so.

We are informed that in 2007, *a leader of the national movement opposing the mine, Mr. Nuruzuman, was publicly tortured by the Bangladeshi military.* In 2009, the National Committee was *marching peacefully in Dhaka* when it allegedly ran into difficulties, with more than 50 of its members coming to harm. *Among the seriously injured was the member-secretary of the National Committee— Professor Anu Muhammad. His legs were badly fractured by police batons.* Two women students who sought to protect him *were badly injured. They are all in hospital now.* In 2011, 200 people sealed off part of a Phulbari highway,

---

* B. Traven, mid-20th century: "All problems are simple when enough land is available, and enough men on the soil who know how to cultivate it and love it; but all problems are complicated the moment men are displaced from the land to which they belong."

demanding that the government finally honor the Phulbari agreement. *The Bangladeshi government deployed Bangladesh's Rapid Action Battalion to intimidate protesters. This Rapid Action Battalion has been denounced by international human rights organizations as a government death squad because of its routine use of torture and extra-judicial killings.* The same year, in a standoff at the Barapukuria coal mine itself, *government-backed attackers publicly broke the hands of a National Committee leaders [*sic*].* In 2012 the police beat 15 protesters, a two-day general strike took place in Phulbari, and the people burned an effigy of the new American Ambassador, Mr. Dan Mozena, on account of those WikiLeaks cables.—All of the events in this paragraph were detailed in another of Nakeeb's typescripts, and I cannot prove that they did or did not happen. But I remember the monument to the three *shaheeds* down in that sweltering meadow by the Jomuna River where the skinny white cows grazed, looking up at the bridge. I remember the look in the old man's eyes when he mentioned that boy and those two men who were shot down for the sake of something indefensible.

BANGLADESH'S BEST FRIEND

And in Dhaka, in a place where the river curved widely toward the horizon of power towers and low concrete buildings, everything crowded together into what my untrained gaze could perceive only as chaos, and people streamed across the tight-packed coal barges. In the foreground lay long rectangular mounds of coal, whose rounded tops were nearly twice a man's height. A woman strode across a many-puddled plain of dirt, bearing a shallow basket on her head. A man guarded his square mountain of stacked logs. Coal waited to go up in smoke.

The lumbermen rarely trouble to put out their fires, such is the dampness of the primitive forest; and this is one cause, no doubt, of the frequent fires in Maine, of which we hear so much on smoky days in Massachusetts. The forests are held cheap after the white pine has been culled out; and the explorers and hunters pray for rain only to clear the atmosphere of smoke.

Thoreau, 1847

The Trend

But this is a time when many things contrary to nature and even against the ordinary course of fate seem likely to happen at any moment . . .

Cicero, "First Philippic Against Marcus Antonius," 44 B.C.

The trend of these chapters has been toward perils of increasing invisibility. We find ourselves proceeding from the earthquake-tsunami, whose horror was all too obvious, to the nuclear accident, whose nightmarishness revealed itself by degrees, through repeated vistas of eerie decrepitude and more importantly by means of instruments; next came the immediate aesthetic, biochemical and climatological harm done by coal, whose soldiers dismissed these matters as political lies and exalted the "heritage" for whose digging up ever fewer humans were needed; we now turn to fracking, whose effects on the climate were once again minimized if not dismissed, whose water contamination occurred discreetly underground, and whose immediate health effects seemed perceptible only by a minority. We who merely *burned* natural gas perceived no ill effects! On a chilly December evening, I, sluggish, sedentary devotee of the fossil fuels I criticized, sat happily by the gas fire, rereading this paragraph through plastic Emirati spectacles whose frames and lenses could have begun as either oil or natural gas, and for a moment I thought about you. Perhaps you have seen the place where I used to live. Does the Central Valley still support human settlement? Perhaps your summers reach 150° or 160° Fahrenheit. I hope your winters are not too stormy. Today began with a clammy fog; I felt chilly all day. Returning home early, I threw the dog a tidbit, meowed back at the cat and aimed the battery-powered remote control at the fireplace, which used to accept genuine logs but was long ago converted, against my wishes, "for the good of the planet." The cat was shy; she had not yet gotten over the disappearance of my daughter, who had trundled off to college. So I sat by the fire alone, with a glass of mescal in my hand, correcting this chapter about fracking. It felt so good to be warm! And since I cannot imagine that you approve of me, just now I don't care about you.

Once upon a time, maybe even while I had guests over and the gas fire was cheering us, an emergency room nurse named Cathy Behr *spent ten minutes with gas field worker Clinton Marshall,* whose head was throbbing and whose gorge was rising after being doused with fracking fluid.* *The fumes were so overpowering the emergency room was evacuated.* And Cathy Behr, they say, came down with *multiple organ failure.* But what if this were only a fable? After all, I heard it from that filthy subversive cabal, the Sierra Club. My intermediary contacted her by phone, then, with Ms. Behr's permission to publish ("go for it," she said), wrote down the following summary: *Can't speak about it—according to her lawyer would risk being sued for libel by the oil and gas industry.* What can we make of this story? When I requested her opinion, the intermediary replied: "She was a very intelligent-sounding woman, very articulate, and, well, her response about being sued was very much laced with sarcasm . . ."—Trump would soon be President. Maybe Ms. Behr feared the new administration; or, if you like, let's say that she and the Sierra Club had libeled the innocent makers of Zetaflow. Perhaps she was merely tired and intimidated—for in Weld County, Colorado, where I soon shall take you, most carbon ideologues appeared to like fracking pretty well. An anti-fracking activist told me: "Sometimes I carry a gun. You have to understand, this is their livelihood. How can I argue with that?"[†]

You from the future had better understand that, too. No arguments, please!

---

* "It was . . . a product with the trade name Zetaflow." According to an advertisement from the manufacturer, "sand problems will eventually plague most wells . . . Sand and unconsolidated particles can erode the near-wellbore and slowly choke production . . . ZetaFlow sand-conglomeration technology . . . increases the attraction between particles . . . This system leaves formations in the optimal wetted condition for increased load recovery . . . The solution is added continuously in both slickwater and borate cross-linked systems . . ."

[†] Bob Winkler, p. 349.

FRACKING AND NATURAL GAS

Greeley and Loveland, Colorado, U.S.A. (2015)

Overleaf: A frack pad in Greeley, Colorado, photographed in July 2015

Fracking and Natural Gas Ideology

Assertions

"Natural gas is considered as a premium fuel. It has a high caloric value and is clean-burning and relatively easy to handle ... Natural gas has the added advantage of adding the smallest amount of CO_2 to the atmosphere per unit of released energy."

George A. Olah, Alain Goeppert and G. K. Surya Prakash, 2009

"Fracking may actually prove to be our best near-term solution to the climate change threat without bankrupting the developed world and further impoverishing the developing world."

Chris Faulkner, Chief Executive Officer of Breitling Energy Corporation and 2013 Oil Executive of the Year, writing in 2014. Although he took pains to share his gaseous ideology with the reading public, this fellow never responded to my interview requests.

"Natural gas emits about half as much carbon as coal and can transition us to truly clean power."

Froma Harrop, syndicated columnist, 2015. She goes on to say: "But the future is truly renewable energy from such sources as the sun and wind."

"Natural gas is one of the most efficient, reliable and affordable sources of energy for homes and businesses."

Pacific Gas and Electric leaflet, 2016

Chevron anticipates that in 2040 "fossil fuel's share of the energy mix" *will be 75%, because oil is so convenient, and because* "one of the advantages of natural gas comes from its reduced CO_2 intensity, which is about half" *that of coal. Hence in 2040* "natural gas demand grows from about 15 percent in 2013 to 394 billion cubic feet per day."

Chevron website, "Managing Climate Risk," 2016

One electronic analogue of the "West Virginia Coal Tree" [p. 10] lists, among other commodities, the following products "made with natural gas": Tires, bandages, insect repellent, crayons, golf balls, perfume (which also happens to dangle from the Coal Tree), allergy medicine, football helmets, cortisone, fishing poles, guitar strings and parachutes.

Embellishments

ECONOMIC APPEAL

"The shale became a lifesaver and a lifeline for a lot of working families."
Dennis Martire, mid-Atlantic regional manager, Laborers' International Union, 2014

"OKLAHOMA JOBS BROUGHT TO YOU BY HORIZONTAL DRILLING."
Ad from Continental Resources, seen in 2016 in the Chickasha Chamber of Commerce Community Guide 2014

"Gas and oil resources extracted through fracking have already added more than $430 billion to annual gross domestic product and supported more than 2.7 million jobs that pay, on average, twice the median U.S. salary."
David Brooks, New York Times *editorial, 2015*

"DRILL HERE, DRILL NOW, PAY LESS!"
Bumper sticker seen in Denver, 2015

LOCAL APPEAL

"As the third largest consumer of gas and diesel on the planet, California has to use oil even as oil slowly decreases in importance . . . A ban on fracking would only lead to higher global carbon emissions, as the state would import more oil by rail and ship, from countries with more lax environmental regulations, and to more unemployment in the state's most economically-troubled region, as well as less revenue to fund vital state programs."
Dave Quast, California director, Energy In Depth, 2015

NATIONALISTIC APPEAL

"The revolution has already begun. It's already reducing our oil import dependency, notably from volatile regions or nations antagonistic to Russia. This revolution may help to defang Russia . . ."

Chris Faulkner, 2014

"Energy: the lifeline of Russia."

Gazprom, the Russian natural gas concern, 2015. In other words, Russia was not getting defanged.

"Effective and sensible utilisation of [natural] gas is . . . of paramount importance to the country."

The Bahrain National Oil Company, 1980

TAUTOLOGICAL TWADDLE

"FRACKING ROCKS!"

Bumper sticker seen in Greeley, Colorado, 2015

"Heat like the rays of the sun."

Slogan of American Gas Machine Co., bef. 1985

DEFENSES AGAINST CRITICISMS

"I care just as much about the environment as you. I have a family to whom I want to leave a beautiful and livable world . . . But, as a pragmatist, here's my main point—we can't have it all."

Chris Faulkner, 2014

"Frack jobs do consume a large volume of water—perhaps 2–4 million gallons of water with each job . . . But let's put things in perspective. That's about how much water a golf course uses in a typical summer month."*

Chris Faulkner, 2014

---

* "In 2011, the EPA estimated that 70 to 140 billion gallons of water are pumped into 35,000 fracking wells annually."

"A fracking ban . . . would utterly decimate an industry that accounts for about 10 million jobs in the US and represents about 8% of US G[ross] D[omestic] Product."

Chris Faulkner, 2014

"State and federal regulations that limit hydraulic fracking will cause even further deterioration in income and standards of living in Weld County and other rural counties dependent on the oil and gas industry."

Barry W. Poulson, Past Commissioner of Colorado Tax Commission, 2013

"Fracking is safe with manageable risks."

Dave Quast, 2015

"There have been no confirmed incidents of groundwater contamination from hydraulic fracking."

David A. Waples, "a communications manager for a natural gas utility," 2012

"A recent Environmental Protection Agency study found that there was no evidence that fracking was causing widespread harm to the nation's water supply. On the contrary, there's some evidence that fracking is a net environmental plus. That's because cheap natural gas from fracking displaces coal."

David Brooks, 2015

"Fracking is like putting tires on a car . . . The process is safer and has less environmental impact than driving a car or flying an airplane."

Greg Kozera, President of Learned Leadership LLC, with "40 years of experience in the energy industry," 2016

About Natural Gas

The whole place was cosy in that it was lighted by gas and heated by furnace registers, possessing also a small grate, set with an asbestos back, a method of cheerful warming which was then first coming into use.

Theodore Dreiser, *Sister Carrie*, 1900[*]

L
ike petroleum, this stuff comes from dead plants and animals, decomposed and crushed for millions of years. Natural gas generally contains, in decreasing order, methane—the lion's, or better yet the dragon's share, hopefully[†] at 85% or more—followed by ethane, whose molecule sports one more carbon and two more hydrogen atoms than its predecessor; propane, which adds still another carbon and two more hydrogen atoms; and a very small percentage of less volatile compounds.

Some natural gas was "naturally" carbonated. Up rose the subject of this book! But of course CO_2 pollution made little impression on us back in 1821, when the good citizens of Fredonia, New York, piped gas from their local well *through hollowed-out logs to nearby houses for lighting.* It must have been especially pure— or else their lamp-flames would have wavered and stank; anyhow, when I was alive we had to "process" it by purifying it of water, hydrogen sulfide and other contaminants—a quotidian industrial task which emitted more greenhouse gases:

GERMAN EMISSION FACTORS IN NATURAL GAS PRODUCTION, 2007,

in multiples of the value for nitrogen oxides

All levels expressed in [pounds of greenhouse gas released for each ton of natural gas produced].

Unless otherwise stated, all global warming potentials [GWPs] are for 100 years. See I:176–79 for a more detailed presentation of these values.

---

[*] This epigraph is probably a cheat, given the date. The gas must have come from coal that was heated in some municipal works which produced methane and piped it around the city.

[†] The proportion might be as low as 60%.

1
Nitrogen oxides [0.029 lbs per ton natural gas]

    × *GWP of 159 = **4.6** lbs of carbon dioxide equivalent*

3.45
Sulfur dioxide [0.1]

    GWP: Disputed

5.52
Methane [0.16]

    × *GWP of 28 [IPCC 100-year value, 2013] = **4.48** lbs of carbon dioxide equivalent*

    × *GWP of 86 [revised 20-year value, 2015] = **13.76** lbs of carbon dioxide equivalent*

931
Carbon dioxide [27.0]

    GWP: 1 [by definition]

Source: Greenhouse Gas Inventory Germany, 1990–2007, with calculations by WTV.

By then our petro-engineers had progressed beyond hollowed-out logs. With its pieces laid end to end, the Gazprom pipeline network would have wrapped four times around the earth! Had anyone inquired, "But what's that work for?," I would have dragged out the explications of my own carbon ideology: selfishness, shortsightedness, metastatic capitalism and my own comfort—of which each tumor was another irruption of the *regulated community.* But I was shallow; I'd neglected to take the long view; being fixated in the sadness of the future, I'd neglected the mindless tragedy of the past, with its forgotten extinctions. *Life is a furnace of concealed flame . . .* wrote Loren Eiseley. In other words, life burns energy as fire does, yet more slowly and steadily. *Fire . . . is a nonliving force that can even locomote itself . . . What if I am, in some way, only a sophisticated fire that has acquired a way to regulate its combustion and to hoard its fuel in order to see and walk?* I liked that. And how could I better hoard the fuel of my own warm blood than to feed widely? The morning was cold, so I buttoned on my soft overshirt of polyester fleece, whose gas- or oil-derived insulation saved me from shivering. I walked up to the bakery and ordered a foam-rich vanilla steamed milk, whose sugars gave up their carbon energies to me, so that I could keep breathing out carbon dioxide, while inhaling the carbon dioxide of women in my herd.

Wishing to extend my biological range with a minimum of effort, I now boarded a bus whose painted sides proclaimed it "green" and "ecological" because it used "clean natural gas." Why shouldn't I see the world? And up in frigid Novy Urengoy (which in your day, reader, must be a methane-bubbling swamp), the Gazprom plant's deputy head declaimed: "Here, we extract 130 billion cubic metres of natural gas annually. This is more than the annual gas consumption of Germany. The gas is of high quality, and consists of 98.8 percent methane. The remnants of propane, butane and nitrogen are minimal."

In my era, our two favorite chemical products derived from natural gas were ammonia and methanol.* (To make a pound of the former we spent 58,300 BTUs—call that 4.7 pounds of coal—for the latter, 64,700 BTUs.) As you know, we loved our fertilizers and plastics! In 1990, with oil still in its glory and fracking's long horizontal penetrations not yet quotidian, such industrial applications consumed the largest single allocation of American natural gas. By 2012 that stuff's highest demand category was boiling water into steam, so that our electric power turbines could keep the lights on.—But humans' *concealed flame,* which Eiseley might have been overkind to call *sophisticated,* yearned greedily for new combustion-expressions, so how could we avoid pressing on with our industrial projects? In Seattle, for instance, a glorious effort to construct *the world's largest methanol plant* promised to *create 1,000 jobs during construction and 260 when up and running.* Had it "created" only one job (for its owner-proprietor), wouldn't that have been worth heating the atmosphere for? Actually, it would provide even more benefit than that—for its source commodity was, yes, American natural gas (I choke up with patriotic tears)—and its intended market was the earth's current number one greenhouse gas emitter. We managed our perceptions of both facts, having taught ourselves how to play carboniferous shell games: A selling point was that *the methanol would replace coal now used for methanol production in China*—and in the previous chapter we learned that coal was bad, bad, bad! Therefore natural gas must be better.—By the way, our Chinese customers would put the methanol to work more or less as we did—*as feedstock to create plastic.*

All the time we drilled, fracked, piped, bottled and shipped. Natural gas was as invisible as radiation; so were its climatic effects. In India, chemical engineers turned clouds of it into fertilizers. *The country was self-sufficient in natural gas until 2004, when it began to import liquefied natural gas (LNG) from Qatar.*

To us "consumers," natural gas's renown lay in its warming power. It kept us

---

* For a brief discussion of ammonia, see "About Agriculture," I:114–17, 126, 176. For methanol, see "About Coal," p. 18.

cozy in winter and powered our clothes dryers. In 2002 we burned 4,995 trillion BTUs' worth; call it 4,995 billion cubic feet. In 2014 we devoured over five times more: 26,819 billion cubic feet—22.16% of global supply. But don't call us prodigal: 487,286 homegrown natural gas wells (as of 2013) produced most of what we used.* Thirty percent of it went for electric power, 29% for industrial applications, 19% for residential, 13% for commercial . . .

The calorific efficiency of North American natural gas was higher than many other varieties. We pegged it at 24,381 BTUs per pound—about 18% more energy-rich than gasoline, with half the kick of rocket fuel.[†] I hope we were all grateful for its high heating value—and for its new abundance, for which we had fracking to thank! You see, we needed ever more power, since we'd dreamed up new work to be accomplished.

Consider the early-21st-century American kitchen. Our electricians certainly had, being perpetually on the case. *The kitchen,* they advised, *requires plenty of well-diffused light for timely preparation of food without accidents.* To be specific, "most kitchens" of the period were equipped with *general lighting from two 40-W fluorescent lamps or two 150-W to four 100-W incandescent lamps . . . The general illumination should be augmented by additional localized ceiling or bracket units.* Let us then imagine a middle-class kitchen with two 150-watt incandescent lamps and another 100 watts' worth of those "bracket units." Perhaps the cook was a working mother who used the kitchen for half an hour in the early morning to make coffee, sipping it under *general lighting* while she checked her latest text messages. Sally's soccer tournament was tomorrow, so she'd better gas up the car. On Saturday the neighbors were expecting barbeque; how full was the propane tank? The children's staggered breakfasts required another hour. In the evening she spent another hour and a half feeding her family and doing the dishes. So there went 1.2 kilowatt-hours (4,095.6 BTUs) each day for kitchen illumination alone.

A month of that routine would have consumed 122,868 BTUs, which a 33% efficient power plant would have spent 368,604 BTUs to produce. Call it 15 pounds of natural gas, or 360 cubic feet—all for one month of lighting one family's kitchen.

What about the remainder of their house and appliances, not to mention the millions of other families who likewise preferred *timely preparation of food without accidents?* Here came fracking to the rescue!—Hence this happy announcement, *circa* 2012: *Natural gas is by far the most important resource in Oklahoma, because it alone maintains a positive state energy balance.*

---

* Almost 96%.

[†] See the table of Calorific Efficiencies on I:216–17.

Gas stove of William T. Vollmann

Moreover, as the ideologues promised, it was good *clean* stuff, giving off far less nitrous oxide, carbon monoxide and sulfur dioxide than did either coal or oil. Germans credited it for some of their reductions in greenhouse gas emissions. Pulling out all the stops, one sermonizer called it *an incredible reservoir of domestic fuel whose extraction can result in cleaner air, reduced carbon emissions, energy security, massive job creation, and unexpected economic gains for business, job seekers, landowners, and governments.*

If that wasn't good enough, we'd make you a deal!—You see, we had to "grow," by increasing demand.—Hence this ad from CenterPoint Energy, accompanied by the depiction of a smiling blue collar man and woman: *When you choose clean-burning natural gas for your facility, our rebates offer you immediate cost savings for equipment purchases and installation.*

An organic chemistry textbook published in the halcyon decades before we believed in climate change* gushed as follows: *Combustion of methane is the principal reaction taking place during the burning of natural gas. It is hardly necessary to emphasize its importance in the areas where natural gas is available; the important product is not carbon dioxide or water but* heat.

Sad to say, one batch of Libyan natural gas contained a mere 66.8% methane, after which came ethane and propane; while a sample from Holland showed scarcely more pep at 81.3% (most of the rest being nitrogen); all the same, even in such unfortunate cases, methane dominated, so let us simplify once again, as I loved to do, and equate natural gas with methane, because what could be more innocuous than that substance? We collected it from underground, and some of it leaked from our pipes, and for that matter from our coal mines, rising invisible, odorless and unmeasured, hence harmless like radiation . . .

---

* But some scientists were already warning of the greenhouse effect. See I:87, 98.

About Methane

And to that end I have recited this fable, which may serve as ensample to warn the youthful reader from attempting the like worthless enterprise . . .

George Gascoigne, *The Adventures of Master F.J.*, 1573

As our most abundant hydrocarbon, methane offers an attractive source of raw material for organic chemicals, such as methanol.*—*Believe it or not, "attractive" was actually what some of us said, back when we were alive!—The Qatar Petroleum Producing Authority used to crow about its fertilizer plant, *with two Ammonia and Urea trains each day each producing 900 MT/day of ammonia and 1000 MT/day of urea. The feedstock for both trains is methane rich gas.*

The carbon black in the ink of this book may well derive from methane. The same goes for the probably carcinogenic formaldehyde in the particle board I bought for my cheap walls and cheap furniture; likewise polyesters—not to mention the chloromethanes from which sprang those wonderful climate-warming Freons in our refrigerators.*

And the news got better. Not only did that colorless gas[†] *(one of the simplest of all organic compounds)* make for an excellent feedstock, it was superlative in its high heating value.[‡] You have just read that its main product was *heat*—and as three industrial optimists explained: *Methane has a H:C ratio of 4, the highest of all hydrocarbons, and contains only 75% carbon by weight compared to some 85% for crude oil and more than 90% for coal. Considering the growing concerns about global warming, a reduction of CO_2 emissions by using natural gas is certainly a step in the right direction.*

Even the gloom-and-doom ecologist Barry Commoner admired methane, once upon a time. His utopia involved communal compost heaps enclosed and

---

* See "About Greenhouse Gases," I:174.

[†] On Earth, that is. Neptune looked blue to us and Uranus blue-green, thanks to methane, "a gas which absorbs red light."

[‡] Heat of combustion: 213 kilocalories per mole. So a chemist would put it. Or, as *Carbon Ideologies* would say, a high heating value of anywhere from 18,057 to 23,861 BTUs per pound. See the table of Calorific Efficiencies, I:216.

cooked by the sun until that wonderful gas came out.* And then? *Electricity needed to operate electrified railroads can be derived from cogenerator power stations driven by solar methane.*

And indeed, if you'd allow me to rank carbon fuels in order of their familiar combustion byproduct, the exhalation of roasted diamonds, wouldn't methane take the prize?

CARBON GAS EMISSIONS OF COMMON FUELS,
in multiples of methane's

All levels (except for gasoline and diesel) expressed in [kilograms of carbon (gas) per calorie of energy generated] in combustion.

1

Methane [<59 kg-carbon per cal.]. This substance is nearly equivalent to natural gas.

1.36

"Refined" petroleum "fuel" [80].
    1 gallon American gasoline: 8.87 kg CO_2
    1 gallon American diesel: 10.18 kg CO_2
        But since diesel-powered vehicles were often more fuel efficient, the U.S. EPA believed that diesel's emissions per mile were comparable to gasoline's.

1.78

Bituminous coal [105].

0

Nuclear fuel.

Sources: The Methanol Economy (2009), U.S. Environmental Protection Agency (2014), with calculations by WTV.

-----

* He also wanted us to have breeder reactors.

And so methane seemed not just relatively but *absolutely* tip-top! When England started making less electricity from coal and oil, and more from natural gas, her greenhouse gas emissions fell significantly. As President Obama enthused to Congress, back in the sweet years when our coral reefs had barely begun to die: *The development of natural gas will create jobs and power trucks and factories that are cleaner and cheaper, proving that we don't have to choose between our environment and our economy.* (You see, we liked to have it all, back when we were alive.)

Unfortunately, as you know, *unburned* methane was a worse greenhouse gas than carbon dioxide. According to the EPA, it exerted *a warming effect on the planet more than 20 times greater* than CO_2. And so that agency suggested new regulations that might, if the *regulated community* were honestly compliant in its usual fashion, bring down the methane emissions of natural gas and oil operations *by 40 to 45% from 2012 benchmarks.*—Meanwhile, down the corridor, our Interior Department authorized Royal Dutch Shell to drill in the Arctic Ocean for natural gas and oil . . . and before he'd gasified the White House for even two weeks, President Trump set about undoing the methane rule.

Our three industrial optimists were more pessimistic than the EPA—but only slightly. They pegged methane's global warming potential at 23 times higher than carbon dioxide's, considered over the course of a century. They wrote: *Methane accounted for about 18% of the cumulative greenhouse effect from 1750 to 2005.* But was swamp gas our fault? You may remember that it came out of septic ponds,* and we could hardly stop pooping! And would you demand that we forgo rice paddies and cattle farms?†

---

* German sewage got the water treatment, and so the bacteria behaved comparably to the ones in flooded Asian ricefields: "Sludge stabilisation is carried out in order to prevent uncontrolled putrefaction. In facilities for fewer than 10,000 inhabitants, [this is accomplished] . . . aerobically, with energy consumption, while in facilities for more than 30,000 inhabitants it normally is carried out anaerobically, with production of methane gas." That comprised a mere 0.01% of Germany's 2007 contribution to global warming (measured at a GWP of 21). The Germans were clever enough to combust the methane for power, or when that wasn't practical, to flare off all they could.

† Although ricefield emissions in their country actually increased, and enteric fermentation from cattle decreased only slightly, from 1990 to 2013 the Japanese still succeeded in reducing their methane emissions by 25.8%, "mainly [as] a result of a 54.8% decrease in emissions from . . . solid waste disposal."

PRIMARY SOURCES OF METHANE IN MEXICO, 2002

"Fugitive emissions [from] oil and natural gas": **26%**
"Enteric fermentation" from cattle: **26%**. This was 84% of all recorded agricultural emissions.
"Waste disposal in sanitary landfills": **24%**
"Wastewater management and treatment": **20%**

These added up to 96% of total methane emissions, which were 6,803 Gg [14,997,893,800 lbs]. They warmed our planet without doing any useful work.

Sources: Mexican greenhouse report, 2002.

Every 400 pounds of kitchen waste gave rise to a pound of methane (and 1/30 pound of nitrous oxide). Again, this was not so very much. We kept nickeling and diming our planet. (In 1755, methane levels were 723 parts per billion. In 2011 they had reached 1,803; like our entrepreneurs, they were upwardly mobile.) And how the stuff did creep out of fossil fuels!

QUANTITY OF FUEL RESOURCE WHICH WILL RELEASE 1 POUND OF METHANE,

ca. *2007, in multiples of the value for hard coal*

Expressed in [pounds of substance required to release 1 lb of methane].
All headers rounded to nearest whole digit.

1

Outgassing of "marketable production" hard coal, when extracted [1 lb methane emitted per 385 lbs hard coal].

65

Incomplete combustion of aviation fuel (jet kerosene) when aircraft are cruising, taxiing, etc. [per 24,943 lbs].

90

Fugitive emissions from extracted oil [34,571].

217

Subsequent outgassing of same "marketable production" hard coal as above, stored [83,448].

1,143

Emissions from crude lignite in open pit mines [440,000].

Source: Greenhouse Gas Inventory Germany, 1990–2007, with calculations by WTV.

These added up, no doubt. And except for the aviation fuel, none of the mentioned substances had even yet been burned!—Then there were our gas pipes.

Germany's natural gas pipeline ran 218,876 miles. In 2007 it had leaked 174,131 tons of methane. The Umweltbundesamt crowed: *Emissions caused by gas distribution have decreased by some 17%* since 1990.

As the table implies, quite a lot did seep from coal mines (recall that it was a methane explosion which killed the 29 at Upper Big Branch[*]). In 2012 *fugitive emissions from solid fuels* made up a mere 0.2% of the European Union's[†] contribution to global warming. However, 86% of that 0.2% consisted of CH_4 *emissions from coal mining.*

Well, so methane's global warming potential was 23. The CEO of Breitling Energy appeared satisfied with that factor. And presently he called our attention to the fact that India's 4,500 reservoirs *emit an amount of methane that is equivalent to 850 million tons of carbon dioxide per year* due to decaying microorganisms in them. His logic entertained me; it was like justifying a murder on the grounds that people had to die anyhow.

The Foundation for Deep Ecology asserted that methane's GWP was 25, not 23 times worse than CO_2's. But what could we expect from those Communist freaks?

The New York Times proposed that methane might be *by some estimates* 80 times more potent than carbon dioxide *in the first 20 years in the atmosphere.*— Well, wasn't it more relaxing not to read the newspaper?

---

[*] See p. 50.

[†] Or, more accurately, the so-called "EU-15" countries which had committed to greenhouse gas reductions.

Presently we were advised that *a more accurate figure . . . is between 86 and 105 times the potency of CO₂.*

Well, so how were we supposed to think about methane? Which time frame was helpfulest, and which GWP most accurate?

I was not the only methane numbers runner. Please try to reconcile the following two aphorisms:

> "Methane in the atmosphere was almost flat from about 2000 through 2006. Beginning 2007, it started upward, but in the last two years, it spiked," said Rob Jackson, an earth scientist at Stanford University . . . "Methane is starting to approach the most greenhouse gas-intensive scenarios . . ."
>
> *The Washington Post*, 2016
>
> Methane emissions in the United States decreased by 16 percent between 1990 and 2015. *(See another claim on the next page.)*
>
> U.S. Environmental Protection Agency, 2017

No wonder my friend Dave didn't believe in global warming! Trained in physics, supporting Donald Trump, he assured me that "all the climate change data sets are corrupted."

I decided on the 20-year GWP. The game might already have been over when I was writing *Carbon Ideologies;* another 20 years of selfishness would surely melt more of the Antarctic ice cap . . .

We used to imagine that in 2012, the United States released only 6,525.6 teragrams of carbon dioxide equivalents. By the more scaremongering ideological calculation the figure was now 8,281.5 teragrams—27% higher. At the old global warming potential of 21 times CO_2's, methane's share had been a mere 8.69%. Now it was 28%.* The joke was on us.

Or was it? Eighty-six percent over 20 years would magically revert to 26% over 100 years, so had we actually let loose only 6,525.6 teragrams "in the long term"? How could I add this up without expert help?

And so I wrote again to Dr. Pieter Tans at the National Oceanic and Atmospheric Administration, inquiring about the perils of methane. He assured me that in comparison to coal-burning utility plants, methane-burning establish-

---

* By the 20-year computation, methane from agriculture alone now made a dispiritingly high 10% of all our greenhouse emissions.

ments kept *14.7 mol[es]* of CO2†... out of the atmosphere per kWh of electric power. How much CH4 [methane] would have to leak from the well, during processing, transport, etc. to match the G[lobal] W[arming] P[otential] saved by not emitting 14.7 mol of CO2?... A fraction "F" of methane is leaked to the atmosphere for every mol burned... [If] F = 14%... it would nullify the global warming benefit of switching from coal to natural gas.*

How high precisely was that fraction "F"? In the immortal words of Stanley Sturgill: "Buddy, you got me."

For his part Dr. Tans insisted that carbon dioxide was considerably more threatening than methane. You have already encountered his grim calculation that *over 2000 years, the cumulative amount of heat retained in the Earth system is about 8 times larger than if we count only the first 100 years.* He explained that in his *accounting[,] GWP/mol of methane is... [1.25] or not much larger than that of CO2. Indeed, CH4 oxidizes in the atmosphere to CO2; 1990 of those 2000 years it will be CO2.*

In other words, methane's high 20-year GWP must plummet due to that greenhouse gas's brief life in the air—after which it became carbon dioxide. Meanwhile, carbon dioxide's steady warming continued, century after century.

By the way, what *was* the lifetime of methane? The Intergovernmental Panel on Climate Change pegged it at 12 years. A certain sulfate-watcher in Wyoming countered: *The atmospheric lifetime of methane has increased 25–30% during the past 150 years to a current value of 7.9 years, implying gradually decreasing oxidizing capacity* of our planet, thanks to sulfur dioxide pollution. Twelve years or 7.9, how could I figure it out? I was getting hungry, so I threw some meat and vegetables into my frying pan and turned on the gas. It burned steadily and nearly odorlessly, with a sweet blue flame. And all that my continued prosperity required was for Dr. Tans's fraction "F" to remain below 14%—a near certainty, because I *wanted* it to be so!

Although between 1990 and 2012, methane releases in the European Union decreased by one-third,‡ American emissions (claimed other ideologues) had *risen* by one-third, *accounting for 30 to 60 percent of an enormous spike in methane in the entire planet's atmosphere.* Where did those originate? According to a

---

* A mole (6.023×10^{23} atoms or molecules of any substance) corresponds to its atomic mass in grams. Since chemical reactions take place according to molar ratios, the mole is a fundamental unit in chemistry—hence in climatology. For more discussion, see the Definitions, Units and Conversions section (I:596).

† No subscripts in original. For a more alarmist "F" number, see pp. 315–16.

‡ The two main reasons: coal mine closures and "reductions in managed waste disposal."

geophysical research letter from 2016: *Major contributions in the U.S. EPA inventory and their interannual ranges are 30–32% from oil and gas,* with which fracking surely had something to do, *31–34% from livestock* (ruminant flatulence), *21–22% from waste* (decay of garbage) *and 10–13% from coal.*

Well, were we Americans throwing out one-third more garbage and raising one-third more belching, farting cattle than before? Nobody ever told me so.* So once again I wondered why methane emissions had increased by one-third. According to that geophysical research letter: *The U.S. has seen a 20% increase in oil and gas production . . . and a ninefold increase in shale gas production from 2002 to 2014 . . . , but the spatial pattern of the methane increase seen by GOSAT does not clearly point to these sources. More work is needed to attribute the observed increase to specific sources.*

More work . . . more time . . . why not?—That benefitted the economy—and we might have generations yet to spare before we got cooked! Methane was up 160% from 1750—so why not let it increase some more while we figured things out? (From the Pope's encyclical of 2015: *We know that technology based on the use of highly polluting fossil fuels—especially coal, but also oil, and, to a lesser degree gas—needs to be progressively replaced without delay.*) And so, in a can-do scientific spirit, we hunted; we made our attributions.

Sometimes we sampled methane emissions from above, grandly swooping from carbon-burning overflights.[†] Meanwhile our on-the-ground experts employed a backpack-sized gizmo called the Bacharach Hi Flow Sampler, whose readings came out reassuringly low.—I was glad to hear that; frankly, I'd been fishing for good news.—Presently we discovered a secret trick of the Sampler, an intermittent failure to auto-toggle from its low-level to high-level sensor. Result: *The missed emissions could be extremely high—perhaps tenfold to a hundredfold for a particularly large leak.* Thank God there'd never be one of those!

In October 2015, seven to nine thousand feet beneath the San Fernando Valley suburb called Porter Ranch, a storage reservoir for natural gas within one of J. Paul Getty's hollowed-out oil deposits grew incontinent. Aerial methane concentrations rose from two to 50 parts per million. Some residents of Porter

---

* On our ghastly worldwide stage, to be sure, the tragedy might be performed differently. Rob Jackson, the Stanford scientist just quoted, asserted that since *two thirds of the world's methane releases come from the tropics, not the temperate latitudes,* "we think agriculture is the number one contributor to the increase." To be sure, "there's been a secondary increase from fossil fuel use, partly because there continues to be more fossil fuels extracted."

† An activist in Denver told me that only these procedures could correctly measure methane outgassings: "The air readings are inherently fraught, because some of the wells are super-emitters and some are not; maybe there's a ton of diesel being burned and stuff that's offgassed; the condensate tanks might be doing other things; the industry really loves doing cherry-picked sites. The more accurate readings come from flyovers."

Ranch got headaches and bloody noses while the rest encountered, as the Japanese government would have put it, *no immediate danger.* This "accident," or "event" (which continued for four months), deserved no public consideration, for we had to respect the tender feelings of the *regulated community*—one upstanding member of which, the Southern California Gas Company, declared in the "Summary" section of its incident report to the Public Utilities Commission: *No ignition, no injury. No media.* As a member of that media I must therefore apologize for writing anything at all about the Southern California Gas Company. And to our great good fortune, only 107,000 tons of methane escaped (quote, *the largest such leak in terms of climate impact in America's history,* end quote)—really the merest little fart!* Natural gas remained America's new best friend.

Fortunately, the *regulated community's* habitual exemplary stewardship must be credited for the good news I got while I was still alive:

> A recent assessment of the potential for a future abrupt release of methane . . . concluded that it was *very unlikely* that such a catastrophic release would occur this century. However, . . . anthropogenic warming will *very likely* lead to enhanced methane emissions . . . On multi-millennial time scales, the positive feedback to anthropogenic warming of such methane emissions is potentially larger . . .

(Once upon a time in 1980, a fossil fuels professional gleefully announced: *Reports of the discovery in Siberia of natural gas in a frozen state—called gas hydrate—have aroused wide interest and have led to the conclusion that hydrates may exist over large regions elsewhere in the world.* He thought there might be 500 trillion feet of them in the permafrost alone, although *the means of exploiting them are not clear.* Reader, I suppose that in your time they must be bubbling up out of the Arctic swamps that used to be tundra, mercilessly accelerating global warming.)

And now I must inform you that in 2016 an American (or as I should say un-American) environmentalist cruelly insulted the fracking industry. I hardly dare repeat his blasphemies:

> If even a small percentage of the methane leaked—maybe as little as 3 percent—then fracked gas would do *more* climate damage than coal . . . somewhere between 3.6 and 7.9 percent of methane gas from

---

* At the 86× 20-year GWP, I calculate it as equivalent to 16% of the nation's entire carbon dioxide emissions for 2012 (the latest year for which I could find appropriate figures). Over 100 years, its carbon-equivalence fell to less than 4%; the future was ours!

shale-drilling operations actually escapes into the atmosphere . . . In January 2013 . . . aerial overflights of fracking basins in Utah found leak rates as high as 9 percent.

Very well—we'd seal our leaky gas pipes! (By 2014 there were already more than 1.2 million U.S. miles of them.) A Harvard research team triumphed over that difficulty as follows: *It sounds like it ought to be simple to make a cement seal, but the phrase we've finally fixed on is "an unresolved engineering challenge."*

Not resolving it, we presently died and rotted. I cannot speak for others, but thus ran my final consolation: *The methane generated in the final decay of a once-living organism may well be the very substance from which—in the final analysis—the organism was derived.*

About Fracking

We will open the gates of our city to those who deserve to enter, a city of smokestacks, pipe lines, orchards, markets and inviolate homes.

Ayn Rand, *Atlas Shrugged,* 1957

nominate *Thanksgiving Day, November 24, 2005, as World Oil Peak Day,* predicted a former Shell Oil Company geologist. Americans would soon be shivering for lack of hydrocarbons!—He was wrong because of fracking. Americans, with a little help from their Chinese and other friends, could go on generously warming the world! ("Well, every oil company does fracking," remarked the retired Chief Executive Officer of Conoco.)

And by the way, the geophysical letter quoted in the previous section *found maximum emissions in the South Central U.S., a region with large sources from livestock and oil and gas production.* This zone coincidentally included Weld County, Colorado, and western Oklahoma, both of which achieved true fame for fracking.

From Oklahoma's point of view (at least according to the thinking in which we humans specialized), fracking was a life-saver. In 2011, 16.3% of the state's residents had endured lives below poverty level, but the *Oklahoma Almanac* summarized a happy new chapter: *The long-term declines in Oklahoma oil and natural gas production have been reversed in the last several years due to the impact of horizontal drilling and completion technology.* To be precise, *Oklahoma's active rotary rigs have increased in 2011 to 175, up from 130 in early 2010.*

Nor was Oklahoma the only winner. In North Dakota *per capita real GDP grew by 11% in 2011–12 and unemployment dropped to 3% (lowest in the US).* In Pennsylvania *per capita income rose by 19% in counties having more than 200 wells and confirmed observations that the more wells in a county, the better improvement . . . on the financials.* Oh, those "financials"! Did I mention that this paragraph's figures derive from my favorite reference source, the *Investopedia*?

The bloom remained on the rose in 2015, when I went to see fracking in Colorado, and U.S. Energy Information Administration announced that it *expects U.S. shale gas production to continue rising through 2040.*

But let us uplift our eyes from merely American concerns. According to the *Investopedia,* the *Countries With The Highest Fracking Potential* were, in diminishing order of *technically recoverable shale oil,* Russia, the U.S., China, Argentina, Libya, Venezuela, Mexico, Pakistan and Canada. Considered in light of *technically recoverable shale gas,* the group changed its ordering, and a few dropped out, while in came Algeria, Australia, South Africa and Brazil. (I regretfully report that several of these slowpokes remained in *initiation mode* or else suffered from *geopolitical effects and unclear policies.*) *Like it or not,* the *Investopedia* concluded, *fracking is practiced almost all over the world.*

Now, why might that have been? Although she reassured us in a television commercial that she would *stand firm with New Yorkers opposing fracking, giving communities the right to say "no,"* in 2010 Donald Trump's "progressive" opponent Hillary Clinton had already informed the right people that *Poland will be part of the Global Shale Gas Initiative,* for which the U.S. State Department would goodnaturedly *provide technical and other assistance.* That was the year when *delegates from 17 countries descended on Washington for the State Department's first shale gas conference. The media was barred from attending, and officials refused to reveal basic information, including which countries took part.* Meanwhile *Clinton . . . sent a cable to US diplomats, asking them to collect information on the potential for fracking in their host countries.* In 2012, the U.S. Ambassador to Romania, a certain Mark Gitenstein, explained that "the Romanians were just sitting on the leases, and Chevron was upset. So I intervened." And so *Romania signed a 30-year deal with Chevron, which helped set off massive, nationwide protests.* (If you notice a certain anti-fracking tone here, consider the sources: the leftwing eco-rag *Mother Jones* and that subversive publication *Intercept.* They based their preposterous allegations on the Freedom of Information Act and WikiLeaks.)

Let the *Investopedia* resume the tale: *Amid continued local opposition* (who gave a rat's ass about that?), *a 2012 amendment of rules allows foreign companies to invest in Algeria's shale gas sector.* This measure must have been both homegrown and spontaneous—for the *Investopedia* mentions nothing about Hillary Clinton or the State Department! There were now *agreements with multinational corporations . . . like Shell, ENI and Talisman, to sell the gas into European markets. This has resulted in increased employment opportunities, an increase in investments and better energy security for Algeria, apart from the positive impact on its economy.*

For that matter, *irrespective of all environmental concerns and geopolitical challenges, the fracking business does have the potential to bring economic respite to Argentina, a country struggling with high inflation and financial stress.*

Etcetera.

Now that you understand how wonderful fracking was, let me try to define it.

At this point, it may be necessary to "open up" the formation to increase the flow, explains a petroleum textbook from 1978. With low-permeability carbonate rocks, *the use of acid will open paths in the rock for the oil.* In the case of sand-stones, *a "frac" job* may be necessary. This involves forcing a sand and fluid mix into the formation to open cracks . . . so the hydrocarbons can flow through.* (Meanwhile a subterranean detonation or two might prove necessary, which benefitted the explosives industry.) In 2012 a natural gas man explained the matter as follows:

> A perforation gun on a 20,000-foot electric line . . . is fired to punch holes through the casing and into surrounding rock. The shaped perfo-ration charges are set off in stages. The small bullet-hole openings . . . will function like holes in a lawn sprinkler as a conduit for "frack fluid" . . .

As for extra accidental explosions,[†] I heard allegations of them, but never saw evidence of one. (I was running out of money, and my publisher still hoped for a short book.) Did these "incidents" ever occur? Why wouldn't they? Wasn't natu-ral gas's virtue inflammability?

But on the subject of disequilibrium, rumor-mongers did begin to claim that in Ohio and in Oklahoma, fracking was associated with earthquakes. (*Time* magazine, 2016: **In 2007, Oklahoma had one earthquake. Last year there were more than 900. What happened? Greed, politics and the biggest oil boom in decades.**)[†]

On the credit side, "frac sand," which had allegedly been *implicated in silico-sis,* climbed happily in monetary value. By 2014, by which time Americans were improving their homeland with 50 new fracks per day, a Minnesota organization called Citizens Against Silica Mining saw fit to claim: "The demand for sand is going up and up. It's a gold rush." A single well might consume 10,000 tons of sand. And so tiny crystals of silica spiced up the air, not to mention our lungs.

But why should silicosis be relevant to you who surely face less trivial

---

* By the time I began studying this subject, "frac" had attached a "k" to its end, like some long-chain hydrocarbon getting longer, although the traditional spelling survived here and there.

† For instance, "an Ohio house exploded after a fracked gas well leaked large volumes of methane into the home's water supply," but I failed to verify this. An abandoned natural gas well allegedly exploded in Bradford, McKean County, Pennsylvania; for all I know, that never happened, either.

† The U.S. Geological Survey put us straight: "Fact 1: Fracking is NOT causing most of the induced earthquakes. Wastewater disposal is the primary cause of the recent increase in earthquakes in the central United States . . . Fact 3: Wastewater is produced at all oil wells, not just hydraulic fracturing sites." But up in Canada, researchers at the University of Calgary concluded that their own quakes "were induced in two ways: by increases in pressure as the fracking occurred, and, for a time after the process was completed, by pressure changes brought on by the lingering presence of fracking fluid."

problems? I was summarizing the fracking process—although, come to think of it, the CEO of Breitling Energy can do so in far superior prose:

> Prior to a frack stage, the near-wellbore area is cleaned up [I positively adore that euphemism!] with an acidic solution, typically to unplug rock pores plugged by drilling mud or casing cement. Then a water-based frack fluid is mixed with a specially designed agent to reduce friction, and thus expedite the flow and placement of the proppant. [Proppant consists of sand or ceramic grit that will "prop" open the frack job.] Then large pumper trucks pump a huge volume of water into the wellbore, along with the finely grained proppant . . . [How much water precisely? Let me quote another source: *Each Marcellus well will require from three to five million gallons of water during the fracturing process . . . In the Susquehanna river basin . . . each well uses about 3.7 million gallons.*] The volumes of proppant are gradually increased while the overall volume of the fluid slurry gradually decreases. Subsequent stages typically contain a coarser proppant. After the final stage is completed, the well is flushed with enough fresh water to remove excess proppant.

As it was meant to do, fracking increased the porosity of what had lain out of reach below us. Not only did oil and natural gas begin their intermingled migrations, but so did less desirable substances.

In 2014, a million gallons of "salt water" from fracking leaked into Lake Sakakawea, North Dakota, which supplied drinking water for the Fort Berthold Reservation. It was only five to eight times saltier than seawater, potentially radioactive,* and could sting the tongue, so why worry? A hydrogeologist

---

* In May of 2013, a company with the bland name of Range Resources attempted to get rid of "two containers of flowback sludge" from its natural gas operation in the shale of Washington County, Pennsylvania. "When disturbed by human activity," explained the newspaper, certain "naturally occurring radioactive materials," to which some spin doctor had awarded the comforting name of NORMS, sometimes turn into *technologically enhanced* NORMS—oh, delicious enhancements!—or in other words, TENORMS. In case the two euphemisms deflected you, the technological enhancements consisted of increased radioactivity. The flowback sludge measured 2.12 micros per hour—the highest reading that my pancake frisker measured in Naraha. Call it 18.57 millisieverts a year—near the Japanese ceiling of "acceptability" in "remediation situations," and, according to the International Commission on Radiological Protection, far into the zone of outright unacceptability. The "level at which Pennsylvania recommends normal landfills not accept drilling waste" was 1.5 micros an hour, which corresponded to Mary Cook's headstone in that cemetery up on profit-flattened Cook Mountain, West Virginia. That was why when Range Resources sought in all good hope to dispose of their TENORMS at the Arden Landfill in Washington County, their gift was rejected on account of "higher levels of radioactivity than Pennsylvania deems acceptable for normal landfill disposal." Fortunately, West Virginia lay ready and welcoming—with opened legs! In the words of Pope Francis: "The violence present in our hearts, wounded by sin, is also reflected in the symptoms of sickness evident in the soil, in the water, in the air and in all forms of life. This is why the earth herself, burdened

emeritus at the U.S. Geological Survey remarked: "You don't want to be drinking this stuff."—Well, he was retired, so what did he know?

Methane ruined the wells of 19 families in Susquehanna County, Pennsylvania. But let's be fair: *Though often erroneously linked by critics to fracking, state officials said the cause was due to improper well casing. There has never been a case of contamination in the state from the actual fracking process.* That surely consoled the 19 families.

In West Virginia, the heartwarming symbiosis of the regulators with the *regulated community,* which we have admired and applauded in my coal chapter, adapted to the nascent opportunities of fracking. Once upon a time in Fayette County, A.D. 2014, an entity called Danny Web (or Webb) Construction happened to conduct business without a permit. (I'm sure that was the merest slip; what form of subversive imbecile would I be, to doubt the *regulated community?*) A year later, Danny Web's injection wells were leaking. And in a minor reprise of the MCHM spill in Charleston, the newspaper reported (although only in an editorial) that *frack chemicals have been found in a nearby tributary upstream from the Oak Hill, Fayetteville and Lochgelly water intake . . . One well goes through a Sewell coal mine which connects to other mines*—which had already improved the porosity of Fayette County so splendidly. *Leaks can be forced up into that mine and, from there, travel through the rivers and aquifers.* At a permit hearing, up stood a certain Mary Rahall, who *to a near-standing ovation* claimed that the injection wells' wastewater had revealed radioactivities of more than 3,000 picocuries* per liter, when 60 picocuries was *the limit for industrial waste, as reported by both the U.S. Geological Survey and the permit application itself.* (Since Americans thought in quarts and gallons rather than liters, these units struck me as obfuscation.)[†] As Ms. Rahall informed me:

> Tom Rhule from Charleston made me aware of the 3000 pico curies per liter number, because he found that info in one of his W V D[epartment] of E[nvironmental] P[rotection] F[reedom] O[f] I[nformation] A[ct] requests . . . [H]e said it was buried in a document hundreds of pages

---

and laid waste, is among the most abandoned and maltreated of our poor . . ." The name of the West Virginian facility was Bridgeport. "Waste Management, which operates Bridgeport, never tested the material's radioactivity levels. The facility doesn't even have the equipment to do so. West Virginia doesn't yet require radioactivity monitoring of natural gas drilling waste . . ." Thus Range Resources maintained its competitive edge.

* For a definition, see I:545.

[†] To know how dangerous this measurement was one would need to be informed (as one was not) which isotope was involved. In the case of (for instance) cesium-134 and -137, the Japanese legal maximum was 10 becquerels per kilogram. Three thousand picocuries per liter would exceed this at 14.64 Bq/kg.

long... [it] was listed on one of the forms, as part of the renewal process of Danny Web's ... permit ...

Then there was the matter of 18 feet of sludge in the wastewater pits.* Mary Rahall remarked that *we are blocked from knowing what's in the waste by Senate Bill 423,* whose protections and benefits you have already learned (see page 178). *And, further*—reader, this is my favorite touch—*the Environmental Quality Board, at the request of the DEP, prohibited USGS scientists from testing the sludge or wastewater* in those pits.

Commenting on this passage, Ms. Rahall remarked that DEP helpfully arranged matters so that the USGS crew

> were only allowed to test the stream down stream from the sediment pits as far as I know. I know it took an act of God to get their study results published to the public and you can now view their results which show an extremely high level of Endocrine Disrupting Chemicals ... as a result of the water being contaminated with hydraulic frack waste.

The permit hearing grew all the more perfect when Danny Web Construction, *scheduled to speak last, was nowhere to be found.* Therefore, let me give Mary Rahall the last word instead, as extracted from her "final statement" for *Carbon Ideologies:*

> Before 2012, I knew nothing about hydraulic fracturing. It wasn't until I contacted the C[enters for] D[isease] C[ontrol] in Atlanta ... to review some water test results ... that it was brought to my attention just how dangerous frack waste can be. As a concerned citizen and parent, I did what was necessary to find the TRUTH about what the drinking water supply in our areas was contaminated with. Regardless of the harassment and obstacles that were put in front of me, I stood strong ... I am extremely grateful to the cancer researchers and scientists, including the USGS team, that took my concerns seriously ... Some state and federal politicians are financially invested in the hydraulic fracking industry and do everything possible to protect it. Putting profit over public health is wrong, and I think it is a real shame that some state and federal politicians enact laws (e.g. WV Senate Bill 423, which prevents the public from knowing the long list of chemicals used in fracking) that

---

* Mary Rahall commented: "I'm not sure exactly how deep the sludge pits were, I don't recall 18'— probably a guesstimate from somebody? ... [They] were closed during the summer of 2014 or 2015."

protect the fracking industry, regardless of the long term health effects on innocent communities . . .*

Thus the West Virginian situation.—Meanwhile, in Ohio, the Center for Health, Environment and Justice labeled that state's Natural Resources Department *a captured agency. It's an agency that's* [sic] *very existence relies on the industry it regulates.* The center had the gall to complain about *D & L Energy, whose deep injection well was the epicenter of more than a dozen earthquakes in the Youngstown area,* and whose owner, in his simultaneous capacity of big cheese at Hardrock Excavating LLC, had been *accused of violating the Clean Water Act by illegally dumping oil and gas waste into a storm drain.* But consider the source: Any organization devoted to health, environment or justice must be suspected of backbiting the *regulated community.*

Of the four kinds of fuel considered in this book, natural gas might have been the least unpopular. As you have now read over and over, it was "clean carbon," as long as it didn't leak. Although it could not match coal for heritage-ideology, its wells and pipelines racked up fewer demerits for unsightliness than mountaintop removal mines. For some reason, its explosive mishaps took place more rarely than oil's—and they were certainly less catastrophic than nuclear's.

But of the various methods of resource extraction, fracking raised the most alarms. Uranium and coal miners harmed mostly themselves (never mind the aftermath of their operations). After offshore oil rigs blew up, the ocean rapidly resumed its virginal appearance, however compromised might be the plants and animals below. But for fracking, well, the most convenience-minded fantasists could hardly imagine that its pollution of aquifers would be anything but permanent. If natural gas had only come to my stove without fracking and without leaks, what a happy carbon ideologue I would have been!

A certain Dr. Cyla Allison, who was President of an organization called the Eight Rivers Council, made an unpleasant case: "Fracking is science without conscience: short sighted, bloated on [sic] greed. Fracking steals irreplaceable water and substitutes poison."—This was hardly fair. Didn't coal's effects on drinking water deserve their own hymns of praise? You may remember Maria Lambert

---

* Wondering what *effect* she had achieved for all her effort and risk, I presently obtained the following response: "Since the permit hearing, more people are aware of the scientific evidence that shows what the drinking water supply is contaminated with, and more people have been advocating in favor of the ban on the disposal of hydraulic frack waste in Fayette County." In other words, there was no ban and maybe never would be.

from the poisoned West Virginia town of Prenter.* Let me quote another reminiscence of hers:

> Probably forty-six out of forty-eight people don't have a gallbladder. I knew a lady from across the mountain ... who was raised in Prenter ... She's on well water ... Their son had to have his gallbladder removed a week or so after his mother did—he's about thirty-five. They squeezed his gallbladder when they took it out, and it was nothing but black stuff come out of it, tarry-looking stuff come out of it.

Fracking couldn't be *that* bad! ...—although I nearly forgot to infect you with a harmful rumor (courtesy of the environmental group Food & Water Watch) that *fracking was exempted from provisions of the Safe Drinking Water Act*. But please do consider the possibility that the Safe Drinking Water Act, like standards of radiation safety, had never been worth following.

Given the theme of *Carbon Ideologies,* I feel obligated to insert the following, again from Food & Water Watch: *Fracking has as much as 105 times the global warming potential as carbon dioxide by weight over the first 20 years of its emission and as much as 33 times the global warming potential over 100 years.* True or not? These figures might simply be high estimates of the 86× and 21× GWPs for methane which I use. Or they might reflect pipeline leaks ...

In this chapter you will read more negative than positive judgments of fracking. This was not my intention; indeed, like any zealous corporate undersecretary, I "reached out" to men and women on both sides of the question. Only one soul even remotely connected to this industry gave me the time of day: the environmental consultant Mr. John Mahoney. I am grateful to him. With the exception of a Community Development Director, who cheerfully made the best of it, everyone else I interviewed condemned fracking.† As for the ones who declined to go on record, if my judgment of fracking is skewed, who are they to carp?

---

* See above, p. 158.

† As you may have noticed, the *regulated community* generally kept clear of my research for *Carbon Ideologies.* Environmentalists were more helpful—although my research assistant, expressing surprise and discouragement, once reported: "Velina Pendolovska of European Commission DG Climate Action (velina.pendolovska@ec.europa.eu); sent 11-14-16, no reply," and: "Sarah Taylor at Cal[ifornia] E[nvironmental] P[rotection] A[gency] (sarah.taylor@calepa.ca.gov); sent 11-14-16, no reply."

4

Among the Most Vigilant Protectors of the Environment

June–July 2015

We now know that we live on top of a supply of hydrocarbons that by some estimates could sustain us for centuries, and in exploiting it we are among the most vigilant protectors of the environment.*

Chris Faulkner, CEO, Breitling Energy Corporation,
The Fracking Truth, 2014

---

* One other group contested the preeminence of frackers as environmental stewards. Fighting the good fight against Pope Francis's encyclical on climate change, the President of the West Virginia Coal Association assured us: "I can say unequivocally that our coal miners are the greatest practicing environmentalists in the world."

"WALK INTO THE LIGHT"

God forbid that you say anything," said the activist Sharon Carlisle, who lived in Loveland. "It's nasty here. Really nasty here. Three blocks from here there's a trainloading area, and last summer all night every night they were bringing those 18-wheelers, loading the oil and gas, loading the bullet cars, crossing over to the north-south tracks. Then the Saudis lowered their oil prices, *thank you, thank you!* Last year, 70 or 80 of those missile cars left Loveland; now it's about 20. If one of them should ever catch fire, we're at the blast zone."

"Do you ever feel at personal risk?"

"I tell people that I am not that effective, that I am not that big a threat. I know that my phone calls are being tapped and my e-mails are being read . . ."

She must have felt anxious. I remember that in her critique of fracking she made certain specific accusations. Upon reviewing my draft of this interview she wrote (which reminded me of Cathy Behr): *If you print the previous statement, stating corporations and individuals by name, I may get sued.* I deleted that part.*

"The way I came about it," she began, "was that I was doing my research about nuclear in this country, for an art installation I am developing. And I started noticing parallels between the way the Cold War was sold to us by Eisenhower; people weren't afraid enough, so they did a really big P.R. push to make people more frightened of the USSR and her atomic bombs, in order to justify, essentially, the military-industrial complex. I started looking at the way they were selling it, and in the meantime I was thinking about fracking.† I don't recall hearing so much about an energy source before! Maybe a little about G.E., or the

---

* Upon receiving the interview draft she dropped in extensive written comments. This explains why the remainder of her testimony reads a little oddly, as a mix of her spoken and written voices. I have let the new combination stand, with abridgments.

† In case you are wondering what drove her to compare fracking to nuclear, let me remark that her father won the Bronze Star and two Purple Hearts and then couldn't get a top-drawer job, so he worked in a uranium processing plant in Apollo, Pennsylvania. By some coincidence he died of cancer. His place of employment was called Numec, which Ms. Carlisle described as "a big black industrial hole looking thing." The Numec employees "went and buried waste in what was once open space and is now housing developments and schools; and just recently they had to bring Homeland Security in because they had not mitigated the site (they found more uranium in abandoned coal mines). It went from they're gonna make it a park to you can't come in here."

big oil companies' products—gasoline—but not like I was finding. As I researched, finding out more and more about Halliburton, about Cheney's part in it all—Cheney had been Halliburton's CEO, and his wife was still involved in the corporation while he was in office—hearing about how all this stuff has been shoved down our throats, was and is still being shoved down our *lungs* . . . ! Halliburton refused to reveal just what chemicals were being shoved down in the wells, claiming it to be a 'proprietary recipe.' Yet a list of what goes in and what comes up and out, it is more than alarming, it is terrifying. You can just go in there and use anything, supposedly.

"My friend Judy and I found out they were going to be putting wells here in Loveland, and we ran to a lawyer to write ballot language to get a moratorium passed for two years in our city, and that was when we found out how much we were in danger of losing our rights, our democracy—while studies were continuing and more information was obtained about the adverse and toxic effects of oil and gas development, and specifically about the health dangers of fracking. Nobody's gonna come and rescue us, nobody's gonna stand up for the Constitution; nobody's gonna stop these corporations from taking our rights away.

"It dawned on us how big and bad this is. When you look at it from the large context, it took me a long time to realize that we haven't had a truly democratic President since Carter. But looking at the big picture, REDACTED Energy and REDACTED Corporation and the big boys, and the Democratic Party and how they're in cahoots, REDACTED's buddy-buddy with REDACTED,* they have children and they know what they are doing. The tobacco industry knew what they were doing. You walk up with a dead child in your arms and they *do not care.*

"You know about the blue ribbon panel that our Governor† put together here? There was going to be a grassroots initiative against fracking put forward by Democratic Congressman Jared Polis. But Polis threw us under the bus. He, they, state Democrats, made a deal with Hickenlooper. If they made sure the initiatives never made it to the ballot, Hickenlooper would establish a blue ribbon panel to study the issue of fracking in communities. Hickenlooper put together a panel, all right; he loaded it with oil and gas representatives and a couple of others so that it would look legit . . . as if we, the people, would have a voice. What a farce. Yet another betrayal!

"I read some testimony presented to Hick's blue ribbon panel on line. What a typical story! Two grandparents moved here to be close to their family. They

---

* Redacted at her subsequent request.
† John Hickenlooper.

bought a house just south east of Loveland, on the east side of I-25*—an area loaded with oil and gas wells close to schools, in shopping centers, in people's back yards. The grandfather said his daughter came down with what the physician called 'environmental leukemia,' making it twice in the last month I've heard that term. The daughter was living with her husband and child in his house. He found out she was sick with environmental leukemia; and his grandchild was sick with an illness that could not be identified; the doctor said to him, do not take your daughter home; do not ever take her into that house again. The well next to their house had caught on fire. Although it got put out rather quickly, a plume of toxic smoke from the well site had blown into the neighborhood. He abandoned his home, stating that he would not sell the house, that he could not, in good conscience, sell it to someone else knowing how his family was affected by being beside that well.

"There was a guy that used to work for the EPA named Wes Wilson. Wes was an EPA whistleblower. He was featured in the very revealing movie 'Gasland.' Just before she died, Theo Colborn, whose DVD about fracking and how the chemicals used and released create endocrine disruption I gave you, told Wes that we are all gonna kill each other before we kill the planet, because of endocrine damage; that will damage our pineal glands and take away our empathy. Reproduction of our species will be disrupted by the toxins being produced, and released, by the process of fracking."

"If someone asked you to describe the fracking process, how would you put it?"

"A continuation of fracking will seal your children's future and they will die a painful death. You know, we're going to go extinct, and it's not going to be quick. Back in the Cold War days when it was duck and cover, my mother said, you know, Sharon, if it happens, you should walk into the light. She knew I couldn't have stood dying slowly in some contaminated fallout shelter. Now with global warming, people don't realize that it's going to be slow and painful. But the big boys must realize it. So why do they do it? You can't wrap your head around insanity."

"When was the first time you saw a frack site?"

"I went to a protest down in Canna headquarters off I-25 down in Longmont, and Shane Davis was going to do a frack tour, so we all got in our cars and we did a little tour of Weld County. And here's a lovely park, and here's a well pad. And they hadn't quite hit I-25 as badly as they have now. I'm terrified we're gonna have

---

* The locality was Frederick. Ms. Carlisle added: "I have included a copy of the above testimony . . . from the Blue Ribbon Panel documents . . ." See source notes.

Ms. Sharon Carlisle

a big grassfire, and when the wells catch it'll look like that big fire Saddam set after Gulf War One."

I asked her: "When you look at one of those facilities, do you understand everything you see? You know, there might be three or five of those round fat storage tanks, and then maybe a tall narrow cylinder, and a shorter cylinder beside it . . ."

"Frankly, I've told people, I'm not interested. I do not feel that I am responsible to know everything. We can educate ourselves. But that's not enough. I always want to take action—*whether it's deemed successful or not.* I've said to so many people, I be will working on this for the rest of my life. My family's sick of it; my friends are sick of it, sick of me being an activist. They don't wanna know. *But if you don't freaking stand up now . . .!* We want you get up off your ass and start working with us. It's gonna go to direct action pretty soon. There's a difference between effect and success. With effect you are nipping at the heels of that monster every day like a little rat. And someday that monster will fall down. Even when there's strategies that I don't agree with, I still think they have a place. We're headed for the streets; it's this close. We're gonna be doing peaceful protests, and they're gonna shoot us; you understand, *they are gonna shoot us.*"

"WE'RE REALLY TALKING ABOUT RELATIVE RISK"

1: "This Is Not All That Different from Other Inconveniences in Our Society"

Twenty-odd miles east of Sharon Carlisle's apartment, the last dozen of these in Weld County, which according to an activist in Denver was right then *the most fracked county in the United States,** lay the municipal offices of Greeley. On the state atlas's double-page spread, Greeley comprised the largest fist of yellow urbanization, with Loveland and Longmont each half that size; anyhow, those two were in other counties. Loveland's moratorium on fracking had failed; Longmont's remained in force. The following year, the state Supreme Court invalidated it.

Weld County was essentially a whiteness desultorily gridded with county roads. The northern part of the county inhabited another page; most of it looked even blanker, being the Pawnee National Grassland, where fossil-hunters sometimes got lucky. Southern Weld likewise continued off the page; no matter how one looked at it, Greeley was the county's metropolis. I admit that Interstate Highway 25 had not deigned to swerve from its near-perfect arrowing north and south; but Greeley did manage to cling like a wasp's nest from the junction of U.S. Routes 34 and 85. It was a flat dry city, sprawling with chain stores and fracking wells. At that time the most ubiquitous energy company in Greeley must have been Noble Energy, whose *mission* was *Exploration Investors Careers:*

> At Noble Energy, we are driven by our purpose—*Energizing the World, Bettering People's Lives* [and this italicized slogan is followed by an "R" inside a circle, evidently to "reserve" it from anyone else, such as Sharon Carlisle, who might have a similar aspiration]. We believe in safely and

---

* Be that as it may, the U.S. Energy Information Administration claimed that Texas was the "top-producing state" for natural gas.

responsibly providing energy to the world through oil and natural gas exploration and production, while positively influencing the lives of our stakeholders. We strive to be the energy partner of choice, a responsible corporate citizen and the preferred employer of the industry's top talent.

So much for "Our Mission." Now for "Our Value":

Founded in 1932, Noble Energy has succeeded where others would not venture—applying global experience to safely and responsibly create new opportunities. The company has proven its ability to move from discovery to efficient execution of large-scale development projects and has additional major projects under development. Noble Energy (NYSE: NBL) is an independent oil and natural gas exploration and production company with a diversified high-quality portfolio of both U.S. unconventional and global offshore conventional assets spanning three continents.

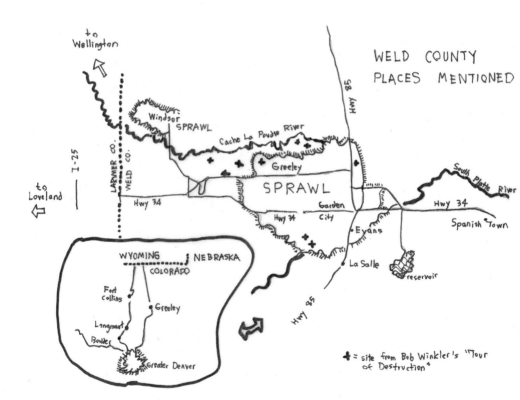

A year later, Noble Energy's fortunes had declined, although unlike Tepco it was not saddled with nightmare liability for a nuclear accident; and unlike Patriot Coal, Alpha Natural Resources, Freedom Industries, West Virginia American Water and other such stars and planetoids of the Big Coal Universe, the company remained uninfected by high-profile bankruptcies and prosecutions.—According to the *Financial Times,* Noble had been suffering *a cash flow problem.* As would any stalwart member of the *regulated community,* the company *defended its accounting and denied any wrongdoing.* Why, until now the thought of wrongdoing had never entered my head! I'd heard only that Noble *continued to burn through cash in the first half of the year and its adjusted net debt, which counts inventories of oil and coal as cash, ballooned to $2.4 b—about $500m more than its market value.* I suppose that weakening coal and oil prices had left Noble's executives plotting layoffs, or packing their golden parachutes, or . . .—wait! Why not consider *the sale of one its best-performing assets?* And so the Noble Group sold *its American energy business* to the equally American Calpine Corporation. That was how I learned that the Noble Group itself was not American at all, but *Hong Kong-based,* not to mention *Singapore-listed.* At any rate, the Noble Group now reassured us that *the divestment of San Diego-based Noble Americas Energy Solutions* (which called itself *a wholesale retailer of gas and power to large customers)* was a big step toward . . . *raising capital this year.*

With natural gas sweeping ever more of the energy market, the company must have been desperate if it chose to sell. But why dwell on unhappy times? Back in 2015, when I savored the petrochemical perfumes of Weld County, Noble Energy was still (at least to all appearances) *positively influencing the lives of our stakeholders.* And that is how I would like you to remember this exemplar of the *regulated community.* Be that as it may, from interviewees you will hear several unflattering comments about Noble, and in the interest of fairness I asked my research assistant to arrange an interview. Here came the expected result: *Noble Energy Inc: Emailed on 1-23-16 to no reply.* So poor Noble will have to take its lumps.

Meanwhile other frackers hung out their own shingles in Greeley—for as an environmental consultant for the gas and oil firms that drilled in that county explained: "We have pretty good quality oil for refining. I'm sure everything here is within a normal or average range. The difference is, our stuff is cheap to get at."

Fracking was not the only business going; the latest issue of *Greeley Unexpected* proclaimed:

WE ARE HIRING! Don't miss out on this great opportunity! Leprino Foods is . . . the **number one producer of lactose** in the world . . .

HONORED TO CALL GREELEY Home
JBS USA is proud to provide great-tasting beef products to customers around the globe, but we always remember our local Greeley community.

And in case lactose or great-tasting beef products ever palled, Greeley, or at least *Greeley Unexpected,* offered precious insights into the goal of human existence:

Making Life Great in Greeley.
Living up to the philosophy of "Making Life Great" is what gets us out of bed in the morning.
Great Western Bank.

That got me out of bed, all right. I felt like burning some carbon.

I remember those lovely July clouds over Greeley, the sky sometimes bluish-grey on one side and then cracked into blue-and-white on the other. There might be a drizzle, then rain and thunder, clearing into a rainbow over a neat new set of frack tanks. West of Greeley half the sky might be purple with rain with a greenish-yellow window of clarity in the middle of the rainclouds, straight ahead on West 34—fresh-cut hayfields, and rolling yellow fields with a nice little frack at the top of the hill, cottonwoods along the roadside, then the Iron Mountain Autoplex; and three white frack towers like chessboard rooks all in one diagonal row right around the time the rental car entered Larimer County; tucked in among cool trees a tall white farmhouse was flying American flags; and here was the Loveland city limit, a railroad crossing, another frack.

"Where do you live in relation to the frack wells?" I asked Sharon Carlisle, who replied: "I am probably in the middle. They're coming in from the south. If you're coming in on 287, they're coming into the valley, right along the ridge by County Road 60; they're all along the ridge. And then they're coming in from the west side. Here in Loveland we're in this bubble, and everything being extracted and produced in Weld County blows over and settles on top of Loveland since we are located up against the foothills of the Rocky Mountains. Pretty bad air. I think the American Lung Association gave our air quality a double F rating: it's from the fracking."

South and north of Greeley the fracks grew more conspicuous along the roads and fields. And slightly east of the city's geographical center, in the municipal offices, the Community Development Director was kind enough to lay out Greeley's situation. He was a pleasant middle-aged fellow named Brad Mueller. We were both Cornellians. The back of his business card read:

> SERVING OUR COMMUNITY
> IT'S A TRADITION
>
> We promise to preserve and improve
> the quality of life for Greeley
> through timely, courteous and cost-effective service.

I have always sought to please others whenever I can, and he might have been even better inclined. There was no reason for him to have made time for me, but he did. The mayor was out of the country; the city manager was just about to go on vacation, so Mr. Mueller got stuck with me, but he never made me feel like the pest I must have been one day before the Fourth of July.

I began by asking which were the most daunting problems in his city, and he cheerily replied: "Well, we were built on utopian principles, so we're just a hair's breadth away from utopia, so there aren't any problems! The city was named after Horace Greeley, and Nathan Meeker decided to follow the vision that was kind of the rage at that time. And because this is at the confluence of the Cache la Poudre and the Platte Rivers, this became a natural place for a city. Western water law actually came out of Greeley! So that kind of utopian attitude is what brought us forward.

"As a medium-sized city we balance the challenge of creating a higher income for as many households as possible. We have a working class, based in part on ag, and our economic development program is based in part on bringing industry that will pay higher than average wages. It really is true that we're really at a wonderful point in our history. About three years ago we recognized that our story wasn't maybe being told based on current realities. There were a lot of stereotypes about being a cow town, a small town, nothing to do, not very sophisticated, and the reality is that we have one of the longer established universities, we've got a thriving art community, a booming downtown, and the largest protein producer in the U.S.: a Brazilian company, and they just announced that they've acquired a chicken company. We're the largest non-fruit, non-nut agricultural county in the United States. So there's just a lot of wonderful things that are happening in Greeley right now, and a lot of the problems that are happening here are the problems that are happening in any middle-sized community."

"What percent of your employment comes from energy?"

"Somewhere between 5 and 10%, even though we're the regional hub for that activity. We've got a wind manufacturing plant here, a Dutch company which is partially located in Greeley, partially in Windsor—and a blade and turbine plant in Brighton, and the solar cells are in Pueblo. We've got some fairly large solar fields. We get about 340 sunny days a year.

"One thing about the ag here is that the water situation has always been a limited resource, and managing that resource and finding new and innovative ways to work with it is a part of our daily living. Water storage is a big part of the solution; conservation is a big part of the solution, and so is reuse. All of these are being looked at. Greeley has some major longterm water projects in the works—reservoir expansion projects, pipeline projects. My entire career in Colorado, these issues have always been there. Up here, along the Front Range, there's the renewable water sources, the river water, and then the Denver Basin is a series of aquifers, a nonrenewable water resource that under state law can be mined. In south Denver they've had to be very aggressive about working away from that for the long term."

He added: "We have a team of water lawyers for the city."

"What are the pros and cons of fracking?"

"You know, I think . . . This is probably my personal observations, not so much city policy. And it's interesting, because these get talked about in public forum. The clear pros are that it provides a job both at the construction phase and at some sustained jobs, and it provides a tax function, directly through severance taxes and through the multiplier effect. One of the things that many residents are proud of is the energy independence aspect of it. So fracking is used both for gas and oil. Much of the drilling used to be on the gas side of things; that provided a bridging energy source; now it is more oil, which it is not so good as natural gas, which can transition a coal-powered plant to a natural-gas-powered plant. But we can transition to cleaner fuels. One of the things that hydraulic fracturing has allowed to happen is, one, I'm familiar with the whole concept of peak oil, and it's universally agreed that whatever the peak time was agreed on has been pushed out now, pushed later. The other aspect is because it is reaching areas that are not depending on the viscous liquid resource and more on the shale formation; from the land use aspect that can be a good thing, because with directional drilling you can in fact site a given well in any of several places. The flip side of that, of course, the negative, is that you're accessing places that might never have been touched there before. But if you subscribe to the fact that we will have to get it out eventually,* then this is a good thing. The negative side of that is that it delays the inevitable of moving to renewables.

"You've got the boom and bust from an economic standpoint. We are buttressed from that, unlike some of my colleagues in North Dakota or in Kern County, California. One of the benefits of being in a more urbanized area is that you can absorb any boom or bust a little better. We've had a slowdown, and economically we've certainly noticed it—there've been layoffs—but as a full economy that hasn't been detrimental to us; it hasn't affected our sales tax.

"The hydraulic fracturing obviously means you've got more underground activity, and the injection wells where you put the produced water,† those if not managed well can lead to seismic activity, but those in Colorado we do seem to have managed. Not true in Pennsylvania or Oklahoma in those kinds of formations! But here it's on rock solid shale a mile deep. All the oil and gas that's being taken out is a mile or more below the ground. In Pennsylvania it might be a couple of hundred feet.

"Under Colorado state law, oil and gas is a matter of state interest. Mineral right, some deference is given to the right to access the mineral. Greeley was a city that in the late '80s tried to ban drilling from within its city limits. That went all the way to the Colorado Supreme Court. Greeley lost, and under the preemption laws, mineral rights owners have to get access to their property."‡

2: Flashback to 1984: The Sensitive Issue

A clipping from *The Greeley Tribune,* which once commanded more paper, ink and local analysis than could its hollowed-out namesake in 2015, begins to tell that tale: *The Weld County Medical Society, a 150-member organization of area doctors, has called for a halt of all oil and gas drilling near county municipalities . . . after a series of oil and gas well drilling and pumping mishaps in Weld County.*

---

* This may remind you of the words of Mr. Rabiul Islam Rabi, the labor union leader in Bangladesh, on the subject of coal (p. 245): "There's no alternative. Today or tomorrow it will have to come out. And it has to be open pit; they have no choice."

† "Produced water" (a cunningly meaningless industry euphemism) is quite simply contaminated water, water that has been used to frack. Sometimes it is disposed of in a reinjection well, but this might cost the frackers more than they care to pay, if local geology compels them to shoot the stuff down deeper. Two oilmen once explained that in offshore wells, "the water may have to be treated to remove contaminant in order to qualify for a permit to discharge it into the sea. Onshore, water may have to be trucked or pipelined to a certified water-disposal facility . . . The cost of water disposal varies from negligible to oppressive . . . from a per-barrel cost from five cents to several dollars."

‡ The anti-fracking activist Bob Winkler, whom you will shortly meet, phrased the matter more bluntly: "Mineral rights preempt surface rights."

The first of these, *a natural gas explosion that continues to puzzle state and county officials, . . . left about $544,000 in damage at Wickes Lumber in LaSalle.* Not long afterward, *about 100 residents were evacuated from their homes when a newly drilled natural gas well about 1 1/2 miles north of Kersey malfunctioned and sent a thick, half-mile long cloud of vapor into the air. Faulty equipment was blamed for the incident.* And since good things come in threes, *late Sunday night a fire at an oil well condensate pit near Mead was quickly contained by firefighters from the Longmont Rural Fire Protection District.**

Please remember that this is but one clipping, retailing the merest trio of incidents. How many others would it have taken to alarm first the Weld County Medical Society and then the citizens and officials of the city itself? How much stress and treasure must have been expended to argue the matter of fracking all the way to the state Supreme Court?—Well, the mineral owners won; we know who they were. And after all, how could any other parties have hoped to prevail? As the *Tribune* had to admit as early as 1985: *Two facts are indisputable: First, the numerous gas and oil well incidents in 1984 are proof that tougher standards are needed to ensure the public's safety . . . Second, stiffer regulations are a sensitive issue.*

3: A Blast from the Past

In 2017, in a Weld County town called Firestone, two men happened to be *working on a water heater* when *the house exploded in a fireball.* Of course they were done for; a lucky wife was *badly burned but expected to survive.* As it happened, the accident occurred 178 feet from an old vertical well now in possession of Anadarko Petroleum. *It primarily produces gas with a little oil.* Several of the neighbors now grew afraid to operate their gas stoves. Meanwhile, in what it called *an abundance of caution,* Anadarko halted production in 3,000 Colorado wells. But why worry? Merrily echoing expert Japanese pronouncements that Fukushima's radiation posed "no immediate danger," the director of the Colorado Oil and Gas Conservation Commission announced: "There is no immediate threat to the environment or public safety."

4: Return to 2015: The Sensitive Issue Resolved

Before returning to the careful sentences of Brad Mueller, let me repeat that in my own day *The Greeley Tribune* had learned to respect the sensitive issue so

---

* This might or might not have had something to do with the fire *at an oil and gas well operated by Petromax Energy Corp. of Denver three-fourths of a mile east of Interstate 25 on Weld County Road 38.*

obsequiously that the yellowed clippings just quoted appear to be not just peculiar but almost impossible survivals of a time when local doctors stood up to industry and the press could report on their reasons in a column whose length was no insult to intelligence. Herewith, the *Tribune*'s guest columnist on July 5, 2015:

By the end of the summer, the Obama administration will likely finalize climate change regulations on power plants *that will have devastating economic effects by driving up energy prices . . . all for a change in the earth's temperature that is almost too small to measure.*

(A national newspaper had just announced that 2015's *global temperatures* were being "pushed"—the actual assertion was a little ambiguous—*to the highest in recorded human history.*)

Stern and sure, the *Tribune* got to the point: *Congress should prohibit any agency from regulating greenhouse gas emissions, and the states need to step forward and reject these regulations entirely . . . Both need to reassert their power before it's too late.*

Fortunately, Donald Trump rode to the rescue.

5: A Bemusing Side-Note on the Reassertion of Power

The anti-fracking activist in Denver was named Sam Schabacker.* He said to me:

"I get a call maybe once a week or once every other from somebody who has a fracking well in their back yard and they don't know what to do."

"And what do you tell them?"

"Unfortunately there's not a lot they can do. It's a rubber-stamping process at the state level."

"So what can you tell them?"

"I try to hook them up with those grassroots activists and leaders, to help them build power in the long term. So for the short term, people probably need to seek out a lawyer."

6: "This Is Not All That Different from Other Inconveniences in Our Society" (continued)

Mr. Mueller continued reassuringly: "But there is still a role for local government under its zoning laws, as long as those laws don't frustrate the state laws. So the illustration I give is that we can get a lot of people with a lot of opinions about oil and gas in the room, and on one end of the spectrum we can agree that the federal government is best suited to handling the casing of wells. On the other

---

* His interview appears on p. 392.

end of the spectrum is this idea that storm water management and traffic routing, that's something that local government can handle best. We also think that things like aesthetic values are things that we know how to regulate and should regulate according to our values. I think that really is the essence of local government and local control. And people hold onto that. They like the idea of being able to call up your neighbor and say, hey, there's a barking dog or whatever. In between that is a grey area of regulation. While we want to assert where the tanks go and how they are configured; we might be going into an area which could be challenged; we might be told we'd gone too far. We're able to assert our own setbacks, which are not more strict that the state ones. (In some ways they're more permissive.) We're able to assert noise regulations for automatic equipment on the facilities, and we can do something about the dust. But we've also said that we aren't able to undertake efforts to handle air quality, which should be on the state. Our attitude and our approach has been more about building relationships, not only with the oil companies but also with the neighbors.

"It might sound simplistic and it might even be simplistic, but land use planners are trained to think about these things, and part of our role in life is to try to mitigate and manage what might be competing usages. In a modern society we know that we have to have substations and cell phone towers and dumpyards, and so forth. Land use planners try to anticipate where uses might be placed. We're able to take that philosophy forward to oil and gas use: The state Supreme Court has said, wells have to be allowed in every zone district—but we can set up parameters, as long as there's no operational conflict.

"One of the things that's really been outstanding about Colorado is to take this collaborative approach, this can-do attitude. Many people across the political spectrum would give Governor Hickenlooper marks for trying to promote dialogue in an extraordinary difficult situation. You'll find people on both ends of the spectrum, but the needle has moved . . . Sara Barwinski is a local activist. She would be the first to say that the needle has moved."*

7: Comment by Sara Barwinski

I certainly tried to foster a collaborative approach, and did help move the needle in Greeley from hostile treatment of citizens to a more courteous dismissal of our

---

* In August 2014, said a local newspaper, she became "one of the 19 members of the new oil-and-gas task force formed last month by Gov. John Hickenlooper to address concerns about boom-time drilling activity as it has moved in and around Front Range cities and towns. Barwinski is in the extreme minority on the new commission—minority of one. She was nominated by her state representative, Dave Young, to speak for concerned Front Range homeowners. She is the only Front Range 'civilian' member of the commission."

concerns. I also served on the Governor's Task Force in good faith—hoping that we could craft some win-win, common sense solutions. We did just that and passed many solid proposals. Sadly, it turned out to be political window dressing. The Governor declared that proposals had to pass with a two-thirds majority, and industry blocked things that could have helped their social license to operate. There was one modest, but potentially very helpful proposal, that did pass unanimously—only to be gutted during the rulemaking process. I do think there was a public relations effort on the part of the industry and the Governor to give lip service to a collaborative, balanced approach. However, because it was not backed up with meaningful action it backfired, leaving more disgruntled citizens in its wake. I moved away from the [G]as [P]atch precisely because the needle did not move in any significant way.

8: "This Is Not All That Different from Other Inconveniences in Our Society" (continued)

"At the end of the day," said Brad Mueller, "the nuanced issue is about what relative risk is, and the land use issue is, when you have to put something somewhere, where do you put it?

"There is an awful lot that can be managed on the surface, such as disclosing the chemicals. And now industry has said, we get why those questions are being asked. But that volatile stuff on the surface, that's the same thing as what's at that gas station where I go all the time. Maybe I should pay attention to those gas station chemicals, too! And so it's been fascinating to me that as our level of public discourse has increased, most people have come down to, well, we're really talking about relative risk. This is not all that different from other inconveniences in our society. I've had this thing come up multiple times in my career: cell phones, and microwave tower emissions, high tension power lines, junkyards, processing plants; so these are things that land use planners have to learn about and then find out how you mitigate, manage, prohibit."

"If I bought property in Greeley, could I buy the mineral rights underneath it or would those have already been sold?"

"It just depends. It's just a real estate transaction. There are instances where an entire subdivision, the rights are severed or in other hands. Most people are pretty savvy about that now, but 10 or 12 years ago, when drilling activity in town was low, it was like, who cares? Same happens with water, too. The industry was savvy enough to price that in. So keep an eye on what you buy, and never say never!"

"What would you do if some company seemed to act in a very predatory way, damaging the environment and harming health?"

"It's a pretty easy answer. If somebody reports something and we witness it

"We Buy Mineral Rights!" The lovely frack pad beside the sign shows what will then happen.

ourselves, we just call the COGCC.* I've literally got a half a dozen of those in all this time—vibration, noise, that, and the general *I don't want these,* which is not a complaint but just an opinion. But locally, we're not afraid of sending our Code Enforcement people. If there are weeds on the property, we're gonna cite 'em. And we recently got authorization for a new fire regulation in the city; we can approach the wells that way. You'll be looking at it mostly from a life safety standpoint. And that's one more way to be more active."

According to the Environmental Protection Agency, *natural gas systems were the largest anthropogenic source category of CH_4 emissions in the United States in 2014 with 176.1 MMT CO_2 Eq[uivalents] of CH_4 emitted into the atmosphere,* that naturally being not the 86× figure, but a middling 25×.—I asked Mr. Mueller: "How do you feel about methane leaks and global warming?"

"My understanding of this is through the rule-making process, through the air quality process. The Colorado Department of Health and Environment regulates this. But here is what I have come to understand. The methane would come out when there are mistakes made. Much if not all of that can be controlled through proper management. Where there used to be 95% compliant equipment, there is now 98%.[†] Where there used to be a mechanical process of opening what is called the thief hatch to get the oil and gas out, now those hatches can be, and in modern cases are, required to be contained."

This might well have been so. For the Environmental Protection Agency had continued: *Those emissions have decreased by . . . 14.8 percent . . . since 1990[,] . . .*

---

* The Colorado Oil and Gas Conservation Commission.

† Not for another month would the EPA advise a 40 to 45% reduction in emissions of that dangerous gas.

largely due to the decrease in emissions from transmission, storage, and distribution. The decrease in transmission and storage emissions is largely due to reduced compressor station emissions . . . The decrease in distribution emissions is largely attributed to increased use of plastic piping . . .

"And it goes on and on," said Mr. Mueller, "which is why Colorado claims that it now has the highest air quality standards in the U.S.!* It is useful for me to understand what quantities we're talking about. To say that it doesn't matter because everyone else does it[†] is not an acceptable answer. But to fix it and not address other sources is also unacceptable."

"If you were dictator of Colorado, what would be your energy policy?"

"I'm not really sure I could answer that. Personally I would take a different kind of 'all of the above' approach. Land use planners tend to be pragmatists. You could impoverish an awful lot of people, and on the other hand, you could impoverish an awful lot of people in future generations. I really don't know how to answer that, but I can tell you that Colorado is doing a lot of things and more of all of those is in the works."

"More fracking, more oil, more solar, more nuclear?"

"Yes," he said.

(That was how most of us reasoned, back when we were alive.)

More of everything! Weld County was a happening place, at least in comparison to another zone that you and I have visited:

TWO COUNTIES COMPARED:
McDowell, West Virginia, *versus* Weld, Colorado

Per capita income*

McDowell:	$13,035
Weld:	$19,447

---

* Here is the spot to repeat Sharon Carlisle's counter-claim: "I think the American Lung Association gave our air quality a double F rating: it's from the fracking." Of course this issue was contentious. An "air division spokesman" at the Colorado Department of Public Health and Environment insisted that the ALA's yearly survey "is both inaccurate and misrepresents air quality in Colorado." (My heart bleeds.) "We also maintain a robust air pollution monitoring network and have added several monitoring sites in recent years." Meanwhile, the Environmental Defense Fund asserted that "a major factor causing worsening air along the Front Range, as well as in neighboring Utah and New Mexico, is oil and gas industry activity."

† That is, releases methane.

Median household income*

McDowell:	$19,031
Weld:	$47,579

Population change, 2000–2009

McDowell:	−19.4%
Weld:	+142.7%

Projected population change, 2009–14

McDowell:	−10.2%
Weld:	+13.8%

* 2008 figures.

The invisible hand of economics had closed its fist upon the West Virginian coal fields. But it extended all its green fingers in a blessing upon the frack pads of Colorado.

"THEY JUST SHUT DOWN THE OPPOSITION"

1: "When Are People Gonna Wake Up?"

Bob Winkler was an activist who offered what he called *the Tour of Destruction*. The name led me to suspect that his tour stops would not be very pretty. He said:

"I live just a little ways from here and I can see the Front Range from my house. My wife called me one night and said: I don't know what's going on; the neighbor's yard is all lit up like aliens landing or something—at two in the morning. I says, call the police and she calls the police and they're just laughing. Then there's all this noise, so I say call the police again and the police say nothing we can do. We live in a subdivision here and they don't even notify us to see what's happening. So I put on my old business continuity hat and try to look at what's happening with an open mind. A lot of what I see in my career is: *Make money at all costs; we don't care what we do to the land and people's quality of life.* We went to a city council meeting and they basically turned off the mikes; they didn't want to hear it. I'm a civil rights guy from the '60s: I went to the Democratic convention, got my head bashed in. So I started petitioning, putting stuff together; I went to the Governor and pleaded to our state representative, saying, this is crazy; they can't just ignore this. My whole thing is health and safety. People are dying on these damn things! Just throwaway people . . . Now I'm involved all up and down the Front Range with Sam* and other people.

"I can't change people's minds; if they want to live in a toxic waste zone what can I do about it? All I can do is educate people. I started reading all this stuff and I'm saying, this is a no-brainer! Our health costs are gonna go through the roof.

"Sometimes it happens in the night; people get their heads blown off from when a well goes off.

"I keep pushing, and the last three years we've tried to put on the floor of the state Senate a health assessment. Just assess the risks to our health! Is there anything different from living in Illinois? It got killed in committee, then passed the

---

* Sam Schabacker, quoted on pp. 339, 348, 356, 366 and (bulk of interview) 392–95.

House, then got shot down in the Senate because of budget constraints. Now, this assessment would have cost $800,000. It would have cost nothing.

"You smoke when you're 20 and then at 40 you say maybe this wasn't a good idea, and then you're dead at 50. Fracking's like that.

"They have exemptions from the Safe Drinking act, Clean Water Act, Clean Air Act . . . When Cheney pushed this thing through, they called it the Halliburton loophole."

"How do you feel about fossil fuels in general? Which is the best, which is the worst?"

"Natural gas has the longest impact on greenhouse gases; it pushes up climate change for 20 years.* Oil you burn and it's over.† I don't think any are good . . . Doesn't anyone realize that if we keep burning this stuff *our kids will not be able to live on the surface of this planet*? This is the beginning of the end. I'm doing what I do for my kids and my grandkids. And I understand, being a business guy, the whole world economy revolves around the fossil fuels. But there's no direction of trying to move away from it. We're getting record heat here like we've never had. The Colorado River is being drained, and the snowmass is a joke.

"Three days a week I can water two hours in my back yard. The industry can open up a fire hydrant and take up to one to eight *million gallons* for *one well*, buy it at discounted prices, and then pollute it. My end game is, we've got to get off this stuff—but in this community," he said, lowering his voice, "it'll never work."

2: From a Leaflet by the Colorado Oil and Gas Conservation Commission (*ca.* 2015)

Q: *Can hydraulic fracturing open up pathways for oil and gas to reach ground water zones . . . ?*

A: The distance between the oil and gas formation and the water formations is substantial.

3: Conclusions of the United States Geological Survey (2016)

Deep well injection is widely used by industry for the disposal of wastewaters produced during unconventional oil and gas extraction.

---

* Evidently this referred to methane's 20-year GWP of 86.

† Too optimistic.

Our results demonstrate that activities at disposal facilities can potentially impact the quality of adjacent surface waters.

4: From a Leaflet by Food & Water Watch (2016)

After fracking and drilling, Chevron and other oil companies sell their toxic wastewater to farms in California's Central Valley for irrigation. Poisonous chemicals, including acetone and methylene chloride, have been found in the wastewater, and many popular food brands—some of which are labeled organic—are being grown with it.

(The leaflet singles out Halos Mandarins, Sunview Raisins and Grapes and Trinchero Family Estates, "which makes Sutter Home Wines.")

5: From the Leaflet by the Colorado Oil and Gas Conservation Commission (continued)

Q: How do you ensure the fracturing fluid, including the chemical additives, don't [sic] escape the oil and gas wellbore and impact nearby water wells?

A: The COGCC requires all wells to be cased with multiple layers of steel and cement to isolate fresh water aquifers from the hydrocarbon zone.

6: From an Article in *Earth Island Journal* (2016)

A salty mystery is brewing in Carbon County, Wyoming . . . Water salinity spiked suddenly in 2009 . . . Wastewater from fracking and failed wells can infiltrate creeks and streams, eventually making them too saline for use on crops or consumption by cattle, wildlife, and people.

7: "When Are People Gonna Wake Up?" (continued)

Bob Winkler was always saying things like: "Now when the injection well blew up, that was news."

"Here's a town here, Greeley, which is the jewel of Weld County, the seat of

Mr. Bob Winkler at a frack pad

government. We've got over 20,000 wells in this county. You think that's enough?*

"We've gone to the Air Quality Commission in this state, which will actually listen to us a little but there's only so much they can do. They've put in setbacks† and the industry's got two to three years to comply, and in the meantime they're drilling and drilling.

"They've got all these roughneck‡ guys out there that are drilling, operating on a shoestring. They gotta keep drilling, because they gotta keep the pipeline open, because last September the OPEC cartel said, we're going to lower the price of oil, so these roughnecks have got to compete! The big guys are not here—not Shell, not Exxon. All the little guys are here. No one's accountable. Everyone's a subcontractor or a holding company. All these little guys, back in September, basically the shit hit the fan, and the wells shut down. My idea is Shell and Exxon are going to buy up all these wells for 10 cents on the dollar.

"An American flag on top of the well, that makes me puke. It really angers me

---

* A few days earlier Sam Schabacker had offered a slightly different number: "The worst place in Colorado is Weld County, hands down. They've got over 19,000 wells in their county. There's a noticeable difference just aesthetically. You can see what's called the *brown cloud*. You're going to see a bunch of fracking wells when you go there."—I did not see the *brown cloud,* maybe on account of favorable weather.

† A setback is a required minimum distance between the frack pad and the nearest home, road or business.

‡ An industry term. The ex-CEO of Conoco, Mr. Archie Dunham, uses this word on p. 457 to describe his first job in the oil fields.

to say, we anti-fracking people are not patriotic. When somebody poisons us, what're we supposed to do, step into the gas chamber?

"I love the ads. We get every night on the ten o'clock and the six' oclock news, we get on TV, *facts on fracking: we've been doing it for 60 years*—but it wasn't there until five-six-seven years ago.* They got all these butterflies in waterfalls and all this other shit they've been shoveling. We have no way to get our voice out.

"All this oil that's coming out, they're selling it to China because China's the high bidder! And they say *we're* unpatriotic.

"It's pretty hard here in Greeley. I've worked in the Democratic Party so I know what a machine looks like. This makes Daley look like a Sunday prayer meeting. There are no liability issues because there's no law to stop these people."

"If you walk around with your anti-fracking button, will anybody threaten you?"

"Not here, but in the well sites they do. Sometimes I carry a gun. You have to understand, this is their livelihood. How can I argue with that?"

We set out on the Tour of Destruction. Everything was flat and hot.

"Water's gonna be the big thing," he said. "Water's gonna be worth more than gold. Why are they destroying it? We've got to have an education in water law.

"All the land around Ault† is owned by Thornton. The farmers sold it because it was a sweet deal. When are people gonna wake up? I don't understand it."

But I think I did. Aldo Leopold had insisted: *The major premise of civilization is that the attainments of one generation shall be available to the next.* We just wanted to make sure that you in the future would remember us for what we did to your water.

"How do most of your neighbors make a living?" I asked.

He burst out laughing.—"Are you kidding? Most of my neighbors are gas and oil guys. Nobody cares about the community. But when 20% of the people are voting, what do you expect? We've just handed over our freedom."

"Do you see the CEOs around?"

"Those fracking bigshots, I heard 'em say: *Oh, I live in Golden; I live in Vail.* I said: That's nice, why don't you live here?"

"Is Greeley's water contaminated?"

"Horace Greeley and Meeker, back a couple hundred years ago, they hand dug reservoirs from snowpacks; they have tunnels and reservoirs; all the water that

---

* But according to "a communications manager for a natural gas utility" whom I enjoy citing on occasion, "since [1949], the process has helped augment production in more than a million wells, producing in excess of 600 trillion cubic feet of natural gas and seven billion barrels of oil."

† A small community not far north of Greeley.

comes to Greeley is snowpack water. Greeley's still fine there. But for how long? It's an unscrupulous business: Get in; get out; make as much as you can. Up in Wellington where my sister was, there's an oil company there; they introduced a water augmentation program. Wellington is on wells. The oil company is filtering their fracking water and putting it into the drinking water!"

8: An E-mail of Comment from the Wellington Town Clerk

The majority of the water the town uses to treat for potable water is surface water but we do have several wells in town we supplement our supply with. There is an old wellfield north of Town [sic] which is still in production but there has not been any new drilling or fracking anywhere close to Wellington that I know of.

9: "When Are People Gonna Wake Up?" (continued)

"Are there certain medical symptoms from being near a frack site?"

"Some people get bloody noses, some people get teary eyes, some people get respiratory problems. A woman I know, she goes to the head of pulmonology, and he won't come out and say this publicly, because he's on contract with the hospital or something, but . . . All they keep talking about is the economy and what a great thing it is. It's wonderful for this town. We're going to put 67 wells behind the elementary school! . . . Windsor is a little farm community and they have no choice there, none at all, and at the city council meeting, they just shut down the opposition."

"What does the Mayor of Greeley say about it?"

"When elected officials are elected to protect our health and they don't do it, they should be recalled. But I got run off the road."

Pulling over, he said: "I'm gonna show you the one that was close to my house, where they got 10 wells and Texas barbeque burning off the emissions. That stack there is supposed to burn off 90% of the stuff. This is the West T-bone. This used to be feedlot . . ."

I did not see any gas flares right there. Perhaps there often were some; I cannot verify it.*

He said: "I'm gonna take you down to the polishing plant, where they take the raw gas and they clean it up. I don't know if we should get out; they know my truck."

---

* From a West Virginia environmental newsletter: "Flaring is used to get rid of impurities before gas is piped. Regulations say companies are not to flare for more than 30 days. I'm concerned about global warming, and that seems like an awfully long time to burn gas."

Driving along the edges of cornfields and then a young onion field, passing a gated community he said: "Now there's some tanks. They're going to make 20 wells right in there."

At a covered open pit identifying its product as 1.45 UPV oil he said: "Don't get too close to those storage tanks. See, it's got something flaming out right there. And here's the water reclamation, right across from the polishing plant: the Platte River . . ."

At the polishing plant I saw varying-sized silver cylindrical towers and then the horizontal lattice of pipes and white storage tanks along the yellow grasses and cottonwoods, and we could hear the sound of an air conditioner, magnified. It was an air compressor. The following afternoon a man named Eric Ewing would point to a frack well a mile away and say: "Sometimes I can hear that air compressor in my living room."*

"See those tanks right up there by the houses?" said Mr. Winkler. "And they don't give those guys respirators; they don't care. And there's the pond where a tank broke off and leaked. Normally, people talk about a metallic taste in their throat. And you know we're on an ozone alert today. Sixty percent of the VOCs are from oil and gas."

---

* See below, p. 369.

About Volatile Organic Compounds

When he rose next a sheet of flame was lighting the sky and the oily reek of burning hydrocarbons tainted the air.

C. M. Kornbluth, *The Syndic,* 1953

They rose up out of pesticides, lacquers and hairsprays, not to mention rubber cement, adhesive tape and diverse other products that we used without much concern. (After all, they underlined our ever so high level of development.) One of them was chloroform, which got to us whenever we boiled chlorinated tapwater. The Japanese government fingered p-dichlorobenzene in "repellent and refresher," but why should we "consumers" care what p-dichlorobenzene might be?

We emitted them during the production of bituminous roofing material—of which Germans made nearly 194 million square feet in one year alone. We emitted more of them when we attached the stuff to housetops. I calculate the German total release in 2007 from that source at 25,363,800 pounds.

In 2006, German railway tank cars carried 459 million cubic feet of petrol fuels here and there, sending up *only 1.4 k[ilo]t[ons] VOC per year,* or not quite 3.1 million pounds of them. The Germans sounded quite proud of that: *The emissions situation points to the high technical standards that have been attained in railway tank cars and pertinent handling facilities.* I send my own congratulations, but I wish they had told us what those *pertinent handling facilities* smelled like.

The Germans must be doing their best;—the European Union had reduced VOC emissions by 58%. I suspect (although I cannot prove) that thanks to fracking and other good deeds, the Americans were going in the opposite direction.

Sometimes they were abbreviated "VOCs," or more sonorously presented themselves as "NMVOCS," the first two letters standing in for "non-methane."

According to the Intergovernmental Panel on Climate Change, they had

atmospheric lifetimes ranging from hours to months. Global coverage of NMVOC measurements is poor, except for a few compounds ... Satellite retrievals of formaldehyde column abundances from 1997 to 2007

352

show significant positive trends over northeastern China . . . , whereas
negative trends . . . are observed over Tokyo, Japan and the northeast
USA urban corridor as a result of pollution regulation.

Commendable academics and officials sought to quantify them by, for instance, seeking (and failing) to discover "shipping amounts" in the ledgers of the
Japan Moth Repellent Association.

When I first heard about them all, in connection with fracking, I supposed
they were newfangled synthetic compounds. Why couldn't we ban them all? But
getting rid of VOCs would be no more practical than abolishing carbon dioxide.
My favorite was *a fugitive emission in alcohol in the process of manufacturing foods
or beverages*. What were we supposed to do about that?

Every 222 pounds of bread gave off a pound of VOCs, whose fresh-baked
fragrance got counted against us! Surprisingly enough, given that wonderfully
penetrating power of the odor which symbolized morning to so many of us, it
required 14,497 pounds of roasted coffee to produce another pound of VOCs.
Exactly three times as much smoked fish would accomplish the same result.

One of them, methanol,* was in window washer fluid (25%, I think by volume). Another made up 76.8% of nail polish remover. Sunscreen contained
83.5% VOCs. The atmospheric emission rates of the last two were 100%; about
the window washer fluid I can't say, although I would imagine that it too volatil-
ized gloriously upward, leaving each window free of residue.

Premium and regular gasoline sent up nearly equal amounts of VOCs.

Whiskey emitted 187.5 times more of them per unit volume than did refined
sake, which in turn surpassed beer by a factor of by 2.3. So we scented our atmosphere.

In case you desire a definition, I now quote:

> Not a well-defined group of hydrocarbons. This group of gases with dif-
> ferent lifetimes is treated differently across models by lumping or using
> representative key species.

And speaking of "not well-defined," consider the proverb *one man's meat is
another man's poison*. Propane was a fuel; cyclohexane came in handy during
nylon production. In "About Coal"† we have already read glowing praise of

---

* Which appears as a fuel in I:211 (header 106).

† Above, p. 18.

Dry cleaning establishment, Abu Dhabi. An excellent source of VOCs.

benzene, toluene, etcetera. What were they, but our dear old friends the hydrocarbons?* In the present context, all these and many others were VOCs.

Around Boulder, Colorado, which lay within the Wattenberg Gas Field, mean propane levels were three to nine times higher than in smoggy Pasadena and trafficky Houston.

---

* From the Environmental Protection Agency: *Non-CH$_4$ volatile organic compounds (N[on] M[ethane] VOCs), commonly referred to as "hydrocarbons," are the primary gases emitted from most processes employing organic or petroleum based products, and can also result from the product storage and handling. Accidental releases of greenhouse gases associated with product use and handling can constitute major emissions in this category.*

Just when I went riding on the Tour of Destruction, the Boulder Atmospheric Observatory was sampling volatile organic compounds. The conclusion: *Overall, the [Northern Front Range Metropolitan Area, which included Greeley] was more strongly influenced by O[il and] N[atural] G[as] sources of VOCs than other urban and suburban regions in the U.S.*

Now let me continue the preceding extract:

> The spread in metric values . . . is moderate across regions, with highest values for emissions in South Asia . . . The effects via ozone and CH_4 cause warming, and the additional effects via interactions with aerosols and via the O_3*–CO_2 link increase the warming effect further.

Carbon monoxide and some nitrogen oxides did similar damage to the atmosphere. But VOCs in and of themselves could increase ozone levels, sometimes above U.S. federal standards.

Their average global warming potentials were 14 times worse than carbon dioxide's over 20 years, and 4.5 times worse over a century.

They blighted vegetation, and accordingly decreased carbon sequestration.

Overweight smokers had higher benzene levels in their blood than trim non-smokers.[†] But smoking actually reduced toluene and certain other VOC levels. As we used to say (we liked to whistle in the dark), "go figure."

Although their primary source was solvents, combustion played its inevitable part. Burning wastepaper gave off a minuscule 2.48 grams of VOCs per metric ton (one pound for every 202 American tons).

Japanese diesel passenger vehicles emitted a pound of them every 272 miles; to achieve the same excellence, light gasoline vehicles had to roll 1,269 miles. If they hadn't been also gushing carbon dioxide, carbon monoxide and nitrous oxide, it would have been an even better world.

Often we made our VOCs on purpose, either as solvents or as reactants. Toluene was both: a solvent in glues and lacquers, a reactant in the manufacture of such useful substances as methyl ethyl ketone—which in turn served as a solvent when we whipped up our printing inks.

When this book was getting printed, its vain pages gave off somewhere between 0.132 and 0.798 pounds of volatile organic compounds for every pound of ink, which had released another 0.03 pound of VOCs during manufacture. As

---

[*] Ozone.

[†] Women had higher benzene levels than men, due to "the increased proportion of body fat in females."

for the glue, each pound had already emitted 0.02 pound of VOCs while *it* was being made . . .

Mostly, when we smelled them, we smelled what we would have called "chemicals"—meaning the chemicals *not* in coffee, beer or burning wastepaper. In 2013 Denmark emitted 239 gigagrams (not quite 563 million pounds) of VOCs caused by "paint application"—by far the most in this category of any EU-15 country. When I was alive, I knew what paint fumes smelled like. My best friend had been a commercial painter, and I sometimes kept him company. He used both latex and oil-based paints. After some years the odor of lacquer thinner began to make him sick. (He got tonsil cancer, then brain cancer—possibly a coincidence.) I myself was an occasional woodcut artist, and I used turpentine spirits to clean my blocks, year after year, until suddenly the smell of them caused me to retch, and after that, the smell of paints and solvents in sealed tins in a hardware store made me queasy.

When a certain frack pad's unfortunate neighbor* likened the smell to a janitor's closet, I knew what he meant, and so did almost anyone who had passed any time in one of our schools or office buildings—so penetrating was the stench of those particular VOCs.

If the smell varied from pad to pad, it must have been because so did those proprietary (I really mean secret) aggregations of injection chemicals.

A study of 21 "symptomatic patients" in Erie, Colorado, who were unfortunate enough to live within 200 to 1,800 feet of frack pads or BTEX† burners measured *specific volatile organic compounds* in their blood at *levels at or above 95% of the general population['s]*. These people suffered from migraines, "breathing difficulties," "generalized malaise," digestive trouble and chronic fatigue. The sample included *a teenage girl* who *has been unable to attend school for over a year due to the above conditions*. One VOC conspicuous in the blood of the research subjects was ethylbenzene, *a compound very similar to benzene in its negative health effects*. In a particularly gruesome case, *the patient's blood itself present[s] itself as a biohazard to herself, other humans and the ecosystem.*—The report concluded: *It is . . . possible that tissue level of these VOCs are even higher in the nervous system and fatty tissues than in the blood.* But this paper, which my energetic research assistant netted from the turbid shallows of the Internet, lacked a dateline or even a journal of publication. So what if it were fake?—Next story: An *Aspen*

---

* Mr. Eric Ewing (p. 369). Sam Schabacker said: "It smells like you stepped into a chemical facility. It depends on how sensitized you are to it. Some people who have been in close proximity, they build a nosebleed or their eyes start watering or whatever."

† A common abbreviation for the group consisting of benzene, ethylbenzes and the xylenes [all considered as VOCs].

Times article looked in on the town of Silt, where a lady, her daughter and her granddaughter were all tested after *showing . . . upper respiratory infections, swollen glands, sore throat, congestion, coughing, sneezing, earaches, shortness of breath and itchy, watery, burning eyes . . .* These tests were not cheap, and the family had to pay for them, which to my mind gives their malaise greater credibility. *The results . . . have indicated that the family has been contaminated by what are known as volatile organic compounds.* In her innocence the grandmother contacted the Garfield County Department of Public Health, whose interim director replied that this entity *does not accept personal health studies . . . If . . . you are concerned with the [study] results, we encourage you to seek medical attention from your primary care provider.*—As the CEO of Breitling Energy has already so wisely reminded us: "As a pragmatist, here's my main point—we can't have it all."

"I GUESS WE'RE KINDA LEARNING TO COPE WITH IT"

1: As If There Were Still Time

We arrived at a badly washed-out road, the blacktop broken off and gulley-traced. Mr. Winkler said: "See that garage over there? That's a chemical warehouse. When Greeley flooded, it was up to the doors."

A strange bittersweet taste settled on my tongue, and I felt a faint nausea. But I was always impressionable when I was alive.

What might that interesting flavor have been? Thanks to the Halliburton loophole I could not find out. This interesting measure, in the words of the CEO of Breitling Energy, *exempted fracking from the reporting requirements of the Safe Drinking Water Act as well as the Clean Air Act and Clean Water Act.* Granting as he did that *keeping secret the chemical composition of frack fluids was a gift on a silver platter for people looking for reasons to target the oil and gas industry on environmental issues,* this "pragmatist and . . . optimist . . . with a signature style" reminded us that *oil field service companies are experimenting with different formulas for more effective frack jobs . . . Keeping trade secrets is legitimate in any business.*

KEEPING TRADE SECRETS IN PENNSYLVANIA:

The Oil and Gas Omnibus Amendments Act, 2012

§ 3222.1. Hydraulic fracturing chemical disclosure requirements. (b).

(11) If a health professional determines that a medical emergency exists and the specific identity and amount of any chemicals claimed to be a trade secret or confidential proprietary information are necessary for emergency

---

treatment, the vendor, service provider or operator shall immediately disclose the information to the health professional upon a verbal acknowledgment by the health professional that the information may not be used for purposes other than the health needs asserted and that the health professional shall maintain the information as confidential ... The health professional shall provide upon request, a written statement of need and a confidentiality agreement ... as soon as circumstances permit ...

§ 3222.1.(c). A vendor, service provider or operator shall not be required to do any of the following:

(1) Disclose chemicals that are not disclosed to it by the manufacturer, vendor or service provider.
(2) Disclose chemicals that were not intentionally added to the stimulation fluid.
(3) Disclose chemicals that occur incidentally ...

§ 3222.1.(d) (2) (i). If the specific identity of a chemical, the concentration of a chemical or both the specific identity and concentration of a chemical are claimed to be a trade secret or confidential proprietary information, the vendor, service provider or operator may withhold the specific identity, the concentration, or both the specific identity and concentration, of the chemical from the information provided to the chemical disclosure registry.

... APPROVED—The 14th day of February, A.D. 2012. TOM CORBETT.

(And in 2017, with one of Trump's most effective foxes in charge of the EPA henhouse, climate change information on the government website was being treated like another trade secret.)

Around the bend of the highway, literally looking down on those tanks, was a sign: RIGHT NOW LEASE YOUR HOME, STARTING AT $89.

That was Prairie View and Cave Creek: modular double-wide houses.

Then came Mr. Winkler's favorite, Freedom Park, where one could walk the dog and admire a fracking well's hundred-barrel crude oil storage tank. He said: "I just had to laugh ...

"The Front Range is in noncompliance for air," he said. "We haven't been in compliance for 15 years. It used to be from automobiles; now it's from fracking."

On Industrial Parkway, we passed a long silver tanker truck emblazoned with the symbol for energy (a capital E in a circle); Mr. Winkler had no idea what that organization might be.

At the next frack pad were separators new and blue, a little like alien fire extinguishers, and a water truck. He said: "You know, they use up to a thousand of those things in one well. Baker Hughes, Upstream Chemical. They put a fence around it now.—I wonder why!" he laughed.—We were at Industrial and 42nd Streets.

Freedom Park, with frack pad on the left

Now we were driving over the Platte, which was wide and brownish-green. He said: "I used to ask the frackers' *so-called engineers* this or that, and they would always say, *oh, we'll have to get back to you on that,* and they never did."

"We just had two tanker derailments," he said.

Farther on, right around the LaSalle city limits, he said: "Now, this is just a staging area here for Ensign, which is a Canadian gas and oil company."

We went on CR 52, with the Centennial Highway off to the right.

At First Avenue and 31st Street he said: "This used to be a fishing hole, and the fish are funny-looking." There was a strong crude oil smell—in fact, a stink; and this stench was literally sickening; it burned our companion's throat. Mr. Winkler was talking about "Mineral Resources, now the Greeley Directional Project. Then a new company came along, since they had such bad public relations, and now they hand out community outreach. Now see this here? The whole thing is gonna be filled up with wells."

Not far past the sign for Double Barrel Oil he said: "This is our university.

They've already sold their mineral rights, so they're gonna drill here in the university. Right in the campus, too! The President of the university is the wife of the Mayor of Greeley, and she used to be an attorney with an oil company. Talk about Chicago politics!"

He told the tale of the instructor in the junior college who got so sick with nosebleeds that he had to move away. I cannot confirm it.

"I want to get the hell out of here," he said, "but I can't afford to move anyplace in Colorado."

He said: "I'm gonna take you to a well site called Spanish Town where the Spanish-speaking people used to be able to work but they were not allowed to live here. You know, around here they're Christian and all that, but it's still real segregated. The people here are very nonconfrontational and very subservient and very religious."

So he drove across the Cache la Poudre River, and then, after showing off a frack pad, back south across the same river, to a pile of huge grey boxes labeled SAND—"to help you get silicosis," he said with a grin.

"And this one makes me nuts; there's homes here," he said. We were on C Street, where the frack site exuded a smell like rotten shrimp and vinegar. What he had promised was true: They all stank differently.

So what's really in the frack fluid slurry? the CEO of Breitling Energy asked himself—and even generously provided the answer! It was *90% fresh water, 9.5% sand and 0.5% chemical additives,* the "representative" ones of which he listed in a facing table. Diluted acid? Think of it as *swimming pool chemical and cleaner.* Biocide? *Sterilize medical and dental equipment.* "Breaker"? (That's ammonium persulfate to you.) *Bleaching agent in detergent and hair cosmetics, manufacture of household plastics.* And so it happily went. *Sure, you wouldn't want to drink antifreeze (ethylene glycol) or swimming pool cleaner (muriatic acid), but there are only tiny, even trace, amounts of these substances injected into a rock formation more than a mile below any freshwater source.*

By the Alpine Veterinarian Hospital, at the gate of Rockwater Energy Solutions,* Mr. Winkler said: "You can see the tower there. When it's up, with those blue and red trailers out, they're fracking." Rockwater, he said, was "just a subsidiary. You can't keep track. Over there, the Best Way Directional was formerly Mineral Resources. You've got the drilling subcontractors, the maintenance companies, and on and on, so you can't find out who does what."

You have seen how typical that was of the West Virginian coal industry.

---

* Address: 131 North 35th Ave., Greeley, 80634.

Interlockings of obscure and ever changing companies, like Dracula's many hidden coffins, evidently served the interest of the *regulated community.*

Again he warned me to be careful of workers. "You don't want to mess with some of these people, because you're messing with their families."

Rolling through Hunter's Corner, we paused at the green athletic field of Northridge High School, whose sunny silence reminded me of that mildly radioactive playground in Naraha, not far south of Reactor Plant No. 2—a spurious comparison because Mr. Winkler must have been one of the few people who perceived anything undesirable about this place. He had photographed a graduation ceremony here, with plumes ascending from the frack pad behind it; I could see no plumes today. He said: "They've sold their mineral rights, too, because the people don't want to pay for a decent education."

The Tour of Destruction continued.—"You see that tank battery there? In the middle of a golf course! Are you kidding me? And these homes that are million-dollar homes and they are putting another well site in there! Their mental outlook, it's that these things are normal."

Our companion needed a restroom, so we pulled in at the Poudre River Learning Center, which was situated on what was supposedly "a National Heritage Site." *Our mission is to awaken a sense of wonder and inspire environmental stewardship and citizenship,* began the center's self-identification; so here among the nature dioramas and stuffed birds I would have expected a certain anxiety about fracking, but the pretty woman at the desk said: "Well, I'm in favor of it. It's my children's lives. One works for X company and the other for Y Company."*— With these wonder-awakening words in our ears, we got back in the car.

"I always thought of America as this pristine area," said Mr. Winkler.

Sharon Carlisle had advised: "You should visit Windsor sometime and see what's going on there. Windsor is on what I call the Eastern Front, because that's the line between Weld and Larimer County. The Eastern Frontline of fracking— the Eastern Front of the war being waged against us."—And as it happened, Mr. Winkler now drove to Windsor, passing the Vestas factory, whose wind turbines lay ready to be shipped out on a certain long freight train: one turbine per car, they were so immense; while across the street lay a frack site adorned with yellow trailers; and crossing another bend of the Cache la Poudre we arrived at the Poudre Heights housing development, which he estimated at about 10 years old. Windsor was well fracked, I could see, although the fact made no special impression on me. But that was how frack pads themselves also struck me: dull, almost minute, drearily additive.

---

* Both entities were frackers. I have redacted their names to make her less identifiable.

There was a hill; I wanted to see the view from there, so we drove up Extraction Road and then New Liberty Road. While the others waited in the car, I climbed through tall grass and up a rocky sandy slope until I could gaze out to my heart's content through the rain and green grass and the new clean frackitecture: I got a nice view of the Rain Dance Sand and Gravel Pit; the air was perfumed by solvents and crude.

The tour went on; you get the drift. Were I not myself an ideologue I would classify Mr. Winkler's commentary as 90% hearsay. How could I know whether some junior college instructor really got nosebleeds, or whether anyone's head blew off thanks to an exploding well? Pennsylvania's Oil and Gas Omnibus Amendments Act was as hatefully dangerous to health, justice and public accountability as West Virginia's Senate Bill 423*—but these enactments of oligarchical absolutism, however much they furthered ignorance, intimidation and corruption, did not in and of themselves discredit fracking. A less negative carbon ideologue than I might interpret the lonely wariness of Mr. Winkler and of Sharon Carlisle as proof of wrongheaded irrelevance. Socrates was equally irrelevant once the Athenians had served him his hemlock. The insipidities of the hollowed out *Greeley Tribune*, the *no comment* of most people to whom I "reached out," and the typical anomie of an American metropolis, whose citizens I rarely saw except in their cars, in retail establishments or at the Fourth of July parade, operated synergistically to create the usual hot wide silence—about fracking, climate change, democracy and every other thing. A certain form of economic development held sway, and that was that. Although if what Mr. Winkler said was true, Colorado was getting hotter and drier, the city government did not yet feel any imminent necessity about that situation. Whatever urgency most of us once felt had been directed toward the bugaboo of peak oil, and, as Brad Mueller had pointed out, the peak was now *pushed later*. Thanks to fracking, there was *no immediate danger*—of running out of fuel, that is—but was there likewise *no good alternative*? Oh, no, there were many! "We can get a lot of people with a lot of opinions about oil and gas in the room," Mr. Mueller had said. And what did those opinions add up to? Which fuel was the best? "Personally," Mr. Mueller had replied, "I would take a different kind of 'all of the above' approach." And the Tour of Destruction went on.

In after days, tooling about Weld County, I discovered that nearly every route was its own tour in each hot late afternoon of low cornfields and dirt roads—another pad of stubby cylindrical tanks tucked into a farmer's field; half a dozen tanks and two cigarette-shaped towers, all khaki-painted metal but for one dark

---

* See p. 178.

tank; more fracks off to the left, and on the right two more pads (environmentally friendly, as Mr. Winkler had sarcastically pointed out, since they were adorned with solar panels); then past them a wheeled irrigation line long and tall, ahead of us a silo, tall fuzzy-leaved mulleins growing imperceptibly toward the sun, then a dark stenchy plain of feedlot cattle. I had come at the perfect time, for the newspaper crowed: *Natural gas overtook coal as the top source of U.S. electric power generation for the first time ever . . .* And let me quote this West Virginia letter to the editor: *There are hotels and motels in Wheeling[;] there are simply no vacancies, due to all the fracking activities. When you do find a room, there is mud tracked into hallways and all over.*

What would Greeley look like once all its gas and oil had been fracked up? (That might be a long time coming, of course.) One summer in West Virginia when I mentioned the strange visual similarities of Iaeger to Tomioka, the mining lawyer Tim Bailey had replied: "I think the parallel with Japan is, no matter what the energy source, you're going to leave behind ugly ghost towns covered in vines. The coal's all gone from McDowell County, at least the good coal is gone, and that's why the black lung is up; I'm representing a lot of coal miners now who are needing lung transplants."—When I imagined a post-frack Weld County, I visualized blue collar housing developments turning into gangland, sickly fields irrigated with poisoned water, bumper crops of cancer cases. Of course, the greater evil of global warming would trivialize all such effects, in which case, since it was already too late, why not frack?—But what if it wasn't too late? (We sometimes titillated ourselves with those false hopes when we were alive.)

Bob Winkler acted as if there were still time to do something. When I praised him for his bravery, he said: "It's the older 55- to 69-year-old people who get involved. The younger ones worry about their jobs." This brought back to mind the security guard at Cook Mountain* who after opining that "nearly a hundred percent of the people in West Virginia would vote against mountaintop removal" added: "I'm not going to say it because I'm with the company." It is no denigration of Bob Winkler, or of his Kentucky coal mining counterpart Stanley Sturgill, to point out that neither of them had jobs to lose.

Turning south on County Road 61, I saw a horse farm whose back field had been decorated with a fracking well. On Highway 34, here came a truck labeled Baker Hughes Oilfield Operations, bearing tanks of hydrochloric acid. On 34th and 95th Avenue, headed toward cloud-crowded snowy mountains, I perceived a frack site in a cornfield, its narrow tower blackened on top and appearing

* See p. 96.

Mr. Stanley Sturgill showing us a good old American double dip:
a frack pad on the bed of an abandoned coal mine

inactive even while emitting a smell as of burning rubber which was followed by a longlasting bitter taste on the tongue.

Unlike coal plants or nuclear reactors, the frack wells were disheartening in their proliferation, each of them a small scale ugliness, all of them monotonous rather than strange, beautiful or horrible; their totality defeated me, for a fact. *Look at their fields,* Thoreau once glowed, *and imagine what they might write, if ever they should put pen to paper. Or what have they not written on the face of the earth already . . .* What the frackers wrote was **MONEY**.

2: A Familiar Theme: *No Comment* Again

Whether the fracking of Greeley resulted from a "choice" made democratically, which is to say, legally and openly, by the accountable organs and entities of Colorado, in a majoritarian fashion that simply left some people in Greeley outvoted, or whether it had been preestablished through crooked, secret and arbitrary measures is a question of significant interest. If we loudly invoked the carbon ideologies while leaping into our graves, then we were fools. Had we gotten steamrollered and blindsided instead, well, that would have revealed a charmingly different set of human deficits.

I suspect that both factors played a part. Bob Winkler would not have been run off the road had fracking not been popular among the locals of Weld County. And Sam Schabacker, who like Sharon Carlisle opined that the forces of fracking were antidemocratic, was quite willing to ascribe a significant degree of agency in the matter: "We're seeing as you do in California places that were agriculturally based, and then there was consolidation, fewer companies controlling the price of the market; the average farmer makes much less and he has to feed his family, so leasing out his land for fracking seems like a pretty good strategy. If you want to talk about the rise of fracking in Weld County you have to talk about the consolidation of the food system."

The way Sharon Carlisle wrote it, fracking was at least partially *imposed:**

> A faux grassroots organization called L.E.A.P. (Loveland Energy Action Project) was created here in Loveland by the industry (during Protect Our Loveland's campaign to pass a two year moratorium on fracking within city limits)—to prevent passage of our moratorium. This group is funded by the Colorado Oil and Gas Association as well as by locals who support their work—as were the lawyers who brought forth the challenge to our petitions—just 10 minutes before the closing of the 40-day challenge period. This was meant to delay the process long enough to prevent us from taking our initiative to the ballot. That didn't work, so our city council majority took over the role of prevention. (A story that could be a book in and of itself.) L.E.A.P.s mouthpiece is B.J. Nikkel—a former state rep for Colorado and former ALEC rep. It gets worse as the story continues: They had such a success here in Loveland

---

* There is a photograph of her in the local *Reporter-Herald,* with her anti-fracking sign beside her, and above her this headline: "Protect Our Loveland organizer Sharon Carlisle plans to sit on a downtown bench every day through Election Day."

that they expanded it into the Larimer Energy Action Project (for fighting pro-health organizations who oppose oil and gas development).

In June 2014, some anti-frackers visited Loveland in order to hear the self-
same *B.J. Nikkel, the director of Coloradans for Responsible Energy Development . . .
CRED is funded by Noble Energy and Anadarko* so that they can continue to have
their way with the people of Colorado . . .* This organization had called a rally
against the two-year fracking moratorium that Sharon Carlisle was hoping to
pass. *Nikkel's charges against the citizen's [sic] of Loveland were her usual canned
aerosol of lies . . . [S]he bestows on herself major credit for the legislation that gave
the oil and gas industry rights that negate those of local citizens . . . Nikkel also told
the crowd, as she always does, that fracking is safe . . . [J]ust last week a family in
Texas was awarded $3 million for damages to their health . . . from nearby fracking
operations.*

In November of that year, B. J. Nikkel co-wrote an editorial for the *Boulder
Daily Camera:*

> Regarding the energy bans that have been passed in local communities,
> we'd argue that anti-energy opponents used fear-based hype . . . to pass
> [them] . . . [T]he fact is, they're illegal.
>
> . . . Fortunately, because Loveland's vote on a similar measure was a few
> months later, we had enough time to adequately inform voters about the
> facts. Thankfully, Loveland voters rejected the poorly-worded and ille-
> gal initiative . . .
>
> Loveland is . . . a home-rule city that has wisely used its own local con-
> trol power to address specific energy issues, including . . . allowing for
> 1,000-foot setbacks from well sites to high-occupancy buildings—that's
> a standard higher than the law requires.

She proposed helping Coloradans to *understand the facts about fracking* so
that they could *thrive . . . without out of state interests interfering.*

Who was interfering with whom? That was one bone of contention between
our pro- and anti-carbon ideologies.

Of course I would have liked to give B. J. Nikkel her say in *Carbon Ideologies.*

---

* In the list of 2011's "top ten energy producers," ExxonMobil was number 1, Anadarko number 3,
Encana number 6, Chevron number 8.

I kept still hoping that the pro–burning carbon ideologues were somehow right. For some years I delayed facing up to the menace of global warming, simply because I *did not want to*. But finally I had to, for my daughter's sake, at least. And as a subscriber to the Golden Rule I tried to believe that most others—B. J. Nikkel, for instance—meant well—which implied that she believed what she said, so wouldn't she please convince me that it was true? After all, she gave speeches and editorialized on just this issue. Like hers, my position was "slanted" precisely because I possessed a point to make. Henry Adams, 1907:

> The historian must not try to know what is truth, if he values his honesty; for, if he cares for his truths, he is certain to falsify his facts ... Yet though his will be iron, he cannot help now and then resuming his humanity or simianity in the face of fear.

So in, for instance, this chapter, I put off as long as I could deciding what was in Adams's sense *true* about fracking. Rather than falsifying my facts, I wished to collect them.—If only B. J. Nikkel had bestowed upon me *her* facts!—Instead I must quote the summary report of my research assistant:

> BJ Nikkel Questions (from 7.30.16 disc): Emailed her at REDACTED* on 9-22-16, which might be an old email. I found an old number, REDACTED that I will try if I don't hear back from the email in a day. REDACTED emailed that address on 10-3-16. She responded on 10-3-16 and wanted to know more about your bias. We corresponded a bit and I am still waiting for her to say yes and then answer the questions. 10-09-16 I wrote her back and told her [of] WTV's offer for her to write at least a page and he will put it right in. If she wants to respond directly to anti-frack quotes she can do that. She hasn't responded yet to offer. On 11-3-16 she wrote back that she was too busy with the election to respond but wished us luck.[†]

Accordingly, the previously quoted (and less than complimentary) judgments of her individual carbon ideology must stand uncorrected.

---

* I removed this contact information for the sake of her privacy.

[†] Sarah McQuitley at the Greeley-Weld Chamber of Commerce (official phone number 970-352-3566, but spelling of her name not verified), the Evans Chamber of Commerce (970-330-4204), the *Greeley Tribune* (970-352-0211) and Basic Energy Services, Greeley, all likewise declined or ignored requests for interviews. What does this say about them?

3: "But That's Business"

Eric Ewing was a human resources man in the libraries business. He lived south of Greeley, on a county road off Highway 85, in the farming community of La-Salle, where through the spray of a great irrigation line one could often glimpse a fine green field, decorated by a beige fracking storage tank. I remember on the way down (just north of County Road 42 this must have been) a frack pad in a field right by the roadside that did not so much stink as sting the inside of my nose. Among the many picturesque regional sights of that drive was the Southern Ute Indian Tribe Gilcrest Gas Plant on the west side of Highway 40. (On the east side was B & J Hot Oil Service, 300 BBL CRUDE OIL, and a small farmhouse whose front window was curtained with an American flag.)

Mr. Ewing said: "My wife was actually on an inhaler when we moved up here from Denver, to get away from the pollution. That was in 2006. You know, it's an oil field, but we rented a house just right over here before this one became available, and it was really un-intrusive. It didn't have a tank battery right on it, not like that one there—we call those superpads. And that big facility down there, it was only 20% of what it is now. This was a good property and it was agricultural.

"One of the biggest problems now is the truck traffic. The tank trucks have four axles; they shake the windows.

"My next option is that I'll flipflop the basement and move down there.

"You know, in 2008 this well was drilled, horizontal; but it didn't bother us;

Eric Ewing, and the frack pad behind his back yard

I have family that drill, and so no big deal. Then in 2010, '11, something like that, they started building up that gas plant, and just seeing that, I thought, oh my gosh, that's not gonna help the resale of my house! Early 2014, we got a rigup notice for that one over there, and so I called them, and said, okay, I got that notice, what are you guys gonna do about the noise? And the guy said: Noise, what do you mean? I said, you're gonna do something more than hay bales, 'cause it's been loud. So that conversation kinda progressed, and even those employees feel left high and dry by the company, and they're saying, what do you want us to do? And the reason the well is there and not farther from the house is to get it out of the circle pivot for the crops. But then they put all the generators and everything right between me and the well, so the setback made no sense, and they ended up breaking the wall they built, and they were gonna move me into a Holiday Inn Express, and I said, a wife, two kids, two dogs, and I said no; so they ended up moving us out.

"The drilling took about two months, and then came the fracking and completion. I didn't agree, and I said, you're breaking the law, which is a big problem. If you're gonna do the fracking part with all of those big generators pushing the stuff down and then the big metal flaring can sends out this vibration, you're definitely gonna break the law. So the whole thing got delayed for three months and then they started mixing sand, and we called the cops, and got another action required, because just mixing the sand was too loud, and so we lived in our camper for awhile, and then in Estes Park, and when I'm staying in my camper in a campground for eight bucks a night they offered reimbursement just for that and I said, no, you're not gonna pay me eight bucks; I don't need compensation on that level.

"I understand it's a business and they have a right to do what they do and all that . . .

"I said: You have my phone number; I'm gonna move out; call me when it's over. I took a leave of absence without pay, because I was really upset. Mark Dickinson the land man said, okay, we'll give you a call; that was July 30th of last year, and come August 15th, I wanna go home, and we went back but then we left again because they were doing the combustion flaring and it was rattling the house, and I never ever got a call from him again to say they were done.

"I'm not anti-oil and -gas, but I don't like being mistreated by anybody.

"You're not gonna stop 'em, so what you want is fair compensation, but they went, no; their lawyer sent me a letter. And so they ended up getting fined, I don't know, $10,000 for breaking the noise regulations.

"That's Noble Energy did this. Noble got busted by the EPA through the use of FLIR* cameras."

---

* Forward looking infrared.

Reader, from this same patriotic event, whose multiple sponsorships by the fracking company in question reminds me of the West Virginia Coal Festival, I have spared you images of the long bus for Weld County School District 6 which was conspicuously labeled *A SMARTER WAY TO GET TO SCHOOL: POW-ERED BY CLEAN NATURAL GAS & NOBLE ENERGY,* and a plausibly horse-drawn and jouncy vehicle from the Newell Stagecoach Ranch, with Noble Energy's name on the door. But perhaps it would give you pleasure to see

Noble Energy at the annual Greeley Stampede

Noble Energy float, Greeley Fourth of July parade

"Working Together for Hunger Relief": Noble Energy and the Weld Food Bank, Fourth of July parade

("Noble Energy is the major supporter of Banner Health," Sharon Carlisle had said sarcastically. "Now, gee, why do you think they're supporting environmental health?")

"Eric," I asked, "do you think that Noble should be prohibited from fracking near your home?"

"First I think somebody can do whatever they want with their land as long as they don't affect other people. You can have a rock band in your garage as long as it doesn't keep the neighbors up at night. I don't know enough to say if it can be done safely or not.

"When they're drilling, I think they use some sort of alcohol base, although some of my buddies say they use diesel fluid. What we experienced was kind of a sweet smell. And then what's actually the fracking fluid has a janitor closet smell. My kids are allergic to it. They're six and four. We ended up going to the doctor about this. And all the sudden it's like I'm congested in my breathing. The toxicologist said, the symptoms you're describing is what you get from industrial toxins. It was all kind of really weird, the way the doctor dropped me over that; I had to say, are you my doctor? Is it from that? Because if it is, I'm going to move.

"I don't know if they comply with the laws.

"Noise is annoying, but breathing stuff like that and having it affect you is kind of terrifying.* My son got a bloody nose but I don't know if it was him picking his nose.

"If they can run the plant better and it doesn't shoot natural gas liquids into the air and drip on my house, that would help. And that flare is because burning the methane and turning it into CO_2 is better than [simply leaking] methane in terms of climate change. But then we get diesel exhaust; my wife's taking a shower and the flare's blowing and the wind's blowing at us. Strange smells, and they're not agricultural.

"They're completing the well; sometimes it smells like burning rubber.

"Sometimes I tell the family, okay, we're gonna take back our yard; we're going outside to play; and all the sudden, baby has a bloody nose!

"My brother that's two years older than me has leukemia. And what causes leukemia? Benzene.† So you have to wonder. The feedlot there, you have to smell that, but nobody gets ill from that! They're not dumping the manure in my living room.

"It's daily when they're doing the extraction. You know what's causing that natural gas smell?‡ It's the pressure valves leaking on those tanks. The lines aren't big enough even under pressure.

"I have a FLIR camera photo that Earthworks made and they took a FLIR picture of that well nearest us, and it was spewing VOCs, but Noble said there was upset flowback, and they have the right to do it when they can't control it. The other day, they can't have visible emissions coming out, and it was smoke, and if it's technically longer than one minute within a 15-minute interval, it's a violation, so I filmed it, and sent to COGCC. It went on for hours and hours on a Sunday, and Monday somebody shows up and fixes something, so COGCC didn't view it, so they didn't break the law. And that flare over there, it smells really bad when it's smoking like that, and I sent a video and got nothing. I think the regulators have been made helpless for a reason. I think when they get really really good, they get hired by the oil companies.

---

* According to an environmental scientist named Wilma Subra, "more than half of the people who live near shale gas extraction or processing operations in Pennsylvania report suffering, alone or in combination, respiratory impacts, memory loss, lethargy and throat irritation." On the other hand, the various waitresses I met in Greeley, the bartenders, hotel clerks and a bookstore manager all assured me that their health was unaffected.

† Of course leukemia could be caused by any number of factors. And was benzene one of the particular VOCs that the Ewings were smelling? If it was, then the following inspirational epigram from the Atomic Energy Commission (1967) might have cheered them up: "Benzene hydrocarbons can permanently damage the very [biological] systems needed for recovery. One severe dose of these may leave the victim anemic for life . . ."

† As we know from Stanley Sturgill (p. 1), "that natural gas smell" must have actually derived from crude oil's VOCs.

"This guy over there, they basically drilled him out of his house because he went crazy; he moved out . . ."

Perhaps the man who *went crazy,* or Mr. Ewing himself, should have made fewer waves for the poor *regulated community.* The *Investopedia* can tell you why:

> A well-structured initiative approach followed by the US for fracking has enabled a decline of around 37.5% in the natural gas imports between 2007 and 2013. Further fracking may lead to the US becoming a net exporter of natural gas and the largest oil producer in 2020 . . . Much of America's oil is in private lands, which enables the private sector to override government bureaucracy. This kind of shortcut is not available in all nations.

Good enough, right?

"See the orange stakes over there?" said Eric Ewing. "I have pictures of tanker trucks blowing stuff out the back onto the field. And they spray the roads with you know what. They just got finished fracking and they were spraying the hell out of it.

"They'll refrack; they're sitting on gold. Out here, it's not exploration, it's production. It's really volatile, I guess; some of the dudes skim some of the volatiles off here and they can burn it in their trucks.

"I work right next to Frontier Academy and my daughter goes to that and someone wants to frack there.

"The same folks that took the FLIR of this one took the FLIR at Northridge, and they caught this huge plume of VOCs and took it to the school board, and the oil company said, oh, that's when we're transferring it to a tanker, so they added the equipment so the VOCs wouldn't be released.

"They do the cheapest thing they can until you complain or catch 'em. But that's business. But we've kind of done that to ourselves.*

"People just want me to stop complaining about it.

"Last year I would not have been able to talk about this hardly.

"That's the worry now, is lightning striking those things, because it does happen and then they blow up.†

---

* A good summation of *Carbon Ideologies.*

† Regarding injection well fires Brad Mueller had said, in contradiction of Bob Winkler: "If these things start on fire, they don't go kaboom, and I can't explain because I'm not a fire expert, but if you have a fire leaking it just starts on fire like a tree would start on fire. We had a very interesting incident where an injection wellsite started on fire. It did have a tank fly into the air."—He showed a picture of

"The blast radius for that thing is really beyond my house . . ."

"How do you feel about fracking and global warming?" I asked him.

"Well, it's probably gonna make it a lot worse, because are we digging the bottom of the barrel now? The first thing is all that methane leaking; global warming is worsening by it leaking. If you take all the politics out of all this stuff, we have to do something and move to renewables."

"How would you rank the various fossil fuels?"

"Well, I don't have a coal mine next door. So I would much rather have coal burning than have a frack well next to our house."

Wouldn't we all! Let me requote the lawyer Barney Frazier:* "I generally, myself, do not oppose mountaintop removal all over West Virginia and Kentucky and Ohio and so forth. But what I am opposed to is *mountaintop removal behind my house.*"

Reader from the future, might even you be one of us? If you could still burn carbon, why wouldn't you? Isn't it merely your own personal suffering, caused by us, to which you object?

Mr. Ewing continued: "If it didn't leak, if it didn't turn the whole farmland into a factory . . . This stuff is in between every single house. This stuff is unliveable. Pueblo has a coal-burning plant and I sometimes come to Pueblo to fish, and once we saw a yellow smoke come out and I said oh my God, and my good buddy who's pretty rightwing is like, well, you're like this because of what happened to you.

"I am what people don't want that want setbacks. They don't want to have happen what happened to me. I don't have a purpose to be out here; the farmers do and the oil companies do, but I don't.

"We can get along, if they try a little. If they flew me to the CEO's mansion and let me stay next to the pool for a month, I might be their champion.

"One day a guy blew the stop sign down and I stopped him and said, hey, you're not out in the middle of nowhere; we drive our kids to school; and his response was who the fuck are you? and I said, I'm the guy who just saw you break the law, and I'm being nice. It didn't improve, so I ended up calling the public

---

the flying tank and smoke, then continued: "And you think, oh my gosh, we can't allow these anywhere, but the thing is that with injection wells the tanks sit on top of the ground. But that was completely consistent with the fire study that was done in 2006, and the setback requirements were followed, and what happened was what the models predicted. The tank did not explode."—Sharon Carlisle saw a different theme in that story: "You know those big holding tanks, the big brown ones that hold the produced water? One of the newspapers caught a photo of one of them that was struck by lightning and it was flying through the air, *on fire*! It's kind of an insult, to even call it water."

* See p. 164.

relations people, because that's how it would work in my company, and that aspect did improve.

"I guess we're kinda learning to cope with it; there's things I can't talk about regarding the gas plant; I'm under an agreement of confidentiality; I can't talk about how shitty the company has been.

"I'm gonna try to sell the house. I'm gonna bake chocolate chip cookies to make the house smell nice."

4: "But That's Business" (continued)

Eric Ewing was an outlier, a brave man of sorts. He dared to challenge Noble Energy, and even won something, that kind of resolution in the face of intimidation that Stanley Sturgill owned in greater degree. Mr. Sturgill was the stronger of the two because his having been first a career coal miner, then a career inspector, endowed him with two kinds of authority: long experience, and power. In the matter under discussion Mr. Ewing owned only a victim's experience. He had been marginalized and traumatized. No government office stood behind him. He could not issue violations and fines. Nor was he a former fracker or oilman; therefore he lacked certainty as to what Noble Energy could legally do in Greeley. Despite these difficulties, instead of simply going away, as I might have done and even Bob Winkler wished to do, he attempted to make Noble Energy in some way accountable, at least to him. "People just want me to stop complaining about it," he said, and yet he still went on record for *Carbon Ideologies*. I hope that he never suffered for it.

He was a peaceable man who wanted to be accommodating: "We can get along, if they try a little." "I guess we're kinda learning to cope with it . . ."

What else should he have done? Unlike Mr. Winkler and Mr. Sturgill, he still had to report to an employer; moreover, he remained vulnerable through his family. I admire him; I wish him peace. Had more people done what he did, perhaps fracking's inevitable after-effects would have been less ugly. The fact that most people, in Greeley, at least, had absolutely no intention of doing what he did, makes him all the more sympathetic to me.

And you from the future, how would you judge him?—"I guess we're kinda learning to cope with it . . . ," and maybe that is how you talk to your hungry children there in the sweltering desert on the edge of the rising acid sea.* From

---

* When I face the possibility that in my daughter's lifetime, and the likelihood that in her children's, if she has any, some horrid divide must be crossed, be it the so-called "tipping point" that climatology's popularizers proclaimed would be reached when the planet's average surface temperature reached this

your embittered perspective it might be convenient to blame Mr. Ewing, and all the rest of us, for merely *learning to cope with it,* instead of stopping it. But how could we have stopped it? And if we could, do you actually expect us to voluntarily live like you? The Ewings had electric lights and a refrigerator. They owned at least one vehicle, and any number of nylon products. As for me, I sit writing this on a peculiarly warm September evening in 2016, listening to my stomach growling and wondering whether I should skip dinner or fire up my gas stove . . .

---

or that number of degrees Celsius, or some pH index for our oceans, or a new threshold of relative humidity, I seem to be looking ahead and down into an age of fear. The loss and helplessness that I associate with your lot may be exaggerated projections of my ignorance; on the other hand, the heat-induced increased activities of various atmospheric molecules will doubtless worsen the unpredictability of human affairs—all the more reason that if I can be sure of any aspect of your character, it is that you are not as I. Since all I can do here is imagine you in my image, of course I have failed. I was as fossil fuels made me. They kept my lights on. Hence I who imagine myself to be open-minded will appear to you as deservedly dead, fossilized in the stratum of my own period's prejudices.

COLLATERAL DAMAGE

My cohort grew up knowing that "there are always winners and losers," because "you can't make an omelette without breaking eggs." In evaluating our four carbon ideologies I have made a point of recording victims' stories, in order to weigh the fabulous energies inherent in, say, methane and enriched uranium against those fuels' collateral damage. A government or corporation which refused responsibility for the immediate and conspicuous suffering resulting from its policies seemed likely to behave similarly in regard to the merely future and hence inconspicuous suffering associated with climate change. In other words, Noble Energy would treat my great-grandchildren as well as it had treated Eric Ewing. Meanwhile, the cancer-ridden town of Prenter, West Virginia, was living (or dying) proof that the coal industry and its political shills and puppeteers *did not care;* hence their angry dismissals, denials and no-comments about global warming only strengthened my belief in the latter.— Well, so what? Which eggs would *you* have broken? As I revised this chapter, sitting warm and comfortable on a bullet train from Tokyo to Kyoto, with fallow ricefields swinging their terraces in and out of sight, Mount Fuji's snowy shoulder fading into clouds, I felt pretty good, I admit; my picture window (which might or might not have been one of the "big five" energy-greedy materials) was extremely clean; the car was silent; my fellow passengers read paperbacks, carefully ate bento lunches or worshipped the screens of their little phones; having paid good money to be here, we were all entitled to speed along, and wasn't I as deserving as everybody else? So which collateral damage should I have comfortably or uncomfortably accepted, in light of the consoling truth that whatever I did made so little difference?—Now Fuji was long gone, and steep little green mountains made up the horizon. Everything between them and me was urban protoplasm. Cloud-glitter rushed across the faces of skyscrapers; an opposing train's blue stripes swished by; all was orderly and bright. The pretty, immaculate conductress came gravely by, with her flat-topped cap just so; she carried her official pad, but found no need to use it, or even check our tickets, because electricity-consuming devices had already informed her which seats were supposed to be occupied. Already we neared Nagoya. That was when I decided to evaluate collateral damage in increasing order of severity, and so here is a continuum for you, keyed in to our first and most dramatic carbon ideology, nuclear power:

DISPUTABLE AESTHETIC OR EMOTIONAL LOSSES

Mr. Osumi Muneshige, formerly in charge of public relations for the town of Futaba, once "promoted nuclear power stations . . . But now I know how unsafe they are." In 2013 he said: "It took several hundred years to create Futaba, yet it does not exist anymore . . . We had beautiful nature, food and clean water. I now realize that it was the best life I ever had."

Mr. Osumi's loss may approximately compare with that of the West Virginian Dustin White, who singled out Patriot Coal for being *currently the company dismantling my ancestral home of Cook Mountain.* Haters of frack pads, or for that matter of wind-turbine farms, might have uttered their own equivalent local objections. In my time it was counter-argued, and in good faith, that satisfying widespread increases in electrical demand justified such involuntary sacrifices, which could be trivialized as "personal" or "aesthetic."

EMOTIONAL STRESS

In the Fukushima chapters you read so many accounts of this that there is no need to insert new ones. Eric Ewing's misery and Sharon Carlisle's anxiety about fracking and the future of our planet were comparable. The stoic resignation of Appalachians about coal-related disasters, which reminded me of Tohoku people's patience after the tsunami, made their worries less conspicuous to this outsider. Perhaps such concerns were more relevant than any others to the main issue of this book.

INTIMIDATION

We now turn to grimmer situations. That emergency room nurse Cathy Behr, who hardly dared to say anything thanks to legal pressure from the resource industry, and again Sharon Carlisle, who as you recall advised me that *if you print the previous statement, stating corporations and individuals by name, I may get sued,* and again Mr. Ewing, with his "I'm under an agreement of confidentiality; I can't talk about how shitty the company has been"—these anxious, cautious people made me wonder: What exactly couldn't they say? Even Bob Winkler sometimes lowered his voice . . .

When I asked my Japanese interpreter whether outspoken victims of the nuclear accident were ever comparably intimidated, she replied: "Some of the activists must have lost their friends, because of the difference in opinion. And those whom I communicate with are all like-minded people, right? So my opinion may be biased. But those people say that they have lost some friends and they cannot talk about nuclear to those people who have different opinions."

"But would the company itself threaten people?"

"That I have never heard, that kind of interference from Tepco. But maybe I don't know well enough. You know, one of the former prime ministers, he used to be pro, and now he's anti although he might now still belong to LDP,* and because he's a big name he's kind of powerful, helping the anti movement. He has no problem. As far as I know, Tepco haven't attacked an individual. But I know the Fukushima prefectural government is going to reduce the number of health examinations so that suspected cases won't show up. That's what I heard from a lady from Fukushima, and she told me that crying! She said, what do they think about us? We are human! I could only say, that's only too bad and they are doing bad things and I am sorry for you, but effectively we can do nothing more for you. That was when I was doing a volunteer project to help give evacuees a good rest, and we did help them a little . . ."

"So would you say that Mr. Ewing is more constrained than a Japanese would be in his situation?"

"That does sound worse to me. Worse! But, well, I don't really read a lot, but at least I get No Nukes Plaza's† e-mail messages every day and read them, and newspaper, and anything like that is around in news, but if it's not in news, because of someone's constraint, of course there's a risk that I wouldn't know."

"Evidently Mr. Ewing has to keep quiet because of some settlement from Noble Energy. So couldn't that happen here?"

"The corporation might solve those complaints by paying money, and then they would be silenced. For those who don't really complain, it seems that the amount is pretty limited and they're unhappy about that. But if it seems that you are too dangerous for the corporation's reputation, maybe they would pay more. Anyhow, that's just my supposition."

"In all the time that we interviewed nuclear disaster victims, did you ever hear someone say, *there are things I cannot tell you*?"

---

* Liberal Democratic Party.

† This was the NGO whose longterm member, Mr. Yamasaki Hisataka, I interviewed for this book; see I:327, and this volume, p. 230.

"These people were basically patient, so they didn't say anything too bad about Tepco. But they never said, *I cannot.* So obviously, when you saw these people, there was no pressure with Tepco."

When I asked Mr. Yamasaki from No Nukes Plaza for his opinion as to complaints and redress, he replied:

> There are a lot of cases filed in Japan as well [as in the U.S.], although almost none of them are filed directly due to radiation exposure, but many [people] claim for damages.
>
> The procedure regarding Tepco's nuclear power plant accident damages starts with sending a claim to Tepco. If Tepco refuses the claim, then you go through ADR (an out-of-court settlement procedure that involves arbitration). If you are not satisfied with the arbitration decision, then you go to the court.
>
> Quite often even the ADR's settlement proposal is refused by Tepco and many people are obliged to file a suit.
>
> In Japan, it is very difficult for an ordinary person to file a suit, which is something that we rarely do. In the case of the Fukushima No. 1 accident, I understand that if you claim for . . . compensation and if part of it is given by Tepco, many people would normally give up [on] going further to get the full amount.
>
> As ADR is an out-of-court procedure, it's not that difficult compared with filing a suit, but still many people would hesitate. There are cases where a municipality was granted some compensation through ADR after their [*sic*] claim was refused by Tepco. Iitate village decided to claim compensation through ADR on behalf of individual villagers, but Tepco's response has been almost total refusal.
>
> The fund for Tepco to compensate [pay compensation from] was assigned/granted by the government. The government paid an advance following the Act on Nuclear Damage Compensation Support Structure. The . . . allowance is currently about 14 trillion yen, of which about 8 trillion has already been expended. Because compensation payment will continue, they would like to minimize the amount they pay for each claim . . . , and they make a variety of excuses for not paying. Under such circumstances, I don't believe that any decent compensation would be made.
>
> . . . In Japan because of this sort of structural problem, Tepco has silenced the victims not even with money but without money.

Once again, I can speak only as an outsider, but I repeat that at least Tepco did meet with me and answered my written questions; at least their decontamination toilers were visible year after year in Fukushima, and, as the interpreter noted, none of my Japanese interviewees was saying, *I cannot tell you.*

The Coloradan and Appalachian situations seemed worse . . .

FAMILY SEPARATIONS

Evidently because of a conflict between employment necessities and *radiation fears,* the mother of the Matsumoto family, who used to live together in Namie, now dwelled with her elder son in Kanagawa Prefecture, while her husband and the younger son removed to Fukushima City. The mother said: "The four of us can only get together for the New Year's holidays."

I cannot find any carbon-powered analogue to their plight. In the United States—if, for instance, some of them worked in Prenter, while the others lived where they could drink less carcinogenic water—we probably would have dismissed their situation as a "choice."

FINANCIAL LOSSES AND POTENTIAL MEDICAL ENDANGERMENT

Next came the matter of Mr. Hoshino Yuji, who used to forage in the mountains of southern Tochigi Prefecture. In 2013 he said: "Everything I've done till now, it's all become no good. I can't collect wild vegetables and I can't sell my mushrooms. There are problems with the fish in the rivers and I have to worry about contamination levels in the wild game, too. That's what makes me the most angry."

And there was Eric Ewing considering baking cookies so that he could sell his house.

Their cases bore some comparison to that of the waitress in McDowell County who wouldn't drink her local water—but when I asked her why it was bad, instead of explaining the root cause—acid drainage from coal mines, which perhaps she didn't know—she replied: "Just everything falling apart one stage at a time."* Was she angry or merely resigned? I'll never know.

Since some of Mr. Hoshino's losses bore quantifiable economic aspects, resource extraction's kinder profit-worshippers might have written him a compensation check. And why shouldn't potential medical considerations have impelled

---

* See above, p. 149.

a progressive government into writing other checks, as well as making those fish, vegetables and mushrooms off limits? Meanwhile, nobody was buying that waitress cases of safe water, paying her damages or even publicly prohibiting residents from drinking from the municipal tap. But couldn't this issue have been reduced to one of "mere" emotional stress? There was no *guarantee* that the water would make a given person sick!

CONTAMINATION-INDUCED SUICIDES

Since she and her husband had just lost their chicken-farming jobs in Fukushima City, and *no one could tell her when she could go home again,* Mrs. Watanabe Hamako, aged 58 and formerly of Kawamata town, which lay 40 kilometers from Tepco's Nuclear Plant No. 1, took advantage of kerosene's high heating value to burn herself alive. In 2014, a court awarded her family 49 million yen in damages, to be paid by Tepco.

A dairy farmer who became insolvent thanks to the banning of his radiocontaminated milk hanged himself, having chalked the following on his new compost shed: *If only there were no nuclear reactors . . . and I have lost the will to work . . . I am sorry for being a father who was unable to do anything for you.*

Perhaps because of their Christian beliefs, Appalachians did not seem prone to this kind of behavior; and in Weld County I met no one whose suffering was so extreme.

CONTAMINATION-INDUCED INVOLUNTARY DEATHS

In that year Tepco also agreed to pay 13.5 million yen to the survivors of an 83-year-old who got promoted from hospital patient to nuclear evacuee, with consequent death from dehydration.

What if some resident of Prenter or of Fayette County had died in cancerous agonies from the poisoned water? We would most likely have never heard about it.

TALLIES AND CROSS-COUNTS

From *The Japan Times,* Friday, March 14, 2014: *Although no one has officially died as a direct result of the [nuclear] crisis, at least 1,656 Fukushima residents have died due to complications related to stress and other related conditions.*

Of course my first reaction to these accounts was sorrow. Then, being who I was, I began to wonder how they stacked up against the stories from Colorado. Had fracking caused as many Weld County as nuclear-blighted Fukushima Prefecture residents to die *due to complications related to stress and other related conditions*? That seemed absurd—but, then, I must allow that Cathy Behr's response shocked me.

In several of the cases just mentioned, Tepco had done the honorable thing, accepting culpability and paying out damages. Mr. Ewing had received some settlement, which was to the credit of Noble Energy, but why couldn't he talk more about it? (Mr. Hoshino could openly express his anger at Tepco in the newspaper.) How correct was Bob Winkler when he said: "There are no liability issues because there's no law to stop these people"? Should Tepco have been liable for the unsold mushrooms of Mr. Hoshino and the end of Mr. Osumi's "beautiful nature, food and clean water" in Futaba? Meanwhile, what moral claim do *you* say Mr. Winkler had against the frackers?

Some of us used to draw ethical lines, and say yes to this, no to that; we could even see the lines we drew because carbon kept our lights on. These lines were as idiosyncratic as we were. I, for instance, believed in women's right to abortion, while others disagreed, because they longed to protect the unborn. Meanwhile, I did what little I could, which was nothing, to protect *you,* my unborn reader, from climate change. So I lacked consistency; maybe I therefore lacked grounds for my disapproval of the drill-now frackers who cared nothing for you.

At the very beginning of this book I stated my belief that what we needed back in our time was proficiency in performing risk-benefit analyses. More specifically, we should have learned how to make them for ourselves, and how to evaluate those proffered to us (or foisted on us) by others.

"SOMEBODY'S GOT TO STEP UP AND
BE CREATIVE"

1: Explications of an Environmental Consultant

John Mahoney, whitehaired and strong, with a red Phillies cap, constantly gesturing, tapping the table, smilingly affable, was the environmental consultant who had explained that all the oil hereabouts was of middling quality, but easy to get at and hence cheap. In him I saw the quality so often and likeably apparent in people who do physical work. It is a kind of confidence, based, I believe, on the repeated experience of getting a strenuous job done, with results that can be seen. Just today I was chatting with a landscaper who sometimes got called upon to clear fields of poison oak. I asked him how he managed, and he said that he got blisters, but just did what he had to do and never made a big deal of it. Mr. Mahoney was made of such unassuming fortitude. He would have been better fitted than I for that hot dark future in which he did not believe.

He said: "I read, our reserves now are greater than what they ever projected they would be!"

I could hardly disagree.

He said: "We got so many people on this planet; we need energy! It's not realistic to say we're all going to go back to live organically or whatever. We have the energy; we want to help the rest of the world. But if we can't be energy self-sufficient, we lose a lot of control."

This, too, I believed. When we were alive, we citizens of developed nations (some of us, anyhow) truly might have given up certain comforts and choices, if they had made a verifiable difference. But we would hardly *go back to live organically or whatever* for the sake of some murky benefit, especially if Chinese carbon emissions, for instance, undid our own emission cuts.* And that matter of the multitudes on this planet needing energy—a point which you will hear in later pages from the mouths of true oilmen—was so evidently, if not true (they didn't *need* it), at least *just* (because why was it fair that we had it and

---

* See Chris Hamilton's assertion on p. 201.

Bangladeshis didn't?), that I wished to accept it; I aimed to please back when I was alive.

Indeed, when Mr. Mahoney outright said that *we want to help the rest of the world,* how could I not be touched by that? The oilmen also said so, and I came to believe that some of them meant it. They pointed out, and were surely right, that we all loved their product! They were proud of what they had done for us, and resentful to be underappreciated and even vilified. I'd call that human nature.

Mr. Mahoney was not an oilman, not anymore. "I don't work for the oil companies; I contract with them. I'm *proud* of 'em."

About fracking he said: "I'm not an expert on this by any means. I had a master's in geology from Ohio University in the late '70s. After that I worked for Macco doing production exploration and geology. That was great. Working for a big corporation, it really straightens you out.

"A lot of us, we walked in one day and our jobs were gone. That was because the energy industry did their job and found a lot of energy. That's where the Deepwater came in. In that mid-'80s time frame, we had offshore lease sales. I wasn't part of that one, but I did the next two or three before I lost my job.

"I'm a basic environmental consultant. I like to do the simple stuff. I do stuff at my age that most consultants don't want to deal with, the stuff that's not glamorous. I traveled around a lot in the oil fields; most of that stuff was in West Texas and the Appalachians, real nice geology. A lot of the stuff I do now is dirt geology.

"One thing I do, a lot of my work involves Phase I environmental assessments. It's basically a property assessment for real estate. The banks don't want to finance loans if there's a possible hazard. So that's what I do. As a side we might comment on things like asbestos in the building. The older the area, the more complex. I have to try to go back to the 1940s. I can't verify dumping, but say there was a garage shop there, and a garage uses such and such type products, and if there's a real problem I recommend drilling to find contamination. Just yesterday when you called I was drilling a couple of holes downtown. The bottom of most tanks are about 14 feet below the ground. I knew groundwater was at 18 to 20 at that area. Well, luckily there wasn't any contamination, so I just went out to 14, 16 feet to get to the bottom of the tank. If you want me to check for any release of contamination I go to 15 or 20 feet. Otherwise I only go to groundwater.

"I also get involved with the oilfield guys, such as when they're selling or buying property with other companies.

"Oil spills, that's pretty easy; you just dig it up and take it to the permitted site.

"The geology here is not very complex. Essentially the rocks are far softer than in other areas such as the Appalachians, and so it's far cheaper drilling.

"There's not any real difference here between gas wells and oil wells. When you have a gas well you get oil production. The oil separates out; you call it condensate oil; well, that almost looks like gasoline; it's very light and clean. In oil wells, it comes out as black crude, but you get a little gas with that. If I get an environmental problem with the oil in gas, that cleans up quick; but crude oil's a different story.

"In this area, you find oil and gas that's pretty shallow in some areas, but on the average, five to seven to eight thousand feet, although it can be deeper. The geophysicist looks at it and says, I know we got stuff at eight thousand, but we need to go below; we need to try that. This is an old basin, the Denver Julesburg basin, or the Wattenberg basin; those are the big names. Most of it comes out of the Cretaceous rocks, but it's not as old as some of the other geological areas in this country; this goes back 50, 60 million years.

"The understanding of how and where oil occurs, it changes constantly, because there's a need for it to change. We can't just stay put, and take it only out of these old wells in West Texas. Somebody's got to step up and be creative, if only to keep their jobs. And now the new technology is saying that the shale can produce the oil and the gas. Before, they've only known it has hydrocarbons. They didn't need to know how to get it out.

"My professors, they were real petroleum field geologists with camping gear and donkeys and going into the jungle or whatever. The person who does that now is trained in both geology and geophysics, which is basically the study of physics and applying it to the rock process. They send sound waves into the ground and bounce it back. Now everything is 3D; they can say, this is what the rocks look like and this is what the structure is. The engineers come in later to decide how to drill the hole."

"How deep can they go?"

"I'm sure they can go give or take 15,000 feet.* Their models are fairly accurate but they drill dry hole."

(A year later a Vice-President of the Bank of Oklahoma in charge of oil loans—his interview appears on p. 486—assured me that the technology was even more superpotent than that: "They can drill 40,000 feet.† It depends on how

---

* In a glass case at the Phillips Petroleum Company Museum in Bartlesville, Oklahoma, lay the drill bit which had pierced through the last 25 feet of a very special shaft in 1959: 25,340 feet!—which held the record for another 12 years.

† "You can see to bedrock," the Vice-President continued. "It's how well you can see. It's what the resolution is. When I tap the table, like that, sound had to travel from here to here. Well, sound travels

hot it is in that area; they have to use chromium and really really expensive stuff. They'd better be looking for something really big.")

Mr. Mahoney was saying: "I believe in the model that I was taught. There's certain environments on the surface that are conducive to the production of sediments that become oil source rock. People think it's from dinosaurs, but that's not so. Coal is plants; coal is swamps—*huge* swamps, stuff that makes the Louisiana swamps look like nothing! Oil and hydrocarbons comes from microorganisms. That's why the shale is so important. The black mud in huge lagoons like some of the Bahama areas and whatnot, a lot of life is going on in there; someday that might be shale.

"This used to be an intercoastal waterway here, from the sea all the way to Kansas, then it got filled up by the erosion of the ancestral Rockies, and the black muds got covered up by thousands of feet of sediment, and they got warm, they got squeezed and eventually they squeezed out all the oil. The black shales are the big thing for the horizontal drilling and the fracking.

"Fracking they've been doing since I got into the business back in the late '70s, and they've been doing it long before it. Here's the deal. When you drill a hole in the ground, you're lucky to get only 30 to 50% of the reserves in the ground. We didn't have the technology. So then we started drilling more holes, and inserting water to push the oil toward the producing well. But you have to be careful because the water can get around the oil and ruin the old well. So they started fracking. Back in the '70s, the common term was an *acid frack*. Oil is in a little hole in the rock, not in a big pool. The idea is to go down under pressure, and get acid in there, to dissolve the carbonous cement, to help it flow better, and it only penetrated so far from the well.

"There was one case in Colorado, before my time, when they actually put a small nuclear weapon in the ground, and, funny thing, after that they couldn't use the oil!

"We're always thinking: What's the biggest thing we can do to make that thing Swiss cheese down there?

"Oil down there naturally has a radioactive component. Everything in Colorado is just like North Jersey; it's a high radon area. Out here, we have thousands of feet of unconsolidated sand. It can be considered bedrock, but it's very soft bedrock, but that radon comes with the sand. Even out here, we can collect

---

much more in this denser material. You can make a sound wave and put it down there, and some of it goes through here and it goes to the next thing and the next thing, and some it reflects back up. With these geophones, you can listen; you get enough of those all around, that's what you use those supercomputers for the mathematics, so you can interpret so you can actually see the formations. Then you get salt, it messes things up. It does something to the sound, it absorbs or it scatters it. So they can now see below the salt."

radon. But it's easy to take care of. You just put in a drain around it if it's a slab, or a fan in your basement.

"But as far as fracking, so, they've always been fracking. The difference is that now, and I can't tell you how they do it, they can drill almost sideways; they can kick it out at a sharp angle, and that increases your production; you can go at an angle and just follow the bedding planes. It's very expensive but it pays off. They drill less holes to get the same amount. They go in when the hole's drilled, and then they frack it to get the stuff out. They were doing the acid fracks back before I started in the '70s. They can use different stuff, not just acid; it all depends on the kind of rock.

"We have a lot of water used to frack a well. Three things come up: gas, oil and water. You've got to get rid of the water. Up until the '90s, they used surface evaporation ponds. That became a no-no, for environmental purposes, so they inject it back into the ground. The truck will go to the tank battery, and they will pump out the water. The wells are pretty consistent: so much oil per day and so much water per day, and they take that to the disposal well; that's far more water than the fracking operation. We've got to get rid of that produced water."

"Would it be economically practical to frack with only produced water?"

"They have projects in the works now where water is recollected; they have to treat it. Anadarko has a huge site with monster tanks, and some people are trying to do that with the brine water. But the problem is the chemistry. You don't want to be putting microorganisms in there that can clog up the well. They have to be real careful with what they've got in holes. These facilities will have *tank, tank, tank,* a handful of tanks. Oil floats on water, so they pump the water out from the bottom and they move that to the next tank. So by the time that water goes back into the ground, almost all the oil is out.

"The basic thing is, can fracking cause problems? There's no real known problems known yet. The propagation of those fractures, it's all happening down at seven, eight thousand feet. The groundwater in this state, it's only a couple hundred feet. After that, it's all brine water.

"Aquifer, it's all shallow. Some places do have shallow oil fields, absolutely. But most problems occur not because of the fracking, but from human error. Gas stations, I've taken out a lot of tanks, but the problem I usually find, it's not in the tank, it's in the piping, or the fitting. A truck runs over that, only three feet under the ground, and the fittings can leak—thousands of gallons, quickly. When you're doing oil field, it's not the fracking, it's the piping. Things leak, things go wrong, when they have thousands of feet of piping in the surface. That's human nature. I can't go through a day without making a mistake; neither can they.

"In the past couple of years in Colorado they passed some regulations that some people might think are overbearing or whatever, but they all are designed to make things better. Piping under the ground freezes, and I've had to clean up spills even from freeze-resistant valves. I've had stuff impacting groundwater, some pretty big deals, stuff they don't know what's happening until they can come across it. If you drive around, you can see all these tank batteries; they can check them every day or every other day; above ground they can monitor it all the time; underground they can't see it."

"Can fracking cause earthquakes?"

"Yes. In Colorado there were essentially two major sites. One was Rocky Flats and one was Rocky Mountain Arsenal. In the '60s they needed to get rid of their bad water, so they injected that water eight to 10 thousand feet deep, and they got tremors, and they stopped doing it. What happens is there's old or minor fault planes down there, and they can shift a bit; they've had a couple in this area, maybe level 4 or 5; mostly it's around 3 or 2. Rarely does it cause property damage. Unless they have problems with the piping, they have minimal impact on groundwater. It's all about how they build the wells, and mostly they do a pretty good job."

"What do you think about global warming?"

"Back in the '70s, the deal was, we're going to start a new Ice Age! We flat don't know. Statistically, they can't prove the storms are more frequent. We watched the Weather Channel this morning. The people that visited us in Missouri, that tornado hit their own small town. Only a shopping center got damaged, thank God. To me it's about the technology. Back when I was at school, the winters were horrible. Science, the weather patterns, they're cyclical, and the sun controls the majority of it. The sun is going through its cyclical periods, and I just think we have more people with cell phones, we hear about more tornadoes and everything; we think that the world's ending. Since Sandy and Katrina, a real hurricane hasn't hit us, and yet we're supposed to have hurricanes all the time. As a geologist, I believe it's more a natural sequence of events. Our lifespan, and the extent of us collecting data, is minuscule compared with that. I would assume this is just a temporary fluctuation. Depending on who you read, it's either a significant change or not. And then it becomes a credibility factor between the news media and the falsification of data. I read a USGS report, and I can't quote the number for you, in the 1990s, and it had to do with data collection for the weather model, and he basically said that all the data pre that report was not usable, because weather stations were put wherever it was convenient to put them, and you might put them on rooftops next to the air conditioning system where they'd get extra heat, either high or low data statistically, and that was a real eye-opener."

"Should the methane released in fracking be a concern?"

"Anything you look at could potentially be harmful. I have a lot of faith and confidence in the Earth that it will potentially redress it, and here in Colorado we have updated our regulations here, and we have constantly tweaked it. And the oil industry here is in my opinion very cooperative, because we don't have the big oil companies in this state. It was Amoco back then. They all packed up and moved up.* Conoco and Anadarko, they're the biggest ones. Noble is still around . . .

"Overall, when you asked about the impact on air quality of methane and everything, let me address that.† I've been in this business since '89. I used to do mining inspections and all kinds of things all the way down to the Mexican border. They did it because of their insurance policies. The way I understood it, they had two columns. They would tell the company, this is what your premium payment's going to be if you follow what the environment guy says. If not, then here's the other premium.

"Since '89 and through the early '90s, I could go to properties and say, my God, how could this facility operate this way? By the end of the '90s, the regulations that we had in effect were starting to be implemented by the companies, how they handled their chemicals and their waste materials, and it really showed. I now see really few properties with that kind of problem. Now back east, that's where they call it the Rust Belt, there's all kinds of *old* problems out there, but out here facilities are relatively new, and properties aren't so impacted."

"Is it a good idea to keep burning coal?"

"Yeah. Most companies that I go see operate pretty well. By the end of the

---

* You may recall Bob Winkler making the same point in different words: "The big guys are not here—not Shell, not Exxon. All the little guys are here. No one's accountable. Everyone's a subcontractor or a holding company."—What level of citizenship could actually be expected of such entities? In 1998 some researchers in Holland, where agreement between the *regulated community* and its regulators was claimed to be the most cordial of anyplace studied, found that "small and medium-sized enterprises," which they perkily abbreviated "SMEs," *were often extremely eager to be helped* to achieve "economic and environmental win-win situations," *but as soon as the consultants left, the motivator seemed to have gone as well. The reasons cited were that SMEs tend to think short-term and do not have many resources to spare.* Since American SMEs would have furiously resisted any such "help," their own modest resources and their short term thinking must have made them extremely selfish carbon ideologues. In this category I count the fly-by-night frackers.

† "There are two kinds of methane generation. The most common one is from vegetation in the ground. And from Boulder across I-25 and farther, there's a huge underground coal field, methane everywhere! That area is prone to subsidence. Those mines are down a thousand feet; my good friend's father was a real wheeler and dealer and it occurred to him that all these mines are full of water and he filed a claim for all the rights, and finally one of the counties figured out what he was doing and he lost out . . ."

'90s and into the early 2000s, at the end of the '70s, the air and water around Philly was horrible, and nobody wanted to drink the water in New Orleans, but by the mid-'90s, those regulations made a significant impact, and I'm just talking anecdotally. So therefore when we go after the coal-fired plants, well, back in the '70s, acid rain was a big deal, but when was the last time you heard that? They improved that. So I get to the point where I get cynical: We're spending a lot of money to chase the tiny percentage to improve the environment."

"What about the coal they burn in China?"

"Our standards are pretty good in this country. We do hypocritical things. We shut down our offshore drilling. The alternative, there's not much alternative in the long run. We'll stop it here, but our federal government will give offshore loans to companies to drill off of Brazil! So therefore you get China and India and real Third World countries that keep environmental regs on the back burner, and we get blamed for that, because our companies are in there, but I would assume our companies are doing better than they once did.

"Look, I don't even think our wind farms and our solar can keep up with our extra energy demand. *We don't have a good alternative to fossil fuels.* I have faith in humans in the long run. The nuclear stuff, that's where I have problems with the nuclear power plants. Check out the same thing with refineries: maybe one new refinery in the last 30 years! It's an outrage. So they make them keep old nuclear plants and refineries operational, and tweak them, and what they should do is shut one down, clean it up, build a new one with state of the art, with underground tanks.

"Every refinery has problems. You can't get around it.

"You can build a state of the art refinery in a safe and secure area, and follow all the standards, although you have to be reasonable because the EPA gets carried away; you could build one of those with the proper monitoring and piping and then go back to the old one and clean it up."

"And what's your opinion of nuclear power?"

"I can't believe that we complain about nuclear energy when we have facilities that are so old. So why don't they decommission them and build a new one in the middle of a shut-down military base?"

2: "It's Like Smoking Around Children"

Mr. Sam Schabacker, whom another activist had described as "leading the resistance," worked for Food & Water Watch in Denver. It was he who had called Weld County "the most heavily fracked county in the U.S."—and, when I immediately decided to go there, put me in touch with Bob Winkler. He said:

"I'm from Longmont originally, and I grew up on this little piece of farm property that my parents purchased in the '70s. My father was a blue collar worker who would fix your gas leak problems if you had them. He'd work in your crawl space. They put all their energy, all their savings into that property, and thankfully they were able to retire with a little bit of dignity. But someone from Chevron or Exxon could come knock on their door tomorrow and say, hey, we own the mineral rights here, and start fracking, and there's very little they could do about it. Their health and their property values would potentially become threatened. Statewide, Colorado has 53,000 oil and gas wells; 95% of those are fracked. We have a Governor who's very pro-fracking, and took over 100,000 dollars from the oil and gas industry for his two election campaigns. He's using state funds to sue my friends in Longmont and my family members who have worked against fracking.

"The preemption law is something that has been on the books on Colorado for the last 20-plus years.

"Back in the '90s, the city of Greeley passed a law banning oil and gas development. It was a citizens' initiative. Well, the industry sued and the state Supreme Court sided with them. So for the last 20 years, until 2012, that was the law. Well, we wanted to challenge that law. One challenge is through the courts, one is through the state legislature and one is through the ballot. The effort in Longmont was to push the state to bring a challenge to the Supreme Court ruling. Oklahoma and Texas have already passed preemption laws to try to stop what we are doing here. So for three years we've been trying to figure out how to stop the food and water disaster in this state. We're knocking on doors, training activists, building leadership skills. So that's what we did in Longmont. It's happening in L.A., in the Fort Worth area, in Pittsburgh, where there's such encroachment from the oil industry. So with a huge amount of work we gathered signatures in temperatures that were a lot hotter than this, and we were in the general election campaign in 2012, and the industry spent half a million dollars; they outspent us 30 to 1; they got the government to threaten to sue Longmont if they passed the measure; and we won the election with 60% of the vote. It was neighbor to neighbor. I personally knocked on over 1,000 doors. So we win, it's a great celebration—and the industry sues, with the governor on their side.

"Fast forward to 2014, where Fort Collins, Broomfield, Boulder and Lafayette did the same thing; they passed the ballot campaigns.

"The Democratic Party said this is too hot an issue to touch, and we're worried that if you put these on the ballots, the industry will come after us. So they found a way to pull our measures off the ballots; they took away our democracy, and the Democrats lost anyway.

"Denver is about 35% of the state population. There are two threats to Denver. One is that there is fracking, encroaching on residential areas, in northeast Denver. A real estate developer who bought the land out there held the mineral rights, and is slowly handing them out to be fracked. The place is called Oakwood Homes. Their CEO, Pat Hamill,* was one of the big contributors to the Denver mayoral campaign. So that's the first threat. It's a community that's two-thirds Latino and African-American.

"The second threat is fracking in our watershed, in an area where Denver gets up to 40% of our water supply. That's in South Park County, the South Platte river basin.

"If there are spills, who's going to pay to treat our water? We have launched a big campaign called the Don't Frack Denver campaign. We want to protect everything from our watershed to the places where people work and play. Our group includes craft breweries, kayakers, social and racial justice groups, a group of mothers, and our equivalent of Ansel Adams: John Fielder.†

"The developers will go in and buy the whole plat of land, subdivide the top, sell off the houses, and then . . .

"Denver is sort of the staging ground for a lot of oil and gas executives in the country. It's typically Denver, Houston and Calgary that are jumping-off places and corporate headquarters. We estimate that up to 30% of business rentals in downtown Denver are from oil and gas, but oil and gas counts for just a little over 2% of the state's GDP and it's only 1% of the workforce. Of course they claim to contribute much more than that. I have a degree in economics, and I learned those tricks on how to change the multiplier and inflate the numbers.

"A lot of the CEOs are very private when it comes to where they live; that's what they want; they want to enjoy extracting the resource and being part of the one percent while the rest us have to deal with climate change and everything else. It's very rare that you get somebody who's not on the take.

"Most of the people who came to Colorado came for the outdoor amenities. So they set up in these subdivisions on the eastern part of the mountains, and suddenly there's a drilling rig in their back yards! So there's a new generation of activists, from nurses to entrepreneurs, who were formerly apolitical. For someone to understand that this is going in his back yard, with the excess risk of

---

* From an article in *High Country News*, 2015: *Pat Hamill, chief executive of Oakwood Homes, the subdivision's developer, has called the [Green Valley Ranch] area "the best-kept secret in Denver." There's another well-kept secret out at Green Valley Ranch: Oakwood Homes held onto many of the mineral rights beneath the houses, and leased them to Anadarko Energy, which in turn sold the rights to Conoco-Phillips . . .*

† A gifted local landscape photographer.

cancer and the birth defects and the loss of property values, and they can see it and hear it and smell it . . . ! Across the political spectrum, people don't want it in their back yards.

"I think that most people understand intellectually that it doesn't make sense to burn old dinosaur bones, but we're locked into a system where all you can do is buy a hybrid car. We need a Marshall Plan for renewable energy. Seeing this in their back yard precipitates action; we view these issues as a way of bringing people into the larger movement. We're seeing more and more that it's the pushing from the bottom that accomplishes something.

"In Colorado they say, oh, fracking's been done for 60 years; it's safe. But this type of oil and gas development is fundamentally new. It began in the early 2000s.

"Unlike the old rocker wells, per one pad you might make multiple wells, and you're drilling horizontally. You'll see condensate tanks that are the size of a VW bus; they're painted this innocuous desert green color; and you'll see a solar panel near them to account for how much of the resource is coming out.

"It's like smoking around children. It's associated with an increased risk of birth defects, increased risk of cancer.

"It's true that in a lab if you burn a hunk of coal and a comparable amount of natural gas, the gas will burn more cleanly. But over the entire life cycle, the extractive procedure is arguably worse from a climate change perspective. Over a hundred-year time horizon, the methane released is up to 80 times more potent than carbon dioxide as a greenhouse gas.*

"So a lot of the mainstream environmentalist organizers, Sierra Club and Environmental Defense Fund, got sold this myth and some people got a lot of money to push it: *It's so easy to transition with natural gas!*

"I don't think the choice is which ones we want to keep; we don't need to privilege one dirty form of fossil energy; we have to make the transition to clean energy.

"Generally speaking, when people are struggling to put food on the table and gas in the car, it's harder to care about climate change.

"I think there's an opportunity to take this crisis to respond in a visionary way and create a whole series of good construction jobs, white collar engineering jobs. *Let's solve the climate crisis.*"

---

* As we know, Mr. Schabacker overstated this particular evil. Methane's 100-year GWP is "only" about 21.

THE GARDEN NEVER WENT WHITE

The complaints of Eric Ewing had little to do with global warming. But since methane, like radiation, could not be seen, smelled or felt, why not cite his testimony as part of the *experience* of fracking? Thoreau once said that *men hit only what they aim at.* And how could we aim at we could not see? In that department a FLIR camera (which I could not afford) might have helped; meanwhile fracking's non-disclosable emissions of volatile organic compounds tweaked our sense of smell in various revolting ways. I never could "understand" methane directly. But I could understand the *extraction* of methane a little, through the effects of that procedure upon the Ewing family.

"And the chickens had black cockscombs . . . The milk didn't go sour—it curdled into . . . white powder." That was what they said at Chernobyl. "The whole garden went white, white as white can be . . ."

Where and when I lived, the garden never went white. So we didn't have to do anything about fracking. It's true that Mr. Ewing's family were not exactly enjoying themselves. But maybe some people just had to get sacrificed for the sake of Gross Domestic Product, or to keep the lights on. They were, after all, in the minority.

3: "A Wonderful Point in Our History"

No, Greeley resembled neither Okuma nor Tomioka nor any other red zone locality. It was certainly not hollowed out like Iaeger or Northfork or those other Appalachian coal towns. The fracking boom glowed steadily, like the blue flames which flowered on my gas stove whenever I twisted the dial. For Sacramento, the weather was cold as I sat revising this chapter—perhaps not as cold as when I first moved here 25 years ago, but we all know that one person's anecdotal memories of old weather prove little about climate change. Rain tinkled down on my roof. It was 61° and I wore a soft heavy shirt, so there was no reason to turn on the heat. On the other hand, how could I begrudge myself a mug of herb tea, with maybe a shot of rum or whiskey in it? Opening the gas valve, I turned the front left burner knob all the way left, clicked the switch on my longhandled butane lighter, and the first petal rushed up tall and yellow, then shrank, turned blue and summoned up its kin all the way around the circle until the flame-flower lived again! Already steam had begun murmuring in my red kettle. Dropping two teabags into the mug, I returned to the counter, to continue denouncing natural gas for *Carbon Ideologies*.—Nearly a year and half had passed since that visit to Greeley. The place seemed better in retrospect, because I was chilly waiting for the kettle to boil, and Greeley had been nice and warm. It must have been the following month when Bob Winkler was writing his editorial to *The Colorado Statesman;* part of it ran: *The sacrifice zones created by fossil fuel dependence are encompassing the Earth. We will eventually all be living in a sacrifice zone as Weld County is.* But nothing in my neighborhood stank like a janitor's closet. As one of my Japanese taxi drivers would have said:* "If someone around here got cancer, I might feel something, but it's invisible, so I don't feel anything." I actually remembered less about Greeley than I supposed. And what *should* I remember? Bob Winkler had called it "a beautiful community." Aside from its sprawl it was a pleasant enough place, with adequately friendly inhabitants, a bookstore of sorts, an excellent brewery; and most of the time one smelled automobile exhaust rather than those frack pads with their volatile organic compounds.—Do you remember what Pastor Blevins told me back in West Virginia? "Here you do see the smokestacks and you know that they do put off the smoke and everything, but it seems to me that the earth is so large and there are so many trees and everything that how could manmade equipment put up enough smoke to make a difference?" Not only did it appear that way; I longed for it to actually *be* that way. And in Greeley, on an evening grey with rainclouds, with the old brick downtown almost silent and the cottonwoods barely stirring, it had seemed as if the way we lived was harmless and could go on forever.

I recall that on the Fourth of July, families turned out for the parade (many of whose floats, as you have seen on pp. 370–71, were the brainchildren of Noble Energy); and then came another warmish evening, with a single car scudding alongside the railroad track, silhouetted birds flying in and out of an old silo, and then finally a cool breeze from the east. How could anything be wrong with Greeley? Let me therefore end with the words of Brad Mueller: "It really is true that we're really at a wonderful point in our history."

---

* See I:345.

But just as instincts may fail an animal under some shift of environmental conditions, so man's cultural beliefs may prove inadequate to meet a new situation . . .

Loren Eiseley, 1957

OIL

Poza Rica, Mexico (2014)

Dana Point, California (2016)

Osage County, Cushing, Chickasha and Edmond,
Oklahoma (2016)

Ruwais, Dubai, Abu Dhabi and Sharjah, United
Arab Emirates (2016)

Overleaf: A PEMEX refinery burnoff over the city of Poza Rica, Mexico, as seen from a sixth-floor window on the evening of December 23, 2014

Oil Ideology

Assertions

UNIQUE OR INTRINSIC BENEFIT

"My inquiries for natural curiosities were not wholly unsuccessful. Ameer Khan sent into the mountains for some mineral liquor, which he told me was collected by dipping cotton into the places where it oozed through the ground. I thought immediately that it was water which had passed through a bed of mumiai, the asphaltum so well known in India, by the natives, under the name of 'negro's fat.' The natives of India believe that mumiai is procured from a negro's brain, and is made to exude by hanging him up by the heels over a slow fire. It is much valued ... Marvellous are the stories related of the cures it will perform ... A person having fractured a bone is immediately made to swallow a large pill of mumiai ... It must act, I should presume, as a very powerful stimulant ..."

British traveller in Afghanistan, 1840

Oil: "One of the leading sources of motive power in the world."*

Encyclopaedia Britannica, *1911*

"[Unlike coal,] liquid fuel may be stored in tanks for an indefinite time without any deterioration whatever."

Ditto

"With gasoline came the different kinds of engines that we use today, all of us, for all kinds of traveling, and for hard work. There could be no automobiles, no airplanes, no motorboats, if we did not have gasoline."

Irmagarde Richards, 1933

---

* "The great modern development of the motor car gives the light oil engine a most important place as one of the leading sources of motive power in the world. The total petrol power now applied to cars on land and to vessels on sea amounts to at least two million H.P. [= 84.6 million BTUs per minute]. The petrol engine has also enabled aeroplanes to be used in practice."

"Greater speed, greater cleanliness—freedom from soot, dirt, dust—freedom from heat of the boilers . . . [Can now omit] expensive and inconvenient coaling at Newcastle, Australia . . . The motorship as we now know it operates at maximum economy at a higher speed than the [coal-powered] steamer. [On one line] the motor passenger liner—6,000 tons larger than one of the steamers—costs 50,000 pounds less per round trip to operate than the steamer."

Textbook on diesel-engined shipping, 1935

"The increased nitrogen supply in soils that have been contaminated probably explains the increased crop yields observed on old oil spills."

Soil science textbook, wr. between 1943 and 1978

"In theory, no booze can be imported [into Saudi Arabia]. However, [Aramco's] chemists can make unusually credible 'gin,' 'scotch,' 'bourbon,' and so on out of petroleum, believe it or not; called 'sundowners'—made yesterday."

C. L. Sulzberger, 1950

"Oil, the only major energy source which is at once capable of providing high temperature heat and is easily transported and stored . . ."

Joseph Barnea, 1981

"Still continues to be the most convenient form of energy available to man."

Asha Han, 1986

"The economic prosperity of our modern society is closely related to the availability of abundant and relatively cheap oil."

George A. Olah, Alain Goeppert and G. K. Surya Prakash, 2009

"More than 95% of the energy used in transportation comes from oil."

The same

"Petroleum . . . is relatively abundant, reasonably affordable, and generates less CO_2 than coal does."

Michael Klare, 2004

"Oil fuels more than automobiles and airplanes. Oil fuels military power, national treasuries and international politics."

Robert E. Ebel, Center for Strategic and International Studies, 2002

"The availability and cost of . . . petroleum . . . will determine our lifestyle and even our freedom. You can't fly a fighter jet on wind power, or run a tank on solar power."

Greg Kozera, President of Learned Leadership LLC, with "40 years of experience in the energy industry," 2016

An illustrated "petroleum tree," whose branches are much less diverse than those of the "West Virginia Coal Tree,"* offers gasoline, diesel oil and other kinds of motor fuel, asphalt, insecticides and munitions. The "Frack Master" Chris Faulkner claims that "more than half of each barrel of oil consumed in the US is used to manufacture things like heart valves, ink . . . vitamin capsules . . . tape . . . candles . . . carpets, sporting equipment . . ." Olah, Goeppert and Surya, however, are satisfied to assert that "worldwide, about 6% of oil is used as the feedstock for the manufacture of chemicals, dyes, pharmaceuticals, elastomers, paints and a multitude of other products."

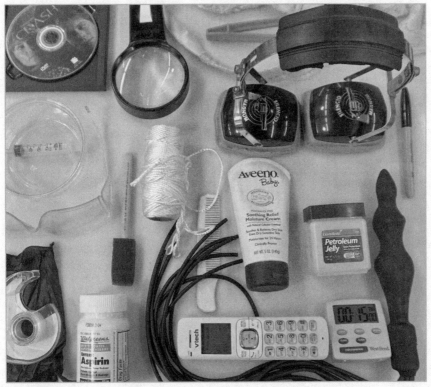

Petroleum-derived products

---

* See p. 10, and for the natural gas equivalent, p. 298.

Embellishments
AVAILABILITY AND EXPEDIENCY

"... President Putin, whose 15 years in power were accompanied by an oil boom that pulled tens of millions of people out of poverty ..."

The Moscow Times, 2015

Oil "for decades has preserved stability in Saudi Arabia and its smaller neighbors, where inflated government salaries and benefits including free health care and education, as well as handsome cash transfers, have largely pacified citizens and fostered a high-spending culture."

The Washington Post, 2016

LOCAL APPEAL

"THANK YOU[,] SHELL ... On behalf of AIO and the communities we represent, thank you[,] Shell for working with Alaska Natives to pursue sustainable economies in the Arctic.* Thank you for investing in our region despite substantial federal roadblocks."

Arctic Iñupiat Offshore newspaper advertisement, 2015

"The city maintains itself with oil. Oil helps with daily labor. Oil products offer more comfort for everyone."

Omar, watchman in Poza Rica, Mexico, 2014

"Without oil, we wouldn't eat."

"Subsidiary" refinery worker, Poza Rica, 2014

"There's instant prestige when you work for Exxon."

"20-year Exxon employee," before 1984. At that time Exxon was "the biggest oil company in the world."

---

* Shell Oil Co. annual report, 1991: "Our mission is to excel ... To accomplish our mission, will be guided by three fundamental objectives: leadership in health, safety and environmental performance; return on net investment of at least 12 percent; and highly competitive earnings."

"The oil and gas operations in the [Niger] Delta have been positive in providing employment and in financing many community development projects in the area. What would have become of those communities without it?"

John Jennings, Chairman of Shell Transport and Trading Company, 1997

"In Markovo, [a Siberian village,] the success, the hopes, and the future itself depended on oil. Oil meant comfortable flats, kindergartens, modern groceries, artists from Moscow, a library, and a hospital."

The Soviet Way of Life, 1974

"The shale boom created a housing boom, a hotel boom and a restaurant boom— and a labor shortage that pushed wages ever upward. The town of Midland in West Texas had the highest per capita income in the nation in 2013 at $83,000. It was a good place to sell backyard pools and high-end trucks."

Time *magazine, 2015*

"The flow of oil into the U[nited] A[rab] E[mirates] and the revenue from gas has sustained Sharjah's development, creating a prosperous and modern emirate, while retaining the charm and tradition of Islamic culture."

Sharjah Commerce and Tourism Development Authority, 2015

"The consumption of 1 [metric] tonne of oil equivalent per person per year is the breaking point between achieving a higher or lower level of well-being."

Mexican greenhouse inventory, 2002

NATIONALISTIC APPEAL

"What might the President do to make that true Morning in America possible? First: Support [the] Keystone [tar sands pipeline] after all."

Amity Shales, 2015

TAUTOLOGICAL TWADDLE

"Gas makes the big difference."

Slogan of American Gas Association, Inc., before 1984

"Gas, the comfort fuel."

Slogan of Philadelphia Gas Association, Inc., before 1984

"See what happens when you start using American ingenuity."

Slogan of Standard Oil Division, American Oil Co., before 1984

"Keep on saving as you drive."

Slogan of Texaco, Inc., before 1984

Statue of *petroleros* in Plaza Cívica, Poza Rica

Oilman statue, Cushing, Oklahoma

About Oil

And then—hush!—they were in the real oil-sands; Dad set a crew of Mex-
icans to digging him a trench for a pipe line; and the lease-hounds and the
dealers in units discovered that, and the town went wild.

Upton Sinclair, *Oil!,* 1926

C*rude oil makes an excellent fuel, being cheap, highly concentrated, and eas-*
ily handled . . . This wondrous liquid proved to *contain about one and a*
half times as many heat units per pound as the best steam coal. Praise carbon, my friends!—These words, published exactly a century before I began writing *Carbon Ideologies,* still held approximately good when I was alive.*

The first gusher in Titusville, Pennsylvania (1859), was only good for lantern
kerosene (10 to 18 carbon atoms per molecule), and although the British Victo-
rians imported that stuff to illuminate their households, profit margins remained
far inferior to high heating value until 1909. But in that year oil began to establish true wealth—by creating dependence. Henry Ford sold horseless carriages
by the thousand, and they all drank an oil product called gasoline (six to eight carbon atoms per molecule). *For pharmaceutical purposes crude petroleum is no lon-*
ger generally used by civilized races, explained my 1911 *Britannica, though the*
product vaseline *is largely employed in this way,* as civilized prostitutes have also
assured me. (In 1984, and probably still in 2018, we blended petroleum jelly into lip
balms, sticks of eyeshadow, you name it.) Regarding all of these, one byproduct
of their manufacture turned out to be carbon dioxide, but what did that have to
do with us?—And we turned out more and more gasoline, because *the employment*
of its products in motors . . . has greatly increased the demand for petroleum spirit.

Since oil supplies often lay in such convenient reach of British steamships,
demand and opportunity stimulated one another like a pair of honeymooners.
The general adoption of the new fuel for marine purposes becomes a matter of ur-
gency for the statesman, the merchant and the engineer. None of these can afford to

---

* For instance, mid-20th-century California crude contained 1.5218 the energy of the West Virginia
Coal Association's value for "average" coal. See the table of Calorific Efficiencies on I:212 [136] and
215 [206].

neglect the new conditions, lest they be noted and acted upon by their competitors. And so they created, too, demand.

In 1901, Oklahoma (which was then still the Indian Territory) produced 10,000 barrels of oil. In 1908 it produced 45,798,765 barrels. In 1970 it produced 224 million barrels. You get the picture.

Ever more sophisticated refineries "cracked" hydrocarbons, first to get more gasoline out of crude, then to create toluene, jet fuel, high grade fuel oil and various higher octane paraffins. (Here I must pause to celebrate the wonderfully eponymous Mr. Carbon Petroleum Dubbs, who filed a significant thermal cracking patent in 1919.) The idea was to organize that long-chain, amorphous stuff called oil into a discrete number of hydrocarbons of lower molecular weight. We cracked butane-rich petroleum fractions into butadiene, and made rubber for tires! We cracked another way to get glycerol—wonderful to save a cigarette from drying out, or to silken up a fancy soap! (Or did you care to buy cellophane, or nitroglycerin? Glycerol was our feedstock.) A children's book explained: *If the oil demand falls, and the gasoline demand rises, the type of cracking is changed to give the new balance of products.* We could pull propane or butane out of natural gas, but we could just as well crack it out of oil. Oh, we cracked wise, we did! We found ourselves ever more on the move, from Sunday drives to strafing runs, so how could we ever get by without petroleum? In 1942, stalemated in the west and losing momentum in the east, Hitler told the Japanese Ambassador why he must resume his drive toward the Caucasus "as soon as the weather allows": *This is the most important direction for an offensive; we must reach the oil fields there...*

His enemies thought likewise. The Phillips Petroleum Company boasted that their fuel added a fifth engine to every four-engine bomber! *All over the world, in trenches, in camps, in planes and tanks, in jungles and deserts, Phillips men are fighting the enemy ... risking everything so that the American way may be preserved.* Bit by bit, the American way was becoming everyone else's.

Natural gas offered the least sentimental appeal, because we could not see or feel it. Next came coal, which did retain, as you have read, strong regional cultural value, but whose extraction sites appeared uglier for longer than frack pads, whose deficits mostly lay hidden underground. Nuclear was the wild card, its radiation being as invisible as the future effects of its wastes; only now and then, as at Fukushima, did it terrify. As for the other fuel, since we had no intention of staying still, oil was the world's default carbon ideology.

Today, which was 1966, *petroleum is the chief source of the enormous quantities of benzene, toluene, and the xylenes required for chemicals and fuels. Half of the toluenes and xylenes are utilized in high-test gasoline...*

From 1978: *Our need for, and our consumption of, petroleum grows daily.* I can almost hear the authors—"two knowledgeable experts in the field"—rubbing their hands.—Talk about success stories! In 1981 a symposium reported that *the world has come to depend predominantly on conventional or light oil, and petroleum products in general, to supply its energy needs.* In 1990, the shortsighted inhabitants of our planet already burned 9,240,000 tons a day—64.7 million barrels. And we kept burning more.

Oil had entered into almost everything! There came a time when the Phillips engineers averaged nearly one patent a day. They turned heptane into the adhesive on the little "sticky notes" that we loved to place here and there on documents, inviting signatures, demanding rewrites. *They took swimming out of the Dark Ages . . . U.S. swimming wouldn't be where it is today without help from Phillips.* (Does it even matter how or why? Let this suffice you: *THE SPIRIT OF INNOVATION.*) Eco-ideologues called us "addicted to oil." Oil stood in for other fuels. We associated coal miners with certain regions: the Ruhr, eastern Ukraine, Wyoming, West Virginia. Oil reminded us of the Middle East, of course, but in the United States we drilled it out of any number of places from Alaska to Louisiana. The Germans pumped oil, and so did the Nigerians. Oil was shorthand for the techno-economic substructure of our lives.

Refinery near Sarykamys, Kazakhstan

In the fracking chapter you read a preponderance of negative assessments, in part because few people other than "antis" made time for me. In most of these oil chapters (in the last, several Emirati guest workers will wax less than enthusiastic about "black gold"), you will not find anyone who damned oil without

redemption. And while oil companies were near about as secretive as their fracking cousins, ordinary people felt so self-assured in expressing and inhabiting this
particular carbon ideology that the interviews were easier, both to arrange and
to conduct. Nobody expressed strong emotional pain relating to this fuel. Although I could only go so far in my questions (I did not care to irritate the kind
and generous Sam Hewes, who had already expressed disagreement with a couple
of matters I "framed"; while Archie Dunham, who also was very good to me, did
not suffer fools gladly), I never felt, thank goodness, that anything I could say
would constitute an outright challenge. My purpose was to learn from them, not
to unload my worries. In the Appalachian coal country it had felt very different.
Sensing that their carbon ideology was "going away," people could be testy, bitter,
aggressive, defensive. In Colorado I perceived an ongoing polarization; the battle
was not yet won or lost, but might swing either way into something ugly. In
Fukushima it was far too late; everybody had to be against what had happened,
but (setting aside a few activists' pronouncements) it was a "tragedy," not an ongoing "crime." In Oklahoma the battle had not yet begun. And in my lifetime,
if all went well, it never would. (Before I died, I did hear once or twice that
oil might not be good for our climate. But we won the victory over such talk.
As Goethe wrote in 1829: *Nothing is more damaging to a new truth than an old
error.*)

A market-load of tires—made from oil, and probably for use in gasoline-burning cars.
Of course the dhow is motorized. Deira quay, Dubai.

Oil remained the greatest cause of global carbon dioxide releases until 2005.* It continued to be *the main contributor to OECD*[†] *emissions and it generated 42.3% of the CO$_2$... in 2009.*

QUANTITY OF GERMAN OIL WHOSE EXTRACTION ALONE WILL RELEASE 1 POUND OF SPECIFIED GREENHOUSE GASES,

in multiples of the value for carbon dioxide, 2007

Expressed in [pounds of oil to release 1 lb of greenhouse gas]

and

<total pounds of greenhouse gas released by German oil extraction that year>.

"German petroleum extraction in 2007 amounted to some 3.414 million [metric] tonnes of carbon equivalents. [See I:585.]"

1

Carbon dioxide [1 lb released per each 48.43 lbs oil].
<155,085,691 lbs released in 2007>.

714

Sulfur dioxide [34,571].
<217,257>.

---

* Like her two fossil sisters, oil warmed the atmosphere from the moment of extraction, and even before. The Marcellus shale in Pennsylvania boasted a leak rate of below 1%, while places in Utah could fizz out invisibly at 11%. In preparing oil for use, the first thing one had to do was separate out its water, natural gas and salt. In Germany, that operation released a pound of CO$_2$ for each 10 pounds of oil. The oil then got pipelined elsewhere, leaking along the way. Some emissions were carbon dioxide, and some were VOCs (see p. 352). (Good news: Oil gave off an order of magnitude less methane per unit volume than natural gas—at least until it was combusted [p. 308] After that, it reached the refinery. Americans spent 1,273 BTUs per pound to extract and refine the crude, then another 16,982 to crack it into what we wished for—18,255 BTUs, all told. You may remember that the calorific efficiency of a pound of this or that petroleum oil is about 18,500 BTUs (I:215). Therefore, it requires the energy in a pound of oil to make a pound of oil—a fair trade, no? In 2012, petroleum refining emitted the sixth largest category of carbon dioxide in the EU-15 nations. But why worry? That added up to only 3%! "Between 1990 and 2012, EU-15 CO$_2$ emissions increased by 6% ... Emissions in 2012 were above 1990 levels in all Member States, with the exception of the UK, the Netherlands, France and Germany."

[†] The Organisation for Economic Co-operation and Development. A consortium of 34 mostly First World countries affiliated with the International Energy Agency. For a list, see I:564.

714

Methane [34,571].
        <217,257>.

1,428

Nitrous oxide [69,143].
        <108,627>.

Source: Greenhouse Gas Inventory Germany, 1990–2007, with calculations by WTV.

Meanwhile, from 1985 to 1995, China's oil demand doubled. As Mr. Zhang Wen had told me in the Barapukuria mine in Bangladesh: Coal is "very important in China. *It needs the coal!*" No doubt; no doubt. And evidently China also *needed the oil!*—just like the rest of us.

In the USA, explained "the New York–based Energy Intelligence Group" in 1997, *oil has re-emerged as a general instrument of foreign policy.* But while America could deploy oil in other nations as a reward, and withhold it as a punishment, for decades oil had been not only an instrument but a cause:—Between 1997 and 2002, American petroleum consumption rose from 36,264 to 38,400 trillion BTUs per year. (That kept the lights on.)—I quote a book from 2004: *At present, petroleum products account for 97 percent of all fuel used by America's mammoth fleet of cars, trucks, buses, planes, trains, and ships.* Furthermore, *the fighting machines that form the backbone of the U.S. military are entirely dependent on petroleum.*

"Internationally, we no longer have the ability to rape and plunder," remarked Mr. Sam Hewes, who was a Vice-President of the Bank of Oklahoma.* "The bidding is much more competitive. The technology deep water and all that, is just . . . a drill ship costs a million dollars and more a day. This is a friggin' *ship,* and you're drillin' in 500 or 700 foot of water before you even get to the floor! The expense to find the oil has gone up tremendously. Onshore, it's all fracking and horizontal drill. The industry is very innovative. The largest use of supercomputers is in the oil and gas industry, for seismic processing. So it's gotten more and more and more expensive. So it's changed exploration onshore to exploitation. All these formations, everybody knew it was there. Before, it would never give enough to make it economic. But with all the technology for drilling horizontally, you can drill maybe three miles. And then you take this piece and you frack it, and then

---

* His interview appears below, p. 486.

another, and you frack it. You make maybe 20 frack jobs, and you're going into really crappy rock . . ."

Advertisement at the Oklahoma City airport

But we had do it. Only one out of 20 Earthlings was American, but out of every four barrels of oil in existence, Americans used one.

(In 2015, they burned 20 million barrels every day—2.5 gallons each. Call it 325,000 BTUs per American per day.)

"Look at the energy use per capita in India and China compared to the United States and Canada," said Mr. Hewes. "It's like 15 times more here! They've got two billion people! All they've gotta do is come up a little bit, and by golly, the whole damn world . . . there's gonna be conflict over energy resources. We use our economic might to make slaves in the way that slaves were bought in the past. Everybody tries to do that. We were able to do that. And we still do that. Why do you think we have the military out there?

"Things are complex. Part of it is definitely due to oil. Why in the hell are we friends with Saudi Arabia? They are the most corrupt; they don't treat their people well . . . The guy that owns our bank, his name is George Kaiser, and he's worth about $12 billion and he owns an oil and gas company too called Kaiser Oil, and he owns a bunch of other stuff. One time he got up at a meeting and he said, I don't understand the United States. The country in the Middle East that is most like us that we should be friends with is Iran.—But our government, what are they there to do? They are there to be sure that they have your mind, will and emotions, and keep information from you.* I can't even tell what's best to do for my own kids . . ."

As Thoreau remarked: *We know not yet what we have done, still less what we are doing. Wait till evening, and other parts of our day's work will shine than we had thought at noon, and we shall discover the real purpose of our toil.*

When I was alive I still felt sure it was noon, when indeed it was not quite evening. You from the future, trapped in that hot dark dusk, after all these centuries can you yet tell us what the work has been for? In your case I suppose the answer will be: *War.* (From the Pope's encyclical: *It is foreseeable that, once certain resources are depleted, the scene will be set for new wars, albeit under the guise of noble claims.*) For us, regarding oil, at least, I would have repeated: *Mobility!* (As our cars rolled down the road, every gallon they burned gave off 20 pounds of carbon dioxide.) In 1990, we spent 72% of our petroleum on transportation. By 2012 it was 80%.

The *BP Statistical Review of World Energy* for 2017 concluded: *Oil remains the world's dominant fuel, making up roughly a third of all energy consumed.*

Oil powered the trucks that supplied the supermarkets and convenience stores that fattened us with petro-packaged foods.—And in the service of that *mobility,* oil frequently generated the electricity used to manufacture our cars, three of whose "big five" components, glass, steel and aluminum, drank in a healthy share of current; meanwhile, as I said, oil even became tires!

For the *substance* of those tires, we liked to distill our crude into petroleum naphtha, which via benzene became ethyl benzene, some of which we turned into styrene and some into butadiene, so that we could polymerize those two latter chemicals into synthetic rubber, which we blended with the carbon black that we had catalytically cracked out of crude—and then we had the soft parts of our tires!

(One American auto tire cost about seven gallons of oil—but whether those seven gallons went for feedstock, or up into power plant smoke, or both, I cannot tell you.)

In 2004, Americans produced 255 million tires for passenger cars alone. That cost us 250.6 trillion British Thermal Units—if you like, 5.7 billion gallons of heavy grade commercial fuel oil sizzling away in utility plants, rotating turbines, wasting two out of every three BTUs, and emitting 71 million tons of CO_2.

---

* Although I struggled not to believe in its accuracy, I knew right away that this would become one of my favorite definitions of government.

Pakistani tire shop

Railroad carriage of oil made its own helpful contribution (didn't everything?). *After the tanks have been emptied . . . , between 0 and 30 litres (up to several hundred litres in exceptional cases) . . . are not normally able to evaporate completely. They thus produce emissions when the insides of tanks are cleaned.* Meanwhile, accidents kept life interesting. In 2013, the derailment of an oil train burned down 30 buildings and incinerated 47 people in Lac-Mégantic, Québec. In 2015, a derailment in the appropriately named Mount Carbon, West Virginia, produced less spectacular results; all the same, the *blaze burned for most of a week.* In Mosler, Oregon, *some residents* (I never learned how many) had to be evacuated when a string of Union Pacific tanker cars containing high-volatile Bakken crude *derailed here* en route to Tacoma. *At least 10 oil-train derailments and explosions have occurred over the past two years . . .* But it was our article of faith back when we were alive that increasing demand might require increasing risk. And if that risk entailed extinction, well, 90% of our planet's previous species had gone that route, so buck up, my hearties! *Washington state,* determined to emulate Oregon's success, *is becoming a western gateway for oil by rail. Future crude-by-rail traffic may increase to 17 billion gallons by 2035 . . .*

Then there were the *ships that burn oil with as much as 5% sulfur by weight and are largely unregulated.* They met their own mishaps, and in 2013, when a fuel tanker misplaced half a million liters, *The Japan Times* proclaimed: **Spreading diesel oil spill paints Manila Bay red**—with unmeasured consequences to our atmosphere.

Wall mural, Cushing, Oklahoma

Well, oil was a fossil fuel, so of course it was dirty! Why should I write a whole book to tell you that? The Weld County Household Hazardous Waste Program brochure will serve well enough: *Coloradans throw away more than a million gallons of used motor oil each year . . . One quart of oil can pollute 250,000 gallons of water.* (But, reader, don't glare at me like that! Wasn't each quart worth it?)— Meanwhile, in its obligatory "Public Warning" leaflet, bundled with my natural gas bill, Pacific Gas and Electric wearily conceded that *petroleum products, natural gas and their combustion by-products contain chemicals "known to the State of California" to cause cancer* . . . Those distancing quotation marks implied that whatever the benighted state of California might have "known," the rest of us could happily ignore—and so I put chili on the stove, then strolled outside to spray my lock with petroleum derivative.

When lunch was ready (all I had to do was stir once or twice while carbon did the work), I opened the electric bill. My usage over the previous month had been (for me) extremely high: 22 kilowatt-hours per day. Usually it was 16 or even 10. Which toys had I overused? I couldn't even say. A guest and I carved woodblocks, in the service of which we ran my air compressor for several hours; that might have been it. Oh, and someone had stolen both steel drain covers in my parking lot, so my contractor friend came to my rescue with custom-cut steel plates which he welded onto fixed chains. How much power did his arc welder draw? He didn't know. (It amused him, my nerdy curiosity about energy.)—Or had I run the fan too long on those unseasonably hot nights? Why should you judge me because I cannot account for my increased energy demand in September 2016?

Not only did I lose track of where my power went, I also failed to discover where it came from. I telephoned the electric company to ask. They promised they would get back to me, and eventually did, with a brevity bordering on curtness, after which I gave up, being very busy back when I was alive.—Hence all I can tell you is that generating my personal 22 daily kilowatt-hours might have burned a gallon and a half of oil in one of their power plants, meanwhile releasing 35 pounds of carbon dioxide.

On the subject of energy-related communications, let me insert a note from my research assistant:

Oil Companies:

Chevron (on 1-23-16 I emailed through the website form under general inquiries).

Texaco (on 1-23-16 I emailed through the website form under inquiry).

Shell (on 1-23-16 I emailed through the website form; couldn't find general so did the media option).

BP (on 1-23-16 I emailed at bpconsum@bp.com).

Exxon (on 1-23-16 I emailed through the website form; used subject category: Marine Fuels).

Chevron is the only [one] to respond, and [only] with a link to the website . . .

Well, even if they didn't have time for me, I made time for them—because in the reverential words of a book from 1938: *Oil has built great cities—greater than gold has ever built, and apparently more permanent . . . Oklahoma I believe offers the most significant illustration of the astonishing effects of oil upon the civilization which produced it.*

Seventy-eight years later I wandered among the decrepit brick buildings of Anadarko and Chickasha, concluding that not all those *great cities* were especially *permanent,* at least in respect to oil. Anadarko's downtown struck me as particularly decayed. Despite the grandeur of the Phillips Petroleum Company edifice with its three dark glass doors all in a row, Chickasha was now defined by its Chamber of Commerce as a "healthcare center," the largest employer being Grady Memorial Hospital. Of all the preeminent businesses in the Chamber's list, none appeared to be energy concerns—although the vicinity did contain six "oil and gas consulting services" (one in Norman, one in Alex and the rest in Chickasha), and 15 "oil and gas operations" (counting Pollution Control Corp.), nine of which had Chickasha addresses. In other words, some Oklahomans still made a living from one or another of those two fuels. *Time* magazine claimed that *the energy business indirectly accounts for 1 in 5 jobs around the state and 10% of its GDP.* Locals worked in the oil fields and sold fuel or equipment to the drillers. One hometown firm was Cates Supply, specializing in oilfield and industrial equipment; it announced itself with a tall pale-painted derrick, at whose base bowed one of those long-necked pumps which the Mexicans called *bimbas* and the Oklahomans *horseheads;* when I was alive we used to see them in ever so many places, nodding their long necks as if their spearlike heads were sipping near the ground. Meanwhile, from the *Thrifty Nickel Want Ads* in Tulsa, I learned that *TOP O Texas oil field service is looking to hire CDL drivers ASAP. Insurance after 60 days. Must have good driving record and be able to pass DOT DRUG TEST. We are a salt water disposal water hauling company.* So oil (this sort evidently obtained through fracking) did continue to bestow its thrifty nickel's worth of local benefit. But were these *astonishing effects* in *great cities?*

In 2015, British Petroleum *agreed to pay out $18.7 billion to settle all federal and state claims* from the Deepwater spill. In the shrewd judgment of "a litigation analyst at Susquehanna Financial Group": *Everyone is getting a significant amount of money. But BP gets to pretty much walk away and close this chapter.** Even BP must have considered that a large figure, since it was *higher than all the*

---

* But the Gulf's ecosystem couldn't walk away. "Surveyed scientists" said that before the spill, the health of the Gulf had scored a happy 73 out of 100; in 2015 it was a 65, thanks for instance to a three-fold rise in dolphin mortalities.

profits it has earned since 2012.—Well, well; so the profits were likewise pretty large. But again, where could one lay eyes on their *astonishing effects?*

That was easily answered—for I must relate the boast of my Oklahoma place-mat that *the state capitol is the only capitol in the nation with a working oil well on its grounds*—and just in case you dismiss that *great* and *astonishing* fact, up rose Dubai and Abu Dhabi, where long cleanish sidewalks ran their barren course through hot nights that glowed with traffic; while couples, families, even a few single women, covered or uncovered, waited together at crosswalks; to be sure, it was mostly young men alone or in packs, some of these sitting on curbstones, chatting with friends; a lady's tailoring establishment glowed open-doored, blaring Arabic songs; the drycleaners remained busy and the fruit vendors still sat in their market stalls. It was 10:00 p.m. (only 27° Celsius*), by which time Pawhuska, Madison, Iaeger and the other long-established American resource extraction towns relevant to *Carbon Ideologies* would have been solidly dark, while here couples sat laughing on benches (although, to be sure, not embracing, which would have necessitated police intervention). In place of Welch's nearly empty drive-in, whose coffee was foul with contaminated water, Abu Dhabi offered fancy Indian restaurants, in one of which I ordered a banana-date-almond shake and an expensive light meal, while headlights and taillights crisscrossed in the windows. On the walls hung enlarged copies of Mughal miniatures. Fluttering their hands at each other, two men in white robes, sandals and red-and-white-patterned kerchiefs leaned together across a table set with many dishes; while at the next table, an American couple in late middle age asked the waiter to explain the menu. The windowsills were richly carven wood. So don't say oil could no longer raise up *great cities!*

Meanwhile, southwest of Tulsa, the local business directory explained that Keystone Gas *specializes in purchasing gas streams from stripper oil and gas wells,* and since that concern was headquartered in the smallish municipality of Drumright, a few more locals had petroleum to thank for their paychecks. (I had my research assistant ask Keystone such apparently dangerous and vicious questions as: *What would Oklahoma be like without oil?,* and in time I heard back: *Keystone Gas (info@keystonegas.com); sent 11-14-16, no reply.*) In the same town, *Blue Flame Gas Company operates the natural gas facilities serving homes and businesses in Yale, Hallett, and Jennings, OK, as well as rural and commercial sales of natural gas for various purposes in Creek, Payne, and Pawnee Counties.* From their descriptions, both concerns sounded so local as to resemble the West Virginian coal company which had employed Pastor Bob Blevins in his youth. You might recall

---

* Or 81° Fahrenheit.

his summation: "Well, we were so few working where I worked; the owner, he was just like one of us; he didn't ever do anything that would make it hard on us."* For a fact, oil had its friendly side, if we are to believe the free pamphlet at the counter of the Phillips Petroleum Company Museum in Oklahoma: *ENERGY PROVIDER . . . Like small neighborhoods within a city, the refineries reflected the spirit of the dedicated employees.*

And indeed the display below (from the Oklahoma History Center) makes it look very innocent and ingenuous, much like the power plant diorama at the Kentucky Coal Museum:[†]

Refinery diorama, Oklahoma City

Wouldn't your little child love to play in a place like this? And from a certain point of view we were all little children.

---

* See above, p. 38.
[†] See I:151.

About Internal Combustion Engines

And the Diesel engines stand like black marble idols, bedecked with gold and silver. Firmly and solidly they stood in long regular rows, ready for work—just a touch, and their polished metal limbs would start dancing.
. . .

The Diesel motors shimmered in black and brass. In the air was a tender singing hum from the pistons and wheels . . . This severe and youthful music of metal, amid the warm smell of oil and petrol, strengthened and soothed Gleb's being.

Fyodor Vasilievich Gladkov, *Cement, ca.* 1920

A s the 2002 *American Electricians' Handbook* wisely expressed the matter: *No machine gives out as much useful energy or power as is put into it. There are some losses in even the most perfectly constructed machines.* Let that be a lesson for people who insist: "You get out of life exactly what you bring to it."

Our engines drank refined oil derivatives, performed work—from which entropy, friction and other realities took deductions—and emitted greenhouse gases. We had created them in our image.

A children's book promised us: *Compared to many other machines, the modern electric motor is extremely efficient. It turns more than 90 percent of the energy fed into it, as electricity, into the energy of motion.*—Ah, but fossil-fueled motors achieved no such results.

Efficiency is the ratio of output to input. Or, if you prefer, output equals input times efficiency.—Do you remember the 3:1 output-to-input ratios of power plants? Internal combustion engines were comparably (if more varyingly) inefficient: To accomplish each BTU of thermodynamic work, approximately three BTUs' worth of carbon had to burn.

COMPARATIVE HEAT EFFICIENCIES OF ENGINES, 1911–2015,

in multiples of the efficiency of a reciprocating steam engine

The date of each figure is the year that it was published.

Dr. Wai Cheng, Director of the Sloan Automotive Lab and Professor of Mechanical Engineering of MIT, calls these figures "OK for the layman. The actual efficiency depends on the operating condition of the engine. The numbers here are probably for the engine at its most efficient operating point."

1
Reciprocating steam engine, 1911, 1972:   12%

1–2.1
"Usual" range of "internal-combustion engine," 1911:   12–25%

1.25
Engine of an "automobile ... driven in the city," 2015:   "About" 15%

1.7
Steam and gas turbine engine, 1972:   20%

2.1
"Best" gasoline engine (piston type), 1972:   25%

2.1
"Peak efficiency" of gasoline engine, 1998:   25%

2–2.9
"Typical" gasoline engine, 2007:   25–35%

2.5–3.3
"Average" diesel engine, 2007:   30–40%

2.8–3.1
Diesel engine, 1972:   34–37%

3.1
"Has been realized" in "internal combustion engine," 1911:   37%

3.2
"Typical" gasoline engine, 2010:   38%

3.3
"Can be as high as" this value for diesel engine, 1998:   40%

≈4.6
"Theoretical maximum" for gasoline engine, 2015:   "About" 55%

5.6
Theoretical efficiency of diesel engine in mathematically idealized
Otto Cycle, 1968:   67%

Thus an intermediate thermodynamics textbook. But Dr. Cheng inserts: "'Theoretical' limit to the diesel engine is 1 since the compression ratio is not limited by knocking. In practice, however, other limits since as friction loss, dissociation loss (associated with the incompleteness of the combustion when the temperature that comes with the high compression ratio is too high) etc. comes in. Also the compression ratio is limited by peak pressure."

5.6
Theoretical efficiency of gasoline engine in Otto Cycle, 1968:   67%

Limited by compression ratio = ≤=10, since more will explode the gasoline prematurely.

Sources: Ellis and Rumely (1911); Zemansky (1968); Stambler (1972); Norman, Corinchock and Scharff (1998); Calder (2007); Environmentally Benign Laboratory, MIT (2010); Prentiss (2015); verified and corrected by Dr. Wai Cheng, Massachusetts Institute of Technology (2017), with calculations by WTV.

All we could hope for was to make the ratio of output to input approach unity.* We could never get there—not nearly. (U.S. Department of Energy, 2017: *The internal combustion engines in today's cars convert less than 20% of the energy in gasoline in on-the-road driving—and that's after more than a century's worth of innovations . . .*) Perhaps we couldn't even reach 55%. But by reducing needless greenhouse gas emissions we might have been able to operate our engines longer before they carried us into your hot dark future. And, as I keep reminding you,

---

* To be sure, it was not invariably so simple. For airplanes and rockets, "high efficiency engines tend to run at very hot temperatures," which released more nitrogen oxides and other pollutants. The solution was to reduce engine temperatures, probably through "new fuel/air mixing processes."

we were always making improvements; "technology" got better and better the whole time we were alive!

The G. B. Selden Road Engine (patent number 549,160: applied for 1879, granted 1895) weighed 370 pounds and could generate a magnificent two horsepower—not quite 85 BTUs per minute. In 1960, on a lake in New England, a paleontologist observed how petro-powered boats made *the shores echo to the roar of powerful motors and the delighted screams of young Americans with uncounted horsepower surging under their hands.* You should have seen how cunningly we'd outclassed even those by the time the icecaps melted!

The Environmentally Benign Laboratory at MIT—site of so many of Professor Gutowski's analytic projects—once decided to analyze a certain 3.5-liter internal combustion gasoline V-6 engine manufactured in 1987. Researchers placed it inside a car, which someone or something drove for an "average lifespan" of 120,000 miles. The energy thus consumed by the engine (which is to say, excluding the car) approached 420 million BTUs, whose production necessarily produced 70,000 pounds of carbon dioxide.

But why pick on engines? Tires were not so efficient, either. Out of every three tankloads of a heavy truck's fuel, one got wasted on the tires' rolling

Internal combustion engine on parade in Edmond, Oklahoma (Fourth of July)

resistance.* For a passenger car the ratio was one out of five. So said the Society of Automotive Engineers.—Actually, the situation might have been considerably less discouraging—for here Dr. Cheng of MIT commented: *The rolling resistance numbers are way too high. For long haul 18 wheelers, the rolling resistance takes about 12% of the fuel energy; for passenger cars, . . . about 5% . . . The high numbers that you have are probably for the share of the mechanical output of the engine and not the share of the fuel energy.* But this must not have been a settled question. Other minds[†] at MIT concluded: *We consider the range of contribution to be 10 to 20% for passenger cars and 15 to 33% for heavy trucks.*—Five percent or 20%, 12% or 33%, what was the difference when we had places to go? In my middle age I began to hear some of my fellow Americans complain that they could not afford to drive their big trucks. But then the price of gasoline went happily down again. As Sam Hewes told me in 2016: "A barrel of crude oil is 42 gallons and it's about $50. You go and try and buy 42 gallons of water from someplace, and you can't do it! I'm trying to let you know how cheap this is!"

If we had to drive, we should have designed our automobiles for high efficiency. But when I was alive, that was not a necessity, but a "choice."

In 2014 any adult with money could own the 12-cylinder two-seater Ferrari F-12, average efficiency a sorry 13 miles per gallon, greenhouse gas rating of one (the lowest possible), premium gasoline recommended. Reading these statistics, Dr. Cheng opined that the Ferrari *probably has an engine with peak efficiency comparable to the other engines. But it is designed to provide a lot of power for acceleration; so most of the time,* this machine was performing thermodynamic work *at a very inefficient operating point; hence the low gas mileage.* Had I proposed to my neighbors that perhaps the Ferrari should have been re-engineered for less acceleration and longer stretches of *peak efficiency;* or, worse yet, had I murmured that maybe, this car should not have been offered for sale, they would, I fear, have cursed me, so I kept quiet.

One day after President Trump announced that he would withdraw our country from the necessary-but-not-sufficient Paris climate accord, I called my old friend Jacob Dickinson, a mechanical engineer with whom in past years I had spent many a happy hour gliding over the freeways of greater Los Angeles at various speeds.—Jake," I asked, "what's the best aspect of driving a car?"

"Options," he instantly replied.

---

* Rolling resistance is "the energy a tire consumes per unit distance of travel" (see endnote for the relevant equation).

[†] Including this book's kindly helper, Prof. Timothy Gutowski.

"Do you think that we should all continue to drive no matter what the mileage and emissions?"

"Fuck, no. I mean, that's gonna kill us."

This slightly shocked me, because neither he nor I had ever expressed such views on those nights with the top down and the city lights below us when we crested gentle curves, headed for delightful places.

"So what about this Ferrari? Should that be made illegal?"

"Maybe the engine could be converted . . ."

"If so, would you be satisfied to let people own and drive it?"

"There's gotta be an accounting," he said shortly. "If you're gonna do that you've gotta plant a lot of trees."

But who had time for that compromise? And why stop living off our engines? In, for instance, 1982, one out of six American jobs was motor vehicle related.

The nearest supermarket lay beyond walking distance, so we drove there. The light rail ran infrequently at night, so my neighbor gave me a ride home; I was grateful. To get to the doctor I must take a taxi, unless some friend could sacrifice his morning for my convenience; either way, I would go by car. (T. R. Nicholson, 1970: *The automobile today is part of the furniture. We take it for granted as we do our other chattels*.) From Sacramento to San Francisco I generally travelled by train, to be a good boy; that cost me $64 round trip—four times more than fueling up an efficient car, so how could I blame the budgeteers for driving? And that train was slow, and often late; once a friend from Berkeley came to my birthday party by train, missing half of dinner; going home cost her seven extra hours, after which she refused to ride the train ever again.

Do not accuse us of doing absolutely nothing about vehicle pollution. In, for instance, 1979, the U.S. government's mandated standards for emitted hydrocarbons were 1.5 grams per mile; for carbon monoxide, 15.0 grams per mile; and for nitrogen oxides, 2.0 grams per mile. By 1985 they had been respectively tightened to 0.41, 3.4 and 1.0 grams per mile, which surely cost us in effort and expense. But if you inquired about carbon dioxide, I would remind you that even after the millennium had turned, our *decider* President still insisted that CO_2 *is not a pollutant under the Clean Air Act*. For all I know, he even believed it.

Those were the years when an American car might well emit nearly a pound of carbon dioxide for each mile that it traveled:

CARBON DIOXIDE EMISSIONS OF THREE NEW-MODEL AMERICAN CARS, 2016,

in multiples of the figure for an "F-Pace" Jaguar.

1

2017 "F-Pace" Jaguar. 4,025 lbs average curb weight. 0.97 lb CO_2 per mile.

1/4,149 lb [1/259 oz] CO_2 per poundweight of car per mile.

1.10

2016 Audi S3. 3,420 lbs curb weight. 0.75 lb CO_2 per mile.

1/4,560 lb [1/285 oz.] CO_2/lbwt/mile.

1.12

2017 XT5 AWD 3.6 "Platinum" Cadillac. 4,333 lb curb weight. 0.93 lb CO_2 per mile.

1/4,659 lb [= 1/291 of an ounce] CO_2/lbwt/mile.

Source: Motor Trends, July 2016, with calculations by WTV.

The Ferrari F-12 of course surpassed that excellence.

What was the work for? Well, to melt the icecaps—I mean, to go from one place to another—and, whenever affordable, to do so in style. Among the accoutrements of, for instance, a 2017 Chrysler Pacifica van were *3-D navigation, HID headlights, LED foglights, power folding mirrors, hands-free sliding doors and a liftgate . . . , the Stow 'n Vac vacuum, and an optional 20-speaker Harman Kardon audio system.* Each of those required electric power, which the car obligingly supplied at an extra cost in combusted carbon. None of us condemned the owner or manufacturer of such a toy, and I would not do so now. I would rather enjoy the happiness of such a person.

The American car has been described as *the ultimate symbol of a modern progressive and mobile society.* It might have been the ultimate; I agree that it was mobile, fanatically and frenetically so, unless there were many other cars on the road. As for progressive, yes, it did carry us in a certain direction.

Having brooded on this story, hopefully with more anxiety than bigotry, I wished to have ended it happily, with the widespread rollout of all-electric magic cars. But in 2017, sober truthfulness brought me back to 1998, when the author of *The Automobile Age* concluded: *There is no alternative in sight to the internal-combustion engine burning a petroleum product.*

What Was the Work For? (continued)

Can't imagine a world without motors. There's nothing finer, if you ask me; nothing that shows better that you're alive and humming and living in this present day and age than when you squeeze the juice and burn up road and there are the signs and the lights and the white lines all so it can happen and everything's moving . . .

Graham Swift, *Last Orders*, 1996

Sun gleamed on the windows, shoulders and buttocks of waiting cars. Through the darkness of a motor-rickshaw's cage I glimpsed a driver's fingers gripping the mesh. My hotel was directly across the street. I might need another 20 minutes and more to be carried there.—Why didn't I get out and walk across?—Too dangerous.

From the airport to the hotel might take two or three hours. Without traffic it would have taken 15 minutes.

This city had won particular traffic-renown; later that month I was in the outskirts of Moscow in a faintly humming van on an eight-lane freeway, beside a huge truck which crept and lurched on our right, dirty trucks blocking out most of the sky ahead, a pretty woman beneath a blue umbrella walking between the lines of traffic, trying to sell Russian flags for Independence Day, the motor idling, drivers staring ahead, another man in a green umbrella trying to sell Russian flags, as the van crept slowly toward a long pedestrian overpass on which no one was walking:—I would rather commute in Moscow than here in Dhaka, Bangladesh.

Inhaling motorcycle-smoke, we all honked at roundabouts. Seldom did any journeys surprise us with their easiness. Pedestrians wandered between us toward the median strip, which was triply strung with barbed wire. A skinny man pulled a rickshaw across our path; his passengers were a young couple holding hands. Now we were all stuck across the street from the Quality Trading Co., in front of which traffic did move in the opposite direction. An ice cream vendor pedaled his purple-umbrella'd stand; now he too was stuck and caught. A busload of sweating ladies in hijabs passed us magically; we could not move. Our

430

Traffic in New York City

Traffic in Dhaka

situation was as stifling as the great debating chamber in Parliament. My driver kept his professional calm. So did I who did not have to live out my life here in Bangladesh, where there were frozen profiles in a striped bus,* each face staring downward or far away, and then a beggar-woman in a green chador was tapping on my window, two vendor-boys next soliciting me, the first grinning while the second tried the mute hard sell; they wore basket-trays of peanuts strapped to their chests; we smilingly waved at each other and made the thumbs-up sign, just

---

* Many Bangladeshi buses were powered by natural gas.

as a man in the striped bus caught sight of me and began to stare; next a one-legged beggar came thumping on my window, the old woman in green watching anxiously to see whether he might touch my heart, in which case she most evidently stood ready to rush in. (A strictly analytical person might call the stalled traffic of Dhaka "beneficial to beggars," and it certainly created a kind of ecosystem; but of course a better world would contain no beggars.) A dark old beggar-woman whose skirt-top had worked its way down her hips, nearly exposing the crack of her buttocks, went slippingly and limpingly away, creeping between buses and cars; soon I saw her no more. In the adjacent lane, the striped bus now pulled two car-lengths forward, allowing the approach of a yellow flatbed truck which bore two skinny horned cows tied by the neck, facing forward as they twitched their ears and defecated, guarded by a man in sandals who stood with crossed ankles, leaning against the back corner—and my driver rushed ahead a full car-length, our motor throbbing. (According to the "Tourist Map of Bangladesh": *Important Notes: Health: Dhaka is known as one of the most polluted capitals and the excessive dust during winter does not help either. Hence, any precautions for breathing problems would include a mask (available at local pharmacies) and other respiratory medicines . . .*) A grey car darted into the breach and almost grazed us, its sweaty driver honking ragefully; the man in the striped bus who had stared at me was now level with me again, so I waved to him and he smiled happily, waving back. An ambulance stridulated, but no one made room. This was terrible to me; someone needed this ambulance and might be dying for lack of help—a problem whose best common sense solution, if one falsely posited that both Bangladesh and the planet could afford it, would be to increase the numbers and widths of roads, so that more vehicles could travel more fluently. We eased ahead of the striped bus. Through the green cage-grating of a motor-rickshaw I could see the gesturing of a woman's slender brown naked arm; we pulled ahead by another foot, and I could see the child on her lap. By now dusk was stealing vibrancy from everything; and as the traffic's colors began to disappear, its purrings and putterings of motors were accentuated. (From up high one better hears the untuned symphony of horns.) The bearded ambulance driver rested his face on his wrist, his siren pleading to indifference, his orange light blinking, and then we turned left onto a boulevard which happened just now to offer some prospect of vehicular motion; the ambulance was still frozen as I looked back, while my driver accomplished the victory of passing a skinny man whose pedaling carried behind him a wooden trailer on which another man sat, holding down mesh sacks of onions. He who pedaled that onion-load was one of the few (among whom I count the ice cream vendor, the rickshaw-pullers, pedestrians and beggars) who when traffic refused to recede before him could merely rest, instead of running another carbon-spewing engine.

Our car hit a dog, and it rolled over and over, squalling, until the next car crushed it. Then three rickshaw drivers were fighting, two against one, the two skinny youths punching their enemy in the back of the head until he snatched up a stone to strike one of them in the face.

What Was the Work For? (continued)

Sleek with lacquer, shining with chromium, their motors machined to the thousandth of the inch, their commutators accurate as watches, they had been the pride and the symbol of civilization.

George R. Stewart, *Earth Abides*, 1949

The preceding section might falsely persuade you that our primary experience of automobiles was traffic congestion—in other words, stupidity. I grant that we too readily resigned ourselves to the disadvantages of internal combustion vehicles whenever there appeared to be *no good alternative.* But you know better than I how often *Carbon Ideologies* wavers on the edge of Luddite hypocrisy. So let me pause to praise the automobile, which solved a problem, cheaply, ubiquitously and sometimes with élan—for more than a century. Now that that "solution" is over, I would not have you retrospectively underestimating our magnificent roads and machines (which even I sometimes found beautiful); nor should I dismiss the proud goals and plans of our engineers, who lived in hope, designing for a future of ever increasing *abundance*—a word which meant *demand.*

It was not only mobility that we won—although that grew inestimable to us. From dependence upon easily wearied muscle power, be it our own or that of our animal chattels, we came into a kind of independence whose limits were merely financial. Commuters drove from the hinterlands where they could afford the rent to the urban centers that paid high wages. Even the Joads in *The Grapes of Wrath* made a pretty good run before they used up their gas money. My friend M. fell behind on her car payments; her credit soured; interest and penalty fees unnerved her—but she hung on to her freedom to drive rapidly and easily far away from the dreary place where she worked. Her family cabin lay at an hour's distance, in a forest on a mountain. There she could sit alone cooling her feet in a stream, or lie down upon the hearth with her latest sweetheart, living her pleasure-life from Friday night until early Monday morning, when she would drive back down the hill, arriving in time to punch the clock in full professional dress.—If I touch on erotic aspects of her freedom, it is because internal-combustion vehicles (in our culture, at least) so efficiently facilitated sex as to be

434

identified with it. Pop crooners dilated on what Diane did with Jack in his car, and how romantic it was to make love in a Chevy van—because mobility had combined with shelter and discretion. Most of the Americans I knew had at least necked in a back seat. Hood ornaments and gleaming metal fins got straightforwardly described as sexy. Moreover, the experience of being carried somewhere by a throbbing engine could be itself erotic, or at least private and dreamlike. We felt it in airplanes. We felt it in trains, as when an Ayn Rand heroine from 1957 *was flying away from* the world, *at the rate of a hundred miles an hour . . . She lay back, conscious of nothing but the pleasure it gave her . . . The words for . . . this journey . . . were: It's so simple and so right!* But most often we felt it in cars.

In a novel from 1962, a man and a woman are in one of those.

> "Where are we going?" she said.
>
> "We're headed toward Santa Barbara, but we don't have to go there. I'm just following the highway. Are you cold? Shall we stop and get a cup of coffee?"

In other words, the point is to be in motion. (Presently he says: "I wish we could drive all night and I could talk to you.") With thousands of BTUs reliably spending themselves beneath the man's foot (which can always depress the gas pedal further, to increase the excitement), quotidian destinations, even undiscovered ones, become gifts for him to offer her. Somewhere there will be a diner if she desires it; or he can turn around; it seems he can take her anywhere. From the standpoint of *Carbon Ideologies* this is waste. And so that standpoint must be idiotically rigid. Why can't a romantic couple wander as they choose? How many millions or billions of such aimless drives would affect the climate measurably? Such ran Chris Hamilton's argument back at the West Virginia Coal Festival: Shuttering all the coal-fired power plants in America would prevent only three paper sheets' thickness of sea level rise.

Ayn Rand again: *She turned to the door of the motor units, she threw it open to a screaming jet of sound and escaped into the pounding of the engine's heart.*

> And once again, in the twink of nothing, I was in another big high cab all set to go hundreds of miles across the night, and was I happy! And the new truckdriver was as crazy as the other one and yelled just as much and all I had to do was lean back and relax my soul and roll on. Now I could see Denver looming ahead of me like the Promised Land, way out there beneath the stars, across the prairie of Iowa and the plains of

Nebraska, and I could see the greater vision of San Francisco beyond like jewels in the night.

Thus Jack Kerouac, in his "scroll manuscript" of *On the Road*. I always loved the sweet joy in that book, when I was alive. His two protagonists sped from A to Z and back again, for no reason. And since I loved that, I too must have loved internal combustion engines.

5

"The Whole World Depends On It Here"

December 2014

The physico-chemical sciences have already reached a point where man is clearly about to become master of matter. But social relations are still forming in the manner of the coral islands.

Leon Trotsky, 1932

THE LIBERTY TORCH

The front page headline said: **Veracruz brings in 41 to 59% of the nation's petroleum and gas wealth,** and in Veracruz, not far from the inlaid stone ruins of El Tajín, in a region which back in 1980 an engineer had labeled one of Mexico's *traditional hydrocarbon-producing areas,* lay an oil city called Poza Rica, whose oil smell often strengthened beneath the humid sky, although in the center one could hardly notice it; nor did one need to, because on the median strip of the highway, old oil valves and other such treasures had been set as ornaments, while the Plaza Cívica offered a great heroic bronze of three oil workers; and overlooking that plaza, the Restaurante El Nuevo Petrolero lived up to its name, drawing in the *petroleros* for *a torta* or a coffee, since the gate of their work facility stood right across the street. Less conspicuous than these emblems, since more ubiquitous, were the concrete superhighways—manufactured, of course, from that "big five" energy glutton, cement, so that internal combustion engines could move us about and heat the atmosphere. In brief, Poza Rica was another *illustration of the astonishing effects of oil upon the civilization which produced it.*

Up on a hill in Colonia Nahua where an old lady was almost motionlessly creeping, assisted by her cane, my oil-seeking taxi rolled past banana trees and then through a gate where a grand facility separated oil from gas. The foreman was sitting with his workers eating lunch. While he was running me off the premises I managed to ask a couple of questions, but his answers were concise, because there was a "problem" just now, a "risky situation" since "it smells a little like gas"; perhaps I should come back tomorrow and ask for someone, but not for him, because in his case it would not be convenient.—I knew enough not to go back tomorrow.

So we rolled away and down along a curving brick wall, behind which two of those heavy-headed machines called *bimbas* swung slowly back and forth, pumping oil, while on the other side of the road, old clothes, grass and animal carcasses lay in the tall wet grass where morning glories grew.

Descending a jungly hill where children's T-shirts hung out to dry from the edge of a tin roof, we came to a watchman named Omar, who wore a grey

Exterior of the Restaurante El Nuevo Petrolero, with the Oil Workers' Union above it

The grand facility with the risky situation

Two *bimbas*

uniform, well-used boots and a hard white PEMEX helmet.* His hand was very soft and moist when I shook it. I paid for his time. He remarked that there were many Americans in this city, working for Halliburton (on the plane there had been two of them, coming in from Dallas), and there were also Venezuelans.

"In your opinion, what is the best thing about oil?"

"The city maintains itself with oil. Oil helps with daily labor. Oil products offer more comfort for everyone."

"Which is best for the world, oil, nuclear or coal?"

"It's a hard question. Nuclear would be the best, but not all countries are prepared for it, and it can be used for war purposes."

As for coal, "it's been discontinued," Omar explained, which would have interested the West Virginians, although from a local point of view he approached correctness. In 2010 (the latest year whose data I could get gratis), coal made up

---

* According to the former CEO of Conoco, Mr. Archie Dunham, whom you will meet in the next chapter, "PEMEX have been a really not very outstanding state-owned oil company. The state-owned company in Malaysia is probably the best, because it is allowed to keep the money it needs to grow. PEMEX lack capital to drill, capital for maintenance, and so historically what they have done is spend whatever cash they can on their own production. They have not invested in their gas resources, but in Mexico the natural resources are owned by the people and the people don't seem to want to develop their gas. Mexico has both heavy and light oil. Nothing wrong with it."

a mere 9% of Mexico's energy-related carbon dioxide emissions*—while oil came in first at 61%.

I asked him: "Do you believe in global warming?"

"It's a difficult problem, and it's worrying. If we all do our part we cannot solve it but we can mitigate it. It's happening."

"How do you know?"

"It was never very cold here ever, but the hot season is lasting longer and the temperatures are getting higher. It gets up into 42 or 43°.† When I was a child it only got up to 40 or so; actually I don't know the exact temperature because I wasn't interested in it then."

Thanking him, I said: "Now please tell me about oil."

He replied simply: "The whole world depends on it here."

Omar could not say how long the supply would endure, but "I guess it would last my lifetime." That was good enough for him.

In 1943 Hermann Hesse wrote: *The wave is already gathering; one day it will wash us away. Perhaps that is as it should be. But for the present, revered colleagues, we still possess that limited freedom of decision and action which is the human pre-rogative and . . . makes world history the history of mankind.* Omar and I both employed our limited freedom of decision in the same way. We said to ourselves and each other: "I guess that the joys of electric power will last my lifetime."— That saved our tranquility.

My taxi driver's brother agreed to be interviewed at the Restaurante El Nuevo Petrolero. He was a patient, peaceful man who would not take any money, al-though I finally persuaded him into a light meal. His name was Jonathan Car-rillo Espinosa. For his work he "took the contamination from wells."—"I'm a manager," he proudly said. "I'm in charge of administration, billing, sales . . ."— Thanks to his high position, he got his own hotel room on his business trips. He traveled to Oriba and other places. Sometimes he was away from home for weeks.

"How old were you when you first started working in the petroleum in-dustry?"

"I was 29. I've been doing it for five years. I was in sales but earning very little, and a friend in sales in a petroleum company who didn't like it recommended that I take his position, and I like it, because you meet a lot of important people and you learn."

---

* Which totaled 416.91 million metric tons. Mexico's per capita CO_2 emissions that year were 3.85 metric tons. The comparable figure for the Japanese was 8.97; that was the last year they could be proud of how "green" their nuclear power was. Meanwhile the United States displayed its usual world leader-ship at 17.31.

† 43° C would be 109. 4° F.

"Why didn't your friend like it?"

"It didn't pay enough for him."

"What would Poza Rica be like without oil?"

"It wouldn't be anything. The best paid work before this was Coca-Cola."—
Then he made a remark which typified Mexican class resentment: "No matter
how much education you have, you have to be the son of an oil worker to work
for PEMEX in Mexico."

He supposed that Halliburton and a certain Canadian company hired locals.
He knew that Baker Hughes was in Poza Rica, and there was also a Mexican
company whose name was something like Balero. He said: "It's thanks to all
those foreign companies that Mexicans in Poza Rica can get jobs. Before that,
everyone had to migrate to Mexico City or Ciudad Reynosa or *maquiladoras* or
even the United States."

"When did all the companies come here?"

"It happened five years ago."

"What was Poza Rica like five years ago?"

"A slow city."

"And now?"

"It's the economic center, the petrochemical center of every place around
here; it's the economic powerhouse! Because the PEMEX offices are here, from
here all the money flows."

He added: "People say: Anywhere you go in this region, you're never more
than 50 meters away from oil."

"What are conditions like in the oil fields?"

"Uncertain," he answered. "It's restricted for security reasons. Our work is
just for certain hours, because of assaults. Mafias steal our trucks even in the
daylight when we are working. Thank God, that's never happened to me."

A year or two before the city had been more dangerous on account of "a lot
of corruption with the state police." The governor replaced them with new officers, "and trafficking has been controlled. You can't get rid of narcotrafficking
but you can control it."

"When was the last time there was an oil accident?"

"Many years ago. I don't remember the exact year. A tank where they kept the
combustion petroleum blew up. I was a child; I don't remember if people were
killed."*

---

* Large-scale Mexican oil accidents seemed to occur only every several years. In 2007 the Campeche
Sound conflagration incinerated 21 petro-workers. In 2013 a PEMEX explosion reduced the population of Mexico City by 37. In 2015 a gas-and-oil-separation platform on the Bay of Campeche exploded, creating "huge fireballs" and taking another four lives.

"Nobody's working on wells right now," he continued, "because it's all gone down.* I'm taking oil residuum; it's already in containers. I drive it to the state of Nuevo Léon up near Monterrey for final deposit. There's another place in the state of Coahuila. These are places authorized by the government for final deposit of oil."

"What sort of places are they?"

"Very well controlled. Big lined pits. It can't get out of there."†

"What does the waste look like? Is it solid or liquid?"

"There's solids and liquids, all kinds of hydrocarbon waste. It's industrial garbage that the guys have used to clean the machines, and the material that sinks from one well into another well, a counter-well. Maybe there's even rags and so forth."

"You drive all this in a big truck?"

"*Sí,*" he said gently smiling, with those *petrolero* paintings above him on the restaurant wall. "It's authorized by the state and by the federal government. The receiving agency checks it against a manifest. If the manifest says two of something, you have to unload two. When I pick up 100 liters, it strictly has to weigh 100 liters."

"Which is best energy source, petroleum, nuclear or coal?"

"Petroleum. But we have Laguna Verde, a nuclear plant, the biggest one in Mexico. I go right by it when I'm delivering waste. Both petroleum and nuclear are important parts of life. There wouldn't be much here without them."—He added helpfully: "There are coal mines in Mexico, but that's not my field so I can't tell."

"Is global warming true or false?"

"True, because all the time I see these illnesses, like skin differences. I don't really know; I can't explain it scientifically; I just know . . ."

"Will there always be oil in Poza Rica?"

"I don't know," he replied again. "Only the authorities know. There may still be oil here . . ."

In other words, he comprehended as much as I.

I liked Poza Rica for the way that the bellies of evening clouds glowed yellow, beige and pink, while the rest of the clouds stayed purple-grey. Come night the streets were perfumed with brightly lit mango stands, and men sat on the sidewalks, sincerely wishing one another good evening. The honking traffic got drowned out by music from the bars and amplified commercial messages in

---

* So much for the "economic powerhouse."

† In other words, there was *no immediate danger.*

Some of the paintings inside the Nuevo Petrolero

certain poor men's cars; and then by 9:00 the city center went quiet, with the exception of certain hard loud prostitute bars.—But the most interesting site in Poza Rica was what I called the Liberty Torch:

Across the mostly flat-roofed city and the diverging rays of the elevated expressway just before the mountains, a grand plume of purplish smoke issued high, very vertical in coils around a vertical axis, like snakes slithering up both sides of an invisible glass wall, then bent like a broken reed, snapping right, widening as it flowed horizontally toward the sunset, like a pennant over all those rush hour horns and evening voices. There was one great golden flare like a Liberty Torch atop a scaffold tower which rose phallic above four white domes and all the various frailer towers of the refinery. At first I thought the flare to be an out of control fire, for it had come from nowhere, and now flamed wide across some flat surface, giving rise almost immediately to smoke. After a quarter-hour, its flames began to burn down, its smoke to thin, although whitish smoke or steam still blew sideways from one of the rightmost parts between various towers; and now from my hotel balcony the air smelled like smoke while the carhorns kept blaring and the heavy-shirted vendors went on selling papayas and mangoes, sitting with their knees spread against the heat, and a lovely dark girl was selling flowers.

I called downstairs to ask if there was any danger.

"Oh, the normal. Just PEMEX doing a burnoff," said the lady at the front desk.

Morning, and the peach-colored flame burning alone in the Liberty Torch now bent leftward. It was nice of PEMEX to show us how the wind was blowing.

A Dutch report informed me that in 2003, worldwide carbon dioxide pollution from flaring "peaked" (from a 2016 standpoint) at 350 million metric tons. Happily ignorant of the total tonnage of our atmosphere, but presuming that it was exactly as unimaginable a figure as 350 million tons, I dined again at the Restaurante El Nuevo Petrolero. *The Russian Federation and Nigeria,* that report continued, *were in 2010 by far the largest flaring nations but managed to reduce the amount of gas being flared by 50% and 40%, respectively. At the same time we see that flaring in Iraq, Venezuela and the United States has increased sharply.*

How much warming would peak flaring cause? Well, maybe not so much. Answering a query of mine, Dr. Pieter Tans of NOAA explained: *For CO2* I can say that for every 1 billion metric ton carbon (= 3.7 billion metric ton[s] CO2†), the Earth retains extra heat over 100 years of 5.9E21 joules (10 to the power 21). Therefore for 1 ton CO2, it is 1.6E12 joules. Per minute that would be 30,400 joules or approx 30 BTU.*

Only thirty match tips' worth of heat per minute, spread out over the entire globe . . . ! Who would even notice?

Multiplying by 350 million metric tons, I worked out that peak flaring would warm our planet by 10.5 billion BTUs—the equivalent of burning 420 U.S. tons of West Virginia coal—minute after minute, for a century . . . after which, as you know,† some warming would continue for two millennia or more. But as Dr. Tans reminded me: *The Earth surface area is 5.10E14 m2, which would be equal to 5.6E15 sq ft*—or as you and I might put it, 5.6 quadrillion square feet. Each square foot would then be warmed by less than 1/500,000 of a BTU per minute. And so once again I fell into tune with the very commonsensical question of Pastor Bob Blevins in West Virginia: "Here you do see the smokestacks and you know that they do put off the smoke and everything, but it seems to me that the earth is so large and there are so many trees and everything that how could man-made equipment put up enough smoke to make a difference?"§ In other words, I could perceive *no immediate danger.*

As for Mexico, she fared inconspicuously in the Dutch report. So why worry about a little flaring? Besides, as Omar had aleady advised us, oil was necessary;

---

* No subscript in original.

† That is, adding in the weights of carbon dioxide's oxygen atoms.

‡ See I:176–77.

§ See p. 193.

"the whole world depends on it here." And so to *no immediate danger* I could add *no good alternative.*

The Liberty Torch kept burning. I wondered what the *petroleros* thought about it. The Union of Petroleum Workers . . . no longer existed, and the plant across from the Restaurante El Nuevo Petrolero, in the *regulated community*'s usual spirit of *no comment,* declined to permit interviews. I won similar success on Christmas day when I approached the local hospital across the bridge which proclaimed: **PEMEX HEALTH SERVICES FOR PETROLEROS.** The hospital's 20th-century plaque proclaimed this institution's half-century of partnership with PEMEX, and on this plaque an oil worker and a squirting, smoking derrick were depicted. I raised my camera, and, in the best tradition of the *regulated community,* a PEMEX guard rushed forth from his kiosk to inform me that photography was prohibited.

The Liberty Torch kept burning. I wondered what the *petroleros* thought about it. The Union of Petroleum Workers . . .

Fortunately for *Carbon Ideologies,* just outside the refinery wall sat half a dozen subsidiary workers eating their lunch. They said: "We wash the trailers; we polish the tanks where the diesel goes, and we leave them like mirrors! We grease the brakes . . ."

I inquired about the Liberty Torch.

"There's a big burn almost every week," a man said. "Often when it's raining they light it, and it's *burn burn burn.*"

"Is it bad for you to be around the smoke?"

"It gives you cancer in your lungs, but little by little, so when you die, no one knows from what."

"Is it otherwise safe?"

"People say at any time there could be a giant explosion, and it could kill half the city. There was one in 1952 in my neighborhood, and it killed many people."

"From the refinery?"

"From the refinery. This refinery is more than 50 years old. They call it the *petrochemica.* There used to be another one called *the point,* but it made too high a burnoff, and so now there is nothing there. There was a big explosion here last January, and they haven't fixed it."

"What do you think about oil in relation to human life?"

"It's necessary, and we maintain ourselves; that's how we eat. That's how the city got the name, of course, because of the wells.* Without oil, we wouldn't eat."

---

* "Poza Rica" actually means "Rich Pool" on account of its sea-catches. There must be fewer of those in your time.

To this another man sadly added: "They say the Japanese will buy PEMEX. If that happens, we think we'll get kicked out. They'll keep the *petroleros*. They're very smart, and they won't keep us."

They wanted to be kept; I could see that.—"But cancer . . . ," I said.

"Cancer is only the smoke but not because of the *petróleo*. Oil doesn't contain anything bad. It's harmless. But over the wall, there's sulfur that comes out of there, and that's really bad.—In Tuxpan," he added, evidently to please me, "there's thermoelectric, and it's killed all the fish. Thermoelectric contaminates a lot."

"Is oil a medicine?"

"The blue diaphanous petroleum would be good for itching, but they don't have it here. Here, they only make diesel, gasoline, and LP"—which must have been liquid petroleum. "Petroleum ran out a long while ago. We used to use petroleum to light our oil lamps. The *bimbas* produce only crude oil. They inject wells so that the crude will come up."

"Is global warming real or not?"

"The climate is the same," he assured me. "It's the same! It's always been hot in summer and cool in winter."

What a relief! I could now stop writing this book.

As for the Liberty Torch, I did not stay in Poza Rica long enough to learn whether that stayed the same. Early on Christmas morning it was glowing in the darkness, then defying the day, flickering from its skeletal tower, while a trace of black smoke and a few low flames rose from the burn-place, after which a dozen bursts of white smoke sprang into flower.* At Christmas dusk that orange flare began glowing more vibrantly against the purple-grey sky, columns of unwinking white light on the farthest of the other towers.

The morning after Christmas dawned overcast. I could see no flame in the tower and the four white smokes on the rightmost tower rose only a bit, with many twists and turns, before widening into the haze beyond.

---

* According to the German government: "In crude-oil refining, excessive pressures can build up . . . Such excessive pressures have to be reduced . . . Flares carry out controlled burning . . . When in place, flare-gas recovery systems liquefy the majority of such gases and . . . [the] flarehead will seldom show more than a small pilot flame." This description did no justice to the magnificent Liberty Torch.

6

"You Will Have Beautiful Lawns and Green Grass"

April 2016

I shall celebrate the chariots that triumph from East and from West . . .

Propertius, before 22 B.C.?

O il," said Mr. Archie Dunham, the retired CEO of Conoco,* "can either be, based on API gravity,† a light oil, which is a more expensive oil, or it can be a heavy oil. Generally speaking it would be by field. Most of the oil in that field would be of similar gravity, and if you were looking at it, I guess I've got most of my oil samples in Colorado, but it would look like the oil you put into your automobile and some would be even thicker. It's gonna smell like rotten eggs in its natural form, much like the mercaptan that they add to natural gas to odorize it. What it is is hydrogen sulfide, which is very dangerous. Some oils have a lot of hydrogen sulfide; most do not. In fact my father who was a pumper back in the '40s and '50s, he would check the quality of the oil back in the big 400-gallon storage tanks; soon as you open the hatch on top of the tank to drop the instruments in, that hydrogen sulfide would come out, and one time he lost consciousness. It's a miracle he wasn't killed.

"Many refineries on the U.S. Gulf Coast are designed to process only what we call the heavy oil. The heavy oils are 'thick' and more difficult to handle. Most of that oil, which we call the traditional oil, comes from Venezuela. That's the way God made their oil. That oil is what's called topped, where they've taken it, usually to Texas, to process the really heavy stuff out of it. You have a real problem when your crude changes. Then you have to spend billions of dollars on your refinery.

"The refineries in California, the boutique spec refineries, all they do is reduce the emissions. It's good for the environment but it's questionable in my opinion as to whether it's worth it. You can't ship a gasoline that's made in a refinery in Billings, Montana, into California, so you have a limited supply, so that's why your gasoline is higher priced in California.

"Most of those countries, they can't afford to care about emissions. They're just trying to produce gasoline, or, more important, heating oil. They'll take the negative qualities just so that they can stay warm."

Mr. Dunham was a gracious host who showed me his art treasures and even let me handle one or two. I wish I could properly describe those statues, ivories and weapons. As a devotee of craftsmanship, I can assure you that his collection epitomized high aesthetic judgment, embodied through wealth—and one would not have needed to see his beautiful home on the edge of the Pacific to know he

---

* In the revised interview transcript he added that he had also been Chairman of ConocoPhillips.
† Defined on I:570, 577.

was well-off. A book at the public library informed me that the Bush-Cheney 2001 Presidential Inaugural Committee got a $100,000 donation from "Conoco's CEO Archie Dunham."

As someone who until death always wondered *why,* I felt interest in his power and experience. Conoco was or had been, according to the same library's admittedly out of date *International Directory of Company Histories,* "a fully integrated and broadly based oil and gas company involved in all aspects of the petroleum business on an international scale." And he had been in charge of it, so his views on oil would certainly be worth hearing—all the more so because another quality I liked about him was his frankness. Unlike those trolls of the West Virginia Board of Education who squirmed out of expressing any position whatsoever on climate change ("I cannot begin to answer . . ."—*Declined comment . . .—Declining*

Mr. Archie Dunham

*to give specifics . . . and hanging up the phone**), Mr. Dunham stood up for what he believed. After all the refusals—from the Heartland Institute, which supposedly had a case to make; the CEO of Breitling Energy, who had "reached out" (as we say) to engage public opinion in his book on the merits of fracking; the P.R. functionaries of several refrigerator manufacturers—to even answer my interview requests, it was a relief to meet this man. And let me say yet again that just because I disagreed with him did not prevent me from longing to learn that he was right.

"Do you believe in peak oil?" I asked him. "My grandfather's *Mechanical Engineers' Handbook* said that our oil would begin running out in the 1960s . . ."

"It's not coming to an end. Look at global demand, say in 2040. In 2010, I think 54% of that global demand, which was probably 575 quadrillion BTUs, was supplied by oil and natural gas. Another 39% was supplied by coal, 6% by nuclear, and 10% by renewables. If you go all the way to 2040, the demand will be like 820 quadrillion BTUs, and instead of oil and natural gas being 54%, it will be 53%; coal will go down a couple of percentage points, and renewables will be maybe 13%.[†] Whether we like that or not, fossil fuels are going to be the primary source of energy for the next hundred years."

Again, carbon ideologues in Germany begged to differ. In a recent greenhouse gas inventory they stated: *On the whole, oil consumption is expected to stagnate or decrease.*

Mr. Dunham continued: "If you look at U.S. crude production, in 2005 we produced 3.5 billion barrels per day. In 2015 our production was 9.6 billion barrels. We went from 3.5 to 9.6 billion barrels in 10 years, all due to fracking. Production will be down this year and next year. We had 2,000 drilling rigs in 2014; today there are 375. That's why we have seen 225,000 people laid off in the oil industry over the last 12 months.

"It takes capital. When prices are where they are today, you can generate cash but you can't generate earnings, so all the oil companies have stopped drilling. This is something you can't just turn on a dime.

"I mean, Russia's dying, Venezuela's dying, Iran is way down.

"It's hurting the lesser important OPEC members. Saudi achieved record production in 2015, but they've used up about 200 billion of their cash reserves in

---

* See above, p. 223.

† Rival ideologues deployed their own percentages. The Sierra Club assured us: "With continued declines in driving rates, continued increases in fuel efficiency and public transit ridership, and continued growth in electric vehicles and renewable energy sources, demand for oil will continue to decline." As for renewables, according to *The New York Times,* in 2015, those already generated 29% of Germany's electric power. (Of course the transportation and petrochemical sectors still required fossil fuels.)

the first 12 months. They can put everyone out of business if they want. If they do that, it would bankrupt most of the industry. And the economy of any Western country is tied to energy.

"For OPEC, even though their cash cost of production is very low, $10 for instance in Saudi, the cost they need to balance their budget is closer to $95 a barrel in Saudi, and that's because everything is free for their citizens. They have a welfare economy, so they need a much higher price.*

"In Venezuela, 95% of the national revenue comes from oil. I think it's 70% in Iran. So all these OPEC countries are in terrible shape."

"How would you compare the advantages and disadvantages of oil with those of coal, natural gas and nuclear?"

"Well, natural gas of course is the premier fuel today. It's what has allowed the U.S. emissions to dramatically drop. Of course Obama takes credit for that. Natural gas in the U.S. is slowly replacing coal in the manufacture of electricity. It's a nice clean fuel. The price of natural gas today is similarly too low to justify drilling. Coal, there's clean and dirty. By dirty I mean coal that has sulfur in it. Traditionally that dirty coal has been exported to China, which now produces 44 trillion tons of coal a year. We're the second largest producer. Coal has been used and has to be used, especially in Asia. They can't afford to care about electricity. Fifteen to 20% of the world has no electricity. Twenty-five to 35% of the human race still use cow dung and elephant dung to cook their meals.† That's not a real sanitary way to cook your meals! India and China, in 2040 they will be one-third of the world population, and there will be something like 9 billion people in the world. They will want electricity. And so that coal, whether we like it or not, will be produced and turned into electricity. Clean coal's gonna continue to be used to generate electricity in the world.‡ Powder River is a higher quality, lower sulfur coal, and also a non-union coal. Union restrictions made the deep coal in Appalachia basically uneconomical."

"Where do you stand on unions?"

"I'm not against unions at all. They were very important in the 1800s when they were first organized against child labor abuses and unsafe working

---

* Two months earlier, that nation's petroleum minister, Ali al-Naimi, claimed that the Saudis could make a profit at $20 a barrel. "We don't want to," he said, "but if we have to, we will."

† This claim was very plausible. According to one 1986 estimate, "household use of energy" accounted for 45% of the energy spent in India and several other countries. Within this large category, the majority of the energy went for cooking. Furthermore, "40 per cent of the total energy consumption consists of fuel wood, crop residue and animal dung."

‡ Stanley Sturgill, the retired mining inspector: "Much of the world is changing its mind about coal, because there is actually no such thing as clean coal. But, to even get close to a clean coal product as could be gotten to, it's just not feasible financially."

conditions in some industries. Over time, however, unions became very detrimental, in the oil, steel and coal industries, and especially the auto sector. They have caused them to be uneconomic, primarily due to the expanded cost of retirement benefits. Look at the cost of benefits for Toyota *versus* General Motors, and you can see why Toyota took over the world market."

"Do you believe in climate change?"

"You know, the climate changes! I'm a geological engineer and I've studied all the changes that have taken place over the last million years. There's a great book, the one by Epstein: *The Moral Case for Fossil Fuels.*—No one can say humans have no impact on the world. There are so many of us, I'm guessing that has to have an impact on the world. Whether or not that significantly impacts climate I don't know."

"So you think that 200 years from now, the world will still be as liveable as it is now."

"Yep. Yep. Look at the stacked layers of rock in the Grand Canyon. There have been thousands of years when Arizona was under water, thousands of years when it was desert and thousands of years when the weather was as it is today. Man adapts. The seacoast changes, and man adapts. Why do we live on the beach anyway?"

"What would happen if we stopped using oil in the next hundred years?"

"In the next one hundred years it would be a disaster. We'd be like the Amazon."

"Could we reduce energy wastage and make a difference that way?"

"We ought to be trimming all the excesses in our society and our planet, but I'm not optimistic that that's going to happen, because we're human. We like a hot shower every day. But more importantly than that, it's those 15 to 25% of the people that have no electricity today, they're gonna rapidly move in our direction regardless of what we want. They want the same things we enjoy today.

"I started going to Vietnam in 1995," he said. "We've been very successful. Ninety percent of the people were riding in bicycles and motorbikes. Now they drive in cars. They now allow capitalism in Vietnam, because they know the benefits of capitalism.

"I first went to Russia in 1998. The same is true there now. More Lexuses and Mercedeses in Moscow than any other city in the world. And all driven by oil, of course, since that's their biggest source of income. The people are not going to go back! The changes we saw in Russia were so dramatic, and I remember telling Bob Gates, the CIA Director, you don't have to worry about them going back to Communism. All these young people with rubles in their pockets, they will never go back!"

"Is there a carrying capacity of the planet, and are we at it?"

"How could you even measure it? It would be interesting to know what someone could calculate as the carrying capacity of the planet. Maybe God takes care of that, with plagues and so forth. But you know, if you look around, we worry about population density, but you go to Mongolia, Africa, and parts of Russia, and you've got a huge percentage of this planet with nobody on it! Sometimes they're not there because of climate, but with technology that can change. If desalinization plants become cheap enough, you will have beautiful lawns and green grass."

"What is the best thing about electricity?"

"Well, I guess I could say, although Bill Gates could answer that better, speed of communication, speed of transfer of knowledge. You can go on the computer and of course not everything's true, but at least the knowledge is transferred instantaneously. "Electricity is what allowed Houston to become what it is today, the fourth largest city in the country. Air conditioning clearly was a stimulus to growth. Well, I remember growing up in Oklahoma and we didn't have air conditioning, and I remember laying down on the hardwood floor, trying to cool off."

"Then would you say that the previous generation that didn't have air conditioning was less productive than yours?"

"I wouldn't say that at all. The generation before us, they were maybe the most productive generation in our history that from a zero start helped us produce maybe a hundred thousand tanks! They were really what we called them, the Greatest Generation. They were the ones who won World War II.

"My grandfather and grandmother both moved from a small town in Oklahoma to the Douglas factory in Oklahoma City, and they worked building airplanes for the war; I remember going there during the war; I was maybe four or five in that time; they lived in a garage apartment, and all they were doing was building planes. My Dad was born in 1909, and his dad was born in 1885. My grandfather was a trader; he would go out and buy hogs; he would feed them and he would sell them; he just traded livestock; he did what he had to do to survive. I think working in an aircraft factory was an opportunity for them to generate cash; but mostly it was a rallying cry. I think they were completely unskilled, putting rivets on planes. If you read some of the history of World War II, it's quite hard to believe. We thought we could stay out of the war, but Churchill kept telling us, you're our only hope. And when we started converting all our automotive factories to jeep factories and tank factories, we started with the wrong tools, and yet it's kind of an unbelievable story; I worry about whether Americans could do that today; we're not tough enough.

"My first job was on a drilling rig in southern Oklahoma; of course I was newly married and we needed money, so I worked 16 hours a day. It's hard work and dangerous but the industry has one of the best safety records. I was a roughneck, and that means you shuffle drill pipe, and when the great big tongs are dropping pipe, you have to put the pipe in the hole properly, and every time you disconnect that 3,000-foot string of pipe, you have to reconnect it, and that's backbreaking work. People lose fingers and toes; they get mashed off. I was very careful; I only did that for three months, to raise money before I went into the Marine Corps. I spent a year in Quantico, and then three really tough years in Hawaii. That was '61 to '64. We were looking for Russian bears. The bears would approach our imaginary property lines and when they were coming we would launch our fighters, and right before they got to the imaginary line,* they would turn back. I was also the top secret intelligence officer, and when the Russians would fly over our carriers and try to take photos, we would fly over their ships and take photos, and give the finger to the Russians. I call those things close calls. It was a scary time in our nation's history—two great military powers fighting over our way of life.

"Well, I left active duty in '64 to go back to graduate school. In January 1966 I reentered the oil industry and joined Conoco as a young project engineer, which is kind of at the bottom, but I didn't have to be a roughneck anymore, because I had a degree in engineering. I got my MBA degree. And I worked my way up. At Conoco our philosophy was *think big, move fast*. When I was CEO, we worked real hard. Conoco was the fast cat, and we would show Exxon as the big fat lazy cat. It just drove them crazy. But we were successful. Conoco was the first company to enter Vietnam in 1995. We were the first in Syria in the 1990s; I signed the contract to go to Iran in 1995—which triggered the U.S. sanctions which were on for the next 20 years!† We'd get things done and we'd get them right.

"We were the first in Russia, too, and the largest at one time, too. It was difficult. Russia's a tough place to operate. This was in the '90s and so basically it was Gorbachev and Yeltsin, right before Putin; Communism had failed and they were now a fledgling democracy, and in a lot of my speeches I used to say, they were probably not a lot different from the United States in 1780. In the area where we were operating in Russia, when they had their first election, they elected a snowmobile mechanic for Governor, and he wanted bribes. I said, we don't pay five dollar bribes, and we won't pay larger ones. I went to Moscow and I told the Oil Minister to tell that governor, Conoco doesn't pay bribes. You know, I think Joe Kennedy was a criminal; he was a bootlegger, but over time he

---

* He later explained: "State of Hawaii boundary."

† "Iran should be our partner in the Middle East, but that will never happen post-1979." (See p. 415 for another oilman's similar view.)

realized that honesty was the best thing. Well, that Russian Governor wasn't happy. Our office got bombed, but fortunately no one was hurt. I think that's why Putin pulled a lot of the power back from the provinces—to cut the corruption. Maybe he liked the power when he got it back and kept it."

"Who is the best President we've ever had?"

"I would say Reagan was probably the best President. He reinvigorated the economy. He also tore down the wall in Berlin and defeated Communism."

"And whom will you vote for in this election?"

"Cruz, he's very rigid, but his values are closely aligned to my values. Kasich is probably the most electable."*

"What would you say about the Deepwater Horizon accident?"

"I'll go on record on this, okay? BP in my opinion historically has been the least safe energy company. They did not have the same focus on safety that most large energy companies had. Partly that was the U.K. culture of being superior: *You Americans are still just colonials.* And obviously there was also some lack of clarity in the governance: Who has authority; who has responsibility? I don't think that was clear.

"And the Russians also do not have the safety culture that we have in the United States. I'll give you two examples. I remember when we were using Russian crews to drill initially. On one of the rigs that were not ours, a Russian roughneck actually drank the methanol that was used to unfreeze the pipes. He died, of course.† We would not allow any kind of alcoholic beverage within 50 miles of our rigs.† Second example: I remember wanting to go out on one of our rigs to inspect pipeline. I remember getting on a Russian charter helicopter, and the pilot comes on board, and he lights up a cigarette. So I'm purposefully waiting a few seconds to see what my managing director does. And he does nothing, so I tell him to tell that pilot to get that cigarette out of his mouth and never smoke around our aircraft again, and we landed in minus 60 degree weather, and Conoco had given me a real good pair of gloves and another pair of gloves on top, but since it was sunny I left one pair of gloves inside the helicopter. When I came back, my gloves were gone."

"Did you say anything?

"No. I figure that pilot needed them more than I did."

"Do you believe in climate change?" I asked again, because I wished to hear more from him on this subject, and he said:

---

* With Cruz out of the picture, Mr. Dunham updated his answer: "The young senator from Florida, Marco Rubio. He has the intelligence and temperament to be President."

† A journalist who visited an Alaskan oil rig summed up the situation there: "There is no alcohol allowed anywhere on the [North] Slope. A person getting caught with it will be run off for life, no questions asked . . . Everything here is flammable. One mistake and you could blow the place up."

"I believe that humans have an impact on climate but I'm not willing to destroy the United States economy based on someone's supposed measurement of the ocean temperature within two-tenths of a degree *versus* the temperature one hundred years ago. I doubt that we could accurately measure the ocean temperature within one degree of accuracy [from] 100 years ago.* Plus, the Paris Climate Change Agreement requires the U.S. to transfer billions of dollars a year to China and India; and neither of these countries is required to cut emissions before 2030!"

Reader, what would you say to this? A minuscule measurement subject to error . . . paying subsidies to other nations who kept on polluting . . . Who could blame Mr. Dunham for his suspicion of these propositions?

"At one time," he continued, "Conoco was owned by DuPont, so I was asked to go to Wilmington, where I managed two sectors including the chemical sector. This was back in 1990; DuPont was the largest manufacturer of Freon in the world. The environmentalists wanted us to stop producing Freon. We shut down all our plants worldwide, and took several billion dollars' worth of write-offs.† Within 24 months, all the capacity we had shut down was rebuilt in India and China, so the world did not benefit from our decision. My point is, unless the entire world signs a binding agreement to do something, all we do is basically punish the U.S. economy at the expense of the U.S. taxpayer."

"And Freon is still in use now?"

"Sure. All over the world."

This staggered me; I must admit that it made me discouraged . . .

In the end I could not find verification for his assertion. On the contrary, according to the respected (and otherwise doomsaying) Intergovernmental Panel on Climate Change (2013):

> CFC atmospheric abundances are decreasing . . . because of the successful reduction in emissions resulting from the Montreal Protocol. By

---

* Chris Faulkner, C.E.O., Breitling Energy: "The possibility of damage to the global economy . . . ultimately could be far worse for humanity than the planet warming by a couple of degrees." John Hofmeister, former President, Shell Oil Company, 2010: "The point here is that debating climate change is a fantastic waste of time and human energy. There is no agreement on what it is or isn't. There is no set of measures accurate enough to be credible to present a clear and present danger. There is no rebuttal for the argument that we have always had global warming and global cooling, and Earth has adjusted accordingly." U.S. Environmental Protection Agency, 2016: "If greenhouse gases continue to increase, climate models predict that the average temperature at the Earth's surface is likely to increase from 0.5 to 8.6 degrees Fahrenheit above 1986 through 2005 levels by the end of this century, depending on future emissions."

† "DuPont was able to develop an alternative," said Mr. Dunham. "HCFC-22 and HFC-134a, those were all new DuPont products."

2010, emissions from O[zone] D[epleting] S[ubstance]s had been reduced by ~11 Pg [petagrams]* CO_2-equivalent per year] ... which is five to six times the reduction target of the first commitment period (2008–2012) of the Kyoto Protocol ...

And the panel went on to itemize some of the chlorofluorocarbons which you have read about in the table back on I:176. All the following were Freons:

CFC-12 has the largest atmospheric abundance and G[lobal] W[arming] P[otential]-weighted emissions ... of the CFCs. Its tropospheric abundance peaked during 2000–2004 ... Its global annual mean mole fraction declined by 13.8 p[arts] p[er] t[rillion] to 528.5 ppt in 2011. CFC-11 continued the decrease that started in the mid-1990s, by 12.9 ppt since 2005. In 2011, CFC-11 was 237.7 ppt. CFC-113 decreased by 4.3 ppt since 2005 to 74.3 ppt in 2011 ... Future emissions of CFCs will largely come from "banks" (i.e., material residing in existing equipment or stores) rather than current production.

Of course atmospheric concentrations of the CFCs' so-called "transitional substitutes," the HCFCs, were still rising, and the panel had this to say about one of their three most dangerous members: *Developed country emissions of HCFC-22 are decreasing, and the trend in total global emissions is driven by large increases from south and Southeast Asia.* In a way this bears out what Mr. Dunham was saying—except that these were not the *old* Freons, and that DuPont and other entities in "developed countries" did not stop making them at all.[†] In other words—and this is one of the few happy outcomes which this book has to report—when "the entire world signed a binding agreement to do something," the agreement actually worked.

But as for those *new* Freons, well: *HFC-134a ... will no longer be acceptable in new equipment in certain end use applications in the U.S. market ... There are no U.S. regulations currently impacting the HFC service market.*

---

* Roughly 24.3 trillion pounds.

† In 2015, DuPont spun off a new company called Chemours, which among its other business handled all "fluro [*sic*] products." According to the Chemours "official page" on our favorite chemical family, "The Freon brand has been trusted by the industry since the first fluorochemical refrigerants were introduced to the marketplace over 85 years ago ... As we began to transition to Chemours, we recognized the value of the Freon™ brand and have since taken steps to amplify it. As a result, our full HFC portfolio will be represented by Freon™ going forward."—The accompanying table following listed various refrigerants by the R numbers of the chemicals they replace, not by their CFC or HCF formula, so I cannot tell you what they actually were, but they are all Freons™—praise profit™!

I asked Mr. Dunham: "Which would you say is most important, coal, oil or natural gas?"

"Well, they're all three important. Coal is going to be predominantly the fuel to generate electricity in most of the world. Renewables will not amount to much because they're uneconomic without tax incentives. Oil is going to be used to heat homes and run automobiles. They're all three needed. But coal in this country is going to be phased out over time and replaced by natural gas, a much cleaner alternative."

"What about solar?"

"I wish it could do more, but it can't because it's just not economic. There've been so many financial disasters in the manufacture of solar! I'm sure it will happen sometime. Oil companies aren't opposed to renewables; I say bring 'em all on. You need all forms of energy to grow the global economy. But it's just not going to happen quickly. Every week I get calls from someone who wants to put solar on my house. They're all subsidized. I believe in free markets. We should not be giving out subsidies for any industry."

"And nuclear?"

"I love it, but it's not going to happen, I think, because of Chernobyl and other problems. People used to use Japan as a great example. Maybe in 50 years it will come back. It should be more economical than oil."

Looking straight at me from behind his desk, he said: "*Oil can't be replaced*. Diesel fuel, gasoline, they're all from oil."

"Can't automobile fuel be made from natural gas?"

"We make gas-fired fuel today, but it's not economic. We have to have subsidies. I just don't think we have to be spending taxpayer dollars to decide what products are going to be good in the future.* I'd much rather be subsidizing some vaccine or cancer cure or . . ."

When I met him it was only April, and so far that year the following headlines had already caught my notice:

Greenhouse Gas Linked to Floods Along U.S. Coasts: Worsening a Certainty: Research Team Reports Fastest Sea Rise in 28 Centuries.

Arizona's Scorpions Get Early Start on Summer: Soaring Temperatures Bringing Swarms to Homes in the Desert . . . Usually dormant until late March, the creatures came out in . . . the second-warmest February on record.

---

* Here he wrote in: "Let's let the free market decide on what products are useful and desired by the public."

Highlight: Near-Record Warmth: Warmth more typical of May or early June will spread throughout the East . . . [*This was in early March.*] So get out and enjoy the weather.

Ice-Sheet Melt Seen Harming Cities by 2100 . . . That is roughly twice the increase reported as a plausible worst-case scenario by a United Nations panel just three years ago, and . . . would likely provoke a profound crisis within the lifetimes of children being born today.

Leaders Meet to Sign a Climate Pact Fraught with Uncertainties . . . Outside experts . . . say the countries' bare-bones plans are still far from enough to keep global warming to tolerable levels.

I therefore asked Mr. Dunham: "What would you tell people who read frightening stories about climate change in *The New York Times?*"

"I would say, maybe try reading Epstein's book for starters, in which he refutes much of what's being written by *The New York Times.* More and more distinguished scientists today are leaving the climate change arena, by which I mean no longer being supportive of that view. Lastly, I would share with them the example of what happened with CFCs. And the same thing's true of unilateral sanctions. Multilateral sanctions like we had on South Africa work. *Unilateral sanctions don't work.*

"I spoke at the 2000 Rio World Environmental Conference; I was one of the main speakers. We never imagined that China and India would immediately rebuild Freon plants. There was no binding agreement! That's one thing you can say about Trump. He's right. We do feel-good things and we just get laughed at all the time.* I've travelled the world for nearly 50 years now. I started travelling the Mideast in 1970s. And people laugh at us; we make a lot of mistakes."

---

* Here he added: "Why do China and India want the U.S. to remain in the Paris Climate Change Accord? They want no restraints to their industrial growth and they want U.S. dollars to be transferred to Third World countries so that the U.S. economy is weakened. We operate in a global economy. China knows this. I'm not sure the media and the environmentalists in the U.S. care or understand strategy or Economics 101."

7

"It's All Economic in the End"

July 2016

The bottom line of my profession: You can do anything. *It's just, how much does it cost?*

Sam Hewes, Vice-President of the Bank of Oklahoma, 2016

"THAT'S FEDERAL REGULATIONS"

When I went to Oklahoma, seeking instruction in the oil ideology, global oil prices were still depressed, thanks in part to fracking, but all was not darkness. The state's gas prices averaged 48 cents lower than the year before, so that as the *Pawhuska Journal-Capital* explained: *A record 558,400 Oklahomans are expected to travel over the Fourth . . . 86 percent will drive.*

In those days a Mr. Chuck Mai happened to be Oklahoma's spokesman for the American Automobile Association. He gloated: "We are well on our way for 2016 to be a record-breaking year for summertime travel. This is welcome news to all of those in the travel industry, from hotels and attractions to restaurants and local shops."

As it happened, 2016 was also a record-breaking year for heat waves.—What a hilarious coincidence!

And Oklahoma did its mite to contribute. According to the *Oklahoma Almanac,* the state's annual oil and gas yields were *more than 3 percent of total U.S. production in recent years.*[*]

In 2013 the Intergovernmental Panel on Climate Change had made the following prediction:

> Human discomfort, morbidity and mortality during heat waves depend not only on temperature but also specific humidity. Heat stress, defined as the combined effect of temperature and humidity, is expected to increase along with warming temperatures . . . Areas with abundant atmospheric moisture availability and high present-day temperatures such as Mediterranean coastal regions are expected to experience the greatest heat stress changes . . .

Well, well. Maybe that would also apply to sweltering Oklahoma. For North America the panel's computer-modeled *warming generally leads to a two- to*

---

[*] Like West Virginia, Oklahoma consumed energy on a higher per capita basis (2012 figures) than the national average: 411 million BTUs *versus* 303 million—in other words, almost 36% more. And like West Virginia, Oklahoma directed her "end-use" consumption more than anything else to a category called "industrial." Something tells me that "industrial" had much to do with oil wells.

four-fold increase in simulated heat wave frequency over the 21st century. For some reason, *warming near the Great Lakes area . . . is projected to be about 50% greater than that of the global mean warming.* Thank goodness I didn't live there! Returning to my darkened motel room in Bartlesville, I ran the air conditioner near maximum, opened the convenient little fridge and drained two plastic bottles of fizzy water which had been trucked in from far away.

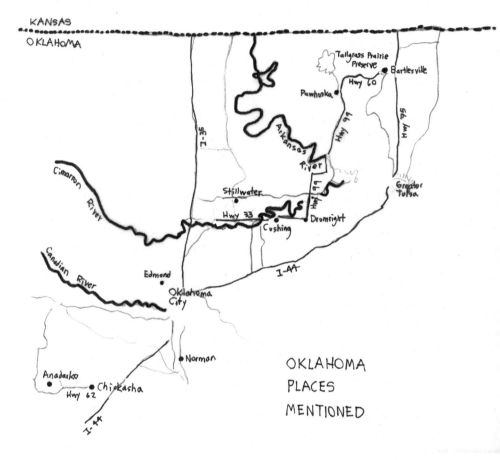

From Bartlesville I remember wide oak trees beginning to silhouette themselves in the sinking sun, locust trees, sparrows, squirrels running in hordes over the grass where people left corn for them by the limestone boulder with an old drill bit on it. There was poison ivy and the slow brown river and a cool breeze like a dog licking me in that spicy-hot evening sunshine. I stepped over the barrier, which from its flimsiness must have been a mere suggestion, and prowled through the replica of the Nellie Johnstone No. 1, *the first producing well in Indian Territory* (1878), whose wooden derrick rose up high, as a rope fell within it to meet a wide spool. Then I went back to town.

The Nellie Johnstone No. 1

Bartlesville was *the energy city.* At least that was what the Convention and Visitors' Bureau called it.

Alex was the manager at the Painted Horse. This sinewy and much-tattooed young man with the physique of a Marine and a handshake of credible firmness had a way of looking a person alertly in the eye. He used to live down in Tulsa but came here for the sake of his "kiddy-o," Bartlesville being "a slow-paced friendly place with lots of swimming and walking."

I asked about oil.

"Well," he replied, "as far as I know just about everybody in this town's connected to oil one way or another. My girlfriend's Dad sells pipes for an oil company. Phillips is what made this town, and all the young people think, if they can get a job with Phillips they've made it.* They get laid off from one building and go apply next door."

The yellow pages for greater Bartlesville listed four local gasoline wholesalers and a propane company—not to mention the Southern Star Natural Gas Pipeline and Oklahoma Natural Gas with a toll-free 800 area code.†

But when I asked around in Bartlesville, no one seemed to know anyone in the oil business.

You asked people how they were and they'd say: *"Blessed."*

You asked about oil, and they couldn't tell you a thing.

Perhaps this was the good old *no comment* so characteristic of the *regulated community.* They looked at me, and maybe smelled environmentalist. All the same, it was strange, because unlike the West Virginians, they saw no need to be militant defenders of their local carbon ideology. Oil's cities might not have been *greater than gold has ever built, and apparently more permanent,* but oil's lineaments remained unchallenged. The Deepwater Horizon explosion was far enough away in space and now time—six years!—to pass virtually unremembered. Oklahomans glorified oil in their materials, place names and even hotel signs, and the "horseheads" kept pumping.

1: "We Don't Comment Publicly" (continued from fracking, coal and nuclear oracles)

Bartlesville lay almost on the eastern boundary of Osage County. Rolling west on 60 (the *Scenic By-Way*) past Osage Hills State Park, one reached State Highway 99 and turned south, quickly coming into Pawhuska *(the headquarters of the Osage Nation),* where three of this chapter's interviews took place. Continuing south clear out of the county on that road, one presently reached Cushing, which after 11 years of drilling had first found itself near an oil gusher in 1912

---

* Here again I repeat the words of Mr. Endo, the nuclear evacuee from Tomioka [I:377]: "In this area, when you say, I got a good job, that means, I'm a Tepco employee or a civil servant."

† Needless to say, none of them returned my phone calls.

Chickasha

Pawhuska

Well pad and farm, south central Oklahoma

"Horsehead" and home, Oklahoma City

(downtown's brick edifices flew up around 1914 and 1915), and which several Pawhuskans had described as active, busy, etcetera, from a petroleum point of view.* In the local phone book I found 14 companies listed under "Oil Producers," 10 of them in adjacent areas. There were two oil marketers and eight oil operators, five of which lay in nearby Stillwater. An advertisement in the *Cushing Citizen* for Plains All American Pipeline ("headquartered in Houston, Texas") announced: *On the average, we handle over 4 million barrels per day of crude oil through our extensive network of assets located in key North American producing basins, major market hubs and transportation gateways.*

"Lots of luck," said the formerly friendly waitress in Cushing, stepping back from my table in one of those repurposed old brick buildings. "They've been told: no talking and no pictures. Everything behind a fence. That's federal regulations. Two days a week I work at the gate, sitting on my rear end, watching television, letting people come and go, and not that many of 'em, I can tell you."

I rolled out thataway and found her exactly right. The road was fenced off at both sides at the Enbridge facility just south of town, presenting massive fat cylinders behind prisonlike fences. Somebody at the library explained that there were concerns about a terrorist attack. I guess there must have been similar worries at Saddlehorn and Blueknight and Magellan Midstream Partners. As for climate change, that hardly seemed to keep them up at night.

---

* "The largest commercial oil storage hub in North America, second only to the U.S. government's Strategic Petroleum Reserve." The former Mayor of Pawhuska remarked: "Cushing, a lot of pipeline..."

And by some coincidence, when I jetted to the other side of the globe, contributing to climate change for the sake of *Carbon Ideologies,* I discovered a near-identical situation for petro-facilities of the United Arab Emirates! As you will see, I collected both talking and pictures over there, through luck and the kindness of others; while in Cushing I quickly abandoned the effort. So once again I had to wonder: What did it say about an industry that most people depended on, that sometimes turned high profits, that was inherently dangerous to its workers, and now arguably threatened humanity itself—that it shut out writers and journalists? Thus coal, oil, nuclear, natural gas. When I began this book I fondly imagined that I would get you photographs in oil fields, refineries, nylon manufactories, offshore rigs, you name it. One might say that my failure implies something about my effectiveness, as it surely does. Once again, what does it also say about *them*?

The Coloradan anti-fracking activist Sharon Carlisle had told me: "The National Chamber of Commerce and every Chamber of Commerce in every damn town is pro-oil and -gas." I would have had to visit every Chamber of Commerce to verify her assertion, and despite some effort I could not even verify it about Cushing, although, not yet knowing that I would strike out, and discovering this municipality's Chamber of Commerce conveniently and conspicuously situated right there at the roadside, I dropped in, ingenuously hoping for a quick interview about energy with anyone, anybody at all! I was always easy to please. The manager, Ms. Tracy Caulfield, was present, but by some fluke (which might or might not have had something to do with my lack of a business suit) she was, she said, too busy just then, so I requested and received permission to try again some other time. As you might expect, my research assistant, who transmitted such questions of mine as: *Does energy use somehow strengthen freedom and democracy?* and: *Is climate change fact or fiction? If fact, is it caused by humans?* and: *Can energy demand continue to grow indefinitely?,* was presently compelled to report: *Tracy Caulfield . . . ; Chamber at 918-225-2400 or email them at* REDACTED. *The email and number go straight to her. Sent 11-14-16, no reply.*

2: "We Don't Comment Publicly" (continued)

It consoles me that I was not the only one to be excluded from the doings of the great:

Once upon a time in 2014, with final approval of the Keystone XL oil pipeline from Canada still in doubt, a Calgary-based firm called Enbridge (yes, the very same whose fences decorated the highway south of Cushing—and by the way, according to the *Federal Register,* Enbridge was *a limited partnership duly orga-nized under the laws of the State of Delaware,* which is to say *a wholly owned subsid-iary of Enbridge Energy Partners, L.P.,* of Houston)—made it clear that a comparable project* required no approval whatsoever! Back in 1967 Enbridge had built its own pipeline, the Line 3, whose fat pipes could carry (*although it's unclear,* as a Sierra Club attorney wrote me) 790,000 barrels of tar sands crude per day. In 2010 the State Department permitted a flow of up to 450,000 barrels per day. An oil ideologue might call this legal limit a tragedy of undersupply. At any rate, En-bridge then constructed Line 67—the Alberta Clipper. In his marginal note on my draft the Sierra Club attorney neutrally related: *It was well-known that Enbridge intended to expand Line 67 up to 880,000 in the future, but the State Department, Enbridge, and the public all agreed that would require a new permit.* Because *the permitting took awhile, Enbridge* in its eagerness to serve us *decided to go forward without waiting by diverting the flow of oil to the adjacent Line 3, which had a larger capacity that was underutilized.* An anti-oil ideologue might have resented this in-tended "diversion." But maybe it was legal; what do I know? It might even have been moral. You see, as Enbridge explained to State's Office of Environmental Quality and Transboundary Issues: *Any failure on the part of Enbridge to provide the requested capacity will cause shippers and refiners to suffer adverse impacts . . . which in turn, may lead to higher domestic oil prices.* Or, as the Sierra Club told it: Allowing this to happen would *give the landlocked tar sands industry access to ports and enormous new overseas markets; and enable the massive, environmentally dev-astating tar sands growth planned by the industry.* But the terminal in Superior, Wisconsin, was waiting! (As the anti-fracking activist Bob Winkler[†] remarked about Keystone: "You put that through Chicago or Atlanta or San Francisco, it'll never happen. They take advantage of the unfortunate people.") And so the com-pany condescended to write the U.S. State Department: *Enbridge intends to con-struct the interconnections and Pump Upgrades, and to operate those facilities to increase the flow of oil on the Line 67 south of border segment, whether or not a new Presidential Permit is issued by the Department.* Fortunately for international com-merce, State did grant the permit. The Sierra Club filed a Freedom of Information Act request with State regarding *permit-application information and State Depart-ment correspondence with Enbridge.* Sierra's lawyer told my intermediary: *The State Department is the slowest of all the state agencies in complying . . . So Sierra Club*

---

* Keystone would have carried 830,000 barrels per day. Enbridge was going to bring in 800,000 barrels per day.

† See p. 345.

filed a lawsuit under [the] F[reedom] O[f] I[nformation] A[ct] and [the] N[ational] E[nvironmental] P[olicy] A[ct]. A year ago, a judge denied summary judgment of the NEPA claim but not the FOIA one. I can say that Enbridge expanded their Alberta Clipper department by diverting to line 3, and we're still hopeful that the administration will continue to review and be transparent. By the time I read that, Trump was President. I had to chuckle over the attorney's hopefulness.*

Perhaps you and I might have felt entitled to learn more of that peculiarly hyphenated *permit-application information*, given its connection with tar sands, which were *described by former NASA climate scientist James Hansen as one of the "dirtiest most carbon-intensive fuels on the planet."*[†] Instead, we were left with the pleasures of speculation: Did Enbridge pay off someone in State? Was State influenced by American investors in Enbridge? Or did the permit need to be denied for compelling reasons of national security? For instance, what if the Sierra Club (not to mention all proponents of human-induced climate change) were preparing some horrendous terrorist plot? (I always liked to give State the benefit of the doubt, back when I was alive.)

3: "We Could Further Enhance Our Systems"

As it happened, Enbridge was not utterly unwilling to comment publicly—for at the Chamber of Commerce I found a newspaperlike **Enbridge Safety Report to the Community,** printed on recycled paper *which is manufactured entirely from wind energy,* never mind the wood pulp, and dated 2014 although it did peculiarly contain a story from 2015. First and foremost, *life takes energy* (I learned to my astonishment), *and our job is to move the energy you need as safely as we possibly can.* How could I argue with that, especially should my life perchance take Enbridge's energy? By 2014, or 2015, or some other time, the company controlled *more than 11,000 employees and contractors across North America*—a robust proffered constituency. It piped in natural gas from Canada, proffered 1.6 gigawatts of solar, geothermal and wind energy (before the

---

* In 2017, asked for a final comment on this story, the Sierra Club attorney wrote waspishly: "This seems to be a pretty minor and irrelevant detail from several years ago compared to everything that's happened since then. We litigated and resolved the FOIA claim, and State just issued a final E[nvironmental] I[mpact] S[tatement] for the Line 67 expansion." In other words, the Line 67 expansion must be going ahead.

† The Sierra Club claimed that "mining for tar sands . . . ravag[es] one of our strongest remaining defenses against climate disruption—boreal forest captures and stores almost twice as much carbon as tropical forest," and that "tar sands extraction and upgrading produce a staggering 220% to 330% more greenhouse gases than conventional U.S. crude."

tsunami, Fukushima Nuclear Plant No. 1 had generated three times more than that) and—hats off!—operated *the world's largest and longest crude oil pipeline system, transporting over 2.2 M barrels per day* at anywhere from 2.5 to 5 miles an hour. (The Sierra Club called Enbridge *responsible for three of the fifteen largest offshore oil spills in U.S. history . . . The U.S. Department of Transportation . . . has fined Enbridge repeatedly for safety violations, including a record $2.4 million fine in connection with an explosion that killed two workers . . . in Clearbrook, Minnesota.* But I'm sure that computed better when expressed as accidents per mile of pipeline.) In 2014, Enbridge had "safely delivered" 2,405,421,468 barrels of crude to the Americans and spilled 2,921 barrels—the merest 0.00012% loss, I see. And in 2010, it had spilled "nearly 20,000" barrels in Marshall, Michigan. As the Senior Vice-President for Enterprise Safety and Operational Reliability cheerfully explained: "Marshall made us realize we could further enhance our systems." In those days I used to be entertained by such creatures.

The citizens of Pawhuska were more down to earth than that Senior Vice-President for Enterprise Safety and Operational Reliability. Not one of them no-commented me or turned me away, and I thank them. Moreover, whenever they told me something about oil, it even sounded as if they believed what they were saying.

4: "Oil Is Everything"

Mr. Roger Taylor had recently been succeeded as Mayor of Pawhuska by a Mr. Brock Moore, but he still served on the city council. Thanks to the good offices of the motel proprietress, he agreed to meet me at Buffalo Joe's restaurant, and here is what he said:

"In this area here, oil is everything. I've been here all my life. That and ranching is all there is in this area. It's all a domino effect. Right now we're in a recession, so it's affecting us. Business I'm in, we sell gasoline, diesel and propane, and we have no drilling going on in Osage County, none at all, and so there's no diesel being bought, no workers working, no frack workers.

"I've lived here all my life. I'm 60 years old. The oil business has been in my life, all my life. I worked for a frack company where a well'd blowed out.* It's messy. They get your attention!

"My oldest son is a geologist, and I'm pretty proud of him. He got into the oil business, and bought into this other company, and this other company hit a

---

* This past participle was in common use everywhere I went in Oklahoma.

couple of home runs. It was when oil was $120 a barrel. And the first check he got was $50,000! A millionaire! At 34 years old! And he didn't quit his job. And then the oil falls off and he says, Dad, I'm gonna move to Colorado. Live off of my interest money."

"Did Pawhuska used to be busier than now?"

"We had a J.C. Penney's; we had a movie theater; we had a drive-in back when I was a kid. We had a hardware store; we had a Turtles, Girdles and Yo-yos. Walmart came to town, and it shut down all them little stores.

"We used to have a railroad through here. The gas used to be brought in here by the railroad. And the railroad cars would drop 'em off and you had three days to pump 'em out. They stopped in Missouri, Kansas and Oklahoma.

"See, when I lived in Pershing, it was an oil camp town. It had a city hall, a jail, a school, and that's where I was raised at. By then the camp had dissipated but the freight train would come through. My mother told me a story: When I was two years old they did have that stop there for just a little while. When we lived in Pershing, the bums, the hobos, they could get off in the middle of those towns, and my mother said I was playing outside when this guy came up and asked me if my mother would make him a sandwich, and I said sure, and she gave him a sandwich, and then she made me come inside and locked the door. The only thing that's there now is the church, and it's a shell. Well, the school is there, and they made *it* a church.

"Okmar Oil Company had a big presence there, and they would drill four wells, and in the center they would pump water into it, and it would push oil out into them producing wells. And then it kind of fell off, and Okmar gave up their presence there.

"There's not many majors left in Osage County. They're all little independents now. The fuss with the tribes was just not worth it.*

"Just a few years ago, there was approximately 300 wells a year. Now we're down to 12, and maybe not that many. They might have drilled five. Two-thirds of those were producers and the rest were disposals.

"I've seen oil at $8 a barrel and $120, $130 a barrel. You know, it goes in spells, and this last boom was a 12- to 15-year boom, maybe a little more, and so now it's back to $48 a barrel."

"To your mind, which is a better fuel, oil or coal?"

"The thing about coal, what can you do with coal, and what can you do with

---

* For more on "the tribes," see the next interview (p. 480).

Mr. Roger Taylor

oil? The thing about oil, you can make more products out of it than you can out of coal. With oil you can make gasoline, diesel; you can fuel your war machine."

"How does your fuel business work?"

"It comes in transport loads and we put in bulk tanks and put in smaller vehicles and take it to sites. To do that we break it down from 85 hundred gallons to 2,500 gallons.

"The company I work for has 17 of these operations. We have to get a special permit going into Kansas. Right now, the most money is in diesel fuel, because when you're drilling, your drilling rigs and frack rigs use diesel fuel. There's a trend now to buy non-ethanol gas. Small engines, people are getting better luck with non-ethanol.

"Most of these drilling rigs will buy probably 3,000 gallons for a week at a time. So around this area, a well is about 6,000 foot. You have to figure out about the rock. Some of is it limestone and some is clay."

"In a week can they drill an entire well?"

"Yeah, you can probably do a well in a week."

I calculated: 3,000 gallons per 6,000 feet divided out to a gallon per two feet. Looking up the high heating value of diesel (my favorite approximation for crude's), I discovered that it cost the oilmen 5,645 BTUs to drill each inch.—Of course those BTUs supported the entire operation, for as Mr. Taylor remarked, "With that being said, they're also using a water pump. And they're circulating water down to the center of that drill pipe, see, so that pump's working, too. They're pulling out the finds. And some's going for the generator to run the lights, too."

He said: "I been in this for 28 years. So when I first started in this, we had leaded gas and we had high-sulfur diesel. And that's a considerable difference in smell. There's a smell and there's a look. I don't have to worry about it too much anymore. From a refinery, I don't ever have to worry.

"We used to have a premium super gas here, used to have a dye in it, and they took all the dye out it, and the only thing that has it now is the off-road diesel, which you don't run on your highway, so you pay no taxes on it. If you put red diesel in it and you get stopped by Internal Revenue, they can charge you a thousand dollars a gallon for a fine. I've heard tell that in several bars where the farmers would sell their cattle, they would pull every farmer over and they would check 'em, and they would pull their records up, and they would go with you to your house and check all your vehicles."

"How many of Pawhuska's citizens depend directly or indirectly on oil?"

"I would say, 60%. I was talking to one of our city managers at Hominy, and they're struggling worse than us. We're the county seat, and they don't have that, and the oil is down. One of the things about Osage County you have to remember is that the Osages have a big finger in the pie here. You have to deal with them. They get a 16th of the oil business. Whenever you produce it, you gotta pay them."

"What do you think about carbon fuels and climate change?"

"I think that everything that's fuel's got a problem with it. I think that we're going have to depend on more windmills and solar power and step back from that. But I don't think we'll ever be weaned off of it. I think we're going to have to step back from it. A few generations down the road, we're gonna see a lot more solar and that stuff.

"But we put up those windmills up there, and those Osages didn't want us to. They don't get a percentage of it. They can't claim the wind. They're just fightin' it tooth and nail. They got 40,000 acres; they could be making money from it."

"What would happen if oil went up to $500 a gallon?"

"Whooh! We could have more businesses, but I don't know if we could afford the gas and diesel prices. Seventy and $80 a barrel, it would be perfect. We sure could keep people working. More people will buy local. People will buy more American-made stuff."

5: The Underground Reservation

In the library of the Wah Zha Zhi Cultural Center, Mr. Vann Bighorse, the center's director, made time for me so that we could talk about oil, and since we are told that oil was *an article of trade among the Seneca Indians,* I asked what his Osage ancestors had thought of the stuff. He said:

"When they talked about it they knew there was something here. Because they hunted here. And the way I understand it is that they used that because it was kinda oozing out of the ground, as an ointment I guess you can say, I think for their animals and what have you. Maybe for themselves for a rash. They knew about that since this was our aboriginal hunting areas back in the early time period; we had this land all the way to the Rocky Mountains: 96 million acres. We owned northern Arkansas, a big portion of Missouri, a big portion of Kansas, and so on all the way to the foothills of the Rocky Mountains. So, yes, they did use some of the crude, as a way to help heal sores or whatever.

"I think they started exploring for oil here in 1898. I think there was a man here from Rhode Island, named Foster. I think he had a blanket lease over the whole county for quite awhile."—And he opened an old book, from which he read out: *March 16, 1896, by Edwin B. Foster from Westerly, RI, then engaged in building railroads.** At that time it was pretty good to get an IP of 20 barrels a day.

"There was a man named James Bigheart, one of our chiefs, and he was one of the ones who had a vision about this place, and once our people made a decision on purchasing the land here off the Cherokees† we went in and put some kind of a rider in the bill where we wanted to have the minerals, which means everything from a shovel's depth on down.

---

* The Bartlesville Convention and Visitors' Bureau preferred to lionize "young engineer H.V. Foster," who "arrived in 1902 to manage his family's blanket lease on all 1.5 million acres of the Osage Lands." I wonder what the Osage had to say about it then?

† Mr. Bighorse had said: "1872 is when we got here"—in other words, when this area became the Osage reservation. From the 1911 *Britannica* entry on Oklahoma: "Until 1906 the Osages lived on a reservation touching Kansas on the N. and the Arkansas river on the W. (since then almost all allotted) ..." According to the Nature Conservancy, the area was once a Caddo and Wichita homeland. "The Osage were permanently relocated from Kansas to Osage County in 1872. They farmed some of the bottomlands and hunted bison and prairie chickens. Eventually, as oil, gas and ranching industries developed ..., the surface ownership shifted from small plots owned by the Osage to large ranches, some larger than 100,000 acres."

"So today we still own that oil, still owned in common, instead of cut up into allotments. I call it the underground reservation. And I don't know what it is right now, but there's some big numbers. And they started secondary recovery in the '50s. That's like pushing water into the formation, and when it started depleting, they started building injection wells around the producer wells.

"There was one over across the street that had the submergible pump on it. There's a lot of 'em that just have the regular rod and tubing. Anywhere, you could stop anywhere and take a picture. Now that one out there, it was a new one where they did that horizontal drilling. And they went two or three thousand feet I guess. I think the IP production was 400 or 500 barrels a day. At the beginning some of those vertical gushers came in at three or four hundred barrels a day. But I think they took a great deal of pressure off back then with those gushers; there's not near as much pressure today."

"And what's each tribal member's revenue from it right now?"

"You gotta remember that it's cut up into shares. We own it in common, but as for the money, well, when we purchased the property, they did an allotment. We held 'em off as long as we could. We kept our land in common until 1906. We were the holdup for the state of Oklahoma! But civilization was here, so what do you say? So they were kind of forced into allotment. And once they did the allotment, well, in 1906 there was 2,229 allottees. After they passed away, and they all had a head right or a full share, time went on, and today it's very fractionated. Some get a fraction and some get a little bit more than a fraction. Some might get four head rights. That's maybe a lone child. And you might say that lone child has it made, but of course that has a bad side: In order to inherit, your folks have to die, and that's the bad part."

"How much do people depend on the oil money now?"

"Well, it's helped supplement people's income ever since it's been here. But in today's time oil prices are so low, I think the head right share annuity payment was only $4,000 in June. So if you get only get a fraction of 4,000, what is it to you?—Once it went as high as $10,000 a quarter. A person can make it on that. So a lot of our elderly people that don't work depend on that money . . ."

"And how does the oil revenue compare to the income from your casino?"

"Well, the gaming money, we don't get a per capita. The gaming money is all for health, education and welfare services. So we have a health card. I think if you're 65 and older, you get a thousand-dollar health card. If you're younger, get a $500 health card. If you go to college, you get up to $3,500 a semester, and you can still get your Pell Grant and whatever else is available. We also have a burial fund, and we get $3,500 for that."

Mr. Vann Bighorse

"What about the old treaty annuities?"

"That's all been said and done and paid for. The last claims from those got paid off in the 1970s."

"Do you believe in global warming?"

"I think it's for real. I think there's something going on in the world. Weather's different from what we're used to. For me it seems to me more hotter. For instance, it's a 115° heat index today. Maybe that's the natural cycle but I didn't know about that. When I was a kid we didn't have an air conditioner until 1950. I think the sea level's rising in different locations. In Louisiana it's rising an inch an hour or a day or whatever. I seen a documentary, and it's coming at 'em."

"In that case, what's your position on the use of oil?"

"I think it's gonna be always needed. Everything that we use and wear, it's made of petroleum. Everything that we have, from food to clothing to our eyeglasses, it has to do with petroleum. It's a medication."

"And coal?"

"I'm not around it, but I know it's used to generate energy, and I think it's something that we need. There's one outside of Tulsa, going toward Arkansas, and there's another one somewhere . . ."

"How about nuclear?"

"I don't know about nuclear. I don't know nothing about it. I don't think it's good, myself."

"You know," he said, "Frank Phillips* started here. His picture's in this book right here. The old chiefs thought he was a good neighbor. They adopted a lot of those guys into our tribe."

6: "I Don't Know Why We Can't Just Make Everything Electrical"

Mr. Keith Lambert was currently employed as a custodian by the Osage Nation. His wife Deanna was a waitress. I interviewed them at Buffalo Joe's.

Deanna and Keith Lambert

Mr. Lambert said: "I did tail rods. They came out of this well and the oil rig pulled 'em out and we had to use tongs to break 'em through, and we had tethered rods and had 'em sit on this rack and line 'em up. There was oil on the socket

---

* From whom sprang Phillips Petroleum.

part. It was a dirty job. We just laid 'em nice and even in a straight line. The rig picks 'em up, the tongs take 'em apart. You bring the cabler down and attach it to the tongs. They also got cables that you bring 'em all the way up.

"It's dangerous. I almost died a couple times. The tongs weigh 600 pounds. You have to pick 'em up out of the back of the truck, and you use a sandline. They throttled it too much and it almost took my head off.

"But they paid me. I made so much money. I did pipeline, too. I made $3,000 a week. Some people made more."

His wife put in: "Yeah, we had just had a new baby when he went on pipeline, so I had him quit."

"On the rigs," he said, "there's supposed to be three crewmen to a rig: an operator, an ace hand and a floor hand. But there was only two of us. When something broke, we're the ones that had to fix it. We were two men, which you're not supposed to do. One time we did four wells . . . Well, I made $2,700 checks every two weeks.

"Pipeline's more dangerous. You can get run over by a bulldozer. You gotta watch it because they can't see you. Soon as it happens, you're dead.*

"I forget how, but the pipes can also blow up on you, and then everyone will die.

"We also had to have this special thing we had on our hard hat. Some kind of gas warning. It was this harmful thing; you can die in minutes if it happens. If it goes off, you sprint.

"I got to work with my brother in law, so I didn't really pay attention to danger. We were very careful."

"What does the crude look like?"

"It's like this. You change your own oil in your vehicle? It's like that. Thicker and nastier. I would have to take showers at work to get it off. We'd have to use that G—."†

His wife said: "They have special washers that are especially for oil clothes at the laundromat."

"There's big wells," the young man continued. "They go up to 30 feet high. Sometimes they don't have ladders and you have to figure out a way up there."

"What do you think about global warming?"

"Got to do something about it. It's real, but they just cover it up and say it's not real. I bet we can have a different resource than oil or gas, 'cause we're just

---

* This reminds me of the death of Wilbert Starcher inside the Pocahontas Deep Mine. See above, p. 160.

† He named an orange-peel-derived liquid soap, which I decline to advertise.

hurtin' our planet. Where are we gonna go once we mess up the earth? Pollution, messing up the ozone layer . . . I don't like any of the sources now. It's just hurting our planet. I don't know why we can't just make everything electrical."*

"Did the other guys think the same thing as you?"

"I think so. Everyone should think the same."

---

* As you will see, this notion that electricity must be some form of fuel-free, self-generating energy was prevalent among the guest workers I interviewed in the labor camps of Abu Dhabi.

"IT'S ALL ECONOMIC IN THE END"

I t goes from black to almost light yellow," said Mr. Sam Hewes. He was a Vice-President of the Bank of Oklahoma, and his purview was energy loans. Of all the people I interviewed for this book he was one of my favorites. Cheerfully self-effacing, hardheaded and slightly cynical, straightforward yet tolerant, he reminded me somewhat of Mr. Endo Kazuhiro, the nuclear evacuee from To-mioka.* I have a feeling that if Mr. Hewes's life were ever ruined he would be as admirably stoic as Mr. Endo, and if Mr. Endo ever got his life back, or better yet struck it rich, he would be as modest as Mr. Hewes. To you from the distant future who may wonder about our long lost carbon ideologies I offer up the words of this kind and attractive personality, whom I would happily have accepted as my friend.

The Bank of Oklahoma, by the way, *continues to be a stable and reliable financial partner. We are part of $31 billion BOK Financial (NASDAQ:BOKF), the largest commercial bank in the country to decline participation in the Treasury Department's Troubled Asset Relief Program (TARP).* I approved of this entity for not accepting taxpayer-financed bailouts.

(One last corporate tidbit: *We're proud to be part of our local communities. At the end of the day, we know it's your money. And that's why we're working hard to be your bank. Long Live Your Money.*)

"In reality," said Mr. Hewes, "there's no such thing as oil and gas. It's all the same. It's just what phase it is. The way they do it is API gravity.† The heavier and thicker it is, the blacker. The closer it is to condensate, which means it's coming off a gas well but it's liquid, when you get up to 55 or 60 gravity, crude, it's all liquid in the reservoir, but it just depends on the chemical structure of the hydro-carbons in the ground."

"Now, what is oil exactly?" I asked him. "Is it mostly fossilized plant materials with a few animal fossils also? That's what I've read . . ."

"Science is a religion," he answered. "It's the new religion: numbers. So, most people in any field other than science, they believe in scientific truth. But when

*See I:374.
† See I:570, 577.

scientists come up with a theory, which means it explains some of the facts, they don't know that it's true. The theory is what you just said. Some people don't believe that. But this standard theory is that mostly plant materials get buried, compressed, and over time they get cooked, and depending on how long they get cooked; it starts out as a coal; it can end up as a gas. Certain types of plants, everyone thinks it's dinosaurs, but it's diatoms. So the classical theory is what you said. But a geologist says, there's gotta be a kitchen. How thermally mature is this? If it's not very thermally mature, it could still be coal. If it's very thermally mature, it's a gas. It's gotta go from high pressure to low pressure.

"Anything that's complex, the answers are probabilistic, not deterministic. I did some studyin' on it. Engineers like to be precise and deterministic. There are some psychologists that did some work on how good are people at giving answers

Mr. Sam Hewes

that actually are correct. So they asked a series of questions and said, within an 80% confidence interval, when did Napoleon do such and such? And lots of people, 63% of the time, people didn't even get it in the range. So they said, well, that's because they're not historians. So, they went to reservoir engineers, and asked the same questions, and it was 62%!"

"How did you decide to go into the oil business?"

"Actually I was programmed to go to work for DuPont, because the University of Delaware is DuPont University. I worked for them in the summers and when I was off in the winters. I graduated in '75. It was a recession when I tried to hire into engineering technical services, and they had already hired more than their quota. So I started interviewing with other people: Sikorsky Aircraft, Procter and Gamble, Douglas, and then, I don't know why, I had an interview with Texaco. When I went on the interview trip and they showed me those things that pump like *this,* I had never seen one! I guess if I lived in western Pennsylvania I would have seen one. But I was from Philadelphia. To make it interesting, my Mom got ill with schizophrenia and my Dad's business was crumbling, so I wanted to get the hell out of there. And I had a good opportunity, since Texaco hired me in Hobbs, New Mexico. I ran to the library and I saw they had a lake there, and I said, oh, this is not bad, but I was not good at looking at scales on maps, because the lake was about the size of this building. But I loved it. The people loved on me, taught me how to brand cattle. And it was an absolute boom, because the price of oil went up and up! So, companies were hiring people that weren't petroleum engineers because there weren't enough of them because the price had been low. I was single, so I would spend all night on the rig, and I would sleep on the top of my desk, and then work all day. The oil business is just, you're playing with such big things, and lots of money. Just fascinating, cause there's all kinds of different situations. At Sikorsky Aircraft I would have been one of 200 engineers and I'd gonna know all about the stress analysis of one rivet on the left rear wheel. In the oil business, you don't actually drill the well, you're the supervisor. You don't frack the well; you . . . It's just *fun.*"

I think that I too could have had fun in the oil business. Here is a description of a certain kind of offshore platform once in use by Phillips Petroleum and Howard Doris, Limited. *More than 200 feet taller than the Washington Monument, the platform is 100 feet above sea level and the helipad 100 feet above the platform.* I imagine how "empowering" it would be to participate in building such a thing.

"As an engineer in Hobbs we had the only building in town with an elevator. People would bring their kids to ride an elevator for the first time. So you as an engineer, you're well kept. The people were just fantastic. I supervised drilling. I would do what they call workovers, where you would take a well that wasn't producing much in one zone, and then you would recomplete it in another zone. So you would write up the procedure and supervise how it was being done. First time I went out, they made me play geologist. I had never taken a course in geology. And they said, how deep do you want to drill this well? I said, I dunno! Things were moving so fast that Texaco couldn't get you to enough schools fast enough, but the service company knew what they were doing. My answer was, what does it say on your contract? It says 13,000 feet. So I said, well, that's what we'll drill to.

"It's really like, you have a log from when you went in, and you have other wells around it, and from that log you can see how other things respond. I would take the samples from that well and the log, and say, okay, there's mica in there, and I would put 'em all in coffee cups, and I compared 'em to the coffee cups of the wells close by, and I would say, we need a little flaky material like that.

"The first time they asked me, okay, I said, we're gonna perforate this zone, like, we're gonna put casing in this well, and then we're gonna shoot it with guns, and they said, how many shots should we shoot, and I said, well, how many did you bring?

"So, I shot every shot that they sent in a 10-foot interval, and then I got a call from the guy that was in charge of drilling, and he said, if I wanted you to blow up the casing, I would have told you, and I said, well, if you had trained me I would have known. And in time I learned.*

"Texaco wasn't the best company in the world. Well, they were really good to me. I got sent to a year-long school in computers for all their well simulations. Computers today were just coming along. I knew how to program, so I was a hero.

"I ran Kerr-Mckee's I.T. department for awhile, I was transferred into their chemical company as a V.P. I'm the V.P. and I'm in a staff meeting and I don't know what they're talking about. But it was fun!"

"Mr. Hewes, what do you think about global warming?"

"When I was in the chemical company—I really like probability and statistics—I was in the forecasting department. I had them do a probabilistic forecast of what we're gonna make in the next year. The President said, Sam, I want to know what we're gonna make next year. I said, well, I can narrow that range if you're willing to sell forward to Sherwin Williams so I can narrow the widest range about what we're gonna get for our product, and he said, Sam,

---

* "There were already in this particular field maybe 50 wells, so from them I could see what to do; if it was a new field they would have never let me do that."

I wanna know what we're gonna make next year. So I did my best, and it come out wrong.

"I draw this curve. This line here, the X-axis, is complexity. The Y-axis is importance. Deterministic answers are down in the bottom quadrant when they're not all that important or that complex. The stuff that's not that important but very complex, in the greater scheme of oil and gas, the only thing I'm certain of is that figuring it out costs way too much money. It is extremely complex and I don't wanna spend any time because it's not my business, so that you have to outsource that and then trust whoever it was outsourced to. In that top quadrant, everything's probabilistic.

"I don't think anyone knows for certain about global warming. We know absolutely that the more of certain chemicals get put into the atmosphere, that should raise the temperature. But we can't predict the weather right here in Edmond, Oklahoma. So we think we can predict what the temperature's gonna be globally? Nobody in the oil business is gonna say, I do a dirty job and I don't think we should do oil anymore! I've come to the conclusion at least that it's really just a battle for your *mind, will and emotions.* If people go in one direction, whether it's for oil or against oil, at least you can get something done.

"The scientific community is trying to be religious, where they're trying to say something certain. I know that with reservoir engineering I could say anything I want. But my wife always reminds me that life is not run by your equations.

"What I'm saying is, it's a battle for your *mind, will and emotions.* So, what's gonna happen is whatever somebody decides needs to happen. *I really don't care!* What I care about is that we come to some agreement about how we're gonna move forward. After awhile, enough people think, well, join the group or you're gonna be out by yourself. In my lifetime, it's not gonna affect me."[*]

"Do you feel that the climate has changed in your lifetime?"

"Every time there are five tornadoes in the Gulf the forecasters can say, see, there's proof. But when you look over the last 200 years . . ."

He was correct there, to dismiss the significance of specific events. Oklahoma's highest measured temperature—120° Fahrenheit, at the Altus Irrigation Reservoir—was recorded way back in 1920. The lowest—31° below zero, at Nowata, came to our notice in the supposedly dire year 2011.

"Is the earth actually warming?" he went on. "I think that's irrefutable. But

---

[*] The day after our interview I saw an article about Tangier Island, Virginia: **The Lost Island: As sea levels rise, the United States will have to decide which coastal towns should be saved . . .** *We can't spend an infinite amount of money defending the coast. And the concept of retreat, which is sort of un-American, has to be normalized.*

somebody *chose* to live on the coastline. Somebody *chose* to live on the San Andreas fault. If something's gone wrong, we want Big Brother to fix it. I run support groups and things like that. I can't tell you how often people come in lambasting the government for their loved ones who are already collecting SSI."

"What do your colleagues say about global warming?"

"*They don't care.* They say, it's not affecting me today!—Well, you've seen this in any industry. They're proud, and I'm really proud of what I've done. You couldn't get here to see me today if you hadn't flown on an airplane powered by petroleum fuels."*

(How could I deny that? I wanted no more or less than you do: for others to conserve, voluntarily or not, so that I would not have to.)

"We're all trying to get people's minds rolling like us on things we don't know," said Mr. Hewes. "The people that I know in this industry, feel like they're very entrepreneurial. They know that oil and gas industry popularity-wise is below Congress. Obviously this state, it's in a terrible trouble right now because the oil and gas prices went down. This state depends on oil. And the people I work with, they think they've done this on their own, they've found solutions to complex problems, and they honestly feel that they're giving a great service. Look at the plastic cup you're using. All these things . . . ! Our modern society is built on having relatively cheap energy."

"What would our lives be like without oil?"

"You can do anything. Can you do it at a point where it is economic to do? You can turn gas into electric and sell it as oil where there's tremendous gas reserves, but how do you get gas from one place to another?

"If you just look at what we get from the oil, you get all your fuels, you get asphalt, you get all the plastics. They're not gonna go away. But I looked at the percent that renewables are of the total energy used in the United States. You know we have all these programs, but the percent hasn't gone up over the last 30 years!"

Could that truly have been so? Not wishing to be ashamed of my country, I decided to look up world statistics instead. Over those three decades the renewables category of total consumption showed improvement—oh, yes, by a good 4%!

---

* As NASA had boasted in 2010: "Every 24 hours, some 2 million passengers worldwide are moved from one airport to the next." Well, good news! Aviation could be blamed for only "3 percent of the potential warming effect of global emissions that could impact Earth's climate."

GLOBAL PRIMARY ENERGY CONSUMPTION BY SOURCE, 1980 AND 2011,

in multiples of 1980 absolute consumption figures

All multiples expressed in underlined numerals (each 1980 absolute consumption figure = **1**). [Example: 2011 "renewable electric power" 42.9 / 18.2 = 2.36.] Absolute consumption expressed in [Q-BTUs].

1 Q-BTU = 1 quad = 1 quadrillion BTUs.

1980

<u>1</u>: World total: 100% [283.1 Q-BTUs]
  <u>1</u>: **"Renewable electric power": 9.96%** [18.2]
  <u>1</u>: **Fossil fuels: 90.04%** [254.9]
    <u>1</u>: Coal: 24.69% [69.9]
    <u>1</u>: "Dry natural gas": 19.04% [53.9]
    <u>1</u>: Petroleum: 46.31% [131.1]

2011

<u>1.83</u>: World total: 100% [519.2 Q-BTUs]
    <u>2.36</u>: **"Renewable electric power": 14.07%, or + 4.11%.** [42.9]
  <u>1.75</u>: **Fossil fuels 85.93%** [446.2]
      <u>2.13</u>: Coal 28.60% [148.5]
      <u>2.23</u>: "Dry natural gas": 23.25% [120.7]
      <u>1.35</u>: Petroleum: 34.09% [177.0]

Source: U.S. Energy Information Administration, 2014, with calculations by WTV.

Meanwhile, as you see, our *absolute* carbon emissions had nearly doubled.* We were using more than twice as much coal as before, well beyond six times more natural gas and a third more petroleum. Sam Hewes's point held approximately valid for the entire world picture!—To be sure, we were also consuming 2.36 more times more renewables, but so what? We continued to warm the atmosphere faster and faster.

And Mr. Hewes repeated: "The bottom line of my profession: *You can do anything.* It's just, how much does it cost? Right now, we have more recoverable oil

---

* Based on world total consumption. You will note that fossil fuel usage increased during this period by a factor of 1.75. Measured carbon emissions meanwhile went up by 1.77.

than Saudi Arabia and the former Soviet Union. It's because the price went to $85 a barrel. When the price went up there, then you could do it. A barrel of crude oil is 42 gallons and it's about $50. You go and try and buy 42 gallons of water from someplace, and you can't do it! I'm trying to let you know how cheap this is!"

"What's your assessment of coal?"

"Very dirty. And when they try to clean it up, well, did you read that story about the carbon capture project in Mississippi, all those cost overruns and whatnot?* Well, that happens all the time. But again, 50% of the United States is powered by coal.† So you're gonna tell China that they can't do that? It's all economic in the end! You can shut it down, but you'd better be damned sure that that trillion dollars you just made them spend for alternatives will be worth it. We are the most spoiled country in the world. When I went from Pennsylvania to New Mexico, I could not believe the attitude of the average Texan. They thought they were better than everybody. I finally decided it was because their whole culture was underpinned by this wealth in the ground. As Americans, we're the same way. We think we deserved it and the country is so blessed, and somehow this all came together just for us, maybe through World War II. We used to be able to rape and plunder other countries, because they needed what we have, but now they too have sent their executives to Harvard Business School.—Coal, oil, I'm for everything, all this stuff; that doesn't affect me."

"So if you were energy dictator, what would your policy be?"

"Try to come up with a policy where people won't riot. That's really the only job I have, because if the masters rise, we're out of office. I think they know America's on a downhill slope; those other people have caught up, and they're just working hard to make sure the masses stay ... It's like being chased by a bear in the woods. We can give 'em enough jellybeans ... and when you run out of jellybeans, the bear's gonna eat you."

"Mr. Hewes, recently I had to put a new roof on my building, and it cost me far more than I expected, because the state of California required me to bring it into compliance with the latest standards for commercial insulation. So my building is now more energy-efficient, and I feel good about that. Do you think that we should require people to conserve energy? In Texas, for instance, I have seen businesses using air conditioning very wastefully, blasting cool air out of open doors to encourage people to come in ..."

"Bill, I don't agree with the way you framed that. I would say that in this

---

* He was referring to the failing Kemper power plant in De Kalb, Mississippi. See footnote on p. 194.

† As you may recall, in 2015 the national figure was merely 36%, although for West Virginia it was still 95%. I am unaware of Oklahoma's share. See above, p. 50.

neighborhood, it should be strictly an economic decision. If there is a decent payoff to putting it on your roof, then, you'd be stupid not to do it. I'd be really pissed off if the government told me to do it. I absolutely believe that the government should be spending a lot of money on alternative energy, because nobody else's gonna do it. I mean, we didn't get nuclear power until we had World War II and the government bankrolled it. Now how you do that efficiently I don't know.—In Texas they *waste* energy? No, they don't! It's their *economic value*."

I asked him about the dangers of methane leaks.

"I'm laughing at you, Bill. Here's what I'm laughing at. Methane may be much worse than CO_2. But the amount of methane that comes up and gets leaked is not even the size of a zit, if you put it all together. So it's these kinds of things that people get upset about. Like about fracking: *Oh my God, it's polluting all the water in the world!* Well, I bet there's less than 25 cases of it's having done anything in the United States. It's the *irrationality*.—Think about autonomous vehicles. An autonomous car could maybe eliminate 80% of all accidents in the United States. But the fact that it's an autonomous vehicle and it has one accident, means, oh, no!

"Look at the statistics. Basically, you just got people on the deep end. Is there gonna be some leak somewhere? Sure. But it's the same thing when I got sued in DuPont, when we got sued for some chemical, and I said, have you never heard of dilution? Have you never heard of some methane seeping out of the ground?"

"Mr. Hewes, what do the antis want to accomplish?"

What he now said is one of most important statements in *Carbon Ideologies*.

"I think that we're all ideologues, and we all have a set of beliefs. I have learned that when you in any way attack somebody else's reality, they're either gonna leave you or attack you. I think, because of uncertainty and not really knowing, there are people that believe fully, one side or another side. So I think, the winner's gonna be the people that are gonna convince the others that they are right, although they don't know either! Those that think that global warming is an unbelievable issue and we're almost at the tipping point where we can't recover, there the model that they have in their head says that we have to do everything that we can to shut down oil. I see this in churches all the time. They make decisions not based on facts and then they call the other side illogical, just because they don't believe the same thing. Well, the other side are not irrational. They're just . . . Science is the new religion."

One of the many reasons I liked him was that he never insisted that his particular ideology of energy extraction was the best. Indeed, he said: "Ten thousand years or so from now we'll probably get it from some other source. There is so much tar sands in Canada that it is unbelievable. Geothermal, at some price, the resource is unbelievable! Think about how much energy's in there! You're using the molten rock ... The city of Edmond, all of the city buildings are heated and aired by geothermal. One of the reasons why it's efficient for them is the police department is open 24 hours a day so they get half the payout.—There's plenty of energy to go around. It just depends on how much it costs to get it."

Indeed, was not cost-benefit analysis the eternal factor? In ancient times, motive power came from slaves and animals. Did it best reward the user to take care of them, work them to death or go without?—Next came steam power, coal, oil.—In a book in the Cushing public library I had seen an old photograph of the Tucker Sand Gusher rising high above the horizon like a tall narrow tornado and beginning to curl into the wind; soon it would begin to rain crude. That book was called *Lakes of Oil: Ben Russell's Rare Photo Record of an Early-Day Oklahoma Oil Boom.* If and when oil ever got too expensive for us, there would be

Gas jockey in Yemen

lakes of something else. Maybe the cost would go up; maybe the cost would be extinction. But I who worried about that was merely one more ideologue. Thanking Mr. Hewes for the interview, I rode in a gasoline-powered rental car back to where I was meant to be: lurking in the motel room with the curtains drawn and the air conditioner on full blast.

ADNOC gas jockey, Abu Dhabi

Nobody seemed to care about the loss of these thousands and thousands of barrels of oil, which soaked the soil and polluted the river . . . The more oil is lost, the higher the price. Three cheers, then, for broken pipes and drunken pump-men and tank-attendants!

B. Traven, 1935

8

"I Am Here Only for Working"

October 2016

But the exercise of labour is the worker's own life-activity, the manifestation of his own life . . . He works in order to live. He does not even reckon labor as part of his life[;] it is rather a sacrifice of his life.

Karl Marx, *Wage-Labour and Capital,* 1849

And frankly, Mr. William, given the conditions I come from, it is not so bad.

Gilbert, security guard in Abu Dhabi, 2016

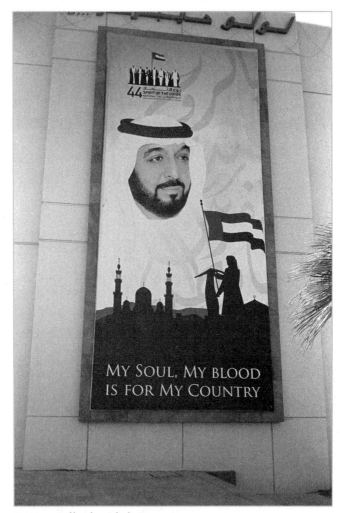

Ruwais Mall, Abu Dhabi Emirate

Mr. Gafar Khan of Rajasthan, in one of the labor camps around Ruwais

LITTLE REASON TO STOP

T he guidebook said: *If you had any doubt as to the oil wealth of the UAE, it'll evaporate when you see Ruwais, an industrial town . . . that exists only to serve the massive refineries set up . . . here. Needless to say, there's little reason to stop.* Wishing to interview petroleum workers, I went straight there.

Ruwais (sometimes spelled Rowais or Al Ruwais) overlooked the Gulf from Highway E11, not far from the Saudi Arabian border. Relatively new productive forces must have engendered the place—for I found no spoor of it on the large-scale national map in my 1976 *Britannica.* Tourists, not that this was their season, passed by, not through, en route to that expensive nature resort on Sir Bani Yas Island. From Dubai the bus ride consumed five or six hours, depending on traffic and counting the brief layover and change of vehicles in Abu Dhabi—which last place-name, if you did not know, referred not only to the nation's capital (said to be *the richest city in the world*), but also to the emirate in which it lay—by far the most sizeable of the seven, and certainly the most petroleum-wealthy. Dubai came in second for oil, followed by Sharjah, whose Commerce and Tourism Development Authority astounded me as follows: *Since the first settlement began, about 7,000 years ago, the Emirate of Sharjah has always been at the core of regional development.*—I wonder if the bright-eyed developers of 5000 B.C. drew up their plans at home or in the office?—At any rate, they must have labored inconspicuously until 1958, when somebody struck oil—in Abu Dhabi, as it happened. The emirates joined together in 1971. And the 1976 *Britannica* went benignly on: *The wealth of Abu Dhabi and Dubai is to be shared among all the emirates, and the continued flow of oil promises a rapid entry into the 20th century.*

Early in the 21st, my bus reached Ruwais, and I discovered a very peculiar place.—But its charms will stand out to even better advantage if I touch on a couple of scenes along the way:

My American neighbors imagined the Emirates as a vacation spot; and those who liked Las Vegas might feel at home in, say, Dubai, which similarly offered air-conditioned venues separated by traffic and hot sand. One hotel in Abu Dhabi was famous for its lobby's automated teller machine, which dispensed gold

GULF PLACES MENTIONED

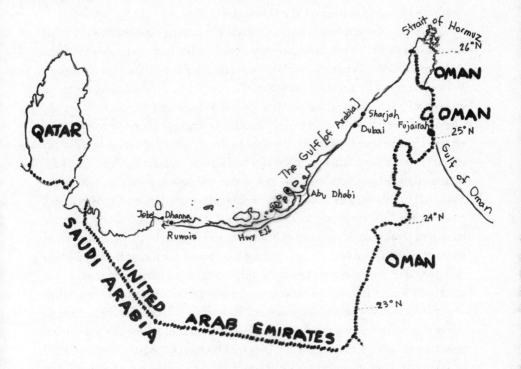

bars; and what my nationality referred to as "family fun" could readily be found in, for instance, aquatic parks. We admired the U.A.E.'s architecture, the safety, and most of all (being Americans) the *wealth*. Being myself a U.S. citizen, I should now relay with extra envy and awe a claim by a certain bellboy with whom you will soon grow better acquainted that gasoline was so cheap here and the Emiratis so rich that when they dined at one of the restaurants at the Ruwais shopping mall they sometimes left the car running. I never witnessed this, nor did the gasoline seem cheap;* although it was true that when the X88 bus to Ruwais paused at a way station, once for five and once for 10 minutes, it left the engine running, the air conditioning on and the doors conveniently open.

In those days my countrymen and I each generously gave 21.2 annual U.S.

---

* One service station offered it at 6 dirhams per liter, or 6.5 dirhams for "super." At that time there were 3.6 dirhams to the dollar, so the U.S. equivalent of "regular's" price would have been around $6.35 a gallon. But the fact-checker for the abbreviated *Harper's* version of this chapter found prices quoted closer to $2 a gallon. (In California I saw prices of $2.50 and $3 a gallon.)

tons of carbon dioxide to the atmosphere, thanks to our flaring* and end use of fossil fuels. The comparable Emirati figure was 47.5 tons per capita—*among the highest in the world,* remarked *The Statesman's Yearbook.*[†]

Expressing my national characteristics, I stayed in air-conditioned hotel rooms every night. (When I complimented one hotel on its quiet peacefulness, the desk clerk replied: "Thank you, sir. You see, we do not allow girlfriends or drunks.") When I went out for the day I turned off the power, like a good Californian; and whenever I got back the chamberman would have restored the chill just for me. The outside weather overwhelmed my brand new laptop and one of my two cameras, so I had to take the former through the dreary sprawl of Dubai to a district called Internet City, where a subsidiary of Petrochina International Iraq FZE happened to be incorporated, so as to mysteriously *support the Petroleum Operations;* and in this neighborhood of traffic-lapped towers, unseen hands in a back room repaired the computer in 15 minutes; as for my point-and-shoot, I learned to refresh its electronics in air-conditioned eateries (thereby discovering the delights of rose milkshakes).—One evening at a Libyan restaurant in Abu Dhabi I dined outside, and although I was the only customer to have made that choice, the kindly waiter insisted on powering up a wheeled Chinese-made fan, which was at least half my height and rated at 290 watts. I inhaled that breeze for 45 minutes, secretly feeding pinches of chicken to the starving alley cats who lurked beneath my table; operating the fan over that interval must have consumed 1/7 of a pint of oil in some power plant. Those were some of the ways I enjoyed electricity.

The American grid still ran mostly on fossil fuels: 73.4%. The Emirati figure was 99.8%.

In Dubai I asked a Pakistani taxi driver: "Is oil good?"

"Of course it is good. Think, sir! This country has no vegetables. It is a very sandy place. This is the good thing for the people."

Well, who could disagree? Having scouted out all seven emirates, the Central Intelligence Agency concluded that their total area contained the merest 4.6% "agricultural land"—from which someone somehow managed to raise strawberries, flowers and dates—fine export commodities, to be sure, but could they pay the air conditioning bill? And since 2000 the population had *tripled.* No

---

* For one egregious example of flaring, see "About Greenhouse Gases," I:171. For a very ordinary instance, see p. 446 of this volume.

† Both of these were 2008 figures—the latest available at the time of writing. American emissions added up to 19.2% of the world total; I felt grateful to China for demoting us to second place in the roster of greedy negligence.

worries! Despite retrenchments the U.A.E. remained the fourth largest oil and gas exporter on earth.

The taxi driver agreed that global warming was already in force; indeed, he even thought he'd experienced the change.

"So is oil good or bad?"

"Good, of course."

Naturally! For without oil there might never have been that touted treasure, the Dubai Mall—*not merely the world's largest—it's a small city unto itself . . .*— extruding klezmeroid Muzak, and a smell somewhere between perfume and air freshener overlying the slightly stale atmosphere as if of an airport. The architecture was grand enough, but as for the merchandise, although a certain Saudi Arabian chain *(one of Dubai's sexiest stores)* offered lingerie, and a branch of my favorite Japanese bookstore was in evidence, I noticed a preponderance of European and American brands.* *With more than 12,000 retail outlets[†] and an unmatched range of world-class attractions,* the mall promised to *delight, amaze and thrill you every time.* What held my interest was the shoppers themselves: two beautiful young women in *abayas,*[‡] one unfurled below the waist like a beetle's elytra to show off her long slender legs in snow-white jeans; three young Indian men in laborers' clothes, the leftmost of whom had his arm around his comrade's shoulders as they wandered without buying anything, one floor above the mall's signature aquarium *(largest viewing panel on earth)* and vertical waterfall. Here came a middle-aged American couple in shorts, and their slender German or Austrian counterpart, also in shorts; then a single lady, black-clad from head to toe, with only a narrow slit for her eyes. One could have called this a triumph of globalism, or syncretism, or something. At the Australian chain restaurant where I ate lunch, my Filipina waitress, still in her first Emirati month, had studied colonic irrigation in Idaho. (The smell of the food court resembled many foul airline meals trapped in concrete.) Two young fellows in sandals and white robes came shopping together, one wearing the typical red-and-white-checked headscarf, his friend in a baseball cap. A pair of Asian girls passed them, one in jeans, the other in a striped skirt which showed her knees.—I went to see the famous skating rink. One could look down at it from a café, sipping a coffee or a hot chocolate while inhaling the chill of its ice. Like the ice bars in Las Vegas, it was

---

* The exceptions tended to be Arabic fashion outlets, banks and exchange houses, pharmacies, supermarkets and a few cosmetic establishments. There was also an attached souk, many of whose perfumes, jewels, etcetera, originated in the Middle East or South Asia.

† This may have been a misprint. The *Harper's* fact-checker found electronic references to 1,200 outlets.

‡ The *abaya* was a woman's black cloak. Her headscarf was a *shaila.* The *burqa* was a "canvas facemask." A man's white robe was a *kandura,* and his white headcloth a *guttrah,* beneath which he might wear a skullcap or *gahfia.*

a triumph of energy consumption—for meanwhile the outside temperature approached human blood's.* Children and occasional parents skated round and round. Two teenaged Middle Eastern girls skated holding hands, one in a T-shirt that said **NO BOYFRIEND NO DRAMA**.

From the mall one could take a moving walkway inside a jointed elevated tunnel, and reach the monorail, or else go a trifle farther, descend an escalator, walk past one more sharp-eyed security guard, open a door and enter another sweltering construction site where a coveralled army swarmed up and down scaffolding, while the world of concrete around me framed untamed zones of sand.

Atrium, Dubai Mall

Everywhere I went I found this national rawness—which must not have appeared raw at all before there were cities on it. But in another decade, if the price of oil reascended, and we could all keep ignoring climate change (please let it confine itself to some other ecosystem somewhere!), Abu Dhabi's branch of the Louvre would be booming, even more splendid hotels would arise, and the urban

---

* 36° Celsius, or almost 97° Fahrenheit. My very rough calculation suggests that keeping this rink frozen required the removal of something like 2 million BTUs of heat (which is the energy contained in nearly 18 gallons of medium grade fuel oil) every *single hour*. And since there is no free thermodynamic lunch (R. J. Clausius: "No process is possible whose *sole result* is the transfer of heat from a cooler to a warmer body"), removing 2 million BTUs would have cost the Dubai Mall far more than 2 million BTUs. To generate that energy, a power plant would need to burn triple: more than 54 gallons of oil per hour.

Multilevel view of
the Dubai Mall

greenery on the highways would appear almost natural, or at least consolidated. As one refinery safety auditor enthused about the latter (he was an expatriate, like every other petro-worker I interviewed; although unlike most of them, he alternated between Ruwais and Abu Dhabi): "They import soil from India and Australia. They use drip irrigation. They're really doing a great job." In 2016 those importations still showed. But the planners and developers kept deploying all "big five" categories of manufacturing energy use*—aluminum, steel, cement, glass and paper—on so titanic a scale that the future of those United Arab Emirates glowed in the night like the new clean skyscraper of the National Drilling Company, whose logo was a red-underlined blue derrick within a circle! That was thrilling, I admit, but the air rarely felt good to me wherever I went in the

---

* As you may remember, these are the five materials which consume a great proportion of all the energy used in manufacturing. For details, see I:134–36. In 2008, 12.1% of the U.A.E.'s Gross Domestic Product derived from the manufacturing sector, whose products included "aluminum, cable, cement, chemicals, fertilizers, rolled steel and plastics, and tools and clothing." Except for clothing and maybe tools and cable, the items in this list were either "big five" energy hogs, or else likely generators of greenhouse gases (chemicals and fertilizers).

Just outside the Dubai Mall

Emirates, not even in Sharjah, which the safety auditor warned me would be "less developed"; certainly not in Dubai's Deira district where the row of high-balconied dhows rode water with their Emirati flags wilting on the mast, while on the quay before them waited box-mountains of Chinese refrigerators, air conditioners, yellow-wrapped tires whose rubber gave off that sickly vulcanized odor in the morning heat, as sweating fellows leaned on hand trucks, and South Asian women helped their men stow ropes in that dreamlike fug of diesel fumes, with minarets and construction cranes for a backdrop on the far side of the creek, whose reflected glare shimmered white and green across this bow and that porthole; and as for Abu Dhabi, no, breathing failed to refresh me in that one-star hotel so chilly, dark and stinking of insecticide, where the poor bellman had to stand rigid and face front at the foot of the stairs, while the reception man manipulated unknown matters on his desk; from there I walked out into the heat and glare of Abu Dhabi, back to the Madinat Zayed Shopping Center for a bottled water or to the Corniche for an interview, tasting exhaust as I awaited my signalled permission to rush through momentarily stopped traffic between strange glass skyscrapers.

What I liked best about the Emirates was their teeming human lives. Behind a window a bearded man's head tilted back as a barber shaved him; next door a grave man in shalwar kameez was ironing; and on the corner, by Light of Beirut Automatic Laundry, beside a woman with long braided hair and hot pink trousers a man in business slacks and a fresh white shirt stood waiting, one hand on his hip and the other on his cell phone, and when the school bus came, a little boy in tartan-patterned shorts toddled off, still wearing his bright new backpack; then both parents stepped into the street to kiss him; while in all those tight alley-canyons, air conditioning boxes dripped and hummed, each hum in its own key; and the hot air was fragrant with solvents, schwarma, fish, roses and sandalwood perfume.

Air conditioners busily warming an alley in Sharjah

Perhaps you have read a parable by Ursula K. Le Guin called "The Ones Who Walk Away from Omelas." Omelas is a perfect city in every respect but one: Its securities and pleasures are predicated on the torture of a single child. Would living there be acceptable to you?—In the Emirates, a quarter of Gross Domestic Product derived from oil and gas, whose extraction required pipefitters, welders, concrete pourers, leak testers, and any number of other laborers. None of the ones I met were slaves. I never heard of anybody torturing them. They came of their own will, and many intended to renew their contracts.

So I rode away from Omelas that morning—or from Dubai, at least; and the furry throbbing of the bus felt almost pleasant. It was air conditioned, cheap and nearly on time, with all of us men packed in the back, leaving a luxuriance of designated front seats for the half dozen female passengers, whose separate queue had boarded first—all but one of them uncovered. The temperature was 31°* at 9:35; and I felt happy to be going out into the world, not knowing what I would find. The hazy skyscrapers, which included the world's tallest building, the Burj Khalifa,[†] were as unreal as the mountains in an ancient Chinese painting. The sun glared on the glass of the New Gold Souk. Now the skyscrapers drew closer. A number were narrow, futuristic and strange, like very angular fountain pens with notches in them. Many of course remained still on the rise, crowned by construction cranes. Then came glittering car dealerships paralleling the highway (Sheikh Zayed Road) for a drearily long time, while the dark and sweaty man beside me, who had ignored my greeting, continued connecting with various distant presences on his laptop. For awhile a metro train kept pace with us. The Mall of the Emirates (renowned for *the wonderfully incongruous alpine slopes of Ski Dubai*— one more way to burn oil) marked our arrival in another clump of urban congealment. Hereabouts rose Silver Tower, the home of Harlow International, which *provides Security Services in Iraq* and was indeed *based in Iraq,* so the only reason I could imagine for Harlow to be in Dubai was that *we also provide construction management* and *manpower support,* the subject of this chapter. The mention of those *Security Services* impelled me to steer clear of Harlow—since corporate police, like the national kind, might put a spoke in my wheel. And we continued west, the city finally beginning to show her bare patches, although there remained such expressions of our civilization as Emirates Wet Wipes (a vast facility), OE Oil & Gas, and Astra Polymers; while through the dust an airplane flew low over transmission towers. Meanwhile I discovered more carbon monuments, signs and allegories: Baker Hughes, Electromec, Electric Way. In 45 minutes we had finally left Dubai, with Sheikh Zayed Road resolving into Highway E11, which ran through flat desert white and tan, freckled with scrub and occasionally pustuled by buildings. On a large screen behind and above the driver, an English-speaking commentator, her pleasant voice amplified for the greater benefit, discussed the American Presidential nominee (a certain Donald Trump[†]),

---

* Not quite 88° F.

[†] At 828 meters (2,716.5 feet).

[†] "One bright spot is [Trump's] agenda for American energy. The President-elect promised to peel away government obstacles, and he will have plenty of work after President Obama's eight-year regulatory

while across the aisle a man in a "Chelsea" baseball cap listened through ear-
phones to some private something on his laptop, and most of the passengers (they
almost could have been Americans!) busily masturbated their cell phones with
both thumbs. Now the desert thickened with low bushes, and a truck passed,
bearing potassium nitrite; one billboard praised the Emirati Aluminum Company (aluminum, you may remember, is the energy-hungriest of the "big five");
another touted the industrial advantages of Khalifa, as we rolled by mosque after
mosque, whose minarets were as white as sand. By the time we arrived in Abu
Dhabi it was 35°.* At the side of the road rose the Petroleum Institute.

I changed to the X88 bus for Ruwais. We continued west past Emirates Steel,
National Marine Dredging Company, more auto dealerships, mosques and
malls, the Abu Dhabi Oil Refining Co., then reentered that flat tan desert, following signs for Saudi Arabia. I passed a profitable hour slowly chewing a sandwich of white bread, soggy vegetables and chicken puree, in hopes of keeping
away my continuing nausea; then on the far side of Tifra rose many tiny identical
concrete houses; with the exception of the great mosques at their center, they
would have fit in at some *colonia* for *maquiladora* workers in the Mexicali Valley.

When my bus pulled in under an awning at Ruwais, at an hour between late
afternoon and evening, I asked the driver if there were a hotel here and he said
yes. Did he know how to get there, and could I walk to it? He said that it would
be left around the corner and then five minutes straight ahead. So I thanked him,
slung on my two backpacks and descended into what the bus's thermometer still
claimed was 35°, although it felt warmer than that. Well, that was business as
usual for me in the Emirates; I stank of sweat wherever I went. The only other
persons to disembark were two African-American ladies, who must have been
oil-connected and now walked out of this story, and a slender, dark young baseball capped man from Manila—a tig welder for an oil company. (All the other
Filipinos I met in the U.A.E. were retail clerks or else worked in the so-called
"hospitality" business.) For the sake of due diligence I asked him about hotels.
He knew of none. His housing was far away; he was wondering how he would
reach it. He said: "If you ever need any tig welding done, let me know and I'll be
there." We shook hands; then I resumed my walk down the hill toward the corner.—On the right was something that would have been a convenience store
anywhere else; no, actually it must be some sort of office. And here was a restaurant or cafeteria; no doubt I could walk to it from my hotel and meet some

---

onslaught . . . More than 85% of area offshore [drilling sites] controlled by the federal government is
closed to exploration. Mr. Trump can unlock this potential, which would be a gusher for global consumers and American economic growth."

* 95° F.

workers . . .—In a concrete bay just off the sidewalk I drew even with an army of young men in orange uniforms. None of them looked at me, so I kept walking. Just as I rounded the corner, the bus pulled away, with all the other passengers on it, as if they knew something that I didn't. And I was alone in Ruwais.

I wish I could describe to you how magnificently mega-industrial it all was. Everything looked either new or unfinished. To the horizon were drums, tanks and such. But the air was hazy and dusty and my spectacles well salted with sweat; moreover, I felt preoccupied with settling myself for the night (for by 6:00 it would be dark), so I did not make out the West Refinery just then. Nor had I seen the concrete honeycombs of "villas" in progress. Instead I found myself in one of those never-never lands of globalism, which could equally well be an industrial park, a hospital, a business college or the sprawling outskirts of some no-frills gridded city.—I was now approaching a new and somewhat grand complex of apartment towers which looked down on a horde of white-awninged parking spaces, most of which were empty. It took longer than I had expected to get past all that, in part because I was hot and tired, but also because the towers were wider and taller than they first appeared; and although of course I could not yet understand what I saw, I began to take heart. Then came salvation: a towering guest house entitled ADNOC. This must be what the bus driver had steered me toward; I would do well in Ruwais after all. Relieved, I turned off the wide street, strolled up to the immense foyer, entered a vast air-conditioned lobby and approached the counter . . . but in place of a receptionist sat a pleasant, bespectacled young Filipina with a security badge at her breast. She explained that the guest house was for workers only; "ADNOC" stood for "Abu Dhabi National Oil Company." Furthermore, Ruwais offered no hotels at all, and no cabs. To be nice and to get rid of me, she led me over to a desk on the other side of that great cave, where a courteous young man, also from one of the Philippines's 7,107 islands, invited me to sit down while he called a taxi from out of town. His name was Richard. Like his colleague he spoke perfect English, and I liked him; he would be free tomorrow, so I invited him to work for me, and he said he would; I wrote down his mobile number and we agreed that I would call him at 9:00 next morning. When I did, he never answered.

Meanwhile the taxi driver arrived. He had a wife and five children back home in Somalia and was now entertaining a wife here, in another city still closer to the Saudi border. Although he could have married two ladies more, I inferred that he had reached his limit. As he drove me down to Jebel Dhanna, I inquired how he liked the Emirates, and he said (along with every other guest worker but the white collar types) that he would rather be home.—"Is life free here?" I asked. "No," he said, "but different. Here they have their Sheikh . . ." and did not finish.

He promised that the hotel in Jebel Dhanna would not be expensive, and

maybe he imagined that for a Californian nothing was, although I was already getting worried to perceive a freeway exit sign devoted only to this one establishment, the Dhafra Beach Hotel, which as it turned out (because the corridors smelled like mold) was actually the cheaper of two adjacent properties, which jointly offered the discerning traveller a private beach whose fine sand had been thoughtfully decorated with bottles and broken plastic, and whose ocean view included, among other petrochemical structures, the first refinery that I had definitely seen in this emirate; it might have been the object of the sign which strictly prohibited zoom lenses, unless the worry was sneak pictures of sunbathing women, not that there were any. Declining to consider my 100 millimeter macro lens a zoom, I employed it on lovely memory shots of the refinery, hoping that I would not be arrested:

Flaring off oil, as seen from Jebel Dhanna

So I took my dusk there on the Gulf shore, which barely smelled of the sea, and the breeze had cooled the air down to 27°.* The sea came in with an almost steady hiss of long and even waves; it had not yet given up being blue, and the sky was a dark dusty purple with two meager wings of cloud across it. To my right a wall of lights, long and low, went far into the sea, which the refinery safety auditor, Mr. Priyank Srivastava, later explained to me was very shallow, necessitating what was actually a spectacular docking facility for oil tankers. And now a man in flowing white robes came toward me, walking rapidly at the water's edge. There was only him and me now. He was silhouetted against his white clothes.

---

* 80.6°F.

He passed me without looking at me. I stood in the breeze. Then I walked east, until I could see over the low dune at the edge of the Dhafra property to the gruesome fairytale lights of the refinery, whose high plumes of peach-colored fire spread out more widely than in Poza Rica; and then, from the rightmost constellation of castles, a long trail of smoke blew parallel to the ground. The full moon was a grimy red ball, as well it should have been to preside over those wonders.

I stood watching the oceangoing row of lights. The moon rose higher. As it did it grew oranger, and more distinct. The sea breeze freshened. I could not smell anything from the refinery, whose doings comprised the most innocuous of light shows.

By day that beach was pink, with reddish and greenish shrubs, and the track of a slender snake, and Navy ships out on the green sea, guarding the border. The Navy must have been doing a good job, for I never saw unauthorized landings.* The low line of lights indeed resolved into a long low bridge, with a long tanker on the horizon. So once again it grew worthwhile to peer across the low and pallid dunes, which were darkened by weeds, at that thing on the horizon which at first could almost be mistaken for, if not a walled city, at least a clutch of skinny skyscrapers behind a long low perimeter; but some of those towers flew flags of orange fire, and some spun out thick threads of smoke that were horizontal in the breeze.

The price for my room fell just shy of 500 dirhams—maybe $142. The calm, pretty Filipina receptionist, less than enthralled by my sweaty clothes and two rumpled backpacks, informed me that I could only stay this one night, since they were booked up tomorrow and after. The adjacent, much more expensive resort would be out of the question, and the nearest alternative was in Abu Dhabi. Not wishing to make that two-and-a-half-hour journey twice each day, I put on my best sad manner, but the young lady could do nothing for me. Fortunately,

---

* I must admit to having been more anxious than necessary about this project. Looking at the map, and studying over that long lonely line of Highway E11, whose main accompaniments were sea and sand, I wondered about getting kidnapped by ISIL or other such benevolent organizations. After all, "UAE forces led the US-led coalition against Iraq after the invasion of Kuwait," which must have made a few enemies. Moreover, after September 11, "the government . . . ordered financial institutions to freeze the assets of 62 organizations and individuals suspected of funding terrorist movements," so there might have been 500 perfectly liquid others. But the U.A.E. was for better and worse a well-watched place, where an *abaya*'d woman all by herself would not hesitate to blackly glide into a dark alley with her high heels clicking; the nation's lack of crime would be extolled by more than one of the petroleum-serving workers I interviewed. It surprised me that the U.A.E. remained so free from terrorism. One cynical hypothesis would be that the militants needed a Switzerland of sorts for their financial transactions, while perhaps some Emiratis were willing to pay off terrorists in order to be left alone. But for all I know, we can credit everything (as did the workers) to the Emirates' security service. While I was in country, one newspaper announced that soon the police would have virtual reality goggles which would display information about who and what they looked at. Money truly can buy happiness.

Ravindra the bellboy was someone who could be reasoned with. On the way up to my quarters we chatted about Rajasthan; he was from Jaipur, where I had once enjoyed the Ameer Fort, as I did not scruple to inform him; I next praised the Lake Palace Hotel in Udaipur, which had been very, very expensive. When we reached the room I closed the door and gave him a large tip. Next I remarked how unfortunate and downright tragic it would be for me to take the bus all the way back to Abu Dhabi tomorrow, and he said he would see what he could do. In 10 minutes the phone rang. Ravindra had wangled me a place at the Danat Resort, just for tomorrow, although the price unfortunately would be 1,556 dirhams;* afterward I could return here. Gratefully thanking him, I asked whether he had time to visit me again. Once he appeared, I invited him to work for me at 500 dirhams a day, finding me petroleum workers while interpreting as needed. He knew plenty of refinery types, several of whom currently boarded at this hotel. And he had a friend who could drive (for that I proposed 300, to which Ravindra replied, as he had in regard to his own wage, "That will be all right, sir"), so that I could tour Ruwais in style! With those details arranged, I set off for the Danat in a golf cart driven by one of Ravindra's colleagues, and by the time I reached the elegant Chinese girl at the immense reception desk, the phone was already ringing, and the elegant girl handed it to me, whereupon I was informed that tonight *and* tomorrow I could stay at the Dhafra after all, oh, yes, and possibly without end. Good old Ravindra! So life was breaking my way.

Browsing the yellow pages for this emirate's western reaches, I found a listing for ADNOC, along with the other national oil and gas companies ADCO and Takreer†; here also appeared a certain pet of a beloved former American Vice-President, which is to say, Halliburton, dually listed under "Oilfield Equipment Suppliers" and "Oilfield Contractors & Services"; meanwhile a company called E-Marine PJSC claimed that it *provides a wide range of solutions to the offshore oil and gas industry,* and then, which encouraged me in my present project, a box in the glossy front matter advertised United International Group—Management Supply Division LLC: *The leading Manpower Supply & Recruitment Services Company in the United Arab Emirates. We supply the following categories of work-ers on contract basis: Shuttering Carpenters / Steel Fixers / Masons / Helpers / Chargehands / Welders / Painters / Office Staff,* with an address on Al Reem Is-land in the city of Abu Dhabi, in a Sky Tower housing six sub-companies, one of

---

* About $432.

† ADCO was "Abu Dhabi Company for Onshore Petroleum Operations Ltd." As for Takreer, "in U.A.E. that name is very biggest company," said Ravindra. The acronym meant "Abu Dhabi Oil Refin-ery Company."

ADCO building, Abu Dhabi City

A sublime refinery view from Ruwais

whom, Mountain Gate International, called itself an *Accommodation Facilities provider at Ruwais Area.*

Here were the ones who served petroleum without glory or riches. One Pakistani taxi driver told me about a friend of his, who worked on an offshore oil rig making 1,800 dirhams a month (an even $500 U.S.): Seven days on for three months, then 15 days off. Those were wages to live up to! And in the *Khaleej Times* one could find solicitations for such worthies as a **PIPE LINE DRAFTS-MAN** . . . in UAE. 3001–5000 salary;* the bosses also hoped to find a **PIPE FIT-TER** at 1,800 to 2,000† dirhams.—Few of the men I was about to interview could hope for more than half of that.

To meet them one made a wide turn down the divided highway that bisected the plain of pinkish sand, with transmission towers half a dozen abreast holding up power wires for their privileged droop high over us; here was a black glitter of windows from the parked white buses waiting to ferry the workers to or from their bedspaces, which were right here where the land began to thicken with sheds, great spools of orange cable and low-budget carports, while the sand grew redder. A mound of strangely scarlet dirt baffled me from some kind of corral, and then came a long high prison-like fence whose outward-leaning top was barbed wire.—"This also a camp," said Ravindra. "All people working here."

"What's the name of the place?"

"Imeco. The camp is Imeco, sir. People are living here. Evening time they are coming here."

There was Almarai, and Artco, which might have been Altco, and other camps whose names I never learned. Now, how would I get acquainted with their residents?

To tell you the truth, my task was tricky. The Emirati press appeared to be controlled. Bypassing that unnecessary commodity called local news, English-language newspapers (the only ones I could judge by) contented themselves with Indian and Pakistani anecdotes, lengthy coverage of American politics (evidently written by and for expatriates) and illustrated five-liners on the itineraries of pop stars or the noble charities of sheikhs, followed by classified ads, a horoscope and health advice. Any prudent foreign journalist (thank goodness I was not one of those!) had better proceed with care.

In my business hotel in Dubai I had attempted to find a driver-interpreter who would bring me to Ruwais, but each candidate demanded a fortune, which I agreed to give, after which he became ever more unwilling, finally pawning me off on someone else. After this happened three times I took the X88 bus. The conclusions I drew were underscored by the atmosphere of the ADNOC guest house, the unavailability of accommodations in Ruwais; and the

---

* About $834 to $1,389.

† $500 to $556.

security restrictions everywhere.—Could I tour a refinery? (In Ruwais there were two of them: East and West. Together they comprised the largest refining operation in the country.) Not a chance!—In fact I could be arrested for photographing one.*

Not an illegal photograph; my shutter must have tripped all by itself.

Ravindra informed me that the refinery workers could not even bring in cell phones unless those lacked cameras. Sometimes they mutilated their smart phones into acceptability. He explained: "Whatever they are doing inside, they don't anyone tell outside. Because in some companies, workers is here just for contact"—I think he meant *spy*. "They are working here only six month, eight month, one year. So then they can apply another job, and they can get information to another company so very easy. Suppose I am working here for some period and take picture, then after that I am getting good job in Saudi Arabia and give my picture there."

More difficult than photographing refineries was photographing bedspaces—which I had first seen advertised in Dubai, spelled as one word or two, on a bright

---

* According to the National Media Council of U.A.E., visitors were not to photograph, among other items, aircraft, docks, military or industrial installations. My American research assistant's wife once attended a talk by a man who said he had been arrested simply for taking pictures of urban Emirati buildings. As usual, my own such camera images are fictions.

afternoon of 36° Celsius.* Switchboxes, walls and the backsides of signs were all posted with fliers offering to rent space to Filipinos only, or Nepalese "executive bachelors" *(4 persons only in one room)*, or "nice couples"; sometimes they bragged about how "nice" the partition was: It came nearly to the ceiling. (Perhaps you from the future remember such situations. I imagine that in your time there are fewer people, but also much less inhabitable land, so that you survivors are as crowded these guest workers ever were. But maybe it is worse than that. I'll bet no one wastes electricity on playing Christmas carols in the middle of October in a semi-air-conditioned Filipino restaurant, whose chicken adobo wasn't bad— nor on air conditioning, nor on restaurants.)—Here were some of the marvelous opportunities in Dubai:

> **BEDSPACE AVAILABLE FOR** female (Indian) in quasias. Rent AED 700.[†]

> **BED SPACE AVAILABLE FOR PAKISTANI.** Rent AED 850.[‡]

> **For Filipino's Only. Big Room good for 6 to 8 and Small Room good for 4 to 6 (with Bathroom). Bedspace (ladies Only) 400/500 & curtin good for a couple 900**[§] **. . . Very close to all Bus pick up point . . .**

> **BEDSPACE FOR INDIAN** Working Ladies near Karama. Rent AED 1250.[¶]

> **BEDSPACE IN STUDIO** apartment all included. AED 2800 only whole room.[**]

All the workers I interviewed on the dirt streets of the labor camps around Ruwais lived in such places, although streetpole advertisements were almost absent because one's employer arranged one's bunking, complete with convenient ethnic segregation. I had expected to see the inside of at least one such room. That, too, was prohibited. One night I dispatched the driver to the camps without my hindering white presence; he had acquaintances there, and thought he could find somebody who for money would allow me in even if for five minutes; he came back shaking his head: The workers were afraid; there was "security"; they would have gotten in trouble.

---

* 95 to 96.8° F.

† About $194.

† About $236.

§ These prices converted respectively to around $111, $139 and $250.

¶ About $347.

** About $778.

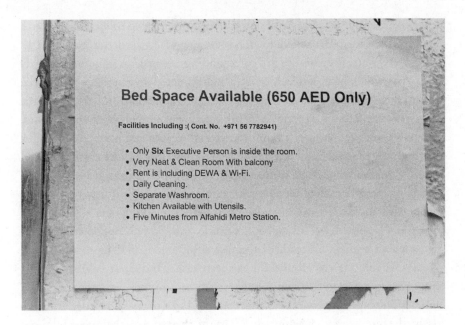

Bed Space Available (650 AED Only)

Facilities Including :(Cont. No. +971 56 7782941)

- Only **Six** Executive Person is inside the room.
- Very Neat & Clean Room With balcony
- Rent is including DEWA & Wi-Fi.
- Daily Cleaning.
- Separate Washroom.
- Kitchen Available with Utensils.
- Five Minutes from Alfahidi Metro Station.

Indeed, they seemed afraid simply to be interviewed. I remember the refinery worker from Pakistan (on the job for three years now), with that checkered head-scarf pulled down over his forehead, peering out with anxious eyes, unable to understand why I would want him in my book, barely looking at me, shaking his head in terror there at the side of the road, happily rushing away. Another one stood white-clad in the pink sand at the side of the road; he too worked in the "plant," as everyone called it. At first he was willing to talk, if through the van window only, but then he took another look at me and my terrifying glasses and hastened off. Immediately after him was the dark red sweating man who shouted: *"Why do you talk to me?,"* then stormed away. Next we coaxed a young black laborer into the back seat beside me; he was very friendly, and I had just shaken his hand when two bearded types in white robes came over very angrily, shouting at the driver while pointing at me: "We saw you with that guy in the back seat yesterday; why do you keep coming back here?"—"Since he can't understand the people," explained the driver, "so we are taking him around to ask for him."—Now they were shoving their fists in the driver's face and threatening to call the police on us. He loyally shouted back, but I put my hand on his shoulder and said that it was okay and we might as well change locales. All this happened in one morning.

We could not interview anyone in a restaurant, as I should have liked to do so that I could have bought those workers something; that would have been too public. But each camp (as a nexus of laborers' bedspaces was called) teemed with dirt roads, and these roads were rich in men who stood or trudged beneath the sun. Accordingly, we would drive slowly along until my two colleagues spotted a prospect; then Ravindra rolled down his window while the driver leaned across him, sweetly pursing up his lips in a commanding *"Tss, tss!,"* or sometimes, as may be, a *"Tch-tch!"* When I was in India and men attracted my attention in that way, I had thought it somewhat rude. But these contract laborers in Ruwais usually came straight up to the window, hopeful, trusting or simply obliging. Ravindra, who now took over, showed a genius for engaging with and persuading people; that morning of utter failure which I have just described occurred when he had to bus into Abu Dhabi on a hotel errand, so that the driver and I (who could communicate with each other only with gestures) tried a substitute.—What was it about Ravindra? He was a handsome, soft-spoken young man who kept his promises. He smiled, was definitive, yet gentle, and showed confidence in everyone, not least himself. He could distinguish a Rajasthani on sight. All the petro-workers we interviewed around Ruwais were either Indian or Pakistani—and here I should mention that the population of the Emirates did not exceed 20% Emirati. During my two weeks in that country, I met to my knowledge only two locals. The rest were overwhelmingly Pakistani, then Indian; there were Filipinas (very rarely men) in retail and at hotel counters; I encountered occasional Bangladeshi or Afghan waiters; and two sweet Russian ladies staffed my business hotel in Dubai. Nepalis whose finances had been ruined by the recent great earthquake staffed warehouses around Ruwais. Cheerful young Africans, especially friendly and perhaps a bit lonely, contracted as security guards at museums and beaches (in Ruwais I never saw them); they often hailed from Ghana or Cameroon. Aside from a few heads of growing families (most of them taxi drivers), none of these people planned to stay on in the U.A.E.

Ravindra said: "If person is local, they pay extra money, because of government rules. If I am from India, my salary is 500–600, but Emirati, his salary is maybe 20,000."*

"It's that different?"

"That different. Because of government rule."

"Does that make you feel bad?"

"That one, sir, I feel bad, too. In the hotel line also. Small hotel is no local person. You can see here Arabic person from Morocco, or Philippines, and no local person here. Because no hotel can pay that local salary. Local is no small person! In U.A.E., local only works for government company like police."

In two years he expected to go home to an arranged marriage in Rajasthan. I

---

* Respectively $139 to $154, and $5,556. I lacked opportunity to verify this.

asked if he would ever consider marrying a local girl here, and he said: "If I looked at a girl for five minutes and she complained, then police would take me."*

As I have mentioned, the bedspaces were ethnically segregated, and the workers seemed to like it that way; compatriots stuck together—which meant that when Ravindra began speaking to our latest victim in the Rajasthani dialect, which not even the driver, who came from a different region of India, could entirely understand, the man would chat back delightedly. And in Rajasthani (or Hindi, which also passed for Urdu, if we needed to cast our nets more widely) Ravindra would then lead the conversation somewhat as follows: "What village are you from? Why, that's near my village! And do you work in the plant? What do you do there? What's the life like? Can you tell us about it? You see, there's an American in the back seat . . ."—at which point the driver's pushbutton magic would roll down my window, so that for the first time the man would see a certain chubby, ageing Caucasian whose head had been crewcut in advance against the heat and who smiled and gave him a *salaam alaykum*— "no, you won't get in trouble; he's only writing a book, a book, a *book*," the reassurances and explanations requiring their own time, and then, more often than not, the man would answer my questions. On two or three occasions only would a worker enter that air-conditioned van to sit beside me, becoming cooler and safer (I would have supposed) behind those tinted windows. Ravindra said that the rest were too afraid of being seen getting into a strange vehicle—and they might well have felt in danger of getting literally or figuratively trapped *by us*.

Whenever we grew conspicuous, Ravindra would advise us to *leave another palace*—meaning leave for another place—upon which we started over again.

I am sad to say that contrary to my usual way of working with poor people, I did not pay these men. The reason was that I was never alone with them, and over the years I have learned not to pull out my wallet too often. I could have prepared a stash of 10-dirham notes in my shirt pocket; but if enough money kept coming out, it might have raised certain thoughts—and as those two threatening bearded types established to my satisfaction, the tinting of our windows was not

---

* On the subject of relationships between two guest workers he said: "Because some girl want money; some people they want only enjoy. Suppose I have girlfriend here. We are boyfriend and girlfriend for one year. And after that if I meet with another girl, then I will leave that lady. All people want to earn money. That's why they come very far. Some girlfriend, she want all the time shopping. I can't do this. Girlfriend is fine, sir. But only if we love each other. One lady, she left from here now. She was Filipino and she had here three year. She made one boyfriend and his salary also good—maybe 10,000" [about U.S. $2,778]. "After two or three years that boy said: I'm going back, so I can't marry with you. She replied, why, and he said, visa problem; he was Canadian. She said: If you marry, no visa problem. He didn't answer. But he didn't go back. He left only for two or three months then came back here. The girl didn't know. She was sad and went home to the Philippines. People like this, that means not good love marriage." | ,

superlative. It was best not to be known as the foreigner who paid and paid and kept coming around. So I never gave these workers the niggling 10 or 20 dirhams that might have meant something to them, and now I feel sorry.

I had prepared a template of questions, then ensured that Ravindra understood them all; his English was self-taught, and I knew none of his other languages. So then I would ask, and he would translate, while I sat clattering busily away on my spare laptop, which in case the U.S. Department of Homeland Security or some Emirati equivalent decided to seize it contained no data excepting these interviews. Once the laptop failed I was back to writing in a notebook, which was also my procedure when we went out at night, so that the blue glare of the screen would attract no busybodies.

At night I also refrained from taking photographs, on account of the flash. In the daytime I always asked each interviewee if I could get his portrait. Oftentimes his agreement required Ravindra's reassurance that this was just a *kitab, kitab, kitab* (in short, a mere book). I had with me two of my Contax film cameras. The T2 pocket-sized point-and-shoot, which on account of its stunning 35 millimeter lens was originally recommended to me by the street photographer Ken Miller, had served me well for more than 20 years, but only as a series of disposable incarnations. Most of the photographs in *Carbon Ideologies* were taken with a T2 or the similar (inferior) T3. Eventually the shutter would fail, or the focusing mechanism grew demented. Then I would have to buy another one. Now Contax was long out of business. This latest T2, bought from a sweet lady who had renounced analogue photography, might be my last. I have told you that in the U.A.E. it quickly grew unreliable. I had first discovered one sweltering month in the Philippines that humidity could disable its electronics: Every display went blank, even the battery icon.

Experimentation taught me that a chilled environment restored these functions. Hence my procedure in Ruwais: Sitting in the back of that air-conditioned van, taking pictures only through a rolled-down window, I could get four or five consecutive shots out of the T2 before it needed to refresh itself. Then I waited. Whenever I wished to photograph a laborer and the T2 had gone on strike, I would uplift the gloriously all-manual, tanklike Contax S2, which, if I ever had grandchildren and if they were perverts who liked film cameras, would continue right on for them, with the same reliability as a pair of scissors. But if I had to pull this monster on one of my interviewees, his face would fill the frame and 30 seconds of twiddling would be required. This made everybody apprehensive. So the S2's deployment might induce my subject to hurry away, picture lost.

Such were the conditions of my research. Now let me tell you what the workers said.

"IT'S A VERY BIG PROBLEM
TO GIVE INFORMATION"

They appear here named in order, and with whatever portraits I could get of them.

MR. RANA SAQIB

Mr. Rana Saqib, one of the few to enter the van and sit beside me, was 23 years old—not muscular, but agile and alert. He had a wide smooth face, a black moustache that curved down beyond his lower lip, and a brief and narrow beard. Mostly I could not look at him because he sat beside me, mostly facing Ravindra, while I bent over my laptop, but when I photographed him, the way he held himself in that hot glare struck me as both wary and vulnerable, with his arms loosely down and away from his sides, and that steady, patient, glittering gaze of his.

"He says, he is carpenter, from Pakistan." I asked from where in Pakistan, and he looked at me, so I rattled off the names of cities I had visited there, and when I got to Multan he nodded and said: "Multan," but later he mentioned an unfamiliar village, so he must have hailed from the mere vicinity of that city.

"He is working in U.A.E. for 39 months," Ravindra continued. "He come here by agent. Because in our country there is some agencies. They take some money, 50,000 something.* They provide visa, they provide job, everything. They came four person together from same village. Now all are working together same company."

(The firm was called something like "Urban Protection," which I could not find in the yellow pages.)

"Are you happy or disappointed with the life here?"

At this, Mr. Saqib pushed away at the back of the driver's seat.

"Sir, he's saying, according his word, when he told this agent I want to go out

---

* This would be an extraordinarily high "$13,889 something" if it were in dirhams. It was probably in Pakistani rupees (103 to the dollar), in which case it would be "$485 something."

of Pakistan, at that time, they told him, you have this job, but once he reach here, this job different. But slowly, slowly, he learning different job. Before, you need to make job same as carpenter, you need to make small small things, but here, they told him you need to make big big things. But about how much salary they told him, the salary the same. That was true.

"Starting time, they took a visa, everything. His brother one children born there in Pakistan, so then this man" (Mr. Saqib) "told them I will go after one month, to be home with his brother, and they cancelled for him, but after one month they said you have to pay for ticket. And he paid one lakh* Pakistani rupees for brother visa, but he didn't get any visa, but after long time, many pressure, they give him back money, give him job.

"Salary is 600 dirham one month, plus 300 they are giving for food, plus overtime also. After overtime he get 1,300."†

"Does he pay any tax?"

"No. But when he send some money home, that cost 20 dirham."‡

"Do you want to stay all your life here?"

"He saying, he was going home. But anything happen wrong. He cut finger this one"—and I saw he was missing the tip of his thumb.

"By mistake, his finger come in machine, and he went in hospital, and he want some insurance and went and no money, so he did police case, but in U.A.E., only if court order company will pay."

"What does the inside of the refinery look like?"

"They don't allow him inside plant. In the refinery there is some office. He went there for some items. They want to make big plant inside; they want to take in beam; they want to cover the steel, and then furniture. They fix cupboard, chair, all. This guy do furniture fixing, and then laborers will come."

"What does he think about oil?"

"When they go inside, at time they give black goggles. Helmet also use. And mask also they use."

I tried again: "Is oil a good thing for the world or a bad thing for the world?"

"Inside, sir, when worker work, to clean things, if they don't put mask on, then it is very dangerous. Then dust and oil will come in, and that is not good. And he's saying, every week there is a meeting regarding safety, but if you don't follow the procedure and have accident, it's your fault."

"So what do you think about oil?"

---

* A lakh is 100,000. At $1 USD = 103 Pakistani rupees, this would have been about $971.

† About $520.

‡ $5.55.

Mr. Rana Saqib

"He's saying, it's very danger for the world."

"So how should people get electricity?"

"He saying, for power, okay, they use generators."

Evidently he believed that the generators somehow made electricity by themselves. A moment later he said: "In company, if they use electronic, that is good, nobody will be injured, because electricity will last. So he said electricity is good, according his mind. No pollution also."

"Is oil a good way to make electricity or a bad way?"

He looked baffled.

"He's saying, according to his word, they use only oil . . ."

"What does he think about global warming?"

He leaned alertly, with arms folded, straining to understand, his eyes wide.

"According his word," Ravindra finally reported, "oil is good, but for labor, it is not good for labor, it's too much no good for labor. Sometimes, we can feel gas. That one's come night time from company. That is not good for company."

"Is the planet getting warmer or not?"

"He saying, it's danger for earth."—But possibly Ravindra was putting words in the man's mouth to oblige me.—"Oil is slowly slowly not good for world because of pollution," he said. "It is slowly slowly danger."

He was off to court in hopes of redress for his injury. Ravindra assured me: "In U.A.E. he will get money, Pakistani one lakh rupees something they will pay." Perhaps he would. But the courthouse was more than 30 minutes away by vehicle, and he lacked a ride, and was missing a day's work. My companions did not want to drive him, so we thanked him and sent him back out into the heat.

Later, while scribbling away a couple of days in Abu Dhabi City, I met Paul the Ghanaian security guard, one of whose duties was to keep anyone from photographing women on the beach. (As might be expected, this zone, the famous Corniche, possessed its own monitored entrances and exits.) Paul had begun considering becoming an oil worker in hopes of better pay. I told him about Mr. Saqib who had lost a thumb-tip and he said: "Well, even driving your own car you can die. This is not the company's fault. He should have been careful."

MR. MAHIVEER

The next one, a longtermer from Uttar Pradesh, was a man of a royal-red sweating round face and greying hair who wore a stained blue shirt with a pen crooked in his breast pocket and identified himself only as Mr. Mahiveer. His eyes looked bloodshot. About his refusal to enter our van Ravindra later explained: "He said, you know, U.A.E. is very danger, so he was scared. It's a very big problem to give information, 'cause in our country it's fine, if we talk no problem, but here, the rules are very danger! So people only do job, then go room. They can't go outside without ID. Because I am local now, I can go out," Ravindra continued, showing me his resident identity card. "When workers first come here, they don't have. Sometimes, if you're going late anywhere, police ask ID. They stop me, because of security reason. In our country we can go anywhere without ID. They don't want here any crime. If I don't have here this ID they will tell me come police station. Then they will call my company. Then my company will send my document, and police will say, it's fine. But if company will not send document, then is big trouble for me."

About Mr. Mahiveer Ravindra reported: "He is working in ADNOC." Apparently he had also been employed at one of the Takreer refineries. "According his word, when they are going inside, at that time, some air problem. When they work inside, the plant throw gas. That time, very danger for the people health. Whatever they need oxygen, they don't get it."

"What does the gas smell like?" I inquired more than once, but never got an answer.

"With the natural gas," said Mr. Mahiveer, "that time oxygen problem come."

And Ravindra added: "Nighttime I will show you some light from plant; some fire is coming. That gas is very danger. That gas is petrol, something."

I remembered the Liberty Torch in Poza Rica. And Mr. Mahiveer stood weary and still, haunting me with the sad and marble-sized pupils of his eyes.

"According his word, when they make any plant, they put a concrete inside. Summertime, there hard work, and too much hot. Humidity also. If that gas stay, work is 24 hours, that time is very danger."

"What made you decide to come here?"

"Before he was in Mumbai long time and there they make only building. He's saying, now company transfer him here. When he get interview, they said they would pay him 650 dirham.* But same position person, not even local, they are also working for different salary; job is same. He said, you know, according his word, like, you are my friend, and you know this guy, and I'm your friend and I will pay you high. You are not my friend; I will pay you low. In U.A.E. he is working maybe 13 years. He's saying, about room is good, with AC, but food is very big problem. The chicken, they don't masala inside; they don't get taste. What kind of taste they need, they're not getting here. But when they came here, this guy doesn't like to work in plant where there is too much gases."

Red and sweating, Mr. Mahiveer spoke in a loud, liquid voice, standing in the sun, touching the front passenger windowsill, and Ravindra interpreted: "This guy, Dubai company hired him, contracted with ADNOC. That gas is very danger, but we need oxygen, but we can't oxygen!"

'What's your opinion of oil?"

"For us oil is not good. For many percent of us working here it's not good. If not good for us, one by one, and not for family, it's not good for world also."

"If we don't use oil, but still we want to use electricity, which fuel source would be good?"

"Electricity is good, but if they use only electricity, how will they make business? Here is too many oil companies, so it's too much problem . . ."—and I could smell volatile organic compounds on the hot air; they offered up a faint nauseous sour chemical smell.

"He said, inside company, when they go inside, they want to finish only work there and come back early. He don't like to work more here, and after two or three months here, he will go back to India, because of gas."

---

* About $181.

Mr. Mahiveer

The man raised his voice.—"Whatever they are working here, not getting salary! One thousand dirham,* and he is working here 13 year!"

"What does the refinery look like?"

"That plant is near the sea," came the answer, so this must have been Takreer's Refinery West. "Gas pipe all, everything there.† You can see from our beach. And when they throw gas, at that time gas will come all over this area . . ."

He stood there, weary and angry, and from the receding hair on his sweaty forehead I suspected that his strength and endurance were lessening; I hoped that in the end he or his family would have something for these years of work . . .

---

* About $278.

† Two white collar types described it more technically. (You will meet them both below.) The safety engineer said: "It's a processing unit, so you have storage tanks, distillation, hydro treatment, cracking, treatments and getting different products. Basically at the gate you see the buildings, and then when you're approaching plant, processing area, the most visible things are the distillation towers, and the big piece of equipment will be the reactors."—Here the medical coordinator put in: "But approaching from Abu Dhabi you will see the flares."

And above the refineries a profusion of power wires sliced up the sky in parallels; these existed in bunches that waxed and waned in density; at their closest-packed they reminded me of the wires inside a piano.

MR. SAHABUDIN

Cruising through the environs of Imeco Camp, we next stopped a Mr. Sahabudin. "He's working with Baraqa Oil Company" (which once again I could not find in the yellow pages).—"Does he want to come inside?"—"No, sir, he don't want."

He was slender, darkish red and handsome, with black hair, and kept holding an empty soda bottle behind his back. He was 36 years old. He wore two shirts and tan slacks, all of them clean. He was barely sweating; only two or three jewels of moisture clung to his forehead.

"They prepare the steel. Salary is 12,500* include everything. The food is good for him. When they throw gas, in that time is problem. At that time, all security things, like the mask and everything. He fixes steel inside. When the people build any plant, they will first put steel, to make plant round, they fix steel round, and the other people come and put in concrete. It's very big big plant."

On the hot breeze came the smell of crude, that half-sickening, half-pleasant sweet smell that exemplified that former luxury and current worldwide necessity called electric power.

"First fix steel, then carpenter will come in there, and then concrete person will come in."

"Can you see the sky there? Can you see any roof? Can you see only pipes?"

"He said, see only plant. He said, sir, they use too much design to plant. Like round also. Whatever they need they use. Sometimes too much plant is there. To that side, they want to go near beach, to oil side, but where tank already there, nobody allowed for security reason.

"He didn't pay anything in India. He applied for his visa here. In India he worked only a little bit there."

"You must be very strong, to work steel."

The man smiled and brushed away an imaginary fly. "He said he will not work here long time. He will work here only for one year and then he will go back. In India too much things is there, but here only plant is there. If he want to see inside something they are not allowed."

---

* $3,472. Since this is so far above the other salaries mentioned, I suspect that Ravindra meant to say 1,250 AED, or $347.

"If you want to go to Abu Dhabi to walk around, can you do it or not?"

"They can't go alone there. There is a trip. Bus trip goes every month."

"What else do they do for recreation?"

"After finish duty they always come room, then come something for food and movie. This area we have only one mall here, very small area and mall is very small."*

"Can I see inside your room?"

"They don't allow that for security."

I asked him what it looked like.—"He said same same this room," pointing at a row of houses.

"His family is in India. He is 36 and wife age is 31. He has two kid also; one is six year and one is eight year."

"What does your wife say about your being here?"

He smiled bitterly. "His wife is saying, she don't want to live alone; she want to live together."

"And what do you say?"

"That's why I want to go back."

---

* That mall had a "courtesy policy." Here were some of the rules: "No overt display of affection inside the Mall. . . Please wear respectful clothing (for example: Shoulders and Knees should be covered). Industrial clothing not allowed inside the Mall." The last provision did not seem especially welcoming toward guest workers.

(Later on Ravindra remarked: "He said, he came here only for see another country.")

The loneliness of these men without their women cannot be quantified, but the following can: The sex ratio of the "total population" was 2.18 males per female. And the "economically active labour force" was 85% male, about which *The Statesman's Yearbook* once again had to comment: *One of the highest percentages of any country in the world.*

"Is oil a good energy source?"

Mr. Sahabudin shook his head.—"You know," said Ravindra, "now he's here, and he's getting no oxygen; same problem . . ."

"Electricity must come from something," I said. "So where should it come from?"

"He's saying energy 24 hours is on. Electricity also is there. We use whatever we want."—In other words, electricity was magic to him.

"Do you believe in climate change?"

He grinned and put his hand on his hip.—"Now a little bit okay, but in future it's very danger."

"So what should we do?"

"Oil is very danger for people, so if we use everything electronic that is good." (Perhaps Ravindra also lacked understanding of where electricity came from.)— Mr. Sahabudin was very patient with me. But when I asked to take his photo, he refused and ran away.

A RUWAIS ALBUM

When I asked why they had come, the guest workers invariably answered, as one might expect: solely for the money. And so Ruwais was built with reference to such people—people who preferred to send or carry their money to their home countries, rather than spending it here. What the city's planners anticipated for the long haul I could not imagine. Was Ruwais someday to be a "destination" aspiring to, say, third-rate Dubai-hood, with another mall or two, and maybe even some "fun," at least for vacationers from Saudi Arabia? Somebody must have had what executives call a "vision," because one saw so much up and coming concrete! And then what? In that latest factbook, the Central Intelligence Agency referred to "collapsing real estate prices" in the Emirates, compounded by the international banking crisis—and of course what gladdened American drivers, the oil glut, was a misfortune for the U.A.E. These three factors surely contributed to the ghostly ambiance of Ruwais. Perhaps Takreer, ADNOC and all the rest had simply overbuilt—because in Dubai, as I have told you, and likewise in Abu Dhabi City, and Sharjah City, and even adjacent to the moldy corridors around my room at the Dhafra Hotel, scaffolding and coveralled, hard-hatted men offered us daily activity with daily perceptible results.

As for Ruwais, perpendicular to one side of a long, bright and utterly empty street ran curving rows of deep canyons comprising those concrete "villas," some blocks of them almost entirely built, while the facing blocks might be higher, lower or maybe exactly the same height but still wrapped in scaffolding; and these had all been sealed off by a wall which loomed behind and considerably above the streetlamps; and along this wall marched slick renderings of the way these build-ings would look once they were finished; but the pictures lacked people. In front of the wall, and occluding the bottoms of those paradisiacal views ran a row of those moveable plastic highway barriers which can be filled with water or sand; they were as long and hulking as tanks. And that was all, except for a traffic island of pale sand, and white sky above. Presently the plastic barriers ended; but villas in progress and the inhuman images beneath them continued on and on.

There was a long, perfectly paved road; and young palm trees on a berm, planted in precisely staggered rows, and there were identical white apartment towers. As I have said, certain parking lots provided individual canopies for each space. Sometimes I saw unmoving cars, and once even a long gasoline tanker truck.

The streets had names such as Avenue A. Each name was in English below and Arabic above.

Sometimes, as in the neighborhood of the post office, the buildings leaned toward each other across the empty street like truncated triangles upturned and windowed, so that they formed an almost-arch with a sky-slit in the middle.

I remember a charming gentle gravel slope inset with a framed rectangle whose stylized flame proclaimed what English and Arabic spelled out as **GASCO**. Light-globes on poles glowed behind it, and the empty parking lot in front of it offered me a special serving of reflected heat.

It was not a place where I should have liked to live—not even if they gave me my own concrete "villa."

From the wide paved plateau on which the Ruwais Mall had been set, one had an expansive view of those villas, which from that distance resembled a crowd of mausolea, and far beyond a silhouetted palm as harshly crisp as a metal pinwheel rose a horizon's worth of refinery towers, whose slenderness, since my eye was accustomed to inhabited buildings, made them seem like alien fabrications on some low-gravity planet.* And as far as I could see, nothing moved.

---

*And why did one establishment's structures so often differ from another's? "A refinery is essentially a group of manufacturing plants that vary in number with the variety of products in the mix."

GASCO's aesthetic masterpiece

At midday Ruwais could be as still as a radioactive municipality in Fuku-shima, or a West Virginian town whose coal mines had closed—although in fairness I must say that even Dubai quieted down during the hottest time of day—but still stirred, if not swarmed, with black-covered women going in and out of stationery shops, workers of many nationalities sitting down for schwarma, bus-tours awaiting their prey; and on every major artery of Dubai, and even sometimes in alleys, that ultimate symbol of carbon-powered human life, traffic: not particularly aggressive but so remorselessly dense that it could take a quarter-hour to catch a safe instant for running across the street—while in Ruwais, as the humidity thickened and sweltered, and from sand, concrete and pallid dusty sky there came glare in equal parts, the silent villas drew together into a spurious concretion of abandoned irrelevance, until even their dark square window-holes faded; the highways seemed to vanish into the sand; those giant transmission towers lost their power wires, devolving into half-seen fenceposts and shipless masts; finally, on the horizon, Refinery East once more pretended to be a city, like Manhattan seen from the window of a distant airplane on a muggy day, a mid-grey-upon-light-grey two-dimensional concretion far more solid and mono-lithic than it actually was.

Once in early afternoon I discovered the silhouette of a moving human being (probably a woman, given the darkness of the figure); she held a white parasol over herself as she went, obediently staying on the curb at that entirely empty intersection; she approached a dual-language stop sign whose stopped pole made it twice her height and whose shadow appeared to cut into the pale pavement; and she kept walking, apparently from nowhere to nowhere.

More grandeurs of Ruwais

PALACES

I was always happy to retreat to my air-conditioned room, shower off the sweat, drink water and lie down while I cooled off; so perhaps these contract laborers also preferred to lurk indoors once their specialized services to petroleum had been completed for the day; certainly Ruwais had little "fun" to offer. Having failed to enter any bedspaces, I had better not rhapsodize about them.* But from the outside, seen through fences, here is how they struck me:

---

* Except to remark: Any place, or worse yet, series of places, that entirely prohibits visitors deserves my suspicion.

The camps tended to be gated off, of course, and in height and thickness those gates would have done credit to a so-called "community" for rich American retirees. But behind them lay shed-, trailer- and barracks-like structures in the sand. I remember a bleached sign for Ewa'a Catering Services & Camp Management, anchored by four of those tall striped poles that Ruwais used for stop signs, and then a swathe of something sackcloth-like on the fence behind it, as if to cover up some other thing on the far side of that mesh fence. Sometimes there would be a windowed, squarish-cross-sectioned security booth, which could have passed for a watch tower in a low-budget prison. Who and what were they watching?

In the sandy yard might lie long cylinders of water tanks with ladders ascending their ends. Sometimes the yards were even paved, with rows of modules inside; I would hazard a guess that these had been fabricated by one of the companies that specialized in "villas" and mosques.

Through a diamond-meshed fence I once glimpsed a long, corrugated-roofed, trailer-like structure with a paucity of windows (and those curtained), and a pair of high exhaust fans; the building was almost white in the almost white dirt, and it gave off the appearance that no one was home. A passing worker assured us that there were bedspaces inside.

Shift change at the "security room" of a labor camp

MR. KAMAL

And on a dirt road in one of those camps, Kamal with no last name closed his eyes and carefully shook his head at each question. He was very clean looking, darkskinned, with white teeth, very gentle and patient, with no objection to being summoned from the side of the road and questioned, but smilingly refused to be photographed.

"He's only for job working. He's two years complete here now in U.A.E. His home is near to Delhi, in Uttar Pradesh. He had some friend over there and his friend went to agent, so he also went to agent. He must pay Indian rupees 30,000.* Yeah, he was scared to come, but his family is more scared than he is. He's married, without children."

He made 600 dirhams† a month—twice what I paid the driver each day, slightly more than I paid Ravindra, which in turn equaled my nightly bill at the Dhafra. "He's a laborer. He remove the grass outside the refinery."

"What do you think about oil?"

"This gas is not good for health. It's bad smell. It's no oxygen. Sometimes smell. Sour smell."

"Is global warming true?" The man shook his crewcut head very slowly.—"He doesn't know."

MR. SHAVAN KUMAR

The Punjabi carpenter Shavan Kumar approached the side of the van with his friend Baljeet. His chin-stubble had just began to go grey; he was 45 years old. (Baljeet, who was only 30 and who withheld his last name, had already been here for 10 years.) They had met in the camp.

Mr. Kumar's workplace was an "electricity room" and he earned 900 dirhams† per month.—"I making villa for ADNOC company for both levels of workers since one and a half years," and if you wish to know more about the villas, which most definitely enhanced the visual splendors of Ruwais, please allow me to quote from another entry in the Abu Dhabi yellow pages, this one for GULFAB Gulf Prefab Houses Factory LLC: *Pre-fabricated buildings made for a wide range of uses . . . Oilfield Buildings and Cabins . . . Luxury Villas . . . Mosques . . .* Yes, all prefab! Reader, how would you like to live in a prefab villa and pray in a prefab mosque?

As Ravindra interpreted him, "I am coming direct here no consulancy"— that is, without approaching a consulate. "I am only here for money or good money; otherwise I am never coming here."

---

* At $1 = 66.7 rupees, this would be $450.

† About $167.

† $250.

Mr. Shavan Kumar

He arrived by plane. "Someone pick us in airport, some company guy." Then he went to work.

"What do you think of oil?"

"Bad. Bad for health."

"And climate change?"

He grinned. "Climate change is good because oil company will close!"

So he could be a jokester, this Mr. Kumar, stubble bristling his chin and cheeks all the way up to his ears. But those vertical creases above the bridge of his nose were not merry, and for a moment I thought that he might be looking at me (which might have also been true for Mr. Saqib) as if this were another procedure which was more easily endured than avoided. Anyhow, Ravindra put him at ease; he liked Ravindra, and perhaps sensed that I meant him no harm . . .

He continued: "Very very bad, this gas! Too many people is sick! Plant is open; anytime gas is outside, very smell. Only air is bad air. You cannot smell this gas in water."

"What is your room like?" I asked.

"One hall and 10 people. Blanket one by one. Three bed on wall, like this:

Top, middle, bottom. Each man one cupboard with lock, but all money you have you must keep with you. Many bathroom is there; no kitchen. Free; paid for by company."

Then he repeated: "I am here only for working, only for eat."

I asked if I could photograph inside this room. He replied: "Cannot allow this camera. Big security."

MR. GAFAR KHAN

Mr. Gafar Khan from Rajasthan, who looked 50 something years old and very professional in his blue ALSA* coveralls, sat boldly beside me in the back seat, while Ravindra reassured and pleased him by speaking Rajasthani; their villages were only 50 kilometers apart. Mr. Khan wore a plastic yellow gas leak monitor on his breast. Watching me through narrowed eyes, he allowed me to shake his ice-cold sweaty hand.

He walked an oil company's pipeline, testing for gas leaks. It was his sixth straight year in the U.A.E.; way back in 1997 he had worked in Dubai.—"He have a money problem that made him come. First time, agent took 15,000 Indian rupees."†

"What is your day like?"

"I get up at 5:00, then duties start at 6:00. Go direct to plant, and half an hour is training. First supervisor is telling you, you go there and this area is safe; and then I go, sometimes to pipeline, sometimes out of company . . ."

"What's the most difficult thing about your job?"

He laughed, showing stained teeth.

"Every work is hard, not easy, but now for me it is all easy. Sometimes is a small gas leaking, but if more than that, this instrument is blinking."

Ravindra put in: "He said is bad smell, same as egg fall down."—That would be sulfur, no doubt.‡

---

* I could not find this company in the local yellow pages.

† About $225, or around one month's typical salary. Mr. Khan did not reveal his own wage rate.

‡ According to Environment Agency Abu Dhabi, for the quarter-year in which my visit fell, Ruwais enjoyed 47% "compliant days" for PM-10 particulate matter, 46% "compliant days" for ozone ("the secondary pollutant typically formed by chemical reaction of volatile organic compounds (VOCs) and NOx with the presence of sunlight"), 100% "compliant days" for nitrogen dioxide, 100% "compliant days" for carbon monoxide (I begin to feel skeptical)—and even, strange to say, 100% "compliant days" for sulfur dioxide, 92 of which were "good days." Never mind that Ruwais's hourly maximum for this latter pollutant [see pp. 302, 412, 454] was 476 micrograms per cubic meter and the federal limit only 350. "All the stations shows [sic] 100% of the operational days were compliant days." In other words, these guest workers must have been malingering.—One more tidbit on sulfur, courtesy of the U.S. Energy Information Administration: "The UAE's natural gas has a relatively high sulfur content that

"What's your opinion of oil? Is it a good energy source or not?"

"Is very bad, this gas, not good for body. I eat nice food and so that's why I am not more sick."

"Is this pipeline for oil or a natural gas?"

Mr. Khan replied, "I have no idea where gas is coming."*

He was the most grizzled and hardened-looking of them all, and yet not brutalized—tempered, tanned, roughened, greying but clear-eyed, watching me with a sort of knowing brightness.

"Do you believe in global warming or not?"

He laughed and said he didn't know, with that slow laugh as if over something that was not for him to work out.

MR. IWAHAR

And now the round sun was half-tamed by our tinted window as we rushed over the empty roads from Jebel Dhanna toward Ruwais, with the transmission towers silhouetted against the peach-colored sky and new-planted palm trees in the white sand on either side. Soon dusk would come. At the roadside waited a long water truck. I requested Ravindra's opinion of employment agents, to which he replied: "Agent, sir, we can't believe him. One friend, he paid 60,000† for a duplicate" (second) "visa; he got in airport and they said, you have no ticket, and agent had already closed office . . ."

Here came those red and white transmission towers in four rows on one side of the highway, with a separate single row on the other side. The temperature was falling, and everything felt new and still, not unlike the way it had been entering one of the red zones of Fukushima.

Turning off toward Slaughterhouse, we passed a long line of long white buses all bringing the workers home to camp.

The half-built houses and their concrete teeth were darkening against the sun as we ascended the hill to the Ruwais Mall so that I could once more admire the outspread scene; and suddenly the wind began to blow cool up onto the parking lot. As I stood on that plateau of asphalt and concrete, gazing down the berm of grey gravel and across the white sandy plain whose main ornament was a clump

---

makes it highly corrosive and difficult to process. For decades, the country simply flared the natural gas from its oil fields rather than undertake the extensive—and expensive—processes associated with separating the sulfur from the natural gas . . . but advances in technology and the growing domestic demand for natural gas make the country's vast reserves an enticing alternative to Qatari imports."

* The contents must have been natural gas, since eyesight would have sufficed to detect leaking oil.

† I presume that this sum was in Indian rupees, in which case the equivalent would have been $900.

of barbed wire which resembled an old bloodclot, I heard a bird begin to sing; and then the formerly lifeless road that underlined the "villas" began to swarm with white buses; while down at the Takreer plant, whose sign proclaimed: **We Refine Right**, a line of lights dully shone from a palm tree's height, slowly gaining strength and beauty against the dusk.

Thinking to give the laborers a few more minutes before we pestered them, we drove down into town, and my friends sat in the parked car while I strolled about taking notes: Ruwais had become a living place!

Evening homecoming in Ruwais

On a blank sidewalk that edged a blank street, a pretty young woman in a magenta top stood beside a car, chatting through its rolled-down window with the young men inside it.

The pond which by day had been turbid and silent within its wrought iron fence, with its ugly sandy shore harmonizing appropriately with the rows and rows of square-cross-sectioned white apartment towers beyond, was now animated by fountains—fountains!—and the pairs of benches that faced each other across the picnic tables, formerly void of anyone and anything excepting empty bottles, were still void, but their existence began to seem less quixotic.

Across the street, a young couple, him in a red uniform, her in green, were bicycling at the roundabout.

Two yellow schoolbuses sat winking their taillights, and two minarets glowed electric green for the muezzin's amplified call to prayer.

At Fifth Street West and Avenue E a lady was following her cat down the street; while a well-covered woman strode rapidly toward the mosque—strangely complex, the way that the hem of that shadow-black *abaya* oscillated around her ankle at each step!—it was as graceful and apparently purposeful as the hovering streaming clenchings and suckings of a predatory jellyfish—and a young South Asian couple overtook her, carrying plastic shopping bags.

I saw a green park where people were playing basketball, two young men in matching blue uniforms were walking toward a street corner, and into the grand parking lot came a smell of fresh and spicy cooking.

A dozen young men sat on their bicycles, chatting among themselves while covered women came out from the mosque. Two young South Asian men in matching coveralls and disposable gloves were grouting some company's tiled square fountain; when I requested to photograph them, they agreed, then prudently turned their backs to the camera.

Probably if I had stuck around it would only have gotten better and better, but Ravindra was dead tired after a long shift and the driver had already sped me around for half the day in the hot sun, so we left the metropolis to its own glorious and stupendous devices, pulling back onto the wide double highway, and passing the many lights of the East Refinery, whose towers and sentry-lights extended magnificently; until once more the double rows of yellow streetlight-globes were shining high above the palm trees of the meridian, with fat cylindrical storage tanks for additional adornment, while in the dusk ahead of us, low lights went on and on as if to mark out runways at some immense deserted airport.

From Ruwais to the Dhafra Beach Hotel was 14 kilometers, and from Ruwais to the camps about the same. Turning right onto the road for Al Gwaifat, we came into still another world of white sand, recently transplanted palms and golden streetlight-globes in their customary long double row, with the transmission towers now beginning to fade into the darkening sky. The land appeared cleaner now, with those streetlamped roads offering us an immense glowing spiderweb. But presently we drew near weaker lights in the dirt, and glimpsed many people passing to and fro among those lights.

"This area is also army training," said Ravindra, and then, passing the sign for **MEGA BAZAR** (spelled thus), we descried many darkly or pallidly-clothed figures going to and fro on the dirt road, there in the familiar realm of Imeco. And we pulled into a huge gravel parking lot. Ravindra rolled down his window and the driver called out through it very softly, saying: *"Sorry!"* and *"Cht-cht-cht!"* in that way of Indian men; at which point we learned that although Ruwais Mall on its little plateau had been cooled by the evening breeze, and the city likewise, down here it remained humid and close.

Ravindra reeled in a very dark man in a camouflage headscarf. Once he had approached the window I saw that he wore an orange vest over his tan shirt. As it turned out, he too was "from near same village" as my cunning fixer.

"I mix concrete for the refinery," the man said. "If any pipeline is going, for support we make a concrete pillar, and also some concrete underneath, both for gas and oil."

"What is the most interesting thing that has happened to you on this job?"

He crossed his arms. "For work," interpreted Ravindra, "if they are working 20–25 person all together, they are feeling good, and here for salary they feel satisfied, sir," smiling, his teeth so white and clean. "He has been working here from six year, and then visa will close and he will open some business back home. His family is mother, father, wife, baby and small boy: six persons all together."

"What do you think about oil?"

He cocked his head, and I watched the rivers of light behind him, in which young male silhouettes kept passing. "Sir, he saying, here all working for money only. All agree, oil is not good, gas not good; for health they are not good. But we are working only for money."

His salary was 1,320 dirhams* for a month of eight-hour days, but everyone worked an extra four hours at an overtime rate to get 250 dirhams more. And it got even more wonderful: "Every week, he have off one Friday. On that day, if he go to work he get double salary."—If I have calculated correctly, that meant that for each month of seven-days weeks of 12-hour days, he took home 1,989 dirhams—about $552 U.S.

"Do you send money back home or keep it with you?"

"Keep all money home. Only a little bit money, like 200† every month, he keep it."

The man stood there in the darkness, with a headlight's reflection flashing behind him on a bus.—"But now no loan," and I could see that the man took pride in this fact. Ravindra explained: "His father is farmer, still alive. What father earned, he gave this one for agent, and now he has paid money back, sir."

Now he was speaking of cropland, and how our civilization improved it. "Sometimes," he said, "oil is down in the earth. If it is down there, they can get nothing. After oil, they can't get anything, because the ground is poisoned. It's not good for us."‡

---

* U.S. $367.

† U.S. $56.

‡ From a Sierra Club leaflet: "Oil companies in North Dakota reported more than 1,000 accidental releases of oil, drilling wastewater or other fluids in 2011, . . . totaling millions of gallons. Releases of brine, which is often laced with carcinogenic chemicals and heavy metals, have . . . sterilized farmland."

Behind him, a slender man in blue coveralls hastened into that light, then reentered the dark.

"In my village we can get good oxygen," the first man said. "But whatever oxygen is here, we can't get. That's why I want to get back to village."

There were four bunks in his room. "AC, TV, everything is there. Any new person is coming from same place. They will give same room. They can be happy together."

"Have you heard of climate change?"

He answered, speaking swiftly, and Ravindra interpreted: "It's true, sure, because here in U.A.E. from six years he didn't see any rain, only December and January, and only a little bit."

His name was Iwahar, and he was 35 years old.

"What do you believe is the best source of energy?"

"In plant they always use small golf cart, electronic, no gasoline engine, because so much danger from big fire! If this is safe, it is good for us. Power we can get from the sun easy. That's good for us."

Another long Takreer bus came pulling in with all windows glowing, and as we talked, the parking lot filled up with more white buses taking over the white sand; their warm exhaust resembled the breath of faithful dogs. Mr. Iwahar was now a silhouette. Through the open window I shook his hand. It was hot and wet. When I smelled my own hand afterward, it stank like a latrine.

MR. ABSAY PRATAP

Then, since some zealous bearded types might have threatened us had we continued trolling that parking lot, the driver put the van back into gear and we went lurching into darkness across dirt roads, toward another pallid row of water tanks where hardly any silhouettes were walking. Low lights lay all around but not near, so we kept going (one of the driver's several virtues was a longterm knowledge of these camps with their people and layouts), infrequently encountering dark figures part-stained with light. We slowly passed the silhouette of a man touching the flank of a long beetle-like truck whose length glowed with many eyes; and then by a metal fence we stopped another laborer who refused us, so we kept on, the semicircle of our headlights leading us into sandy hummocks, with three figures suddenly illuminated: all young men who came toward us; their occupations were unconnected to oil, so we drove back out onto the tarmac,

---

From a farmer in Ecuador: "Just below the surface the soil is oozing crude. At that time all this was planted with fruit trees. Little by little every one of them dried out and died as the oil contaminated the earth . . ."

returning at last (for by now our presence might have been forgotten) to the so-spelled **MEGA BAZAR** where our headlights picked out a crowd of silhouettes standing around fires in a wide sandy lot; I think they were eating but cannot be sure because the driver thought it inadvisable for us to get out of the van, so we left them. Just inside a Christmas-light-hung doorway an automated teller machine shone upon a man who was using it, while a horde queued behind him; and behind them, on the concrete steps descended to the parking lot, sat four or five others, chatting; perhaps their bedspaces were some distance apart, or mutually excluded by "security," or maybe they were simply enjoying the night air, which smelled only slightly of industrial chemicals.—Many blue light-strings decorated the adjacent food court; that added a jolly touch.

We called over another silhouette in the sand between buses; he was holding a white shopping bag.—"Sir," said Ravindra, "he's from Nepal and he's working in the plant and getting food, getting everything very good . . ."—then came a glare of lights across concrete and through chain-link fence; the Nepalese was tired and maybe scared, so he had to go, as accordingly did we; and there were many more people than before along the dark roadside, passing each other in twos or threes, going home to their bedspaces, I suppose, or to the **MEGA BAZAR**; a few (barely distinct) were sitting in the sand; only one man did I see holding any kind of flashlight, and his beam was pale. As they went, the men cast long shadows. Our headlights combed the corrugated wall behind them. There came a dip in the earth with many evenly spaced lights in it, going all the way to the western horizon in promise of ever so many dreary possibilities.

Finally we scooped up another Rajasthani named Mr. Absay Pratap. I could barely see him in the darkness but he was young and unshaven, in a striped shirt.

"He makes in Hyundai factory aluminum and they bring the pipes"—for oil or natural gas, he said—"there to coat when one pipe, when they fix that one, they put another thing upstairs and cover it. First they prepare aluminum. Then what size if round or whatever. Then they cover that things."

"Is he married?"

"They are seven persons in family. He is single. He saying, after he gets some money then he will do marry. So January-February last year, then three-year contract is finished and he don't want to come here again. Because visa problem. Company has some contact with oil and gas companies and if in future in that company want to bring him here again, he's ready to come again.* His salary is

---

* I did not pry into this contradiction.

1,500* and duty hour is 10 hour, but some type company says you can work extra, but only sometimes. So he go to room, take rest.

"They are six person in room, big room. Bed is three person" (I think Ravindra meant there were three bunks in one height of wall), "and they have air conditioning but not TV. All people are from India but from different state. They always live happy."

"Is there ever any stealing?"

"There is separate cupboard, so nothing missing and all are honest person, so everybody living happy."

"So his life is good?"

"Whatever oxygen they were getting in India, they're not getting here, because of oil pollution."

"What about right now in parking lot? Can you smell any gas?"

"In India it is better, sir"—as might have been expected, for the Pratap family lived in a rural village.

"Does he know about global warming?"

"According his word, this car own also too much pollution, and oil and gas company also. Now it's fine, no problem, but in future big danger."

"What does he think will happen in the future?"

"In future, if we don't stop now these things, in future we will not have whatever oxygen we need, and we will be sick. If we try to make all electric car, we can control . . ."

In company with others, he could not see that all electricity, even in electric cars, must come from somewhere.

"What is his favorite part of his day?"

"Duty time, we need to follow rules, but not happy. But now he is free. Now is happy. Now he is talking with girlfriend in India. She is 28 years old . . ."

By now it was half past seven, with even more men passing slowly through the darkness; I saw one group of three sitting in the dirt by some white bags; then came the slanted red cross of **UAE EXCHANGE**, and standing silhouettes held glowing cigarettes between the white buses, other buses creeping almost silently in and out, winking their lights, no longer dropping off passengers, merely pulling out past the men in pale shirts who swung their pale plastic bags as they walked quietly alongside the road. This place was Camp Partel. To me it did not look like the best place to come home to, but what did I know? As Mr. Pratap had said, "They always live happy."

The next morning at 9:00 a.m. I left Ruwais, finding several gleaming white

---

* $417.

vans at the bus station, and I took my seat across from a man in flowing white robes and a red-checked headscarf who was texting on his cell phone with dark brown hands, then looked around me, waiting for this X87 bus to depart. The houses were still; the sidewalks had gone dead and empty again but for one man in lemon-yellow coveralls who stood reading a clipboard. Then we departed, briefly heading toward the refinery, then turned back onto Highway E11, back in the world of white sand, transmission towers and recently transplanted palm trees.

A COMPARISON TO CAMEROON

The average salary of all these men* worked out to 13,894 dirhams a year, or U.S. $3,859. My sample was small, and each worker must have made his own decision as to whether to include his overtime pay and food allowance in the number that he gave me. All the same, these salaries lay within a credibly narrow range. Let us assume, then, that a typical annual wage for petroleum-related contract work around Ruwais was $3,800.

According to the CIA, in 2014 (the latest year for which this data was available) the per capita GDP of the United Arab Emirates was $66,300.

What should I have made of the fact that the second number was slightly over 17 times higher than the first? Remembering that other striking fact, that Emirati citizens made up only 20% of the population, I concluded that the Emiratis were servicing their petrochemical machine at bargain rates.

Would you call these laborers exploited? Mr. Rana Saqib's missing thumbtip, and that common complaint about not enough "oxygen," not to mention the fear they so often showed—of meeting me, going on the record, getting out of line—made me thrilled not to be in their shoes. To work six or seven days a week from at least dawn to dark, and then go maybe to that **MEGA BAZAR** for cigarettes, fast food or the automated teller machine, after which the destination would be some dirt road, followed by an assigned bedspace; to accumulate money dirham by dirham, so that if all went well one could start a business back home, or even marry—or, if the man was married already (which seemed to be the case more often for the taxi drivers I met in the busy Emirati cities than for these contract laborers), he might go home when his contract was up, to father another child, then return here to pay for that privilege—was that a decent life?

But of course they always lived happy.

As for the "oxygen" problem, when I raised this with the refinery safety auditor, Mr. Priyank Srivastava, he said: "Actually, in any area of the world, if you go in some industrial area, ammonia or nitrogen or whatever they are not producing, they are venting something that is not good for health. If you are living in a house

---

* Excluding Mr. Sahabudin, whose wage figure Ravindra might have mis-expressed.

surrounded by trees, the atmosphere will be much more healthy than here. In your country and my country they try to put as much green around refineries as possible, to absorb the toxins. If you're not planting so many trees and you have this concentrated industrial area, you will breathe the area that is not fresh, and you may get sick; you may feel some choking or suffocating feeling. But here they take care. If you are below 40 years old, you must go at least once every three years for medical evaluation. Between 40 and 50, once in two years. After 50, once a year. What are those people noticing? Maybe carbon dioxide or hydrogen sulfide excess or maybe just the dust, over an approximately 10-kilometer area. You are in the refinery, so you cannot get the quality of air that you would be breathing in the jungle."

And, as it happened, those laborers could even be envied:

At the Corniche of Abu Dhabi was a beach, and at the entrance a sign warned against "unruly behavior," which activity was graphically prohibited by means of a slash through the silhouettes of a boy and girl holding hands; but on the sand, accompanied by her tote bag, sat an hourglass-shaped single Asian woman in a bikini, and behind her gamboled some European or American family, the young father going shirtless in and out of the water with his fairskinned little boy, who was well covered up against the sun, while the young mother lay filling out a bathing costume that should have been inexpensive considering how little material went into it. So here was an Omelas of sorts—patrolled by African security guards in uniform. One midmorning (34°*) I started chatting with these men; they were friendly and lonely. Gilbert from Cameroon shook my hand hello, goodbye and in between. He said that he had come "just for income," on a one-year contract which he would renew if he could find nothing better. He lived in a two-room rent-free bedspace with nine other men. He proudly said: "My bed is mine only, until the day I quit. And frankly, Mr. William, given the conditions I come from, it is not so bad. Most of us, you live in your own bed and you don't know more than three people. In one hour my relief comes, and I shake his hand, go back in the company bus to the camp, and I'll be the only one in that room. It will be all mine."

Posted in the Deira district of Dubai

---

* 93° F.

He said: "I have never touched the water. People ask why. You know, where the work is, there is no fun."

I remember looking out at two skyscrapers through the almost lowered blinds of the second level of a two-storey restaurant in that city, where men in industrial coveralls or cheap clerk-suits ate a cheap and delicious meal of phyllo dough cooked into flat pancakes that one dipped into a stew of beans, peas and spicy ground mutton. They would walk over to the sink to gargle and then wash their hands. Presently a woman, fat or very pregnant, lumbered up the stairs and passed through the narrow slit in a painted metal partition to the family room, where she sat alone. A clerkish type held his cell phone to the side of his head while he dipped bread into his curry. Another man ate hastily with both hands, while a black cord dangled from his earbud. Within a few steps lay a dozen other places to eat, each one good, cheap and unique; for that reason alone I might rather have been a guest worker in Abu Dhabi or Dubai than out in Ruwais—but Gilbert, for instance, almost never went out. Oh, once in a blue moon he and some other workers might drink together some evening, but that came at the expense of both income and sleep. Hence he mostly preferred to keep himself at work, in his bunk or on the company bus.

He was awed and admiring of the U.A.E. He said that once upon a time the Emiratis had asked Nigeria for an oil loan and got stupidly mocked. Now this place was the wonder of African countries.

I never inquired into his wages. For his part, he asked how much those Ruwais petro-workers might be earning, and when I gave him the estimate that I have just given you, he said that he might consider a change of career.*

Well, in that case any number of kindly entities stood ready to place him in some new dream job! For instance, a short bus ride away was MCT International Dubai, *a global player* with *offices on five continents*. That thrilled me, I had to say. MCT had been *established in 1988 . . . with a mission to become the regions* [sic] *leading supply house for seamless procurement services.* The company's *target industries* included *the global oil/gas industry, NGO's* [sic]*, world milatarise* [sic]*,* etcetera. Among its *diverse lines* were *camp supply/operation.* For all I know, there were

---

* He was not the only one. Three days later, a Cameroonian security guard at the Sharjah art museum expressed his desperate desire for a higher-paying job. I related the salaries of the petroleum workers, which seemed good to him; he said that he would start seeking such a position if he could, but how could he? I advised him to find the camp of Sharjah's petroleum workers and ask around there. Upon being warned about the lack of "oxygen," he remained keen.—"What do you feel about oil?" I asked him.—"It's very, very good, of course, but it pollutes everything."—"Do you know about global warming?"—"Yes, I've heard about it, but only God knows."—The other museum guard said much the same. Even Ravindra possibly envied the refinery workers, at least in one respect. "And, sir," he said, "Takreer, in one week they has two days off. We have here one off."

MCT camps in Ruwais. And if Gilbert decided against MCT, some other good Samaritan might offer him an even more stunning opportunity.

Given the conditions I come from, it is not so bad. What might this mean? Well, it could be at least superficially quantified:

PER CAPITA GROSS DOMESTIC PRODUCTS AND UNEMPLOYMENT RATES OF SOME U.A.E. GUEST WORKER COUNTRIES,

from Central Intelligence Agency data,

2001–14,

in multiples of the GDP for Cameroon

All GDP figures expressed in 2014 dollars. Unemployment statistics dated from 2001–14 inclusive. All data (including the ancient unemployment figure for Cameroon) were the latest the CIA could scrape together. Individual dates can be found in my source-notes.

All major interviewees and a few barely mentioned people appear in this table. Names in **boldface** are of white collar workers.

1

Cameroon [$3,000 per capita GDP, 30% unemployment]. *Gilbert, beach security guard, Abu Dhabi.*

1.1

Kyrgyzstan [$3,300, 8%]. **Utkir Umarov, refinery senior process safety engineer, Ruwais.**

1.6

Pakistan [$4,700, 6.8%]. *Rana Saqib, refinery "furniture fixer," Ruwais; various taxi drivers, Abu Dhabi, Dubai and Sharjah.*

1.9

India [$5,800, 7.3%]. *Mr. Mahiveer, refinery concrete mixer (?); Mr. Sahabudin, refinery "steel fixer"; Kamal, refinery groundskeeper; Shavan Kumar, fabricator or assembler of "villas" for refinery workers; Gafar Khan, pipeline tester; Iwahar, refinery concrete mixer; and Absay Pratap, refinery pipe-coater—all in Ruwais—and Priyank Srivastava, refinery safety auditor, in Ruwais and Abu Dhabi.*

2.3
Philippines [$7,000, 6.8%]. *Unnamed tig welder, Ruwais; Richard, reception desk, ADNOC guest house, Ruwais; various female hotel clerks, Jebel Dhanna.*

8.0
Kazakhstan [$24,100, 5%]. *Marat Sagimbayev, refinery medical coordinator,* Ruwais.

18.1
United States of America [$54,400, 6.2%]. No guest workers interviewed.

22.1
United Arab Emirates [$66,300, 2.4%]

And so if those petro-laborers in Ruwais ever felt sorry for themselves, all they had to do was thank their lucky stars they weren't seeking work in Cameroon! Really they had a wonderful life.

SPOONS AND CHOPSTICKS

At 6:00 and 6:30 in the morning (Fridays and Saturdays generally excepted), the Dhafra Hotel's restaurant filled with men in navy blue Takreer coveralls whose reflective silver safety stripes at cuff and seam and whose elbow-patch of Emirati flag* enrolled them into brotherhood. By 7:00 the buses had taken them away.

Of course their service to petroleum was considerably more remunerative than that of the men in the labor camps who slept seven or 10 to a room. To tell the truth, I would not have sought them out, for it came more naturally to me to impose on a poor man on the street than to trap a fellow guest of my hotel: Easier for the poor man to walk away!—But even these better paid men were still guest workers, and perhaps that was one reason why they were as generous to me as the others, petro-laborers or not.—One evening in Sharjah City I stood looking across the diesel- and fresh-smelling river at a line of steel dhows, one of which bore cars, and on the far side of the river, up a forest of steel: skeleton-trunks towering crowned and branchless into the dusk while the muezzins' wavering calls echoed from electrically glowing minarets. Men on benches watched me, and when I smiled or waved they touched their hearts. Sometimes a man would leap up to shake my hand. And if I could have walked alone in the labor camps of Ruwais and communicated in Urdu or Arabic, it might have been like that for me; as it was, whenever Ravindra called to them they would usually come, and stand around in the hot sun answering some foreigner's personal questions without resentment or even perceptible impatience. So because Ravindra was friends with several white collar men, and introduced me, I succeeded in imposing yet again: Three of them, all kind and interesting souls, consented to answer my questions in my room at night.† In contrast with those furtive camp

---

* The national colors were green, white and black, with a red band along the left side (or, depending on orientation, the top); and you will be pleased to learn that the black represented "petroleum resources."

† And here also I must guiltily confess to letting down my interviewees. I had meant to take them to dinner at the Danat, or at least do something for them—and this time I lacked the previously plausible excuse that it was imprudent to whip out money. In fact I was ill for most of my stay in the U.A.E.; I could barely do what little I did, and then each night I lay down in tired misery. So let me once more say, to them or about them, that I am sorry.

conversations of a quarter-hour or less, interpreted into extreme simplifications of my language in an atmosphere of interruptibility and anxiety, these final three interviews went on for a reasonable length, without constraint, and in English, so that the only information losses occurred when I could not input their words as rapidly as they uttered them.

Of course the reason that I wanted to meet oil workers in the first place was a certain little problem that we had, called climate change; you from the future have probably solved it, by starving or choking to death. But in our time we were not as advanced as that; we still hoped for solutions that would be cheap and maybe even pleasant, as we meanwhile kept right on manufacturing, air conditioning, travelling and illuminating exactly as we pleased! (You see, it was our human right.) We tried to believe that, no matter what our socioeconomic level. Remember for instance Mr. Pratap: "In future, if we don't stop now these things, in future we will not have whatever oxygen we need, and we will be sick. If we try to make all electric car, we can control . . ."—How marvelous it would have been, if only switching to electric cars had halted the warming of our atmosphere!

You have seen that the laborers rarely understood where electric power came from. They knew that oil was harmful to human health, that petro-extraction could damage farmland and that a sulfur-smelling "gas" impaired their ability to breathe. All in all, they might have thought along Mr. Sahabudin's lines: "Now a little bit okay, but in future it's very danger." That was how I liked to think; that way it wasn't my problem. (Mr. Sahabudin I excuse for proceeding on that basis: He was a poor man without many choices. I do blame myself.)—Mr. Kumar had made the wittiest statement: "Climate change is good because oil company will close!" But most likely our children (and maybe our grandchildren) would live with both climate change and oil companies.

In any event, I was curious to ask the white collar men about this issue.

Two of them were post-Soviet men the shattering of whose birthplace had flung them into separate countries. Mr. Marat Sagimbayev was a Kazakh, probably ethnically as well as formally; like many of his countrymen, he was Korean-looking to me. He said: "I am working here as a medical coordinator, which is totally administrative job, so I'm not using my skills in practice as a doctor.* I

---

* "I need to say that I'm not going to refinery by itself. I'm dealing mostly with security guys. In Iran as well as Kazakhstan I was an emergency doctor on site. Here it's a strictly administrative job."—But what was a refinery worker? If Mr. Sahabudin the steel "fixer" and Mr. Iwahar the concrete mixer were, as I say they were, petroleum workers, then so were these other three, none of whom was operating the controls of some catalytic cracking tower. For the Dubai Mall's skating rink to stay frozen, and my ailing laptop to stay charged, hordes of specialists needed to accomplish tasks that at first sight had little to do with oil.

graduated medical university and worked six or seven years as a doctor and a teacher, so because of hard economical situation in Kazakhstan I am here. I have a family which hopefully can come here."—So his domestic aspirations differed considerably from those of the laborers—for which his white collar status might have accounted. He valued the U.A.E. for its safety and stability, and I could see his point. He was 48 years old.

Mr. Utkir Umarov, the senior process safety engineer, was a trim 37, and more European in appearance, with an almost military bearing. "I started in Kyrgyzstan," his homeland, "as a safety officer, spent 10 years in the refinery there and then got an offer from a firm in Abu Dhabi doing consulting work, and have been doing some jobs for Takreer. I just joined the oil refining company one year ago." He must have been extremely capable to have escaped from the economic conditions of his country, whose per capita GDP did not much exceed Cameroon's.

At my request we conversed about the old days, which reliably fascinated me and which I never saw, having visited Russia and Kazakhstan only after *glasnost;* to Kyrgyzstan I had never been. They told me family stories of bravery and cowardice in the Great Patriotic War, which (being unrelated to oil) I reluctantly omit from this record. Mr. Sagimbayev said: "We had really good things in old Soviet Union, real friendship and good things with neighbors. The Soviet ice cream is something you miss!"—But Mr. Umarov remarked: "You know, I had a good memories about my childhood, but I would say that it was in the past so we should not be going there back."

In Kazakhstan I had been haunted by mentions of a region called the Polygon, where the Soviets conducted nuclear testing (*the equivalent of one Hiroshima for every inhabitant in the village* of Sarjal). The inhabitants were said to develop cancers at a rate three times the national average. Unfortunately, in those days I had lacked my pancake frisker, and its use would surely have been prohibited anyway. The KNB secret police kept their comic opera eyes on me, so I skipped the Polygon.—And Mr. Sagimbayev had been there.

"During my army service," he said, "I was on service in the rocket testing area of the Polygon. The testing area was actually crescent-shaped, and all of this crescent was in our part of Kazakhstan. I was a part of the special research team sent by Karaganda University, so it was not inside the Semipolygon area but nearby. We have made some research and it was mostly statistics, but a higher amount of psychological abnormalities was found, which is normal due to stress, and a higher amount of genetic abnormalities, and a comparatively big amount of endocrine disturbances."

I asked him how high the radiation had been, and he said, "In general, a big amount." But he declined to give me any numbers because the data were so old.*

As it happened, Mr. Umarov had also been near a hot zone. He told me: "There is a city called Mailuu-Suu which is close to the town where I was brought up, and they have wasteland,"† evidently meaning *contaminated place;* "it was closed city in my time but now it is open." He did not elaborate on the present-day charms of Mailuu-Suu, but they might have informed his views on nuclear power, which ran as follows: "It is probably good to provide boost to economy of single country, but in longer run you will have nuclear waste and radiation leaks, and then all the gains will be eliminated."‡

Mr. Sagimbayev contented himself with saying: "In my opinion, this is something that can be considered a really clean source of energy, but definitely a security issue."

It frequently turns out that a subscriber to some given ideology of energy use will be likewise sanguine about other fuels, while a naysayer (like me) may well be negative wherever he goes. And in the case of these two refinery men, each one's position on nuclear power approximated his judgment of oil.

I asked them: "Which energy source is more important now, coal or oil?"

---

* Perhaps, as would only have been reasonable, he also felt security compunctions. In its report on nuclear mementoes in Central Asia ("a leakage or catastrophic failure due to a seismic event at an un-remediated site in one country could result in the radiological and chemical contamination in another country"), the International Atomic Energy Agency did not get around to mentioning the Polygon by name. So let the following serve: In the western part of the country, in Koshkar-Ata site by the Caspian Sea, uranium tailings from the Caspian Hydrometallurgical Plant emitted gamma varying from "10 µSv/h to several Sv/h." Ten microsieverts an hour would more than qualify a Japanese locality for red zone status (I:250); some of my measurements in Tomioka and Okuma showed comparable values. "Several sieverts per hour" would definitely cause radiation sickness. "Unfortunately, this same part of the tailings pond is also a popular scavenging ground for scrap metals." The IAEA had a sweet name for all such areas: *uranium legacy sites.*

† Mailuu-Suu, Kyrgyzstan, whose uranium was dug out from 1946 until 1967, doubtless for all the right reasons, won third prize on "Blacksmith Institute's Worst-Polluted Places list for 2006." [Blacksmith was "an independent environmental action group."] Poor Chernobyl merely placed ninth for horribleness. According to the International Atomic Energy Agency, in due course Mailuu-Suu's nuclear workers "left behind mining and milling waste deposited in 23 tailings and 13 mine waste piles ... some of which are within the town's boundaries." But speak no evil, please! You see, "after closure of production ... the waste piles were abandoned in accordance with former Soviet standards." Near the village of Min-Kush, one pile of tailings (and places within the village itself) emitted 5 microsieverts per hour, which would be in the upper limit for a Japanese yellow zone; another pile gave off as much as 12 micros (comparable to a certain drainpipe I frisked in Tomioka; I:420). "Potential exposure ... is associated both with geotechnical instability of the tailings dumps situated upstream ... and with radiological risks for the local citizens living ... in contaminated houses (where tailings materials and contaminated material from the mill were used for home construction and insulation)." Now for the good news: "Regarding the potential for further growth it is important to note that approximately 20% of the world uranium reasonably assured resources are located in Kazakhstan."

‡ For more about this "wasteland," see I:185.

Mr. Sagimbayev said: "Definitely oil is much more important in the current moment than coal, because the coal era is ended. The most appropriate for me is a book of one German writer, *Ausgebrannt,* or *Scorched,* really describing the importance of that fossil fuel for mankind."

Mr. Umarov said: "At present I think oil's competing with gas, which is being liquefied and used worldwide. From an environmental point of view, gas is more better, but it is not quite practical to switch to gas, economically."[*]

"And what's your take on global warming?"

Mr. Sagimbayev said: "In big perspective I think we should also understand minor Ice Age which is now approaching. All things which Mr. Gore mentioned[†] should be evaluated properly. But definitely humans can affect the environment. But it requires more to cause global warming. It's been proved: It's a minor Ice Age that's been starting, something that comes periodically every 700 or 500 years."

Not looking at his colleague, Mr. Umarov said: "Looking at the poles, sea level is going up; it's a clear evidence of climate change."

Mr. Sagimbayev then said: "I want to say that all of these effects including global warming need to be properly calculated. We cannot say how it will be in the next 20, 30 years; the intensity of the Ice Age will be gradual, but it will compensate. I mean, it's necessary to have proper calculations, by independent sources."

When I asked him whether he thought that the human race had worked up too great a demand for energy, he replied:

"To me, the world is right now more or less balanced for that issue. The problem will not be with us. The problem will be for dealing with the next generations. Before dealing with this, our generation should start preparing for them. Maybe it will come from self-regulation . . .

"You know that energy demand is reaching the critical point where it will go down afterwards. Energy of energy-consuming devices is reaching the point where it will go down. But the number of these devices will go up. That's a different story.

"Africa and these countries will be really problem, not when America becomes satisfied with this.[†] Those countries want to be more or less equal to what American has for right now. In Russia it was one anecdote: When the world will

---

[*] He said nothing about coal, which surprised me, since even back in Soviet times it was said that "Kirgizya ranks first in Central Asia in coal reserves (about 3,000,000,000 tons)."

[†] In the movie *An Inconvenient Truth.*

[†] That is, when American energy demand finally reached some satiation level, and stabilized.

Storage tanks and transmission towers, Ruwais

meet total starvation? Only when Chinese start eating with spoon and not chopstick!"

EIGHT-FOOT OR FOUR-FOOT MEN?

Mr. Priyank Srivastava will have the last word. This outgoing, studious-looking workaholic, who would have put me up in his shared flat in Abu Dhabi, became my friend. About his routine he said:

"My day start, getting up at 6:30 in the morning, because the offices in the Middle East are working like 7:30 to 3:00. By 7:00 I'll be ready for the office and at 7:00 I'll be at the office if I'm in Abu Dhabi. I'm working here at Ruwais for the past few months: various audits. If at Ruwais, by 7:30 I'll be in office, and after that, the office work, and after that, if I have some other work, maybe in Abu Dhabi, I work until 7:00 or 8:00 p.m., and then I'll be free. But when I'm doing the fieldwork in audit, tomorrow I'll get up at 5:30, I'll be downstairs for my breakfast at 6:30, my bus will be here at 6:55.

"Refinery East, it's like 12–15 kilometers from this hotel. It will take approximately 15 or 20 minutes, and then some kind of security check, another 15 or 20 minutes, and then after 30 or 40 minutes I begin my work. So we are reviewing the manual, checking data, talking to owners, finding some loopholes in the process; we are just highlighting those issues, those observations, and further discussing with the process owners, the Senior Vice-President, the Vice-President of the refinery, etcetera: *This is the scope of improvement that we have. This will save you money, this will save you energy.*

"I am reading manuals, then asking for data, then reviewing data, then asking for the process owner some kind of justification, and then I'll give them my kind of clarification and justification, and if it's a genuine observation they will definitely agree.

"In between I will have my lunch as well. And then I'll come back to hotel by 7:15. and then if I have some office work from the refinery I'll work on that, and possibly, if I have the time, I'll go gym. Because my tummy is coming out!"

"When you come into the refinery, what do you see?"

"You see," replied the polite, bespectacled young man, "Ruwais refinery is a big complex. Many companies are working here, oil, gas; there's even a liquid nitrogen plant! When you leaving from here, what you will see is a huge structure of metal. As soon as you enter the refinery, what you see everywhere is huge pipes and columns of metal standing up, and so many things. And if you are in the middle of refinery and you are standing up, you will only see steel structures, and you will hear hammering in the pipes. This is the first thing that I see, and then I go inside sitting area. Then again, I may be going to see physical verification, for instance in the warehouse, to see if they are taking care of the specs, and how the chemicals are going in and going out."

"Do you understand everything that you see?"

"No. Though my background is from downstream, I have done my master's in applied oil and gas management, but if you ask me, what is this unit, I may not be able to tell you."

"Do all the refineries look different from each other?"

"Every refinery same, more or less same. But if you are middle of refinery you will only see pipes, walls, columns, pumps. And some buildings, some offices."

"What made you decide to get your education in petroleum?"

"In India, many youngs are engineers. So people always will do engineering, some chemical or some mechanical. Then if they're not satisfied with their job, they will do MBA. I passed my 12th standard. But am I going to computer science or I.T. or chemistry? But one thing I know: I am good in chemistry. Have you heard of the University of Petroleum and Energy Studies, in Dehra Dun? I came to know about this college; I took my test and I certainly passed it, and I was enrolled there in a five year course, bachelor's plus MBA. I chose and this was how it started. I got a campus placement in New Delhi, and studied what is market of oil and gas. I was a senior oil and gas analyst, and then I joined Productivity Middle East as an auditor, and now I am here.

"For the past three years I do the analysis and all these research, all these things, from the market side, not the chemical side. Here you don't have to be hard core technical. But if you have a basic understanding of refinery business and process unit, then that's good. This job I'm doing, it's sort of a blend of consulting work."

"How does oil rank as an energy source?" I asked—thinking, I suppose, thermodynamically and environmentally, but since I did not make that clear I got what might have been more standard in my time, an economic answer:

"In 2012 and 2013, when oil prices were above $100, people were always running for renewables. But now at $50, of course, it's better for me to use oil or gas in my car than renewables."

"And how do you feel about renewables?"

"Despite the price of oil and gas are hovering very low, still we are pushing forward for the renewables. Dubai and Abu Dhabi are all pushing places like Masdar,* for renewables. It's obvious if the oil price will remain high, definitely the renewables will get a better share. I believe renewables is a future, even though oil and gas will be there until 2040 or longer."

"Mr. Priyank,† I understand that coal, oil and natural gas are more or less interchangeable; apparently you can make anything from one that you can from another."

"From the oil and gas, yes; oil is the mother of everything, like the fridge, the watch and laptop, your specs,‡ but from coal, yes, you can convert coal to gasoline, or if you can convert coal to natural gas and further to your petroleum products, but most of the coal is used as a heating process. So I will keep oil and natural gas on top of *everything,* the mother of *everything*! But indirectly you can convert coal into a feedstock."

"It seems that many products come from benzene, and benzene can come from coal or oil . . ."

"This is for sure," he insisted, "that benzene feedstock is oil."

(Here was another indication that King Coal might finally be approaching senescence—for in the previous century, benzene had been derived from both of those fossil fuels—and natural gas also.)§

---

* A "carbon neutral" district in progress in Abu Dhabi.

† I called him this out of respect, since people often called me "Mr. William."

‡ Spectacles.

§ See above, "About Coal," p. 23.

"What would you say about global warming?"

"What if I say that there is one emirate, Fujairah, and last year there was a snowfall there, there, *a snowfall,* for the first time, and how you will explain this? In the northern part of India, in Agra and so far, if you go back 50 or 60 years, the intensity of heat was much lower than now. And this is because of deforestation. But on the other hand, we are producing tons of carbons, tons every day, and we are creating a huge nuisance. Maybe in New Delhi you are getting rainfall just for one month instead of two or three months. Maybe there are many factors, because of population outburst. I am 27 years old, and after 30–40 years where will I be? Gone. We must preserve our nature for the next ones."

"Will humans go extinct?"

"There will always be evolution. A snake used to have very little hands and feet, if you are talking many thousands of years back. So they shed them because they are useless to them. They cause resistance, actually, when they are moving. So, actually, if you see this pinky finger, it's still somewhat useful for gripping, but over 1,000 year or something, this will be gone. So maybe, yes, we human being, they can do anything, so I believe that of course we will survive. Whatever will be the circumstance, even as in zombie movies, we will always survive. We will evolve over the time! If we need eight-feet man, we will evolve like this! If we need four-feet man, we will evolve over the time."

Thanking him, I said goodnight and prepared to enjoy the accomplishments of petroleum. No matter which hotel I laid myself down in, those were more or less the same, except at the Dhafra, where (at least in that season) there was never any traffic once the refinery buses came back. In Dubai or Abu Dhabi or Sharjah it was less quiet, even in the predawn hours, when no sound came from outside but a very occasional far away motor, and I, unable to sleep for jet lag, finally switched on the bedside lamp and lay listening to the faithful hiss and throb of the hotel's air conditioner, which it seemed could never fail. I looked around me. Electricity controlled my room just as I wished, and steadily. The plastic telephone shone like ivory, ready to carry my voice back to the other side of the planet had I been in that mood. Down on the street a motorcycle passed; after awhile the cars began to come. The dark sky turned greyish-blue.

About Batteries and Fuel Cells

*... Pour into the water surrounding the zinc an amount of sulphuric acid
equal to <u>5 thimble full</u> Then the battery is ready for business*

Thomas Edison, 1885

O il was bad, decreed the bigot who wrote *Carbon Ideologies*—bad like
coal, natural gas and nuclear fuel! Leave 'em all in the ground ...!

Departing Cushing on Highway 33 West, past red-white-and-blue-
striped oil tanks and the crimson-brown of the Cimarron River, I sat in a seat
made mostly of petroleum products, while an internal combustion engine conveyed me on my travels. Touring Tomioka in an immaculate taxicab, I powered
down the windows and tasted radiation with my pancake frisker, whose disposable batteries never seemed to wear out. For your sake, reader, I rolled past
Drumright's oil worker statue and derrick mural. Having flown to Bangladesh
and Mexico, I now viewed South Linwood Avenue, Magellan Midstream Partners and those fences with their three outward-leaning strands of barbed wire.

Combusting gasoline with supernatural evenness, a rented engine took me to
the Tallgrass Prairie in the shadow of blue-grey clouds, with a cool wind blowing
along sand-creeks whose white reflections shimmered and trembled. But I didn't
permit fossil fuels to do all the work: I walked two miles or even three, observing
a mother turkey with her brood, over which a raptor swept low (another unseen
bird suddenly feathering its wings very loudly in the grass). There came a pelting
of rain, lightning, then cool wet silence and swallows bumbling low over the
prairie; having now done my part, I returned to the rental car, and my companion keyed on the engine. On we drove, until it came time to gas up. In some years
a full tank for a vehicle of this size cost $30; sometimes it was $50; whatever the
price, it invariably came in cheaper than thinking.

Overnighting in Chickasha, I departed on a lovely cool Saturday morning, the
sky packed with cirrus clouds, some of which were reflected in the asphalt-framed
puddle from last night's rain, as occasional cars rushed to and fro above the speed
limit, accompanied by many birds, and the sound of a power screwdriver from the
tire shop across the road.—What if I could have seen all that without combusting
anything?—"If we try to make all electric car, we can control ...," said Mr. Absay
Pratap in the darkness by those Emirati labor camps. He meant: *control fate.*

The electric car which he imagined would have been powered by magic batteries which required no energy to manufacture and somehow gave off energy without letup, cheaply and cleanly.

I wanted what he did: something for nothing. We all loved the phantom of "clean electricity." Those few who were both educated and honest admitted the sad fact that more thermodynamic work must be done to generate power than could actually address demand—the vast losses of the Rankine Cycle in utility plants being my own favorite hobbyhorse.[*] What about solar electricity and the other so-called "renewables"? I tentatively believed in them . . .—but the market for those remained insufficiently "robust," explained the neighbors. Maybe by 2098 or 2231 we could make them pay.

Until then, why shouldn't batteries bail us out? In 1999 the *Concise Encyclopedia of Chemical Technology* touted them as *providing a direct means for energy utilization without going through heat as an intermediate step. As a result, electrochemical reactions can be significantly more efficient than Carnot Cycle heat engines.*

Do you remember the "plutonium batteries" of our two *Voyager* spacecraft?[†] Which isotope they were I never learned. Suppose that they had been plutonium-239, which warms its surroundings with 816 BTUs per pound per second. Over the half-century lifetimes of those probes, which as I completed *Carbon Ideologies* were still continuing their lonely service, a single pound of Pu-239 would have emitted nearly 129 billion BTUs of heat from atomic decay—9.2 million times more energy than inhabited a pound of pure carbon. And their nuclear waste would pass clear out of the solar system!

Compared to those, our workaday batteries fell short:

COMPARATIVE BATTERY ENERGY DENSITIES

in multiples of the lower range of a lead-acid battery

All figures expressed in [BTUs per lb]

Pu-239 energy density calculated from [BTUs/grams/sec], 50-yr lifespan assumed as a result of *Voyager* space probe information. All other energy densities calculated from [watt-hours/kg].

---

[*] See I:151.
[†] See I:236.

All energy densities rounded to nearest 5 whole digits, per original. (The Pu-239 datum was originally expressed with greater precision, but I conformed it likewise to +/– 5.] Final header rounded per my usual procedure with huge numbers.

For reference I have included the inherent energy [high heating value] of pure carbon. See the table of Calorific Efficiencies, I:213.

1–1.7

Lead-acid, prevalent in automobiles and forklifts [45–75 BTUs per lb]

1.6–2.8

Nickel-cadmium, often used for power tools and two-way radios [70–125]

2.1–4.1

Nickel-metal hydride, "mainly used for satellite applications" [95–185]

2.8

Reusable alkaline, used in various small appliances [125]

3.4–4.4

Lithium-ion polymer, used in cell phones and laptops [155–200]

3.8–5.6

Lithium-ion, used in cell phones and laptops [170–250]

311

Pure carbon, calorific efficiency [14,000 BTUs per lb]. *Actual power output would be lower due to efficiency losses.*

2.86 billion

Plutonium-239 [128,739,728,160]. *Again, this ignores efficiency losses.*

Sources: Mechanical Engineers' Handbook, 1958; Concise Encyclopedia of Chemical Technology, 1994; batteryuniversity.com, 2017; with calculations by WTV.

Pure carbon still looked awfully good at up to 311 times more watt-hours, BTUs, joules, whatever, than the lowly car battery. How then could we possibly jilt our beloved fossil fuels?

To be sure, solar cells might enable continued demand. You might remember from the previous volume that in 2017 a solar ideologue had assured me* that photovoltaic cells operating at 17% efficiency could feed Americans' current energy-hunger three or four times over—106 daily terawatts!—if we only set aside 120,000 square kilometers.

I did my arithmetic: Each square foot of solar cells would generate 268,882 BTUs, day after day—19 times more than that pound of carbon's one-time best.

How should one compare feet to pounds? I threw up my hands.

For consolation, a chemical engineering expert at Purdue Polytechnic Institute now assured me that the *Lithium Air battery has the highest theoretical energy density and actually exceed[s] the energy density of fossil fuels.*[†]

And let me touch on fuel cells, which if continuously fed hydrogen (for instance from hydrocarbons) and oxygen could go on producing electricity indefinitely. *As of 2009,* said the U.S. Department of Energy, *more than 200 buses and several hundred cars powered by fuel cells are navigating cities around the world . . . A fuel-cell vehicle running on pure hydrogen produces only water vapor.*

Wouldn't it be a relief if Mr. Absay Pratap proved correct? All electric cars, all electric everything!—But how much carbon must we spend to manufacture their energy-storage treasuries?

In the summer of 2017, my intermediary, the unwearied Mr. Jordan Rothacker, *reached out* to the U.S. Department of Energy on my behalf. His first communication explained:

> *Simply[,] we need to know:*
> *What is the most efficient battery?*
> *What is the most efficient fuel cell?*
> *Is it possible to make a battery without using fossil fuels?*

The G-man replied: *Could you please give me more information about the book and the author? Is it this book?*—indicating some electronic announcement or advertisement for *Carbon Ideologies,* with *this book* helpfully highlighted in blue. *What angle is the author taking, etc?*

---

* See I:168 ("About Solar Energy").

[†] He added the following smidgeon of local color: "Batteries were made before we knew of fossil fuels. It just takes the creation of a reversible corrosion process. Metal reacts to give electrons and accepts electrons to plate out the metal to return. Any metal could be a battery, and you could use any electrons from muscle, solar or wind power sources if not from the fossil fuels. There are reports and unsubstantiated speculation that there were batteries in the great pyramids or ancient world made from terracotta and metal electrodes . . . that would behave very similarly to the old acid jars on a model T car."

Mr. Rothacker, hopeful and guileless, confirmed that it was *this book,* then elucidated my *angle:*

> *The battery info will be in volume two I believe.*
> *His angle is that he's addressing the reader 100 years from now and show-ing how much we knew now about climate change and the environmental destructiveness of our energy choices.*

I myself would have been cagier—why, that almost made me sound like an ideologue!—but Mr. Rothacker must have imagined that American government agencies still served the citizenry, giving out tax-funded information upon re-quest. So he kept on building me up: *He has interviewed people in energy, and activists* (this latter a sinister word, perhaps) *and people living around fracking sites and coal mines and near Fukushima . . .*

The G-man wrote back: *Thank you for your follow up and sorry for the delay. We have to notify the D[epartment] O[f] E[nergy] of these kinds of things and I have not heard back yet. When do you need answers by?*

Mr. Rothacker told him. Oh, I was *in the pipeline* now! And on the very next day my queries got addressed in full:

> *Unfortunately, we will not be able to participate in the interview, etc., for the book you're researching . . .*

Perhaps President Trump or his creatures considered *Carbon Ideologies* to be *fake news.*

(And once upon a time there was a fellow named Elon Musk, who proposed to help us out by means of giant storage batteries. My friend Greg was devoted to him; he insisted that Mr. Musk would be part of your future. But our prospec-tive savior declined even to acknowledge my ever-so-nice questions. In other words, he belonged to the *regulated community.*)

Left to my own resources, I shall now soldier on.

A battery is a device whose chemical energy can be converted into electric power. A fuel cell can continue making that conversion for an indefinite period, assuming a continued supply of fuel and oxygen. The latter has been praised as *a remarkable and valuable device because* (and what follows much resembles what we have already been told about batteries) *it can convert chemical energy directly into work, thus bypassing the wasteful intermediate conversion to heat.* That was lovely—but alas, both batteries and fuel cells, like every other creation of human-ity, still waste some portion of their energy inputs.

In one-time-use "primary batteries," corrosion can cause *waste of material;* oxidation can interfere with the flow of electrons. In the Edison-Lalande battery *the electromotive force may be about one volt initially, but in practice only about three-quarters of a volt can be relied upon.* My 1911 *Britannica* considered primary battery cells *a convenient means of obtaining electricity for laboratory experiments, and for such light services as working telegraphs, bells, &c,* but overly expensive for *electric lighting and traction.* Better to rely on steam, or maybe hydro power!

Exactly half a century after that edition went on the market, *a hydrogen-oxygen fuel cell system can be purchased from a number of U.S. manufacturers. These engines or power plants range from hundreds of watts to kilowatts and will operate thousands of hours with negligible maintenance at energy efficiencies of the order of 60 per cent*—comparable to a steam power plant.* More than another half-century later, when I was writing *Carbon Ideologies,* a very few hydrogen-powered cars, several buses and a commuter train, all powered by hydrogen fuel cells, operated in Germany. Their only emission was water.

In 1999 fuel cell efficiencies ranged *from 40–60% based on the lower heating value*[†] *of the fuel,* which was *higher than that of almost all other energy conversion systems.*

But how much energy did hydrogen production require? I wish I knew. If it were easy and cheap, why then, demand for everything might maintain itself indefinitely. Then I would have loved hydrogen fuel cells.

As for non-hydrogen fuel cells, they were hardworking emblems of our carbon ideologies.

From the economic standpoint, the hydrocarbons have a clear advantage over other fuels suggested for fuel cell use. By some coincidence, the writers of that sentence hailed from the Esso Research and Engineering Company. They continued: *It is not adequate simply to obtain current when hydrocarbons are changed to fuel cells. For maximum efficiency the reaction should be complete, resulting in carbon dioxide and water.*

Carbon dioxide, eh? Well, who had a problem with that?

A saturated hydrocarbon, in particular, ethane, can be electrochemically oxidized to carbon dioxide at relatively mild temperatures and pressures. I accomplished comparable atmospheric good whenever I lit my gas stove.

---

* Other versions, "using natural gas, etc.," attained an "efficiency of 50 per cent or higher." Meanwhile, the previous volume of this compilation advises us that for fuel cells "it should eventually be possible to attain over-all efficiencies exceeding the 40 per cent now realized in the best steam-turbine alternator plants."

[†] See I:535.

Meanwhile, following his own professional bliss, a fellow from the Gas Technology Institute proposed a *methane fuel power pack.*

And why leave out our all-time favorite hydrocarbon? *In the simplest fuel cell . . . coal is supplied at the anode, where it interacts with oxide ions to form CO_2 and releases electrons to the external circuit.*

Summing up: *Only a fuel carbonaceous in part can be low enough in cost for the central-station fuel cell.*

Those words date back to 1963. By 1999 we had progressed to the point where *fuel cells operating on pure H_2 and O_2 provide a useful power source in remote areas such as in space or under the sea . . . On the other hand, fuel cell power plants operating on fossil fuels and air offer the potential for environmentally acceptable, highly efficient and low cost power generation.*

In reality, degradation or malfunction of components limits the practical operating life of fuel cells. Therefore, fuel cells were defined by *high initial cost and short operational lifetime.*

And in 2017 the man who installed my new steel door could plug his battery-powered drill into my wall, and revive its energies at demonic speed. That was wonderful; all the same, he must have drawn more current into the drill than he actually got to use. And from whence did that current derive? Perhaps wind spun the turbines, but in my day the prime mover was more likely coal or heating oil.

HAPPILY EVER AFTER,

or,

THE MARRIAGE OF CARBON IDEOLOGIES

Overleaf: Construction in Moscow

From a Newspaper

Companies to develop floating reactors for drilling oil . . . "This partnership will push forward the organic integration of the offshore oil industry and the nuclear power industry," China National Offshore said . . .

AND SO,
AND THEN

Overleaf: Construction in Sharjah Emirate, with lavish employment of "big five" materials

Fear of global warming (or "climate change") led to the adoption of taxes, regulations, and subsidies aimed at reducing carbon dioxide emissions. But now we know the threat of global warming was grossly exaggerated and reducing emissions would have an insignificant impact on temperature or weather. It's time to repeal those laws and give the American people an annual global warming "peace dividend" worth hundreds of billions of dollars.

Heartland Institute, 2016

LAST DAYS OF THE ASSYRIANS

Opinions are spread abroad on a quantitative scale and . . . the leading position always goes to what is easier to grasp, that is, whatever is easier and more comfortable for the human spirit. Indeed, the man who has fully educated . . . himself . . . can always reckon to have the majority against him.

Goethe, 1829

When I began to look around me in this world, what I could see unnerved me; what I deduced from what I saw began to alarm me.

Why shouldn't we go on forever? And why did the climate ever have to change? *There will be no end to time and the world is eternal,* insisted Aristotle. We tried to believe him. But even had that been so, we should have lived more carefully, because *if rivers come into existence and perish and the same parts of the earth were not always moist, the sea must needs change correspondingly. And if the sea is always advancing in one place and receding in another it is clear that the same parts of the whole earth are not always sea or land, but that all this changes in the course of time.*

In one of the so-called "lost notebooks," Loren Eiseley marveled that *life . . . demands a security guarantee from nature that is largely forthcoming.* In other words, *there is change, but throughout the past, life alters with the slow pace of organic change.*

Now organic change was quickening. The results might not necessarily be survivable—but as Mark Mooney had told me in Twilight, West Virginia, "that was the way God intended. It's designed for perfection, and He's gonna do what He's gonna do."*

Although prospective eco-catastrophes had struck a chord of urgent sadness within my heart ever since I was a young man, I had never loathed myself sufficiently to craft the punishment of full understanding—and how could I? No one person could. You from the future must have found ludicrous stupidities in this book; some are mine and the rest come from the "experts" in whom I believed.

---

* See p. 88.

Their most sinister prophecies refrained from proving themselves when I was alive; even in your time the worst of them might be never-nevers. For instance, perhaps you saved the oceans from becoming stinking dead zones. (Between 1750 and 2011, they grew 26% more acidic. What if that trend magically reversed itself after I died?)—Or maybe climate change and radiocontamination are the least of your troubles. What about disease?—I repeat: Could I get one wish, it would be that *Carbon Ideologies* was mistaken to the point of irrelevance. Of course, being pessimistic, I expect to be right—for it appears that *people,* like gods and corporations, do what they're going to do.

Strange to say, this affords me consolation—because we irradiators, window-glass-buyers, paper-powered scribblers, refrigerator manufacturers and airplane passengers were not the only villains who ever lived. Consider this epitaph for the ancient Assyrians: *Amazing as was their intelligence and culture, yet here was a race that was so purely and solely destructive, so utterly devoid of the slightest desire to make any real contribution to the welfare of* humanity.—My epoch was superior; we touted ourselves for improving everyone's situation—didn't we keep the lights on? What we lacked was stewardship; we couldn't be bothered about "some ecosystem somewhere."

> If one can imagine a man with no small amount of learning, with all the externals of civilization, ... fine taste in certain aspects of art and science, and a tremendous aptitude for organization, discipline and learning, and then imagine such a man imbued with ... an absolute delight in witnessing the most ghastly forms of human suffering, one will have a fairly accurate conception of the average Assyrian king...

Our kings of industry (and their satellite politicians) were not cruel. Nor were they necessarily refined. Culture was rarely emphasized in their annual reports—although I hope to have shown you that the carbon ideologies they upheld could be entertaining in their own right.* They declined to raze cities, rape children or flay rebels alive. They erected no graven images of themselves. Their goal was *profit.*—How can I be against that? I too grubbed up shekels when I could. That way, once my air conditioner died I could afford another one. Other demands of mine effortlessly enlarged themselves. Hence if profit were inherently evil, my self-interest would avoid admitting the fact. But *profit without limit,* well, I could never aspire to such luxury, so why not fulminate against it in *Carbon Ideologies?*

---

* For instance, let me hum a few bars of this Mexican swan song to the *regulated community:* "In countries where there is greater reliance on consumption of commercial fuels, more CO_2 emissions are produced and there is a higher level of human development."

How much is enough? Who would dare say, "Having provided for all contingencies, I may safely rest?" Hurricanes, unplanned babies, bank failures, business competitors, enemy agents, not to mention the beautiful burgeonings of our own ambitions, couldn't any of those abrade somebody's cash reserves? And so a great novelist once had his most disillusioned character say: *There is no peace and no rest in the development of material interests. They have their law and their justice. But it is founded on expediency and is inhuman . . .*—Remembering from West Virginia that long slanting forest of Cook Mountain, and then, on the far side of the unseen valley where Twilight lay, the naked terraced mazework of strip mining and mountaintop removal, I saw that expediency (which I, alas, instead of dismissing it as inhuman, must call human), and felt disgusted.* (The already cited Manoranjam Pegu, *circa* 2010: *Capitalism in Bangladesh is characterised by corporate land grabbing and dispossession of people.*) Yes, I named it "human," because to what may as well be called "ordinary human nature" it seems not only natural but obvious that a person (for instance, an Assyrian tyrant), or that agglomeration of persons called a corporation, may chase self-interest wherever it leads, right up to the boundary-line drawn by law to protect the self-interest of others. And once this boundary-line has been painted across our desires and possibilities, self-interest will howl and slaver to overrun it, even if only by an inch. No matter how well or poorly that line may be respected, any talk of moving it *outward* upon us, the *regulated community,* so that self-interest loses that inch, will transform the howling into roaring; thus the animals we are. (B. Traven: *Not only for the development of the human body, but for that of a capitalist enterprise, the possession of an excellent appetite is of vital importance . . .*) In short, self-interest and the common good cannot help coming into collision. And yet these animals (in which category I emphatically include myself) were not invariably blameworthy for doing what they did. From Phulbari I remember a white-bearded, earth-colored man in a grey mud-colored shirt, sitting on a cement step, sweating, scratching his feet; and I remember a sputtering, painted open truck packed with bony cows staring outward and on its tailgate two men of comparable skinniness sitting side by side; then a ewe and four lambs strutted briskly down the street. Didn't they all deserve to prosper? More food, more rest, company and pleasure—I suppose that is what each strove for. Somewhere else, other members of my species turned on the gas thermostat, because that was how they stayed warm; meanwhile kindred bipeds engaged in industrial processes that might incidentally acidify the neighborhood— because as my high school chemistry textbook used to say: *The consumption of*

---

* See the photos on pp. 91, 92, 94, 95, 100, 106, 117.

sulfuric acid is an index to the state of civilization and prosperity of a country—as was the manufacture of adipic acid, without which Sears and Roebuck could not advertise more nylon-lined brassieres in their 1956 catalogue. In our day, we acidified, fracked, cracked and carbonated in order to enrich. Above all, we gasified carbon. We spun turbines for profit; and in that spirit a mutual fund manager assured me: *As one of the most attractively valued sectors in the stock universe, energy is an area that we are monitoring closely for opportunities.* Meanwhile I stood by a frack pad in Colorado, writing in my notebook (Bob Winkler was saying: "When are people gonna wake up? I don't understand it"); then reentered the car; the engine burned carbon, and off we rode, so that I could see more for *Carbon Ideologies.* (Later on, feeling thirsty, I bought a plastic bottle of electrically-chilled water.) The realities of processes and byproducts which determined our ways of being remained as invisible as any greenhouse gas. We all did as we thought best, and whether or not the coal came out from beneath our feet, not to mention how much the gas pipeline leaked, mostly depended on somebody else.

To most of the world, wrote Steinbeck, *success is never bad. I remember how, when Hitler moved unchecked and triumphant, many honorable men sought and found virtues in him . . . Strength and success—they are above morality, above criticism.* Perhaps that was why the *regulated community* remained likewise above criticism:

ACCIDENTS AND THE BOTTOM LINE: COMMON TESTIMONY ON THE *REGULATED COMMUNITY*, 2010–14

Oil

Culprit: *British Petroleum [BP]*
Disaster: *Deepwater Horizon explosion, Gulf of Mexico*
"The immediate cause of the Macondo well blowout can be traced to a series of identifiable mistakes made by BP, Halliburton, and Transocean that reveal such systematic failures in risk management that they place in doubt the safety culture of the entire industry."

Report of the Presidential Commission

"The focus on controlling costs was acute at BP, to the point that it became a distraction. They just go after it with a ferocity that is mind-numbing and terrifying. No one's ever asked to cut corners or take a risk, but it often ends up like that."

Former BP engineer, testifying to the Commission

Coal
Culprit: *Massey Energy [later Alpha Natural Resources]*
Disaster: *Upper Big Branch mine explosion, West Virginia*
"This . . . confirms information that we have heard time and time again that the practice of advance notification and the 'War on Safety' at Massey was . . . condoned at the highest level."

Tim Bailey, Charleston lawyer

Nuclear
Culprit: *Tokyo Electric Power Company [Tepco]*
Disaster: *Plant No. 1's explosions, Fukushima*
"The major problem . . . for Tepco . . . was only a matter of cost that prevented them from building a dike or siting the emergency power generator in a higher place."

M.Y., former Tepco engineer, in interview with WTV

Moreover, thanks to the innate secrecy of the *regulated community*—another utterly predictable manifestation of untrammeled self-interest—the casualties of its mishaps got first denied, then minimized, finally tabooed or else brightly dismissed as "lessons learned."*

Oh, but surely their better natures would come into play in the matter of climate change! They wouldn't cover up *that* accident in progress; didn't they love their grandchildren?

A wise woman once wrote: *As a man's real power grows and his knowledge widens, ever the way he can follow grows narrower, until at last he chooses nothing, but does only and wholly what he* must *do.*—Yes, she was wise, but was she right? William Blake insisted: *He who desires but acts not breeds pestilence.* And electric power fulfilled so many desires; it even created new ones! As our power grew from watt to megawatt, and our knowledge widened in tune with each "big five" manufactur-

---

* Allegations of the organization Beyond Nuclear, regarding the Vermont Yankee Mark I boiling water reactor: "Beginning in 2008, Entergy management officials made false representations [some under oath] to the review panel, the Public Service Board, and the state legislature that there was 'no underground piping carrying radioactive water.'" In 2010, Entergy dug up two pipes leaking "up to 2.7 million picocuries per liter" of tritium.—In 2014 the facility was closed.

ing triumph, we found more ways to extract carbon and even split the atom; for awhile our way did widen before us: We could do anything, because the future was ours! And the ones who lacked what we had made rarely sang their own self-denying virtues; instead, they bred further pestilence; because Archie Dunham and Sam Hewes were correct: *Everyone* wanted climate-controlled cars and illumination on demand. Why shouldn't each Bangladeshi laborer come home to an air conditioner?* (Fancy this: Their country was still above water when I was alive.)

Not long ago, factory smoke, like that stinging odor of sulfuric acid, was a welcome emblem of progress. As an esteemed socialist realist novelized: *He went to see Artem down at the engine sheds. Here in this grimy, smoke-blackened building Pavel felt at home. Hungrily he inhaled the coal smoke. This is where he really belonged and it was here he wished to be.*—Believe me, reader: Pavel imagined that he was helping *you*!

Mural, Cushing, Oklahoma

---

* They exemplified *relative poverty,* the Marxist notion that people who possess the basic means of material subsistence may still think of themselves as poor, and be called poor, in relation to others who have much more than they.

Let me remind you what Lucien Lucius Nunn said so long ago: *To raise man's efficiency—to reduce man's toil—to give him time and means to love his family, his country, and his soul is the work to be accomplished.** Who could be against that? Pavel tried to in his way to be Nunnian; in fact he aimed higher; still more than family and country he loved the future, for whose sake he proudly *increased* his toil. The chemists who built up our treasuries of adipic acid likewise raised efficiency, sparing seamstresses from ill-paid mendwork, shrinking housewives' washdays, easing budgets, increasing glamor-allure! We were all in it together. I pretended to come out against "waste," but can I honestly claim that prepackaged foods trucked to me at ever greater distances, with their energy-intensive plastic shells then gloriously thrown "away," did not give me more time and means to love whatever I wished? I remember happy picnics, and cafeteria food eaten quickly so that I could rush back to the library to take more notes for *Carbon Ideologies,* not to mention magically warmed airplane meals—all predicated on the value of my life's moments. At the dawning of a certain cloudless Sunday in Tokyo I showered to my heart's content, rode the elevator 11 floors down to the lobby of my business hotel and strolled round the corner to the convenience store, where happy music entertained me as I selected two piping hot cans of unknown-flavored coffee, took a plastic tray of tonkatsu and sesame-seeded rice to the clerk, who microwaved it for me, after which I paid, exchanged bows with him, took my plastic sack of food back to my room, and in 10 minutes had created a good pound of trash. Until I began to write this book it never occurred to me (so occupied was I in loving family, country and soul) how much petroleum and electrical power such meals embodied. I wish it weren't so.

In 2010 four Democratic Socialists wrote: *Capitalism must maximize profit to survive. The requirement of ceaseless capital accumulation, no matter how efficient, is ultimately incompatible with the health of the planet.* This struck me as true—but why blame capitalism in particular?—The planned economy of the Soviet Union had dried up much of the Caspian Sea for the sake of an unsustainable cotton monoculture, and poisoned the "Polygon" zone of Kazakhstan with nuclear tests. In short, *there is no peace and no rest in the development of material interests.* A late-20th-century article on "energy management" summed up the principle best: *A saving of millions of kilowatt hours or gigaJoules is only communicable when converted into dollars, or into a ratio of dollars saved per dollar of incremental investment.* Hence this result:

> By 2012, more than 365 billion tons of carbon had been emitted to the atmosphere from fossil fuel combustion and cement production.

---

* See I:28.

Another 180 were added from deforestation and other land use changes. Remarkably, more than half these emissions occurred *after* the mid-1970s—that is, *after* scientists had built computer models demonstrating that greenhouse gases would cause warming.*

In 2016, as I finished writing this book, we were busily sending up not quite 33 billion tons a year. Meanwhile, a reporter finally pinned down a Sacramento County sheriff on a peripherally related topic: *Does Jones think human-caused climate change is real? "Well, there's a body of evidence for both sides of the argument."*

NOT GOOD ENOUGH

I loved our shell games. In the days of President Obama (President Trump would prove more honestly dishonest) we closed down coal-fired generating plants, and all the while kept selling coal to others! I had to grin when I read the following newspaper item from 2014: *When the Prime Lilly, a massive cargo ship, set sail from Norfolk [Virginia] recently, its 80,000 tons of coal were destined for power plants and factories in South America. The 228,800 tons of carbon dioxide*[†] *contained in that coal disappeared from America's pollution ledger.* What a good trick!

Another shell game, played to advantage by Great Britain and several other nations, was to outsource manufacturing, and reduce declared carbon emissions accordingly. *The UK's consumption-based emissions are now more than 30 per cent greater than reported production-based figures. This difference is entirely due to . . . offshore materials production . . .*

We reassured ourselves that if anything untoward ever did occur in some ecosystem somewhere, we could someday somehow emit somewhat less! If push came to shove, we could even (temporarily, all the while crying God forbid) *stop.*—Like the cruel doctor who refused to promise that my father would live forever, the Intergovernmental Panel on Climate Change informed us that even stopping wouldn't be good enough! *The commitment to past emissions is a persistent warming for hundreds of years, continuing at about the level of warming that*

---

* From the Purdue University professor whose team investigated the MCHM leak in West Virginia (see p. 176): "Many of the problems that occurred once the spill was detected were not due to poor planning. They were due to people making decisions without understanding science, engineering, or public health principles . . . Failure by many people at the local, state, and federal level not understanding the consequences of their decisions is a root-cause of the incident." Well, who gave a damn about those scientists and their computer models?

† Of course it was carbon, not carbon dioxide, that was "contained in that coal." I remind you that as long as mass and energy were conserved, a pound of coal obviously could not generate 2.86 pounds of carbon dioxide without outside help—in this case, from oxygen in the air.

has been realized when emissions were ceased. And in place of "hundreds," consider substituting "thousands."*

The thousandfold and millionfold deaths might have already begun, but if so, the decedents were obscure marine organisms or unfamiliar plants that nobody who was anybody (that is, a politico-corporate carbon ideologue) cared about. There might be floods and droughts, but we could still go on building better lives for ourselves. In 2009 an ecologist in Vermont remarked: *Our local ecology, our seasonal traditions, and our weather-based economic infrastructure are in the early stages of flux. In this way we are not much different from others in northern climates, living with the paradox that the short-term gratification of warmer weather may well be the preamble to a miserable fate.* (Call her a sourpuss.)

In short, it seemed very possible that within the time granted by the laws of physics to delay, prevent and alleviate global warming, nothing could now be done. Some doomsayers even began predicting that nothing *would* be done. If that proved to be so, then the question of resistance might demand consideration. Of course, back in the time when I was alive, that question was almost never raised.[†]

When by some chance or mischance the regulators did act, their actions often looked to undo themselves, for they went no further than to punish the *regulated community*'s serfs—who accordingly voted for Trump. In 2015, wicked liberals in the Obama administration finally compelled American Electric Power to shut down six coal-burning power plants. Three in West Virginia had provided jobs since the 1950s. The *Herald-Dispatch* simply noted: *More than 250 workers . . . affected.* I never read that any federal authority then stepped in re-employ these people in, say, making solar collectors.—Who *should* have helped them? This question left me blank, because when I was alive I was as incapable of imagining regulators who were not foxes in the henhouse as were our ancestors in envisaging a round Earth.[‡]

---

* See remarks of Dr. Pieter Tans on I:173.

[†] In another book of mine, called *Rising Up and Rising Down,* I once attempted to consider when violence is justified for our collective protection, and for the defense of other species. The case here would be what I call *scientific imminence*—that is, a roof's impending collapse, as certified by a structural engineer, or a planet's dangerous warming, as predicted with an extremely high degree of certainty by a league of expert climatologists, who, lacking financial or other incentive to state any particular view of the matter, were among the few of us who were *not* carbon ideologues. If this group established the scientific imminence of global warming, then and only then, if there was to be any future at all for us, even the miserable one which you, reader, presumably inhabit, the time might come for a concerted conflict between those who respected their children's interests—and therefore yours—and those who cared most about shoring up the convenient and profitable religion of unfettered carbon-burning.

[‡] West Virginia was "poor." But why? The heavier the social-environmental impact of resource extraction, the greater the corporate tax rate should have been. West Virginia got pillaged for decades, leaving contamination and poverty in commemoration—and I never heard the county commissioners demanding greater compensation from the coal companies; they contented themselves with bemoaning

Mr. Kawama Tatsuhiko of the National Railway Mito Motive Power Union once remarked to me in Iwaki: "Normally, when we are arrested, we will be blamed for what we did in the past. But those who decided in my childhood to build the nuclear plant will never take responsibility."

"So you feel that these people should be punished."

"Yes."

No decent person who visited the red and yellow zones could have lacked sympathy for Mr. Kawama, and the logic of his position was plausible. But as I began to forget the sad and eerie things I had seen in Fukushima, I grew less outraged at Tepco's negligence in declining to pay for that more expensive backup system which should have been situated higher and farther from the sea. Was Tepco (toward which I harbored better feelings than, say, Noble Energy: at least they'd answered my questions!) any worse than the *regulated communities* of coal in Appalachia and fracking in Oklahoma? Perhaps they were better! A pie chart from 2012 showed greenhouse emissions for the EU-15 countries. One busy developed country had come close to escaping blameworthiness. Her slice of responsibility was a mere 5%. *The relatively low share . . . from energy industries in France can be partly explained by the use of nuclear energy for power generation.* Couldn't a case be made that the near certainty of another nuclear catastrophe sometime (humanity and the *regulated community* being what they were) was preferable to the almost absolute certainty of a climate change catastrophe soon? If I opposed nuclear and all the other carbon ideologies, what should I advocate *for*?

So, as usual, I forbore and empathized wherever I could. (My fear of the future resembled the way that in the red zones of Japan the asphalt sometimes fissured like a dropped dinner plate, with weeds more or less radioactive beginning to grow up through the cracks.) Hence I repeat: Those who found themselves compelled by economics to be complicit in the production, distribution and consumption of harmful energies may not have been noble and scarcely deserved to call themselves victims, but they were not especially at fault. For them, fossil fuels constituted sheer subsistence. (David Kennedy, "business owner and magistrate" in Harlan County, Kentucky: "My little town is dying. Do you think I'm a fool and going to support somebody that doesn't support coal?") What about the two executives I interviewed for the oil chapter? They were open and good with me, and quite naturally took pride in the industry through which they had made their living. I cannot fault Mr. Sam Hewes, who knew what he stood for and was at peace with his accomplishments. Mr. Archie Dunham was another active intelligent man who believed that

---

their withdrawal. A higher business tax on coal mining, whose revenues could have been used to pension the miners left unemployed—as sooner or later so many of them inevitably were—to me this seemed the height of obviousness.

he had brought good to the world. I myself have surely benefitted from his oil. As I keep saying, I *wanted* them to be right! Even less could I accuse those who had not been educated to understand the almost invisibly approaching misery.

However, I began to believe that those who selfishly, maliciously or with gross negligence did harm ought to be singled out, shamed and maybe even, as Mr. Kawama had proposed, punished.—What constituted gross negligence? A parent who left a loaded gun in reach of a baby was surely responsible for the result. Those West Virginia officials, Colorado lobbyists and Oklahoma Chamber of Commerce types who publicly advanced the agendas of their chosen fossil fuels but refused to even acknowledge questions about global warming stood convicted, in my mind at least, of authoritarian partisanship. I would have heard their side; they were not even willing to *tell me* theirs, much less ask about mine. And they had power.—The man from Boilermakers' Local 154 in Pittsburgh who held an American-flagged placard reading: **STOP THE WAR ON COAL**, when I saw his portrait in the newspaper I excused him, on account of his legitimate if parochial self-interest—and, more importantly, his lack of institutional authority—how many underlings could he himself compel? Call him only one of *more than 2,000 union workers and others organized by the coal industry in Pennsylvania, Ohio and West Virginia who joined top state officials Wednesday to rally against proposed stricter federal pollution regulations for coal-burning plants.* Their enemy: Environmental Protection Agency regulations intended *to cap carbon pollution 30 percent by the year 2030.* Even you from the future may possess the magnanimity to write off the apparatchik from Local 154— but if you look back into the past, when we were alive and busily injuring you, you will see before a row of American flags Pennsylvania Governor Tom Corbett in a short-sleeved white shirt that matches his hair (he has ever so eloquently explained that *this is about are we going to have enough electricity in Pennsylvania with these regulations*—and, oh, yes, it was he who signed that freedom-loving bill to protect fracking's trade secrets from potentially blabbing doctors*), smiling wearily, clapping; while Ohio Lieutenant Governor Mary Taylor, a longhaired, executive sort of glamor girl, watches through her sunglasses, bringing her painted nails together not far above her crossed legs, and then good old Governor Tomblin from West Virginia (who might or might not have been less vile than Governor Jim Justice), dressed for a wedding or a maybe a funeral, holds his hands the highest and straightest as he claps! *These* are the ones, my friend. These are the ones who laid you low.

Could it be that they, too, really, *really* didn't know and couldn't understand? In that case, the best I can say is that they were unfit for office.

From another newspaper: *Wyoming Gov. Matt Mead is asking the White*

---

* For the text of it, see above, p. 358.

House to disregard pressure from the governors of Washington and Oregon and re-
fuse to evaluate the effects of greenhouse gases that would be emitted by exporting
U.S. coal to Asia from ports in the Northwest.

Refuse to evaluate! What does that say about him?

And here is a pair of headlines from Japan, two years after the Fukushima
catastrophe:

6 of 8 panelists who voted to phase out atomic power by 2030s axed: [Prime Minister] Abe purges energy board of antinuclear experts

and

India closer to getting technology: Abe, Singh ink statement on nuclear deal.

No one could accuse Mr. Abe of not knowing what he wanted!

I presumed, although naturally I could never prove, that Governor Mead and
Mr. Abe were puppets of the *regulated community,* whose mission partook of
ruthless self-interest of a type so strangely narrow that it excluded not only their
own grandchildren, but their employees.

Consider a Koriyama subcontractor named Build-Up, which got in on the
decontamination bonanza, dispatching its peons to grub about in the hideously
dangerous ruins of Nuclear Plant No. 1, for which task no less than five bosses
instructed several workers to fashion 12 covers out of lead sheets that had been dis-
posed of in the plant. These covers were not intended to protect anyone from
gamma rays; on the contrary, their function was to block dosimeters! *Build-Up*
executive Sagara [Teruo] . . . claimed . . . he only wanted to reassure workers . . .
when the dosimeters beeped. What a sweet boss! I wonder how many developed
cancer from his kindness.

In the construction of the future that we left you, was Build-Up the rule or
the exception? In other words, did our carbon ideologues more resemble Governor Matt Mead from Wyoming, who sought to *refuse to evaluate the effects of*
greenhouse gases, or Roger Taylor from Oklahoma, that open and I believe honorable man who straightforwardly allowed, although he made his living from it,
that "everything that's fuel's got a problem with it," and that "I think we're going
to have to step back from it"?*

---

* See above, p. 479.

The answer is sadly clear. As the translator of an oral history of the Chernobyl disaster noted: *Much of the material collected here is obscene.* I feel the same about many accounts compiled in *Carbon Ideologies.* (In particular, the anecdotes from West Virginia possess for me a cumulative nauseating power.) Of course the *regulated community* saw matters differently.

One of my favorite dioramas at the Phillips Petroleum Museum in Oklahoma had to do with technocrats at their beakers and switchboards, *maintaining a competitive edge.*

To *maintain a competitive edge* requires amassing advantages for oneself and disseminating disadvantages to others. That edge gets blunted by taking care of neighbors, grandchildren and the future. Some people would do so regardless. As for the others, such as Don Blankenship, well, in a way, their motives were pure. One can hardly fault a crocodile for doing what it does. But when the supposed regulators put the *regulated community* ahead of everyone else, that beggared belief.

"Maintaining a Competitive Edge": display at the Phillips Petroleum Museum

The Union of Concerned Scientists details how even after Fukushima the U.S. Nuclear Regulatory Commission, while voting to require the installation of containment vents, declined to require that whichever radioactive gases might someday emerge from those vents be filtered! (As the Commission explained:

The NRC considered revising its regulations through the rulemaking process to include strategies for filtering or otherwise confining radioactive material that gets released as a reactor core is damaged. In August 2015, the Commission directed the staff not to proceed further with the rulemaking.) Indeed, the NRC seemed to prefer a gentle vagueness to any specific rule. In 2012 it called on nuclear plants to *provide reasonable protection for the associated equipment from external events,* without defining what "reasonable protection" might be. The UCS likened this to a posting a highway sign saying **DON'T GO TOO FAST** instead of **SPEED LIMIT 55.***

The actions of the West Virginia legislature following the MCHM spill were worse. As you may remember, once the water supply of 300,000 people had been compromised, those callous hacks deliberately exempted almost every aboveground chemical storage tank from oversight.

But it improves nobody's mood to grant these details undue significance. I prefer to quote a pro-nuclear Assyrian from 1971, when climate change remained happily hypothetical: *We . . . admit to being energy optimists who see in unlimited energy a means for supporting many more people than now live on earth.*

LET ME COUNT THE WAYS

Now I must finally rank the four carbon ideologies in this book.

Nuclear would have been best, had there been no risk of accidents, and no longterm radioactive waste. From an economic point of view, this fuel had unstintingly given to the cities and villages of Fukushima, and its thermal pollution appeared insignificant in comparison to the climatological effects of carbon combustion. Although at the end of 2016 our planet clung to mediocrity at a mere 449 reactors—only one more than before the Japanese nightmare—China had just built another five, *the largest ever annual increase in China's nuclear history,* with another *more than 20* on the way!—My own country continued to be

---

* When I brought this passage to the NRC's attention, a public affairs officer replied: "In the years following the Fukushima nuclear accident, the NRC carefully examined the available evidence with detailed analyses. The agency used these analyses to determine appropriate actions for enhancing the already acceptable safety of U.S. nuclear power plants . . . The plants have put in place additional resources and most, if not all, of a suite of additional procedures to reduce the already very low possibility of an accident releasing radioactive material into the environment . . . In short, the NRC certainly imposes well-defined safety limits on U.S. reactors, without prescribing a single solution for complying with the limits. The reactor owners must then prove to the NRC's satisfaction that their methods will meet those safety limits. To use the car speed limit analogy, the nuclear power plants must show their brakes will prevent the reactor from exceeding the limit, or that their engine (reactor) cannot exceed the limit, or that another set of systems will keep the reactor under the limit."—I am grateful that he answered me at all.

the number one nuclear electricity producer in absolute terms . . . at 839 terawatt-hours,* or 20% of the global total, because the US-EPA Clean Power Plan of 2014 *aims to reduce carbon emissions.* How noble of us! That same year, 42 of Japan's 54 reactors were again *operable and potentially able to restart, and 24 of these are in the process of restart approvals,* there being no immediate danger.

Coal would likewise have been wonderful, had it only not been poisonous, and its emissions anthropogenically disastrous—and had its seams remained inexhaustible, with its local payouts as fair, generous and continuous as the Japanese nuclear industry's. While revising this paragraph of *Carbon Ideologies* I met a lady whose sister lived down in a shady West Virginian holler where solar would never, ever work! The sister had no good alternative. And on that same day I learned that *coal is the dominant fuel in the Asia Pacific region, accounting for 49% of regional energy consumption.* So the Bangladeshis might have been quite right when they insisted that today or tomorrow it will have to come out.

Natural gas would have been the most perfect of the three carbon-burners (which is to say, still dangerous), if we could only have prevented methane leaks. The pleasant Community Development Director of Greeley, Colorado, had reminded me that "it provides a job both at the construction phase and at some sustained jobs, and it provides a tax function, directly through severance taxes and through the multiplier effect. One of the things that many residents are proud of is the energy independence aspect of it." The same advantages and more were preferred by the Japanese nuclear industry. Were these fuels locally convenient to hand, I could well imagine Coloradans going nuclear—doubtless with inadequate radioactive "setbacks"—and the Japanese fracking their islands, but paying community subsidies of un-American generosity, meanwhile beautifying or at least kitschifying each wellhead.

As for oil, without that, what would have become of our mobility? *Oil remains the dominant fuel in Africa and the Americas, while natural gas dominates in Europe and Eurasia and the Middle East.* As I wrote this paragraph I was sitting upstairs on the Amtrak "Chicago Zephyr," smelling diesel and watching frosted cornfields give way to closed-up brick buildings, then to suburbs; how was I supposed to be carried across the country, if not for oil? No good alternative there!

So all four fuels won the race, first place, hurray.

---

* In our units, 2.864 Q-BTUs.

ABOUT IMPROVEMENTS

Truly we needed them all, like a diabetic who meant to someday get healthy, but couldn't possibly give up whiskey, fruit juice, bagels and chocolate cake— because there might be a magic pill, or even an operation, to save the patient from reducing demand! In our case the pill was called *technological innovation*.

The efficiency of Thomas Edison's light bulb in 1879 was 1.5 lumens* per watt. The efficiency of a standard warm white 100-watt fluorescent Mazda lamp, *circa* 1958, was 52.0 lumens per watt—an almost 35-fold improvement. If our scientists could only keep increasing efficiency more rapidly than we increased demand, we'd be home free!

On lucky occasions, heightening efficiency might in and of itself reduce greenhouse gas pollution. From 1986 to 1992, improvements in engine design lowered "global airplane transport" emissions per unit volume of fuel by 16%. Other advances reduced the energy needed to accomplish certain work, as when we invented Bainitic steels, which could achieve 3.6 times the strength of ordinary steels, *without a significant change of energy required in processing.* That meant we could get by with less steel in our next skyscrapers and bridges, reducing energy demand still further. That was why increasing Gross Domestic Product had begun to decouple from increasing power consumption.[†]—So why not trust in perpetual future improvements?

Indeed a report to the Parliament of the European Union happily proclaimed that

> energy efficiency should be the first energy source as it plays a key role in speeding up the clean energy transition . . . and contributes to the EU's security of supply. Energy efficiency . . . has become a sustainable business model . . . EU-28 primary energy consumption dropped by 206 M[illion] t[ons] o[f oil] e[quivalent] in 2005–2014 mainly due to an improvement of energy intensity. In other words, primary energy consumption would have been 23% higher in 2014 without the energy intensity improvements made since 2005. This level of energy saving has reduced consumer energy bills and . . . reduced greenhouse gas emissions by around 800 million tonnes[†] [of CO_2 in 2014].

---

* A unit of what we "consumers" called brightness. For the definition, see I:579.

[†] See I:54.

[†] 882 million U.S. tons.

Unfortunately, the most ingenious efficiencies must eventually reach their thermodynamic ceilings. As Gutowski and his colleagues wrote in 2013:

> The number of people in the world with basic needs still unmet is enormous. Of the current seven billion people on the planet, only approximately one billion are in the high-income category (i.e., gross national income per capita greater than approximately $12,000), and approximately three billion people are below $3,000 . . .

Manufacturing's *future improvements,* they warned, would be *limited* because its *efficiency has already improved significantly. In fact, the basic processes to make* the "big five" materials

> have been in place for a long time [~80 years for some plastics (the newest materials on the list) and more than 200 years for iron and steel] . . . During this time the primary processes have been improving, and the very best are now approaching their thermodynamic limits. For example, the best available smelting processes for iron and aluminum are now in the vicinity of 55–65% efficient.

Well, why couldn't they become 100% efficient?—For much the same reason that our power plants wasted two out of each three BTUs they burned. (Gutowski disliked my calling that waste, because it was inevitable; it could not be helped; all the same, waste was what it was. Friction is waste, and so is entropy. They are as inescapable as death.)

With great cunning and luck we might someday approach 80% efficiency in making the "big five," which would reduce their energy inputs by 37%.* After that, progress would dead-end against the ceiling.

Comparably sad limits applied to all other forms of thermodynamic work.

Fortunately, we had carbon to burn. In 1980, Earthlings "produced" 59.6 million barrels a day of crude oil. In 2013 we "produced" 76.1 million barrels a day. There was plenty more left!

---

* Gutowski et al. proposed that rising Third World demand might cause global production of the "big five" to double, while absolute energy inputs must fall by half in order to hold the line against climate change. Hence, "we have looked at the possibility of reducing the energy intensity of material production by 75 per cent by 2050, and found that this appears very unlikely."

COULD SOLAR HAVE SAVED US?

In the previous volume* I touched upon the solar ideology. Why shouldn't the expert who had helped me there solve all our problems forever, right here? Again I "reached out" to him. He answered:

> Regarding the question "If the whole world went fully solar tomorrow, how safe would we be from global warming?": . . . I'm not sure what you have in mind when you say "the world went fully solar," since solar is primarily converted to electricity and electricity . . . accounts only for a portion of the different energy sources required by current technologies for heating, transportaton, industrial applications, cooking, etc.
>
> However, if (this is a BIG IF, e.g. could we electrically power a transatlantic plane?) we could move all our technologies to be electrically powered, and [if] all this electricity came from the sun, then perhaps . . . we could indeed go back to pre-[I]ndustrial [R]evolution greenhouse emission levels . . . [—n]otwithstanding that greenhouse emissions from agriculture, forestry, etc . . . . would not be curtailed by moving to solar energy† . . .
>
> On a more realistic scenario, I can only point that electricity [demand] . . . accounts for around 20% of . . . total world consumption of energy, with oil (40%) being the most consumed resource (around 64% consumed in transportation) . . .
>
> If you note that the Paris agreement calls for a 20% reduction of greenhouse emissions in the short term (from around 50 G[iga]t[ons] to 40 Gt) to contain global raise in temperature to about 2°C[,] . . . it is clear that if we could stop using fossil fuels for the generation of electricity, and move to green renewable sources (solar, wind, etc.), then we would be [i]n a very good position (according to current models) to contain climate change in the near future. In the long term, it is important to note that as a species, the goal really needs to be zero net emissions (even perhaps negative net emission using carbon sequestration schemes).

Ah, yes, the Paris agreement! Rolling the rubber band off my morning newspaper, I read:

---

* I:161.

† Here he more optimistically inserted: "Although [these] greenhouse emissions by agriculture, . . . etc. account for around 24% . . . EPA . . . points that . . . 'sequestering carbon in biomass, dead organic matter, and soils . . . offset[s] approximately 20% of emissions from this sector.' This is really important, from an environmental impact perspective."

TRUMP ABANDONING GLOBAL CLIMATE ACCORD

"IF PEOPLE WANT US TO LOOK AT SPECIFIC QUESTIONS"

So we fought thermodynamics, maintained our *competitive edge* and fulfilled demand. Call us heroes—although in eastern Kentucky certain public meetings about health suddenly revealed what the newspaper hilariously called *a surprising concern:* mountaintop removal. Who would have guessed?—"It was something that people were afraid to talk about," said an eyewitness. By a fluke the director of the Centers for Disease Control was present, and somebody even inquired as to whether he might consider investigating the morbid effects of that practice. He replied: "CDC only goes where it's invited. So if people want us to look at specific questions, we can look at them."*

When I lived, these cases made me sad. You from the future who read this will surely smile with your cracked lips, for the tragedies I proffer have to do with single integers—or a mere thousand or two. You who must have seen death everywhere, and who may perhaps wonder if your line ends with you, if you laugh at me, I'll laugh right back, not to mock you but to show my joy that I lived and died before you had to.

A ROSE MILKSHAKE

Indeed, I tried to live well! As Archie Dunham had remarked, "We're human. We like a hot shower every day."

When my daughter, home on break after her first semester of college, agreed that it might be fun to replicate one of the rose milkshakes I had enjoyed in the Emirates, she turned the key in the ignition of the little white car, and off we happily drove to the ice cream parlor—two miles each way, so that four pounds of carbon dioxide got released from 17,000 BTUs' worth of "regular" motor fuel. Her driving had improved, her parking was perfect and she knew the way. I felt lucky to be with her, just the two of us, after several months of missing her, which I had done my best to hide, in order to help her get free.

It was two days before Christmas, and unseasonably warm (for which we could not blame the jihadis whose oil we considered worth fighting over), so we

---

* In case you are wondering how that issue turned out, let me once again quote the newspaper: "The Interior Department has ordered a halt to a scientific study begun under President Barack Obama of the public health risk of mountaintop removal coal mining."

found a long line at Gunther's: parents and children, boyfriends and girlfriends, girlfriends and girlfriends, all happily licking ice cream cones and taking selfies. In default of rose ice cream we settled on vanilla; then my daughter picked out a pound of mint chocolate chip for her mother. I enjoyed watching the strong, healthy counter girl scoop out our ice cream from the freezer case so that her biceps and shoulders swelled. My father had once worked at such a place, before I was born. In his time the employees were called "soda jerks." The owner allowed him to make himself any milkshake he liked so long so he drank it. That had been one of his life lessons.

When my daughter was a baby, I sometimes pushed her stroller down the sidewalk, slowly, peacefully bound for ice cream. Once she could walk, if my parents were in town, we all sometimes strolled down to get a scoop of this and a cone of that. My father would watch the crew of soda jerks and smile a little, remembering what he used to be. They were local high school and college kids. After my father died I felt more sentimental about this aspect of small town American culture, especially as I got older and visited such places less often for my own sake than for the pleasure of some child or young adult such as my daughter now was; in short, I generally forgot that while the "next to nothing" of ice cream's price surely addressed the economic cost of the electric power that had first mixed and frozen the sugared cream, then kept it frozen until the soda jerks scooped it out, the cost in greenhouse gas emissions remained unpaid. How many BTUs got thus consumed, and how much carbon dioxide, methane and nitrous oxide rose up I could not begin to calculate, knowing neither the volume of ice cream, nor the rate at which it froze (improving a kilogram of water into ice in one second cost 431.37 BTUs, but they surely chilled theirs down over hours)—not to mention for how many days it must consume freezer-BTUs until it sold.

So we got our ice cream and I stroked my daughter's long brown hair. Then we drove home, stopping at the grocery store to buy vanilla, nutmeg and milk—all shipped or trucked there, and the milk then refrigerated. How many BTUs had *those* eaten up? I already had the rose extract. We pulled out our electric blender and happily mixed, pulsed, liquefied, added and repeated, maybe for as much as five minutes: about 215 BTUs.

It was a delicious milkshake—all the more so in that during her childhood I had rarely passed as much time with her as I had wished, and now it was nearly too late; at Christmas she and her friend went off to an animated movie, to the tune of I don't know how many BTUs, and although she said they would not have minded had I invited myself along, it seemed more loving to leave those two girls to their so-called "alone time." Soon enough, if all went well, she'd be half of some couple; then maybe there would be children; and then how often would

I have her to myself? Hence from my point of view, making our rose milkshake justified its modest expenses.

When I think of you in the future reading this, my error is to imagine that you are similar to me. Being nurtured and pampered by electric power, I owned full leisure to emote, consider your point of view and even feel goodwill toward you. I wish to understand you, and to help you understand me—as if that could somehow distract, console or at least enlighten you about the hot dark world in which you dwell. If you could end up saying, "well, yes, we might have made the same mistakes as you, if we'd been lucky enough to live when you did," I'd feel that *Carbon Ideologies* had accomplished some of its purpose. How you judge us can mean nothing to us who are dead, but to *you* it might mean something, to accept that we were not all monsters; and forgiveness benefits the forgiver, so why wouldn't I prefer you to call our doings mistakes instead of crimes?

But weary common sense, and my own experience in war zones and other desperate places, drags down these narcissistic expectations. Most likely, you are a hard, angry person. By my standard you must be somewhat uneducated, since we expended most of the magic that would have kept your lights on. How well can you read? For that matter, how well can you see? Do spectacles still exist? Beset by floods, droughts, diseases and insect plagues, unable to jet around the world as I did, but probably aware nonetheless that humanity's habitable islands keep shrinking; engaged in wars about food, and united merely in tribal hatred (your strongman against theirs), radiation-tainted, fearing for your children in the face of multiplying perils, how can you feel anything better than impatient contempt for my daughter and me, who lived so wastefully for our own pleasure?

If *Carbon Ideologies* travels all the way into your era, it may exist only in scorched or water-damaged pieces. If you are reading this fragment, perhaps you light yourself with a carbon-fueled lamp. How could you possibly possess the luxury of worrying about *your* future? You reduce demand by murdering your enemies. Like me, you hide from reality (in your case, from sunshine) in a cave.* Having skimmed bits of this, you might now understand some of our why and how, but does that appease or merely inflame your hatred?

---

* If you read this at all, it may be to distract yourself from your dingy terrors. Béla Zombory-Moldován, before 1968: "One could get used to catastrophe, too; in fact, one had to get used to it. The past was gone. If I had the time for it, and if I was safe, I might daydream about it."

RHONDA REED

Death, ere thou hast slain another
Fair and learn'd and good as she,
Time shall throw a dart at thee.

William Browne of Tavistock,
"Epitaph for Marie, Countess of Pembroke," 1623

You from the future, who presumably would go back in time and wipe us out (if by so doing you could avoid extirpating yourself), in this book I have tried to tell you "how it was" for us. Our self-serving ploys, and the low cunning of certain high executives, the apathetic greed of the horridly named "consumers" and the cynical short-sightedness of those who served them, the nationalistic loyalties of irradiated Japanese and much-abused patriotic religiosity of West Virginians, the ignorant desperation of Bangladeshi activists, the unhealthy toil of guest workers in the Emirates and the gleeful stupidity of adversarial political systems under which a politician would squander a planet in order to distinguish himself from the incumbent, all these I have now presented to you, leaving (I suspect) a pervasive impression of our depravity. You will not wonder that I exerted no influence! Who is grateful for another's disapprobation? I scolded and exposed follies, all the while burning carbon or fissioning uranium. I must have undertaken two dozen airplane trips for this book! Then I died, leaving the world worse than I found it.

Whether or not we deserved the extinction now swooping in on you, I would like to proffer a less unpleasing picture of us. (You might have brought your own death upon yourself, had we been unavailable to do it for you.) If you can pity the child who, not understanding what he does, shoots his brother dead with some adult's carelessly stored firearm, why not compassionate my kind? Some of us who possessed both power and intelligence did behave with malice. The rest had fewer capabilities for action than it may suit you to believe. And even though we raped it, our world was beautiful to us!

I remember the city of Aizu-Wakamatsu, whose radiation levels, despite its greater distance from Nuclear Plant No. 1, were slightly higher than Iwaki's—mostly 0.18 microsieverts an hour, although one fallen leaf on the gravel walkway to the castle was 0.24 micros, probably on account of all the exposed rock where

I stood measuring—what trivial readings! By the way, are gingko trees extinct in your time? I remember the rich stench of those gingkos, and the steep V-shaped paired steps in the stone walls whose parapets were now carpeted with grass, a persimmon-colored leaf face down in the gravel (there was my 0.24 micros), and the pagoda roofs of the white tower; on display was a lovely black helmet with lacquered wings, from the era, not unfamiliar to you, of wars without quarter; there was also a scroll depicting dragonlike lions, peonies and a waterfall.—Do you still have peonies? Your lions must be gone. We had a fine planet, back when we were alive.

I remember the railroad track in shadow by the Wharton Church of God, which wore a six-pointed Jewish-style star; and then we rolled into Barrett, where on the right across the bridge stood some of the best-kept lawns I had seen in the county; and here were company houses, all different colors, some with grand pillars, many flying American flags. The section down the road was more dilapidated, but still it had on offer the Bald Knob Methodist Church and a dark shady creek way down in the forest. Soon we arrived at a huge establishment of America's best friend, where from a vertical cylinder powdered coal kept spilling out onto a pyramid of powdered coal, mimicking sand in an hourglass. Lovely orange lilies bloomed along the road. Driving south alongside the railroad tracks in the steady cloudy lush hills which could almost have been Hawaiian, we reached another mountain of ground coal, which loomed up so black as to be almost blue: This was Eastern Associated, which bore some connection with the Harris No. 1 Mine and Gateway Eagle. The shedlike office of United Mine Workers Association number 1503 was closed like all the other union establishments I saw in West Virginia. Ascending the hill to an overlook, we found maples outspreading from a low cliff by the junction of highways 85 and 99, the turnoff to the Bolt Mountain, from which coal trucks now came snarling. Old Bill White from Bim* had once seen a peacock on Bolt Mountain. He'd told me: "You don't see no mountaintop removal from up there. What you see is roads from auger mining." He had remembered Cook Mountain from the days when it was still a *mountain* with forest on it, and he had visited "Chap" Cook's grave. He said: "Used to be able to go up this way, but the coal companies may have it blocked off."—Bill White had been a coal miner, but I could never blame him for what the coal companies did. (I do wish I had thought to ask how he envisioned you and your time.)

I gazed back down across the valley and it seemed to be all trees! Just then West Virginia appeared to be the loveliest place on earth—hot and humid, to be

---

* See p. 157.

sure—even the daisies, yarrow and lush green insect-eaten leaves somehow brightly humid!—I suppose that you in the future can no longer survive in West Virginia; it must be a sweltering hell like Dubai, Karachi and whatever is left of Florida.—Both my hands were sweating. Peering across into the green atmosphere of trees, I thought of Jefferson's wilderness, with Virginia not yet subdivided, the hollers and rivers mostly wild, coal barely a promise. I could not see south into the next valley, so I kept looking north across the first green horizon and down into the valley, at whose far northern end a V-shaped notch lay below the other horizon, which was bluish mountains. I inhaled clouds, my arms sweating, the breeze barely moving, and everything seemed perfect, even the coal trucks convoying down from the Bolt Mountain road—but especially the flowers bursting out of all this hot wet greenness. Although it would soon be July, the raspberries were still hard red buds like a young woman's nipples. There came a brief breath of breeze, and then it got hot again there among the water hemlock, yellowjackets and vetch, with wetness shining upon miniature staircases of slate, and everywhere the outspread yarrow whitely reaching up from their dark green skeleton-stalks.

Continuing our drive, we came immediately over the pass into Wyoming County, where on the right the rich hill had been carved away as if for a ski lift, two power wires running up it because there was another coal mine, followed by Kopperston's one- and two-storey company houses, and then a long winding row of trailer homes, addresses such as Twenty-One Row Hollow Road. I wish I could boast that my life had been as long and narrow as Kopperston! Certain lawns reminded me of nice-kept Barrett. On the right we passed slightly more prosperous houses, then a few opulent ones just before the end of town, followed by the Pioneer Fuel Coal Mine.

After that we rolled into Oceana, Wyoming County's oldest city. It was 93° that afternoon, with the Sportsman's Grill unfortunately closed, as was the funeral parlor, and the next UMW office—a fine brick one, number 764.

Just past 5:00, with the temperature declining into the sweet 80s, clouds and mosquitoes comprised the main attractions of Pineville, although on the lowest terrace of that town's steep-sloping courthouse lawn, whose grandeur was almost shocking, the Ten Commandments rivaled them, having been pricked out most conspicuously upon a granite stele. Frisking that Judeo-Christian monument, I got my highest exterior West Virginian radiation reading: nearly half a micro an hour*—twice as much as that fallen leaf in Aizu-Wakamatsu. Not yet having measured the Japanese red zones, I thought this reading was really something!

---

* See I:247.

The black sheriffmobile followed other shiny cars and pickup trucks across the river bridge. Castle Rock was darkening down in consort with the trees upon it and the forest behind it; mosquitoes shot into my forehead and neck; birds sang along with the cooling breeze.* The lone streetlamp by the courthouse steps now began to shine with surprising brightness. The windows of the Pineville Pharmacy (**Rhonda Rose, RPh**) had gone dark; likewise ATV Parts & Accessories beside it and the three-storey brick building on the corner where Lambert Law constituted itself. The streets were nearly empty; thus another hollowed-out West Virginian locality—more hollowed out in your time, I'd bet. Pineville's lovely humid twilight was especially marked by mosquitoes crawling on my arms and the Family Dollar sign slowly gilding as if it were a church steeple. I asked the hotel clerk where to get moonshine and he said, "Rite Aid. Will do you good." At Rite Aid the teetotaler checkout girl was patriotically disappointed that I chose Georgian rather than West Virginian moonshine, which was mostly adulterated with vodka. Now came time to settle in with the mosquitoes and old cigarette smoke in the motel room.

Next morning I met a man in that town who said he often used no electricity.—"We had it off here for a few of them summer months. One white oak about *that* big fell down, and I had to get two big men to help me cut it up. It took awhile, with hand tools."—You would not object to him, I suppose.

Just outside of downtown, one curving hill-road met another which presently led to the forest-enclosed home of the Reed family. Rhonda Reed, who had not yet reached the appearance of middle age, was there visiting her parents. "Where we're at now," she said, "if I was to throw a rock I could hit my house."

The greenery was so wonderfully lush, I couldn't see any other house.

"In this kitchen," she said, "Dad made the cabinets. The baskets, that's Aunt May's; that's something she really got into. And we kept that phone; that's from an old mine."

I wish that you from the future could have met her. In default of that, I will tell you some of the things she said. She knew her world's roots, birds, rocks and hills. I cannot find her complicit in their loss.

---

* I once investigated the birds, mosquitoes and cloudy humidity on that noon at the lichened, layered chimney called Castle Rock, from which one could look down at the courthouse on one side and the river on the other. To be sure, there was garbage down by the river, which nonetheless looked pretty and hissed nicely, and there were some nice steep-roofed two-storey houses, some of them maybe from a coal company, down in a holler. On this high mossy rock-island strewn here and there with dead leaves, the air dose was 0.18 micros. There was grass in some parts, worn down to soil in others, and then the chimney rose straight up out and up toward the lovely white sky, narrowing and then widening like a mushroom, with trees growing out of it.

About the MCHM spill* she said: "I was like, for real? Because the company gets too greedy, don't do what it's supposed to. It's about money. Too much money is a bad thing. What happens if the system shuts down? Money will be worthless."

It was not in her to say that the company should be punished. I'd guess she was a good Christian who would never cast the first stone.

She had kind words for almost everybody. She never professed to know what she did not understand. What she did understand was her home country. She said: "If your digestion's out of sorts, pick up some ginseng and gnaw on it, 'cause it settles your stomach. Now yesterday, my man's Daddy done the gardening finally, and there was some pokeweed. You snap off the stalks, and you can deep fry it. Now I've got pokeweed all around my house for the sole purpose of you can eat the leaves and shoots for vitamins; and the berries, the birds will eat those. There's a lot of stuff in these mountains that will kill you and cure you, and it's the same plant. I grow rhubarb, strawberries, three kinds of beans, six kinds of tomatoes, four kinds of corn, six kinds of cucumbers, squash . . ."†

"It sounds as if you eat very well," I said. "Why do you think in Pineville there are mainly low-quality chain restaurants?"

"Laziness," she replied. "Society, as each generation comes on, they get actually dumber. Their knowledge of the world, of how to fix things, a lot of kids, they have no clue as to how to fish, how to hunt, how to take care of animals. You got to have respect for your elders, like Mom and Dad. Each one steps up as we go. We're a hierarchy. We're a clan. We keep it going."

She kept praising all the generations of her family. "Now my grandson Jesse," she said proudly, "the first time he went deer hunting, when he was 15 or 16, he got a buck—beautiful, first shot!"

She said: "Every chance I get, like with the kids, I try to teach 'em old ways. In the spring, you get the dandelions, pokeweed, ramps; it's the greens you pick in the springtime for the sole purpose of getting your body ready. Wild animals, I have blackberries growing around the place for the sole purpose of putting 'em off. Leatherbritches, you take your green beans and put them up on a string until they can dry. In the wintertime you wash 'em off one time and then you boil 'em, and they're tough, but you cook 'em, and you have that with cornbread, and it's that good. You can dry 'em within four or five days.

---

* See p. 169.

† Her two lists of plants had been longer, but several names were localisms rendered more unintelligible to me by her beautiful Appalachian drawl. Lacking opportunity to meet her again, I simply omitted what I misheard.

"I don't like buying new stuff. That just grinds me something fierce. It's like buying our land. I sweated for six months, I got viciously sick, but then I had $12,000, and it's paid for. Work for Cyrus* has been sporadic, because I said to him, you're not going in the mines—because, you see, there's the *fear* of that. He does gas and oil.

"My grandson and them, they're cohoshing and such. My grandson makes more money with his cohosh.† Out of the freezer come deer, squirrel, peppers, apples, peaches, pepperoni. Everything but the pepperoni come off of my land. What he hunts, he guts 'em in the woods; that way it goes back to the woods. The tails are hanging back on my back door.

"My grandfather, Dad's Dad, he taught me the weather. I know when it's gonna rain, and not from my bones aching. Mom's Ma taught me quilting, gardening, all kinds of things. When I was young, I would have muslin cloth and dye it and make pocketbooks. My grandparents never had to whip me. It's like I know the old ways. That's what they call it. I took a little bit of nursing, but I got too involved emotionally, in the blood part . . .—but it was not wasted, because the knowledge I did have, I can use it for my family.

"Some of my elders, they still had the Irish and Scottish accent, so when they spoke, you had to know 'em to understand 'em. They did mining. When it first started—consider where they had come from!—it was a very good life. But with our age group, it can be very good but sporadic, because of strikes coming on and so forth. When a company works good, it can be good. But I've known families because whatever they were paid, they were paid in company scrip. But a lot of families did well and were happy.

"We're Scots Irish. You notice our songs are sorrowful and sad. And that was right, because when you kissed your husband goodbye in the mines, you never knew if you'd see him again. Life was so rough. Granddaddy was born in 1899. Let your imagination go back to 1899. You hoped your woman was going to grow what you needed, so you could get through the winter. But in my 56 years, I've watched too much change, and it's not for the good.

"Coal mining's not as much as it used to be. A lot of that's Obama, the rules and regulations—and some of that is good: It just tears your heart when they take the tops off the mountains. It would be different if they could use it for a purpose, fix it up where animals could come and eat, but it's not like that. Now we've gone to somewhere we used to go four wheeling, and it's changing; the

---

* Her man.

† A generic name for any of several medicinal plants.

Boone County, West Virginia

soil's not the same. I can remember as a young teenager going way up in the hol-ler, and them big dumping trucks had filled it up a mile and more. I cried; I said, just take me home. The coal in the ground, I said, if you got the guts, go for it.

"Dad worked on the outside of the mine. When I die I want to be outside.

"Most of 'em they seal off, due to the danger. When I was a little kid and we lived on the other house on the hill, there was an old mine that I researched and found it was from the 1920s. We went down far enough in there that we could stand up.

"Now something that went through for awhile, that was the gas fields. That was what my man did about seven or eight years, and that's not here anymore. They done tapped that out."

"Do you believe in global warming?" I asked her.

"I think it's happening, but because of different factors. The Earth has cycles regarding whatever man does. China is doing so much to pollute! But the Earth breathes, and it does its own thing. I think we're in one of those circles. Every so often the earth gets shook off and that means things will change drastically. You had to have the one where the dinosaurs were. If that hadn't happened there might not be us. As far as evolution and such, it all starts with stardust. God says man was made in seven days. God created Earth, and then there was water; it took awhile, and then the ocean, and the sky, and then us. It was supernovas and stuff. That's what gives us all the elements, everything we need.

"The way I figure it is, the climate does need regulating to a certain extent. But also, coal does not last forever. Mankind has been here long enough. We need to change. I would like to be able to have to the money to unhook myself from the grid, with solar and wind. We've got a wood stove that we could use if we need it; we don't need very much gas, either. It's not bad being a little cool."

"Have you heard about the nuclear accident over in Japan?"

"I didn't really follow that."

So I let that drop, and asked: "What would you say about coal *versus* nuclear power?"

"I think coal. Nuclear, where it is, atomic energy, it's really not been here too long. To me the nuclear is not a good thing. It don't make sense. Coal has been here long before the Roman times. And coal *works*. You have to work with it. West Virginia has the best burning coal! But coal's gonna run out."

We had invited ourselves to the Reeds' house upon five minutes' acquain-tance. The whole family was sweet with us; we could have stayed as long as we wished. And Rhonda Reed, what else shall I say about her? Change was upon her; she would age and perish like the rest of us; but that was ordinary; how would it be for her when the land changed? Perhaps in her lifetime it would not be blighted like those greed-ruined mountaintops; for all I knew the greenness would for a time intensify, even supporting more animals in that humid lushness; those acres might still be home to her when she died. But change was in the carbon-laden air; she would perceive it sooner than I, who had become more of an indoor type. Yes, the coal would run out, but that was just the beginning.

I remember the sun coming out after an April rain, the grass going greener, the leafless trees just now beginning to bud and the bungalow houses of Mayberry along the road; I remember driving from Madison to Cook Mountain at 7:30 on another June morning, with the deciduous hills in a still blue haze. Past Madison the river turned deep and narrow and brown. The houses were less structurally beautiful than in McDowell, but better kept, with gardens and lawns, and more vehicles parked by them. A tanned shirtless man was already working his cornpatch. I remember the Quinland Freewill Baptist church, and the sun striking the wall of tree-haze ahead, so that it almost resembled blue dust. There were three fat hens by the roadside. The steep winding valley led past the Coal River Safety Center. Driving along the shady side of the valley, I looked across to serpentine twists of sunlight. I remember a settlement of double-wide trailers, and one nice brick house whose porch was heaped up with garbage; I remember the three-quarter moon in the morning sky. Here again was the town of Van, with a lady in a long grey skirt weeding her garden; a sign said Gospel Sing Tonight, and light came glaring through the glades. Cook Mountain Road felt cooler than before, probably because the morning was still early. On that walk I smelled grass and leaves; I heard birds and sounds of heavy machinery somewhere neither close nor far. Presently I came around the bend, where the ruination of Cook Mountain saddened me. But this had never been my place, so how could my heart be broken? By the time I got back to Madison it was not quite 3:00 in the afternoon, and 96° and probably 97% humidity. A young couple and another young woman, evidently a sister of the wife or girlfriend, were sitting outside the pharmacy, which also sold ice cream. I was hot; I was weary; so I went to the hotel and turned on the air conditioner.

THE ETERNAL CARBONIFEROUS

*Brownouts, blackouts and electric rationing reversed the "safety above all"
response . . . The people in great majorities prevailed and the industries
serving them stimulated their senseless choices . . .*

Philip Wylie, The End of the Dream, 1972

I n his less educated days, the forester-ecologist Aldo Leopold twice took well-
meaning part in eradicating predators. *I had to learn the hard way that excessive multiplication is a far deadlier enemy to deer than any wolf.* He was a kindly
sort who controlled the rage and grief he must have felt in regard to the doings
of his fellow humans. In an essay entitled "Game Management," he approached
but turned away from the logical consequences of what he had learned, writing:
*The hope of the future lies not in curbing the influence of human occupancy—it is
already too late for that,* and the sentence continued, but his hope was too vague
and gentle for me; mine, and it saddens me to say so, was for a massive decrease
in human population, by plague or war if birth control still refused to serve—
better the cruelty of biological decimation than the still more pitiless operation
of heat physics. You from the future who try to hook whatever sickly dinner
while rowing over your grandparents' house, wouldn't you be better off without
grandparents? Then you would never have been born into the world we be-
queathed to you. (Thoreau: *If we were indeed always getting our living, and regulating our lives according to the last and best mode we had learned, we should never
be troubled with ennui.* So you must be too anxious and tired to be bored; my
congratulations.) As for those few of us who wasted a thought on your miseries,
we might have enjoyed higher self-esteem had the race ended with us.

In our day, we busily denied the obvious fact that *excessive multiplication* was
our enemy as much as any deer's. As you know, we multiplied in two ways. First,
we increased our population. Second, we increased our demand for goods and
services, and hence for energy. (How many times must I repeat this? I kept hoping that someone would listen and "do something"—but not me, Lord!) Sometimes we improved efficiency or reduced consumption here and there. But just as
a dieter who keeps eating his daily fill of cheese, pastries and ice cream will gain
weight despite the laudable fact that he put broccoli on his lunch plate last

Thursday, so it was that we with our occasional energy-frugalities could not save ourselves. Consider this datum for Canada: Between 1990 and 2008, total carbon-dioxide-equivalent emissions increased by 24.15%. What if her consumption of fossil fuels and feedstocks had decreased marvelously per capita, and only risen absolutely because the population had grown? That made no difference. All that mattered was that emissions had increased.

But didn't we have the right to improve our standard of living?—Citing two studies, Gutowski concluded: *Household energy use grows in proportion to disposable income, and this pattern appears to be independent of householders' stated ethical values.*

Why shouldn't that truth about householders have applied equally well to real estate developers, corporate executives and municipal planners? Unoccupied by honeybees, larvae or even guest workers, the concrete honeycombs of that petro-city Ruwais grew nonetheless—in anticipation of demand. Meanwhile, two emirates northeast, Sharjah kept improving its "built environment":

Improving the "built environment," Sharjah

Some people there reminded me of Rhonda Reed. I remember a Pakistani driver who picked me up in Dubai's "Internet City" district; he was bearded, gentle and softspoken with those lovely liquid consonants in his speech. We chatted about his home region; he was a Pathan from south of Peshawar, and I fondly remembered that region—where they had now set to building coal plants, he said. Soon, I supposed, the air would be unhealthy, and various watercourses poisoned and acidified. He had heard (although of this both he and I lacked firsthand knowledge) that in India farmers were being dispossessed of land, for the greater good of industrialization. In his area, farmers were still "a little in good condition." But "everyone says they must *rush to develop,* because people have only six hours electricity!" And what would electricity's work be for? He remarked: "My grandfather didn't need cell phone. And you know, we were happy. One hour watch black and white TV, then all talk together. And now when I go home, I have only 15 days in a year to go back and talk with my cousins; all the time they are looking down, texting on social media, and I feel sad."

But we were helpless victims; what could we do, but multiply demand?

In 1875, more or less, Americans burned the most fuelwood ever. The calorific efficiency would not have impressed a nuclear engineer; the air sometimes grew smoky; but logs were abundant enough to come out near-perfect in that era's cost-benefit analysis. Presently we began to use less wood and ever more coal and oil. Pound for pound, these fuels did better work for us: From white pine to commercial fuel oil, the heating values almost doubled!—"The peaceful atom" contributed its carcinogenic, mutagenic mite after World War II, and in the 21st century we went wild for natural gas. When I was alive we exchanged one carbon ideology for another. We never reached the peak; the eternal carboniferous drew us on!

Some got what they wanted, while the rest could hope for equivalent riches. I myself, an American born in the mid-20th century, enjoyed the best life that carbon could give. My parents raised me to believe that if I worked hard enough I could acquire much of what I longed for, and then manufacture new desires. Our social planners preached that a healthy economy must grow forever.—If only it could have been so!

Storage tanks in Cushing

In the first pages of his celebrated *Elements of Chemistry* (1789), Lavoisier remarked that for a young child investigating the physical world, mostly by means of sensation, *want and pain are the necessary consequences arising from false judgment; gratitude and pleasure are produced by judging aright.* And indeed, a hand in the fire will infallibly lead to suffering.—But what of those phenomena whose exploration requires reason instead of touch? We could misjudge any number of them and still escape unpleasantness. For centuries, clinging to the errors of Aristotle—who dared doubt that genius?—furthered our scientific careers. And for kindred causes, even when I was alive it still paid to deny climate change. I myself longed to reject it. *Mass extinctions,* ran a technical article from 2009, *typically involve runaway greenhouse warming, major changes of acidity of air and water, dramatic increases in light carbon isotopes, and anoxia over hundreds to tens of thousands of years.* I hated that; I preferred not to worry about "some ecosystem somewhere."

In my corner stood Mr. Dave Gillihan, *an old-timer with a thick handlebar moustache who's been working for different companies for 42 years at the proposed site* of an coal-export terminal meant to serve China and suchlike irrelevant ecosystems. To hell with China—it might pay *us,* right here in Longview, Washington! According to the newspaper, *the project's operations and the combustion of its coal would increase global carbon emissions by 37.6 metric tons* over a 20-year period.* But what were a few puffs of invisible gas on the other side of the world? "Just ask the community," said Mr. Dave Gillihan. "Talk to the local guy that works. The taxable income alone is going to provide so much for our schools and our roads . . . I don't understand all the opposition, and all the people coming in from the different communities to oppose something that's going to happen in my back yard."

One might accuse Mr. Dave Gillihan of "lacking imagination," of not being able to see the general interest.—For that matter, a carbon ideologue might well attack *my* alarmist selfishness. What about the Chinese? Didn't they deserve what we had? Calculate back from 37.6 metric tons of carbon dioxide, plug in some good old West Virginian bituminous coal, and you end up with 420 million BTUs! That would cover the energy bills of not quite 17 Swiss citizens, but in China it would stretch further, I'd bet; how could I begrudge our fellow Earthlings so modest a demand?—But I was stubborn; I insisted that for China to get what I had would be against the general interest. Turning my attention to Mr. Dave Gillihan, I demanded: Why couldn't he recognize that what happened in his back yard might contribute to *anoxia over hundreds to tens of thousands of years?*—But what if his affliction were actually *too much* imagination? After all, continued Lavoisier (who predictably got guillotined for his projects and associations), *imagination . . . which is every wandering beyond the bounds of truth,*

---

* 47.51 U.S. tons.

joined to self-love and that self-confidence we are so apt to indulge, prompts us to draw conclusions which are not immediately derived from facts; so that we become in some measure interested in deceiving ourselves. So, if you like, Mr. Dave Gillihan simply "imagined" that the incomes and profits of Millennium Bulk Terminal would outweigh the cost. As for me, my own self-love directed me to keep spending electrical power; my self-confidence urged me to hope that I could do so indefinitely—because it was so wonderful that way!—Billions deployed kindred wishes.

"How's the drought here?" I asked a longtime resident of Redding, California.

"Shasta Lake, it's been 10 feet from the top. You get one good hard rain and it'll fill up. This is the first year people's wells have run dry."

"Is that from climate change?"

"Yeah, but I don't know if it's due to what we're doing. I could take you out to creeks and you could see dinosaur tracks. I was out there at Bella Vista goofin' around, looking for obsidian, then I found a wide deep hole, and then another 10 feet away, and then another, and finally I understood what I was lookin' at . . ."

Although even Chevron eventually admitted it,* my new friend *didn't know* whether climate change was anthropogenic; he believed in dinosaurs, had seen their footprints for himself, accepted that they had gone extinct and was willing to consider that prehistoric climate change might have killed them off, in which case we bore no provable responsibility for the current warming trend—all reasonable so far as it went, and I admired him as I do all persons blessed by stubborn self-education.—But his reasoning ended there. If climate change had wiped out the dinosaurs, why couldn't it get us, too? And if he *didn't know if it's due to what we're doing,* how should I feel about his apparent lack of interest in finding out? In short, his observations and analyses served to mask his preconceived conclusion. *Want and pain are the necessary consequences arising from false judgment*—but in this case they might well pass over the judger (who was old), and afflict his grandchildren.

The Intergovernmental Panel on Climate Change now warned us:

> While previous long-term droughts in southwest North America arose from natural causes, climate models project that this region will undergo progressive aridification . . . Because of the very long lifetime of the anthropogenic atmospheric CO_2 perturbation, such drying induced

---

* "Chevron . . . recognizes that the use of fossil fuels to meet the world's energy needs contributes to the rising concentration of greenhouse gases . . . GHGs contribute to increases in global temperatures."

by global warming would be largely irreversible on millennium time scale.

They might be wrong, of course. Wasn't everybody else?

So we kept right on wishing—and our engineers sought to satisfy us. They were never wedded to carbon, as proved by their triumphs at Fukushima and Chernobyl.*

Let's revisit that matter of improvements; what if thermodynamics *could* be tricked? In 1927 they sent their vision-rays down into the earth. *A cubic mile of rock cooled 100 F would yield as much heat as 180,000,000 tons of coal.*[†] *Nothing practical has been suggested in this direction.*

They even did their best to improve the efficiencies of solar power. Three nuclear-loving Democratic Socialists opined: *The problem with renewables is that they are intermittent, diffuse, and fluctuating forms of power. Consequently, they require hundreds of times more space to gather this diffuse energy than coal or nuclear, which offer far more concentrated forms of energy.* In the winter of 2010–11, they said, there were days when a certain German solar array *never* at any point[†] *got to 1 GW—or 6 percent [of] capacity factor.*—And then a book on "the promise of efficient technology" assured us that *on one bright summer weekend day in 2012 the majority of the electricity generated in Germany*[§] *was provided by solar power.* So I wouldn't be surprised if you in the future had worked out efficient solar energy generation. Perhaps your solar-powered pumps have not yet failed to keep the ocean from overtopping your diked-up cities.

The Japan Times proposed that sodium-cooled fast reactors might able to *burn up uranium's most long-lived radioactive waste products, reducing the need for deep storage.* Who was I to say that they wouldn't be even more safe and efficient than Tepco's Reactor Plant No. 1?

We kept hunting miracles. Why disbelieve? Hadn't carbon been one? Maybe we could keep burning it and then somehow "sequester" it. Oh, yes! We remained *in some measure interested in deceiving ourselves.*

Belatedly we mentioned conservation. But there was ever so much against it! Back in 1974 the Organisation for Economic Co-operation and Development

---

* Even the Phillips Petroleum Company, so adept at "Commercializing Creativity" in "The Spirit of Service," considered branching out; in Oklahoma I have seen an ore sample from the Phillips Uranium Corporation.

[†] This figure equates to 4.5 quadrillion BTUs, which could have kept the Montréal subway system in motion for more than 4,700 years at the latter's 2009 energy requirement, or powered the 1997-era intensive care of nearly 1.9 million American coma patients for a year.

[‡] Italics in original.

[§] This was 22 gigawatts.

(Paris) had proposed slowing down driving speeds from 115 to 80 kilometers an hour.* This would have reduced fuel consumption by 25%. Once upon a time, even my nation attempted such a thing. In 1975, a study of American truckers foresaw catastrophe: *A shift in speed from 65 to 55 miles per hour might result in a reduction of miles generated per month for long-haul owner-operators of approximately 13 percent.* And of course the trucking companies got paid by the mile.— Well, we did try to be good, like the old lecher who was proud of having used a condom once. Although most drivers disobeyed it, we kept that law on the books for 15 years. First the states exerted their rights; finally the federal government gave in.† As I write *Carbon Ideologies* I try to remember just how and why it got repealed; just the other day I had the pleasure of being carried at 75 miles an hour on a California freeway . . .

When I was in high school, the Organization of the Petroleum Exporting Countries raised the price of oil. That was the only reason we concerned ourselves with driving speeds. Then the price of oil went down again. NASA had been attempting to make airplanes more fuel-efficient, but *the "extra mile" in fuel efficiency the advanced turboprop provided was no longer required.*

No longer required. Doesn't that bear repeating?

And so much new thermodynamic work awaited accomplishment! *High levels of illumination have been found to result in an actual saving of dollars and cents owing to the increased production and decrease in spoilage produced by their use.* Well, well. If there was an *actual saving,* not just a projected one, who could argue against it? We'd better illuminate the whole world!

In 1990, Canada's "forest land" still sequestered 84,000 kilotons of carbon dioxide. By 2008 only 25,000 kilotons remained there. Well, I'd call that pragmatic conservation! Were the Canadians supposed to idle their forests forever?

Whenever we had to actually *support* conservation, we kept our goals realistically low. That way we never got tired. In 1981, an American report called *Auto Transit and Cities* vexed itself in trying to figure out how to save energy through improving mileage, encouraging mass transit and raising fuel prices. The authors concluded: *All these efforts might save a quad‡ or two or even three, or roughly 2–4 percent of total U.S. energy consumption at mid-1970s levels.*—You from the future will probably shrill out that we should have reduced our energy consumption by

---

* 69 and 48 miles per hour.

† As for the truckers, one journalist concluded that "the era of the independent owner-operator largely ended when fuel prices began going crazy in the 1980s . . . Now it's just a shitty job . . . Since pay is by the haul, the incentive is to break the law, fake your logbooks, which everybody does."

‡ The reader may recall that a quad is 1 quadrillion BTUs. At the average gasoline mileage of that time, a quad would have powered 200 million American cars for about 455 miles apiece.

50 or even 75%, but we just couldn't. Instead, we advised people not to turn their heat too high or their air conditioning too low, which proves that we cared. We raised the price of water. Sometimes we even nudged the *regulated community.*

"The trouble is, you'll never regulate your way into a clean environment in my mind." The speaker was Tom Jones, who had described so vividly the effects of coal-mine-induced acidification upon the waterways of West Virginia.* Even if he were wrong, and we could have diminished disagreements between regulators and industry, shamed the wastefullest carbon-burners, helped companies reduce regulatory compliance costs while improving their "competitiveness" against whomever had not yet reduced emissions, still, unless some plague came to the rescue, increasing demand[†] would inexorably undo all our good:

THE TRIUMPH OF CONSERVATION:
ANOTHER AMERICAN ACHIEVEMENT

Change in energy consumption, * *1997–2002*

Per capita: 1.8% less,

but

Total: 3.3% more.

Almost all our energy was generated from fossil fuels.

---

*Defined in terms of energy flows [see I:522].

How comforting per capita breakdowns can be! *India is the world's third largest emitter of greenhouse gases,* remarked the Australian government. *However on a per capita basis it is much lower down the world rankings at around 140.* Well, then there was nothing for Indians worry about!

Tom Jones continued: "You'd end up having to be a Communist or socialist or a dictator. And it's weird, because we don't talk about it much in history, but in some cases, dictators have had a lot [of] benefit . . . Probably the best example was the Dominican Republic and Haiti. The Dominican Republic had two long-serving dictators who were pretty hateful people, but one of the dictators, an older

---

* See above, p. 152.

[†]Emblematic was this feature in *Investor's Business Daily:* "Oil rose after stronger U.S. payroll growth allayed concern that the economy of the world's largest crude-consuming nation is slowing." A senior market analyst expressed "a little bit of optimism," remarking that "there's been a lot of concern about the economy and what that will mean for demand."

guy in his 70s when he came back into power, he realized that everyone was cutting the forest down, because people cook their rice with wood and they were over-harvesting the forest as Haiti had already done. Haiti has something like 95% cleared land, so there is less than 5% forest on their third of the island, and the other two-thirds is the Dominican Republic; if you go to Google Earth you can look down on the island: This side's brown and that one's green. And so the Dominican dictator said, you're not cutting any trees down. And he actually sent soldiers into the forest so if that you cut a tree down they shot you, which is a horrible thing but it worked to save the forest, and then he turned around and said, well, I understand that people have to cook, so the government subsidized propane, and now in the Dominican Republic virtually everybody cooks on propane."

But when I was alive the rest of us declined to shoot people for burning carbon. And cooking propane still emitted carbon dioxide. As for other solutions, some we rejected, and the rest were inadequate—because, as a Japanese nuclear advertisement depicting a slender female silhouette with chopsticks reminded us: *Our future life needs electricity.*

Abu Dhabi

TOMIOKA, 2017

Energy is Eternal Delight.
William Blake, 1790

The taxi driver was roundfaced and darkhaired, with many moles. He wore a black vest.

"What is the situation now?"

"Almost no change. Originally the radiation's effect was not that strong—not here, anyway. I was worried when the reactor exploded, yes. About one-third of Iwaki people evacuated."

"It's been such a long time. What do people say now?"

"Because it's *invisible . . . ,*" he chuckled.

"Do you ever talk about it?"

"We sometimes talk about it . . ."

"What is the situation in Tomioka? Have they returned?"

"No, no; they haven't come back. The next town, Naraha, to that they can come . . ."

"When I was last in this area," I said, "people could come to Naraha only sometimes. And children under 16 could not come."

"Now you can stay over in Naraha," he said. "In Tomioka there is only a part where the radiation is high. And in Okuma . . ."

I waited, but his sentence ended there, so I asked: "Is nuclear power good or bad?"

"Personally, I don't think that it is really good."

"And what do you think about global warming?"

"That's too large a topic, and I really don't know. Nothing can be just totally good or totally bad."

Although it was early January, midday in Iwaki had achieved a springlike 13° Celsius.* On my previous trips, the ones in February and in October, the city had been colder. At least February had; about October I no longer remembered

---

* 55.4° Fahrenheit.

exactly. So we reasoned, back when we were alive. Forgetting that in Japan February tended to out-chill January anyhow, I made our customary mistake of treating weather as climate, and inquired: "Isn't it unseasonably warm?"

"Iwaki can't get too cold anyway. Today's not unusual."

Consoled to learn that we all still remained in excellent shape (that night on the Hitachi Limited Express back to Tokyo the crawling glowing display of Japanese-language news announced that 70% of Okinawa's coral reef was now dead), I settled back, fiddling with the pancake frisker, which like me grew crotchety with age, and took a timed measurement of the taxi's interior air dose: as expected, 0.12 microsieverts per hour*—quite comparable to home, and to last time.

The driver was Iwaki-born. "So I have lived here for more than 50 years," he said with that patient, understated Tohoku pride.

Since 2014 a second beach had reopened for swimming. The tourist information office at the train station no longer offered that pamphlet combatting harmful rumors. Having taught myself to hunt, sometimes even craftily, for whatever was worst, I had to admit that in Iwaki the worst had become as *invisible* as climate change.

"Are the fishermen still suffering from being unable to sell certain fish?"

"They are still taking the fish for a test," he answered.

"Do you worry when you eat sushi?"

"The sushi that we are served comes from other places," he assured me. So once again all seemed right with the world.

"What about mushrooms?"

"The mountain produce is not good yet. Those mushrooms and vegetables, when they will be safe I don't know. The 500 becquerel standard in the past, now it's more strict."† Then this bringer of good news added: "Probably the radiation level all around here has reached the previous standard."

"Please tell him that I very much doubt it," I said, forgetting the rudeness of conspicuous contradiction in my desire to helpfully warn, all for nothing—an impulse I so often indulged in *Carbon Ideologies*. I strove to inform him that in three decades the cesium, wherever it might reside, would still retain half of its peak virulence; and just now only six years had gone by!—although scraping immense quantities of surface matter into those famous black bags, which we had just now begun to pass, undoubtedly reduced local concentrations. But the driver

---

* Unless otherwise stated, all frisker readings in this chapter are one-minute timed SCALER counts, multiplied by 60 to express microsieverts per hour.

† See table "Their Standard Is as Arbitrary as Ours," beginning at I:506.

was not interested in my pessimism; nor was I, anymore, so we rolled in pleasant silence deeper into that springlike day with the beautiful ocean leaping to the east, then flashed through the first tunnel, on whose far side uniformed construction workers were erecting new breakwaters; while sea-spray leaped up before them; the driver said that this "tsunami protection wall," five or six meters high, would go "all up the coast from Onahama." So clear was the afternoon that I could see far to the west, where the mountains (as locals liked to call them) hunched brown and green, overlain by a long narrow cloud like the outline of a flock of birds. Here came Hisohama, more tunnels, Yotsukaya, a Family Mart, and then the sign: SEE YOU AGAIN IN IWAKI. At 12:47 p.m., with Hirono's incinerator towers ahead, I checked the dosimeter: 589.7 accrued microsieverts. Over the last four days in Tokyo it had averaged the usual 1.1 micros per day.

J Village bustled rather less than I remembered from three years ago; the driver remarked that it would soon resume its prior role of soccer facility; nowadays the decontaminators were deployed closer to Nuclear Plant No. 1—one more reason to let radiation here be *invisible*. While the interpreter engaged in her customary pre-radioactive pee, I rolled down the taxi window, extruding the frisker to sample an air dose of 0.042 micros a minute: 2.52 micros an hour—a trifle high, but perhaps the decontaminators had tracked in some dust; thank goodness that too was *invisible*.

At 13:04, with the dosimeter still at 589.7, we entered Naraha, which had been modernized with many new rectangular bays of black bags, some of which must have achieved capacity, for they lay out of mind under greenish-blue tarps. At 13:10, I took a stroll on the chilly forested hillside. Since 2014 the decontaminators had moved on, and the place now looked parklike. I remembered overgazing the Kido River from this spot and seeing black bags. They were still down there. At my waist the air dose read 0.24 micros an hour. Setting down the frisker in a pile of dead leaves at the edge of the path, I let it measure for a minute and got a reading of 1.2 micros. A day spent lounging there would have earned me a month's worth of Tokyo radiation—far less than the 10-hour flight back to Los Angeles.

Naraha had become almost busy. I saw traffic, cars in parking lots and a lady out walking in a short navy-blue skirt. At 13:21, the air dose remained 0.12 inside the car, while the dosimeter had accrued a tenth of a micro. At 13:24, real-time readings* continued to jitter around 0.1 micros per hour as we drew level with the entrance bridge to my friend Aki's† former workplace, Nuclear Plant

---

* That is, unaveraged instantaneous NORMAL measurements. See I:399.

† See I:448.

No. 2; then as we rolled past, with the windows still closed, the frisker stridulated up to a trivial 0.3 and 0.2; call it local variation.

We drove down that familiar hill. At 13:25 the interior air dose was 0.18 micros. At 13:30, with the dosimeter at 589.8 micros, we entered Tomioka.

The town now, as had Naraha back in 2014, looked rather more quiet than decrepit. Our level inside the taxi rose to 0.36 micros an hour. And presently we were making that sharp backward turn off the highway, parking by the weed-grown pachinko parlor **TSUBA**; and it was just like old times to return to the nightclub Sepia, whose door had been trimmed of creepers; in 2014 this spot had read 4 micros an hour; today it was a mere 0.96.

A hundred yards down, the rusty grating and the pavement which had given off one of the highest radiation levels I measured in Tomioka—21 micros an hour—now varied between 3 and 4 micros. A certain drainpipe which had won the prize at 32 micros had been removed. I took real-time readings on a formerly "hot" culvert: 2.23, 2.32, and so forth in an almost innocuous string of chittering digits; this would approximate 20 millisieverts a year, the lower limit of a former "residence restriction" zone, and the upper limit of the new "safe" standard. Strolling down the street, with the frisker at waist level, I measured 0.78 micros an hour, which had been 3 and 4 micros before.

At the house which in 2014 I had measured before and after decontamination, I frisked the drainspout, whose level had declined from 12 micros to 1.98.

Just beyond it, hard-hatted decontamination cadres swarmed aggressively about. One of them rushed over to prohibit me from photographing a house I had photographed before, "because it is private property," although I somehow accidentally took my picture just the same. In keeping with the neighborhood trend, it looked better than before. Once my new friend turned his back, I frisked the front door, whose level was only a micro.

Many homes, some in apparently excellent condition, had been marked for demolition by double red lines. That was new. The decontaminators were on the move in Tomioka; in another district we even discovered an automobile parts shop praising progress with fluttering red welcome banners of plastic; and one chain convenience store had reopened in July; at the entrance, a poster with an idealized image of the cherry trees of Yonomori proclaimed: **TOMIOKA will never die! standing up against 311 *MY HOMETOWN***. I went in. The barrel-shaped, middle-aged clerk commuted from Iwaki.—"Are you worried about radiation?" I asked her.—"Not at all."—Here on these side streets by **TSUBA**, where red zone warning signs persisted at the meadow-edge, life had not yet become as bright as in her establishment; but the decontaminators were certainly doing their mite.

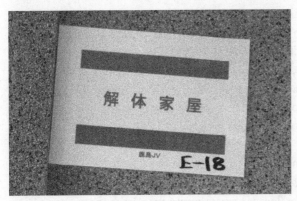

解体家屋

福島JV E-18

Tagged for demolition

A turn away, down another familiar road where yellow tape used to warn me off from those shabby little near-identical homes for workers, the tape had gone; no house bore the double horizontal red stripes marking it for demolition. I remembered snow in a certain doorway, and a shovel leaning as if its owner would soon return, while two whisk brooms had fallen before it; these implements were out of sight, but in the yard of a facing yard now lay a rusting bicycle. Here the pavement read 2.46 micros per hour. A square drain varied from 2.48 to 3.03 real-time micros. Peering farther down that side street into the angular shadows of other workers' houses, I read 2.88 micros—nothing—only 24 times the Tokyo dose—25 millis a year . . .

Frisking another Tomioka grating: 0.048 microSv (2.88 micros per hour)

Exactly here I once bypassed a barrier and stepped into the forbidden zone, where the weeds rose high and a traditional Japanese house had allured me from some half a kilometer's distance; having only the dosimeter to protect me in those days, I had refrained from proceeding very far. The house appeared more decrepit now; the roof was caving in. As for the weeds, something akin to the impression they now supplied had been conveyed quite prettily by Lady Murasaki nearly 1,000 years ago:

> Amid the desolation of the autumn garden only the pampas-flowers stood out, their long stems beckoning from the borders like waving arms . . . They were not flowers of great beauty, yet now . . . the scene was not altogether devoid of charm.

The air dose was 0.84 micros an hour.

In place of the *robot girl,* a much less crackly *robot boy* now echoingly instructed us: "Please do not use any fire."

The interpreter and I turned back toward **TSUBA** and the taxicab. Approaching that busybody decontaminator, I inquired what his accrued dose might be, but he, adopting the West Virginian strategy of the *no comment,* suddenly had no more time for me, so we left him to his own affairs, and soon reached that gap in the laurel hedge, with the forbidden meadow through and behind it, into which I had insinuated the pancake frisker in October 2014, and had watched the levels nightmarishly rush up to 7 micros before the device had even come out the other side; today the hedge was trimmed back, the hole so much less distinct that I doubted we were in the right spot until the interpreter convinced me. I pushed the frisker all the way through, and the real-time measurement flickered between a mere 1.8 and 2 micros. Tomioka had improved from horrible to dreary.

Finishing up at **TSUBA** at 14:14 o'clock; with the dosimeter at 590.1 (which implied that we were accruing half a micro an hour), we proceeded north on the highway, well beyond the former checkpoint, passing pads of black bags in their rectangular bays, reminiscent of frack pads. I wanted to frisk Yonomori's empty street of cherry trees one last time. Our interior dose with the windows rolled up was 0.6 micros an hour. Both western exits to Yonomori had been gated off, with a helmeted sentry at each, so I asked the driver to proceed to the ruined Japan Railways station where Mr. Endo had taken me in 2014.*

---

* See I:377.

Three dark-jacketed men in respirators and hard hats huddled at an intersection. We rolled up toward the pale glow of a shed which spanned the entire street view ahead as if it were a horizon, and construction cranes emerged around it. Once again I could not recognize the place. The driver proved to me on his shining map-screen that here indeed was where the station had been. Men in hard hats swarmed everywhere, along and under the long skeleton of what must be the new station. Across the street, by the tsunami-smashed Chinese restaurant whose fare Mr. Endo had impugned, an army of helmeted, respirator'd men marched with rakes and bags around some building marked for demolition.

We drove across the tracks. There was definitely no station anymore, just black bags, and then a large berm which must be the commencement of a new seawall.

I got out of the car, stepped over the rope that prohibited one from entering, then took a one-minute timed count right there, seven or eight meters from the black bags: only 0.24 micros; then, before any of the hard-hatted types could come closer, I touched the frisker against a black bag, which read surprisingly safe at between 1 and 2 micros.* A sign labeled their contents *cut branches, etcetera from forest.*

I asked the driver: "What do you think of all these black bags?"

"Ah, we are scared. When I first saw this, I felt horrible. Now, of course, I don't like it, but I somehow got used to it."

In that quarter-hour we had accrued another perfectly innocent 0.2 micros, so I directed the driver into the business district, and we parked around the

---

* NORMAL mode.

corner from the eerie garment shop I always visited. The hole in the front remained, so the interpreter and I walked in. It was as dark and clammy as ever, but
over the past two years the air dose had fallen from 0.36 to 0.24 micros. Someone
had moved that haunting kimono'd mannequin and turned it so that its profile
caught the light. From not very far away it still appeared to be a real woman, a
pretty one, standing and waiting, looking out at the street as she had been doing
for years—but her waiting must soon end; this street too had become active as I
had never seen it; on the next block the decontaminators were toiling. They saw
us go in and out of the garment shop, and I half expected them to order us away,
but no one said anything, even when I kept frisking. At the corner, the pavement
by the rain gutter read the same as before: 1.32 micros.

It was now 15:08, and the dosimeter had recorded another tenth of a micro. So we drove to Yonomori, where I strolled about, taking waist-high air measurements which unfailingly read 0.6 micros an hour. When I had frisked this place with Aki and Mr. Kojima, the dose had been 4.08 micros.

The dead leaves at the edge of a bamboo grove read 0.54 micros. Finally I found one patch of pavement that offered me a respectable 6 micros. That was the only spot in Tomioka that I could call truly dangerous. For the most part, the radiation levels had fallen by three-quarters and more.*

Surprised and pleased, I telephoned Dr. Lyman from the Union of Concerned Scientists, and requested his opinion.

"I do see that they have done a lot of remediation," he said. "So Tomioka is one of those areas where they say they have completed the decontamination process in residential areas, and almost completed it in the surrounding areas; they aim to complete everything by March. They're doing a pretty good job, but it remains to be seen if there are reservoirs of contamination."

"So do you think that it's now acceptable for people to move back there?" I asked, and he replied, as he always had: "Depends on the local geography. I think the issue is that you have low-lying areas that are gonna get the runoff, but, well, you know, there is weathering over time; topsoil becomes buried; and the cesium's soluble, so it will get spread. As you know, the half-life's 30-odd years, so it's still there; it's just become more dispersed."

In short, the Prime Minister, Mr. Abe, could declare victory, maybe restart more nuclear power plants and perhaps even sell a time bomb or two (let's call them reactors) to India . . .—because our species was saddled with *demand,* which Mr. Abe most heroically sought to fulfill.[†] I myself, meaning to polish this chapter in a coffee shop in my home town, felt as annoyed as every other customer when the power failed the entire block, so that our lives fell dark in there, as in the garment shop in Tomioka, and they couldn't make coffee; soon we all left, wandering away, hunting electricity.

Wanting a happy ending, I asked Lyman: "Based on that dramatic reduction in radioactivity, are there parts of Tomioka that could be kept safe?"

"It's a localized problem, but there are hot spots; you can't survey 100% of the ground, and one area may be safe, and then get recontaminated. People who

---

* To quote a sentence I wrote about Iwaki in February 2014: *The time to one thousandth of the 2011 contamination value, which you have just heard Ed Lyman calling inadequate, would still be three centuries.*

[†] As of June 6, 2017, five Japanese reactors were on line. Tepco's famous ice wall kept working miracles; perhaps it might even reduce the flow of radioactive water into the ocean from 130 to 100 daily tons!— As of July 11, fewer than 200 people had returned to Tomioka.

move back should have their own dosimeters . . . ," he said; he sounded weary, and the supposition did sound dreary, to be logging my exposure every morning and night in my home, and every time I trimmed the front yard's bushes, or went to the river—and most people would at best possess only dosimeters, pancake friskers such as mine being largely unavailable, so that if the child tracked in a dollop of cesium on her shoes, which then stood waiting at the entranceway for her to wear to school, all the better to keep irradiating her little feet; or if a storm from Futaba or Okuma recontaminated first the roof and then the drainpipe and thus the bricks below it, irradiating the wife once she began scrubbing them clean, we would never learn exactly why our dosimeters were accruing more than the usual risk of death; and this obscure poisoning would go on and on.*

Fortunately I was only a Tomioka day-tripper. Enjoying the long shadows, I walked into a cool headwind, holding the frisker before me to verify that little gamma-flavored harm[†] would come of inhaling. Then the *robot boy* commenced echoingly reminding us that it was now 15:30, and at 16:00 the gate out from the "difficult-to-return area" would be closed.

And so we began leaving Tomioka, rolling toward the mountains, with everything cool and clear, and rows of black bags down in the neatly cleaned gulley below the bridge to Nuclear Plant No. 2; at 15:47 we departed the town limits, with the dosimeter at 590.7. Call it 2.25 hours at 0.4 micros an hour, or 9.6 micros a day—8.72 times the Tokyo dose[‡]—3.5 millis, or 3.5 times the international safe standard. And the full moon hung high over the afternoon pines.

---

* "Injury by exposure to an atomic bomb first occurred in August 1945 at Hiroshima and Nagasaki, and its effects are still in evidence" 26 years later. As of 1971, cesium-137 body concentrations in the residents of Nagasaki's Nishiyama district, which had received high fallout, were "nearly twice that of other districts."

[†] After my failure at Hanford, I rested less faith in the frisker's alpha and beta detections.

[‡] Given the short time I was in Tomioka, this average must be quite approximate.

WHAT WE SHOULD HAVE DONE

By 2007, Germany had already fulfilled a large part of its obligations within the framework of the aforementioned European burden-sharing, amounting to a reduction of 22.4% with regard to the base-year emissions [of 1990]...

German Greenhouse Gas Inventory, 1990–2007

Total U.S. emissions have increased by 7.4 percent from 1990 to 2014... at an average annual rate of 0.3 percent.

American greenhouse gas report, 2016

There had only been one hope for us: To reduce demand.

I've told you that we could have done it through birth control or genocide, but the second strategy appealed to no one whom I cared to know, and the first would have been angrily resisted in the name of freedom, or of religion. Because the warning of Thomas Malthus had been temporarily invalidated through astounding increases in food production, we could pretend that the fate of rabbits that bred unchecked on little islands need not be ours. Reader, we might not have loved *you,* but most of us took joy in our children whose descendant you are; wasn't that cause enough to beget them? In his brave, eloquent and otherwise subversively sensible encyclical, Pope Francis still found it necessary to say: *Instead of resolving the problems of the poor and thinking of how the world can be different, some can only propose a reduction in the birth rate ... To blame population growth ... is an attempt to legitimize the present model of distribution ...* I caviled, disagreed, was outnumbered.—Well, a third way remained: Through changes in policy, mores (not to mention improvements in technology and education, which might buy time), we could have deliberately reduced consumption.* Had we leavened *the present model of distribution* with a sprinkle of decency, the affluent would have reduced their per capita use of energy, while the

---

* To his great credit, the Pope did reiterate that point. For instance: "A person who could spend and consume more, but regularly uses less heating and wears warmer clothes, shows the kind of convictions and attitudes which help to protect the environment."

poor could have consumed more—even as *total aggregate* consumption fell.—
Carbon Ideologies has told you why it did the opposite.

In 1829 Goethe remarked: *Mankind is conditioned by needs; if these are not
satisfied, there is impatience; if they are, there is indifference.* Obviously this rule
applied in equal measure to demand: **For the Bearded Man In Your Life**:
Bluebeards Original, the #1 beard care company, has everything your loved one
needs to grow (and maintain) healthy facial hair.—Well, since he needed it, how
could anyone deny him the fulfillment? Never mind what it was made and pack-
aged from, and how much energy that work required. And once he possessed it,
how dare anyone take it from him?

Demand is vitality, and vitality, like this book, is forward-looking. We set
snares for the future, then check our traps to learn what gifts we have caught. The
baby crocodile (like the mining corporation) needs to grow; in other words, it
must keep eating, in order to enlarge its appetite. The human counterpart to
reptilian voraciousness is called "hope." Won't tomorrow be better than today?
Believing so helps make it so.—In your time, optimism will truly be the best
counsel. As for my generation, pessimism would have served.

Had we better valued what we already possessed, and distinguished needs
from desires, demand might have lessened, if not of its own accord, then through
education in what was necessary and what was well-made, through emulation of
seeking, searching heroines and heroes; and through supplications to harmless
vanity. Love of family (without excessive multiplication), pride in place, and, for
those who cared, happiness in learning, seeing, worship or meditation of any
sort, would have cost our planet little. One January morning at the Fushimi
Inari-Taisha, a temple well-famed for its 10,000 torii, snow was melting loudly
into rain through the vermilion windings of rectangularly puddled stone steps
that wandered from shrine to shrine up the mountain of snowy-shouldered trees;
here and there we won views of Kyoto; in place of loudspeakers we heard crows;
and although, to be sure, the construction and maintenance of this masterpiece
required thermodynamic work, we who walked those ways in our crowds drew
down no electricity (and, for that matter, spent no yen) in the "consumption" of
that experience. What we claimed to cherish most of all (these appeared as placeholders in our action movies, and "accessories" in our purchases) included the
following: love, sex, reliable, friendly affection; happiness, excitement, personal
attractiveness, novelty, distraction; safety, both for ourselves and for those who
depended on us; freedom from fear and pain; and, yes, acquisitions of things.
Many such aspirations could have been indulged in at the cost of fewer greenhouse emissions. I have slept in homeless camps whose inmates took pleasure in
sunsets, breezes and even visiting skunks. The obvious fact that I (and most of

them) preferred not to be homeless did not invalidate the other fact that I could have pleased myself more frugally than I supposed.

Of all the things I owned, I valued most my books. I hoped that someone would use them after me. In 2013 some materials technologists thoughtfully noted:

> A garment from a "fast-fashion" chain can be discarded after one outing as cheaply as it was bought, whereas something made or enhanced by a friend at a time of crisis cannot be discarded at all, as it has become "emotionally durable" and part of an individual's life-story. The emotional permanence of buildings and objects has been a constant of many societies, and only under the mass availability created by industrial production has it been lost.

What sort of society would invest meaningfulness in durability? People who understood scarcity in their bones certainly valued the lasting function of objects. In Madagascar, in a waste called the Burned Land, a man once asked me for an empty plastic bottle of the disposable kind; it was not disposable to him, and his gratitude shamed me. I suppose that you in the future would take better care than did I of aluminum foil, reclosable plastic bags and other "big five" materials. Again, I would not have chosen to live either his life or yours, but why shouldn't I have *voluntarily* done a portion of what extreme impoverishment compelled you to do? That essay from 2013 proposed that *the "heroes" or "role models" . . . should be "local" rather than being exemplars of high mobility and consumption.* My own heroine was Rhonda Reed, because she had said: "I don't like buying new stuff. That just grinds me something fierce."—That localism has its dark side glares out at any student of Balkan hatreds; but patriotism must be healthy to the extent that it instills pride in living where one is, and in caring for people and things pertaining to that place.—Some years ago, on a certain mossy island, lay an Inuit town a couple of miles beyond which I had raised up my tent, and since I made a point of attending their church, people began to know me a little, and children and teenagers visited. We swam together in icy water, and I sat sketching while they caught baby birds. They felt sorry for me because I could never be from there, which was self-evidently the best place on earth. How their lives turned out I cannot say, but during our three-week acquaintanceship they mostly struck me as social, contented, healthy—at a considerably less than American cost in power consumption.

One category of Japanese beauty is *sabi,* or "beguiling decrepitude," as exemplified by a half-decayed scarlet leaf in a mound of yellow ones, or the face of a lovely old woman. Noh mask-carvers (and some Noh actors) assured me that the beautiful masks associated with their professions did not "come alive" until they had been worn on stage, breathed through, sweated in and scuffed for a good century or two. A renowned carver told me: "You use nice color, and then on top of it you put something else to make it look more subdued, so it doesn't look too shallow. After 50 or 100 years, if you've used something unsubtle, it's really going to show; it'll start looking worse and worse."* This did not mean that a visitor would see ordinary Japanese walking around in hand-dyed rags; nor did Japanese convenience stores use plastic bags other than extravagantly; all the same, in Japan I sometimes got hints of "emotional durability." What if more objects had been manufactured with this as an end?

Like my books, my eyeglasses frames should have lasted all my life. If somebody had to swap lenses in and out whenever my vision changed, so be it; and whether the greater good called for those to be glass or plastic I leave to the calculations of Professor Gutowski; for the frames themselves "big five" materials might have been justified: aluminum or steel for at least the lens sockets, to minimize warping and degradation over the years. I can envision their other parts as metal also, or maybe some kind of wood. They could have been ornate or not, to satisfy my taste and budget. (Being "artistic," I might have decorated them myself.) These eyeglasses would have gained "character," like a strong walking stick, an heirloom firearm, a well-built musical instrument, a wedding ring, a family Bible, a dependable all-manual camera, or a leather briefcase. Had what I imagined actually been made, someone, even you, could have used my eyeglasses after me.—Needless to say, there was "no demand" for any such. My own demand, expressed to optometrists, inevitably met with silence. Once my nearsightedness stabilized in middle age, my storebought frames inevitably failed long before their lenses, which, aside from a scratch or two, would have served as well as ever—but a screw began to loosen in a sidepiece; after only two or three re-tightenings, the flimsy ring it inhabited would break; then the frame jockey would calmly inform me that my only way forward was a new pair. Of course this remained my "choice": I could and sometimes did get by with epoxy or even masking tape, but because the glasses now looked decrepit without any *sabi* I got treated like a poor man, which sometimes inconvenienced me in my work; more importantly, functional durability was lacking. While I was interviewing oil workers in the Emirates, my right lens fell out day after day, sometimes striking concrete and accruing more scratches, until, preferring to "fight other battles," as my father would have said, I gave up and went to the Ruwais Mall. The frames

---

* This discussion derives from my book *Kissing the Mask*, which deals at length with Noh.

were "worn out," they explained. By the time I finished *Carbon Ideologies,* aged 58, a good 20 pairs of glasses might have flowed from oil wellheads or natural gas pipes onto my nose, and then . . . Well, it seemed wrong to throw them "away," so I kept them awhile; perhaps I could "donate" them to an NGO, but if I couldn't get the frames fixed, how would some poor man manage? After I died, they and their embedded manufacturing energy must have become junk.

Of course I was a freak. My thoughts about spectacles were trivial and strange. And so another study concluded that even though *the evidence on increasing product lifespans suggests that . . .* life extension is beneficial and technically not difficult,* durability *is not currently wanted by consumers or producers.*

Well, was one of Tom Jones's eco-dictators supposed to rein us all in, impoverish us, limit our choices and keep us goddamned grateful for whatever patched and grubby cardigans we rented?—Now, *there* was a nasty concept! Better to focus on easy, boring little improvements . . .

For example, we could have double-glazed our windows and insulated our walls. Thanks to these tricks, the required furnace output of a certain two-storey house in Saint Louis fell from 118,800 to 67,800 BTUs per hour.—For that matter, a certain German company had begun producing boilers with built-in communications systems. If this improvement truly reduced needless fuel consumption, then why not install only those boilers from now on? In both cases the cost might have pained us, but once Goethe's indifference set in, we would have met an inertia not of consumption, but of conservation.

But again, trusting in such individual and voluntary choices (which would be avoided by most, in order to satisfy more glamorous demands, and which anyhow only the affluent could afford) reminded me of that leaflet I'd seen in Charleston, West Virginia, shortly after the *regulated community* had poisoned the water supply: we were merely to TURN UP THE TIPS! to struggling waitresses in order to somehow BEAT THE SPILL! . . .—while Jim Justice continued to rack up his 23,693 water pollution violations.

Yes, I did my part to ruin our future. But Jim Justice did worse.

We could have burned fossil fuels at higher temperatures, which would have more completely combusted them, thereby emitting smaller quantities of methane and nitrous oxide.[†] But the "we" in this case would consist of the Puerto Rican utility officials who secretly sulfurized the air,[‡] the obliging souls at the

---

* "Unless the product has a high use phase impact." That is, unless it consumes a high amount of energy during its period of use. For consideration of the use phases of various household appliances, see "About Power," I:67.

[†] See above, pp. 310 (second table), 424.

[‡] See I:37.

John E. Amos facility in West Virginia who bought Jim Justice's coal—without previously soliciting his or any other bids*—the "Public Information Specialist" at my local utility company who when I asked about climate change would not even mention that phrase in his meaningless reply,[†] and those Bangladeshi brick kiln owners who could afford only to burn the cheapest coal, cheaply:[‡]—Let us call them all the *regulated community*. No doubt a few did care about what our planet would become after their deaths. As for the others, if nothing else, I hope that *Carbon Ideologies* has showed off the sterling virtues of those individuals.

Hence we should have begun by truly regulating the *regulated community*. But we couldn't, for two reasons: Firstly, they would have screamed; and secondly, we all would have paid more.

(In my homeland we believed in economic "checks and balances," and even in an adversarial legal system. So did we also need an adversarial regulatory system? As a self-employed businessman, I would have hated that. But what do you from the future care about what I might have hated?)

Yes, we could have regulated ourselves before nature did. It has not escaped you that we decided to wait. Biology imposed its penalties. Species died; epidemics and insects took their commissions; our food supply, let us say, simplified itself. Well, since 90% of all species had already come and gone in that long pre-human time of trilobites, mastodons and dinosaurs, why take responsibility for anthropogenic extinctions? Happily agreeing with this line of reasoning, the *regulated community* sold us more carbon—because as the Executive Vice-President of the National Corn Growers Association indignantly demanded: "Are we going to reduce greenhouse gas emissions today because we believe there's an economic benefit 15 years from now? That's way too hypothetical for a family-owned and operated business that has to make a payment this year. The banker doesn't get paid in hypothetical dollars." I hope he enjoyed his real dollars.—Since we (to tell the story in his terms) postponed payment, the fine grew at compound interest, with sterner punishments presently imposed by physics. We were all safely dead by the time the oceans boiled.

To stop Jim Justice, and apprise the Executive Vice-President of the National Corn Growers Association of the existence, believe it or not, of non-economic benefits, we should have continued tracing and quantifying our emissions and energy flows for the products we made; then made, and enforced, socioeconomic policy based on that information. (We could have retired petroleum jelly from

---

* See I:103.
[†] See I:152.
[‡] See pp. 278–86.

lip balm, even if that necessitated going back to goose grease. We could have phased out the coal tar derivative colors which I loved so well in certain high-quality paints.) How much radiocontamination, chemical poisoning* and greenhouse gas pollution was each "demand" worth? How many more tons of cement *must* we produce? To determine that, we had to answer the question *what was the work for?* In my time we never did.

Had we lived up to this elementary duty to our children, and to you, *then* we could have usefully gone about the dreary business of setting permissible standards by the hundreds. (We should certainly have established *and enforced* permissible limits for each significant greenhouse gas.) Could steel be manufactured at lower energy expenditures? Then enact those into standards. Could plastic wastes find any new beneficial use? Why, then, enable—and require—the *regulated community* to put them to that use.

But the future can never trust the present; it must unceasingly (often unavailingly) plead its cause. Ignoring and denying the compound interest rising up against us in an invisible greenhouse cloud, we expected to rub out our names on every proof of obligation, bequeathing the debt to you. So we had always done before, with our dispossessions, blood feuds, invasions, rapes and robberies, as had been done to us; the past necessarily stained the present, so we passed on the favor, kicking it down through time.

Case in point: In 2014, Infinitus Energy established a recycling center in Montgomery, Alabama. A year later the establishment closed, and the press announced: *Montgomery's recyclables are now going to a landfill.*—A merely local failure, perhaps, for meanwhile Waste Management in New Jersey was still managing to sell bales of flattened recycled bottles at $230 apiece. By the beginning of 2016, Waste Management could only get $120 for them. With oil now retailing at less than $30 a barrel it was *cheaper for the makers of water bottles, yogurt containers and takeout boxes to simply buy new plastics.* It could only be cheaper thanks to what Pope Francis called *the current system, where priority tends to be given to speculation and the pursuit of financial gain. As a result, whatever is fragile, like the environment, is defenseless before the interests of the deified market . . .*—to which Mr. Jeb Bush, a Catholic—and Republican candidate for President—replied: "I don't get economic policy from my bishops or my cardinals or my pope." It showed him up ever so well, that he characterized the Pope's

---

* Both regulators and regulated should have been required to know a great deal about all chemicals used, *before* their release into "some ecosystem somewhere." You might have supposed this was obvious, even for us. But it wasn't: "Crude MCHM is more toxic than previously reported by Eastman Chemical Manufacturing."—"That's in a way a difficult thing to say, because everybody has a different definition of safe."—"We don't know that the water's not safe. But I can't say that it is safe."—My stomach turns.

remarks as "economic policy"! I can almost see him and Jim Justice on the same sinking ship, prising out waterlogged boards to sell each other.

I say that we should have paid Infinitus and Waste Management a guaranteed base price for recyclables. Moreover, the price of manufacturing recycled plastic should have been lower than the price of manufacturing virgin plastic.

Those measures might have required more drastic enactments. We could have raised the price of motor fuels, whether or not we spent the surplus on renewable energy. A fraction of that increase could certainly have subsidized the recyclers. Another fraction could have double-glazed poor people's windows, and funded research on climate change mitigation. If we had nothing better to spend it on, we could even have given some to the oil companies. I had nothing against them; they had kept my lights on. The main thing was to get the cost up, in order to lower demand.—U.S. Environmental Protection Agency, 2016: *Changing fuel prices played a role in the decreasing emissions. A significant increase in the price of motor gasoline in the transportation sector was a major factor leading to a decrease in energy consumption by 1.2 percent.*—In Germany, *continual increases in fuel prices* likewise reduced transportation emissions—although one apparent fraction of that reduction actually constituted its own shell game: Resisting high local fuel prices, Germans drove across the border to gas up—a phenomenon beautifully called *Tanktourismus*.* Therefore, we *all* should have raised fuel prices—which would have necessitated some form of global authoritarianism.

Believe me, I never wished to think in this direction. If only we could have quickly and voluntarily paid down our carbon balance before the interest mounted!

We could have captured more of the methane that outgassed from municipal wastes† and coal mines, and burned it for fuel. If this did not prove "cost-effective," well, we could have made cost-*in*effective the generation of municipal waste and the digging of coal mines. As the Germans already had done, we could have prohibited the landfilling of *any* organic wastes. Other methods in the war on methane might have included reducing the portion sizes and the number of food choices at supermarkets, in hopes of decreasing the mountains of uneaten

---

* See I:147.

† The Germans accomplished this through "intensified collection of biodegradable waste from households and the commercial sector" and "of other recyclable materials, such as glass, paper/cardboard, metals and plastics; separate collection of packaging; and recycling of packaging. In addition, incineration of municipal waste has been expanded, and mechanical-biological treatment of residual waste has been introduced. As a result of such measures, amounts of landfilled municipal waste decreased nearly to zero from 1990 to 2006 . . . Over half of municipal waste produced in Germany today is collected separately and gleaned for recyclable materials . . ."

food that had warmed the atmosphere in landfills.* I, a happy eater, would have despised such measures to the extent that they deprived me of my favorite fodder; in my insistence that "something else" be imposed on us instead, I would resemble the man who tranquilly accepts that he must die—but not of lung cancer, and by no means from a stroke, while Alzheimer's would be of all fates the worst . . .!—In such terms we considered the prospect of regulation. We were better than Jim Justice only for lack of opportunity.

Our regulators should have assessed the future as gloomily as did insurance accountants. We should have prepared for the worst. What were the various dangers *and certain longterm costs* of our carbon ideologies? Enumerate those; expect them; make standards—which had to be *simple, reasonable* and *public,* hence subject to public criticism and debate.

Well, who *would* debate them, aside from the *regulated community?* Which "consumers" would take the trouble to learn their own interest, and gather the courage to say, "I may be wrong about this, but it seems to me . . ."? (Almost none, you know.) And to whom would the regulators and their experts be accountable?

Many of the science fiction books I used to read advocated a technocratic future. If disinterested slide rule types were only in charge, how much safer, wealthier and happier we would all be! The authors made the same mistake as Plato in his *Republic.* Guardians, philosopher-kings and the others were never disinterested. Meanwhile, the technocrats did not own the means of production; they were wage slaves or grant-seekers.

Do you remember the technocrats at Oak Ridge National Laboratory who admitted back in 1982 that fossil fuel combustion was increasing atmospheric carbon levels, which *could become a problem a half century from now,* then reassured their own generation that *even in the high CO_2 case, a level of 700 ppm (which may prove to be acceptable) is not exceeded in 2050?*† They disqualified technocracy.

And weren't the regulatory commissions made up of, or informed by, just such experts?

Blinkered by my own time, I could not imagine how to address these objections. But suppose that I could have. Suppose that the standards had been transparent, that the public cared, disinterested knowledge was cherished and the government listened. As you have read in this book time and time again, the *regulated community* could not even be compelled to follow the most elementary safe and decent practices—but suppose that we had figured out how to compel

---

* See "About Waste," I:36.

† See I:88.

them, or better yet persuade them, for the sake of their own posterity. These suppositions appear preposterous even to me—but please remember that like you and everyone else, I vainly wished to keep my case from getting worse. The dinosaurs must not have liked the way matters were going. Why couldn't the climate stay tropical for them? And so I offer my list of should-have-beens, to distract myself from impending change.—Let me then imagine accountability:

In 2014, Ms. Nakanishi Junko, *a leading expert on chemical risk assessment,* addressed the Japanese government's ongoing decontamination of 11 Fukushima municipalities whose contamination surpassed 20 millisieverts a year. Judging that level (as would I) "unacceptable," and one milli evidently unrealistic, she proposed a target dose of five millis. Her cost estimation was 1.8 trillion yen, and 65,000 nuclear evacuees could go home in a year or two. *A resident would be exposed to around 38 millisieverts over 15 years, a risk that, when compared to the average risk of exposure to a chemical like dioxin, is not high* . . . (Well, was it or wasn't it?* And was this mention of dioxin a helpful comparison? "Somebody" should have decided. Did "somebody" include the same members of the public who rejected talk of global warming?)—After those 15 years, continued Nakanishi-san, radiation levels would have finally reached one milli per year. Whatever you or I might think about the merits of her plan, it did at least offer specific figures which could be debated; moreover, the levels at each place could be monitored by citizens with friskers and dosimeters; even I knew how to do that. If some district in Iitate or Tomioka measured above the statutory five millis, people might have some chance of holding the government to its word. "Some chance" was as good as it ever got, back when we were alive.† Not quantifying, debating and disclosing while setting public policy (or using shills and ignoramuses to set it, as in West Virginia) meant no chance.

No, "some chance" was barely superior to none; yes, my should-have-dones *were* preposterous!—Meanwhile, the longer the regulators and the *regulated community* continued to fail the rest of us, the higher the temperature would rise, and hence the more severe any corrective action would have to be. Hence some of the should-have-dones farther down on this list take on what in my lifetime

---

* As you know, Ed Lyman from the Union of Concerned Scientists rejected it.

† Consider the infamous doings of the Chisso Corporation, which released mercury into the ocean in Minamata, Kumamota Prefecture. In due course, local people ingested this interesting liquid metal. W. Eugene Smith's heartbreaking portraits of the sufferers are to my mind among the most powerful photographs ever made. Although Minamata Syndrome was *officially recognized* as early as 1956, I learned in 2012 that *how many people have been affected remains unknown . . . , as intensive medical checkups have never been conducted in and around the affected areas.* Compensation for the "unrecognized Minamata victims" was not enacted until 2010. That was "some chance," all right.

would have been considered an unpleasantly authoritarian character; perhaps for you in the hot darkness they failed to go far enough.

Instead of constructing new homes and offices, at considerable expense in "big five" materials' energies and emissions, we could have staggered occupancy into day and night shifts. I can almost hear the real estate developers raging over lost revenue opportunities, business owners worrying about competitors going through their desks at night; ordinary citizens complaining about sharing their bedrooms with third parties. Would it have been better to dragoon them into it, or to simply watch the temperatures rise?—No, we didn't like that; we preferred a heart attack to cancer.

All right: We could have tried to *shift the burden of taxation away from the renewable resource of labour, and onto the non-renewable resources of materials and fossil fuels.* This drastic alteration would have inflicted considerable economic suffering. Only the rich could afford aluminum-frame bicycles; car ownership would grow impossible for almost everyone. Perfect! As we know, declines in business reduced emissions.

Meanwhile, since the energy-hungry transportation sector was a poster candidate for remediation, we might as well have limited the manufacture and possession of private internal combustion engines, and built more passenger trains and buses. Come to think of it, we could have penalized rather than encouraged travel of every sort. No more Hawaiian vacations! (I wouldn't have liked that at all; my business *and* pleasures would have suffered.)

With ferrous products now growing expensive (although not perhaps in the same league as gold), less steel would find buyers—and many steelworkers would lose their jobs. We would learn to hoard our "big five" materials; "big five" recycling would suddenly more than pay for itself. The replacement of inefficient refrigerators, air conditioners and the like by less power-hungry models might become not merely sensible but *profitable* to "consumers," not to mention to those scrapyards where, in the unlikely event that our system proved kind and rational, former steelworkers would be reemployed at taking old appliances carefully apart.*

Steel products are most commonly replaced because a subset of critical components are degraded. These critical components typically account for a small share of the steel mass within products . . . Our new taxes would most certainly facilitate

---

* To me this sounds so practical that your hot dark future may well emulate it. One post-apocalyptic science fiction novel details the procedure of "looking for the bits of rusty steel which could be found by shattering the larger sections of columns and slabs to extract the ancient strips of that metal, mysteriously planted in the rocks by men of an age almost forgotten to the world."

redesigns, to allow those *critical components* to be swapped in and out of each still perfectly serviceable *steel mass*. This source continues:

> On average, one-third of all material use could be saved if product designs were optimized for material use rather than for cost reduction, because downstream production (and design) costs are generally dominated by labour and not materials.

In other words, once upon a time, reducing material use increased labor costs. No more.

I have mentioned eyeglasses, and asked why shouldn't there have been well-made pairs that not only served their owners through life, but also reflected their tastes. But there I revealed my own dreamy impracticality. In fact we should have standardized eyeglasses, silverware, sofas, houses and all the rest for maximum durability, reuse—and therefore, perhaps, uniform dreariness. But we had long since grown pretty standard anyhow, in the "choices" that we made; and perhaps if the fast food chain that served equivalently foul hamburgers in Thailand, Russia and the United States were emulated by an entity that sold everybody the same ugly, permanent kitchenware, only a few killjoys like me would kick.

(Some technologists once asked: *A key driver of profit in product companies has been increased differentiation, yet design for re-manufacturing and re-use would favour standardisation. How can these needs be resolved?*—By eliminating profit. How would you and I like that?)

Meanwhile we could have tightened the screws on the hardworking people who fed us. Here is how the Danes promulgated *measures to prevent loss of nitrogen from agricultural soils to the aquatic environment*:

> A ban on manure application during autumn and winter, increasing area with winter-green fields to catch nitrogen, a maximum number of animals per hectare and maximum nitrogen application rates for agricultural crops. All farmers are obliged to do N-mineral accounting at farm and field level with the N-excretion data from . . . the Faculty of Agricultural Sciences . . .*

Our self-reliant farmers wouldn't have stood for that—or would they? Should we have thrown the dissenters in jail? I hated this line of reasoning; sometimes I

---

* Such micromanagement actually could have reduced nitrous oxide emissions significantly. Dutch experimenters concluded that the quantity of fertilizer actually required might be cut by nearly half "simply by applying the fertiliser only in the required location, at specified dosage."

even pretended that it wasn't too late for massive birth control! But since it was, well, here came another pleasant measure:

We could have treated our cattle even more like machines than we already did in our factory farms. I have read a proposal to *optimize* each cow's *productive life,* which entailed back-to-back pregnancies (the first one as soon as she was breedable), and then, the instant barrenness set in, off to slaughter, to avoid unnecessary flatulence! That this was methane-frugal I had to confess; as for its cruelty, addressing that issue in a bracing spirit of post-carbon realism, the European Union concluded that kindness to livestock (which is to say *greater compliance to standards and requirements for animal welfare and the housing of animals . . .*), *may contribute to increasing emissions.*

Actually, between 1944 and 2007 we had already reduced by 63% the carbon dioxide equivalents *required to produce 1 billion kilogram[s] of milk*—a miracle accomplished by making our dairy cows more "productive" (I suppose with hormones), which in turn allowed us to reduce the number of belching, farting members of our herds.[*]

Ramping up the fight against methane pollution, we could next have prohibited unlimited floodings of ricefields. We could have learned to eat insects instead of beef.[†] As our centralist tendencies matured (in our quest to squeeze more out of people we might have resembled the coal baron Don Blankenship), we could have rewarded vegetarians and penalized meat-eaters, thereby further decreasing cattle populations.

The processing of sugar into sugarcane in Australia took 4 times less so-called *cumulative energy* than the same operation in Mauritius. So what we should have done was outsource production of energy-intensive food preparation to the nations that did it most economically, with due correction for transportation energies.—But what about the relative poverty of Mauritians, and their proportionately greater need for sugar production revenues?—*Carbon Ideologies* has no answer to this.

Growing corn products required 31 times fewer BTUs per final food calorie than did growing watermelons. Wheat was merely 4% less efficient than corn. So were we supposed to consume more corn syrup and wheat bread, and eat fewer

---

[*] However, meat-loving Americans had slightly increased their nation's enteric fermentations between 1990 and 2014.

[†] According to an Oxford University study, *by 2050, food-related greenhouse gas emissions could account for half of the emissions the world can afford if global warming is to be limited to less than 2°C.* Fortunately, if we all became vegetarians, *food-related emissions* of greenhouse gases would fall by 63%—or 70% if we became vegans.

Planting trees, Japan

watermelons? Being pre-diabetic, I didn't like the sound of that. Fortunately, I'd be long dead before we commenced such measures.

Meanwhile, Iowa corn and Kansas wheat, being heavily fertilized, released far too much nitrous oxide, so should we have banned those crops?

As you can see, a whole new department of rule-makers could have fattened itself on these considerations, comparing apples and oranges as usual—and, worse yet, enacting compulsions bereft of simplicity.

Well, then: We could have required every able-bodied person to plant so many trees, under penalty of the law.

Perhaps injecting more aerosols into the upper atmosphere would have delayed and reduced global warming. That might cause some previously unimagined ecological problem; such measures always did; but we were running out of time, so we'd force it through, for the sake of the greater good.

We could have contracted our economies, thrust people out of work, so that do-gooders like me could get by with one pair of glasses forever. Oh, yes, we could have crushed businesses, increased homelessness, spread suffering.

At the beginning of 2017, Mr. Abe, the Prime Minister of Japan, announced that *to accelerate Abenomics and have the economy steadily grow is the mission given for us.* Our mission would have been the opposite.

Cement *emissions decreased significantly between 2008 and 2009, due to the economic recession and associated decrease in demand for construction materials.* The European Union reported that *in 2012, emissions decreased in particular in the Member States that experienced persisting economic downturn or recession such*

Child scrap paper vendor, Pakistan

as Italy, Spain and Greece. And consider the good old days of East Germany,* which landfilled 42% less waste than the rival system:

> Food waste was collected and used as feed; feeds tended to be scarce during certain periods of time. Paper was collected; it was also a scarce resource. Wood and paper were often burned in ovens for purposes of heating and cooking ... Deposit systems were operated for glass ... All in all, the former GDR's economy was subject to scarcities of resources, and this led to efficient waste recycling.

So we could have taken East Germany as our model, which would have made for a wonderful life.

---

* It hardly need be said that the Communism of this state was defective. As for the ideal version, "the Soviet Union's entire history is one of incessant constructive work. Today, thousands of young volunteers rally to the Party's call and go to construction sites." Our eco-despots would have put a stop to that.

DELAYED CONSEQUENCES

*Near noon, in shadows, objects probably could be seen to glow by their own red heat.**

Encyclopaedia Britannica, 1976, describing the greenhouse effect on Venus

Not wishing to build ourselves a new East Germany, we, as usual, put off paying for our "choices." And why not? The lender had extended our grace period, and there was always the chance that she might forgive the loan. She'd destroyed the woolly mammoths for no reason; why wouldn't she spare us for the same? More likely, interest kept accruing—but when one borrows against the environment, a delightfully typical experience is delayed consequences.

Back in 1768, a Pennsylvania lawyer named John Dickinson sought to drum up alarm against British taxation. One of his warnings ran: *A people is travelling fast to destruction, when* individuals *consider* their *interests as distinct from* those of the public.†—But when we borrowed carbon trouble, our destruction appeared to be a long way off—because the more ignorant the borrowers, the longer the delay

---

* But now, just to show that I am capable of bearing good news, let me inform you that when I imposed again upon the kindly Dr. Pieter Tans at NOAA, this time to inquire whether we Earthlings might warm ourselves into Venusian conditions, he replied: "If humanity is reckless enough to burn a large fraction of fossil fuel resources, atmospheric CO2 might perhaps peak at four times pre-industrial levels *[= 1,100 ppm.—WTV]* at the end of this century and eventually remain about double preindustrial CO2 for a few thousand years ... The Venus atmosphere is almost entirely composed of CO2, at a surface pressure of ~80 bar. On Earth almost all of that carbon is in rocks, in the form of solid Ca/Mg carbonates. So, on Earth a run-away warming in the next few thousand years would have to come from a water vapor response ... As the ocean surface warms, the amount of water vapor in the atmosphere is expected to increase by ~7% per degree C ... Currently, the added water vapor approximately doubles the direct effect of increased long-lived greenhouse gases *[since water vapor is a potent greenhouse gas]*. If we assume that holds into the future, we have *[radiative forcing of]* about 3% of all absorbed solar radiation. [See I:592.] The surface would have to warm 2 degrees C [3.6° F] to send that extra 3% to space, in a new long-term equilibrium situation. This does not look like a run-away to me. However, I have not considered the effect of clouds—how do they respond to a warmer surface—as well as other potential effects. Therefore I cannot really answer your question, but I do suspect that a run-away is unlikely. We will have significant ocean acidification from the high CO2 regardless of the Earth radiation budget."—In other words, maybe the clouds *would* get us ...—and for all I knew we'd keep burning a mere "large fraction" of our fossil fuels. Finally, please remember that just about every aspect of global warming was now proving worse than our predictions [see p. 651]. But why be a killjoy?

† Italics in original.

before they are forced to see. Like a callow couple who decide to be homeowners, no money down, so that their monthly mortgage payments cannot even cover the ballooning interest, we enjoyed our possession of carbon- and nuclear-powered lives, unceasingly expanding demand. While revising this chapter I sat in one of a row of chairs in the Japan Railways office, and everybody on either side of me held a plastic bag of something, or a plastic-skinned electronic camera, or a glowing cell phone; so far as I could tell, nobody worried about foreclosure. Then I took a ride on an express train to Shin-Osaka, enjoying through immaculate plate glass windows the sparkle of passing rails, the massive exoskeletons of triple bridges, purple evening clouds in the pale winter sky and the various electric-powered glows of urban humanity. We paused in a station of concrete, steel and long light-tubes. On the platform, a uniformed official bowed to us. Then the train, clean, punctual and deliciously warm, hummed like a harmonica and gently, carefully, brought us back up to speed. Looking out the window, I saw children playing soccer, and a semi-wild belt of parkland, after which we crossed another low wide river, and rejoined the sprawl. An unknown shrine beguiled me, and I wondered how it would be for me to wander from temple to temple, shrine to shrine; thus I added to my own list of carbon-powered demands. Why couldn't such a life go on just a little longer?

And maybe it could. As Sam Hewes had said, we were all ideologues; no doubt if I asked around, that question would eventually be answered as I wished.

"WE MAY ENCOUNTER ANY UNFORESEEABLE HEALTH PROBLEMS"

Three years after the Japanese nuclear disaster, I wrote to the Fukushima Kyodo Clinic requesting information on visible and anticipated health effects. The following is extracted from the clinic's reply:

1. Anticipation of health problems

While 26 years have passed since . . . Chernobyl, health problems that are suspected to be effect of radiation have not been resolved and onset of new cases are still prevalent. The reality of Chernobyl is that out of 10 then healthy children only 2 are healthy now. We believe that what we need to do to anticipate the effect of [the] Fukushima accident is to capture the ongoing reality of Chernobyl.

Cases of childhood thyroid cancer showed significant difference depending on the patients' date of birth, before or after the Chernobyl accident. This is the only disease [the World Health Organization] was obliged to acknowledge as . . . radiation related . . . Other health problems, however . . . have been ignored as "causation [from] the accident cannot be proved."

Health problems due to internal exposure, which [are] of late onset . . . require long term health management of individuals . . . As [the] large explosion of Reactor No. 3, which had been run with MOX fuel (plutonium is considered . . . the most toxic material in the world) . . . was added to Hiroshima and Chernobyl, we always keep in mind . . . that we may encounter any unforeseeable health problems.

. . . [A] residents' health management survey conducted by Fukushima Prefecture . . . revealed that as of the end of November 2013, one in 10,000 children has thyroid cancer. The prefectural and national governments have denied . . . causation [from] the accident, despite [the fact that] the onset is 100 times more than the onset of one in 1 million that has been considered . . . normal.

. . . Learning from the experience in Hiroshima and Chernobyl, we realize that we need to carefully watch [for] the following symptoms: bura bura disease (feeling tired, not wanting to do anything; thus others see you as being lazy), fatigue, sluggishness, mental instability, defects of memory, ophthalmologic problems (cataract, detachment of the retina), circulatory organ disorders (cardiac infarction, cerebral stroke), ear-nose-throat problems (dysfunction of thyroid gland and others), gene damage, defect of reproduction functions, miscarriages and stillbirths.

As you may be aware, the entire Administration has been [conducting a] return-to-home campaign to encourage people to go back to the contaminated area. "Authoritative" medical doctors join this campaign saying "up to 20 milli per year is safe" or even "up to 100 milli." Our position is opposite to these. We are determined to listen to and stand by the victimized people.

2. How much effect has already appeared?

. . . What we can say at this moment is only about childhood thyroid gland cancer. We consider the current situation as . . . "abnormal." We, therefore, are afraid that the prefectural checkup schedule (once [every] two years) is not frequent enough. We recommend a checkup every 6 months [as in] Chernobyl ... We regard this spate of [cases] as representing internal exposure of radioactive iodine due to the nuclear accident.

Meanwhile, Wade Allison, professor emeritus of physics at Oxford, declared the danger from Fukushima to be "even less than the risk of crossing the road." The World Health Organization spoke of *one additional percentage point of an infant's lifetime cancer risk.*

Did these disagreements come about simply because it was still too soon to know anything? But even after 20 years, people kept disagreeing extremely about the Three Mile Island accident in the United States. While the Pennsylvania Department of Health could not discover any spikes in cancer mortality in people who had lived within five miles of that reactor, a certain Arne Gundersen asserted that there had been 10,000 extra cases of lung cancer. He opined that the Chernobyl incident might lead to "1 million extra deaths from cancer." And presently we got to read this emblematic headline:

Fukushima thyroid cases grow but nuke link denied.*

DISAPPOINTING, BUT NOT SURPRISING

In the autumn of that same year, the Pentagon began placing global warming alongside extremist Muslim terrorism in the short list of urgent national security threats, to which Senator James Inhofe of Oklahoma responded: *It is disappointing, but not surprising, that the president and his administration would focus on climate change when there are other legitimate threats in the world.* As for me, I likewise found it disappointing but not surprising that after sending out my questions about climate change, my research assistant finally reported: *Sen. James Inhofe: Emailed/Website Form on 2-4-16 to no reply.*

We can hardly deny that the future is semiopaque, rendering consequences unknowable and therefore susceptible to good faith disagreements. Delayed consequences thicken the opacity into real darkness.[†] This phenomenon enriched Senator Inhofe's promises, not with added credibility, but with diminished verifiability—perfect for him! We went on mortgaging the future, leaving you to pay for it.

---

* Eight more children came down with the disease, but "the panel of mainly doctors and other medical experts" distinguished the "types and sizes" of the new tumors from those in Chernobyl's children.
† Edmund Spenser, 1590: "Worse is the danger hidden than descried."

A RAY OF HOPE

Our desirable future includes a per capita GNP growth rate in the range of 2% per year, which would maintain or increase our present level of amenities. Adequate energy is needed to keep this future possible.

Robert S. Livingston et al., 1982

By most accounts, said the newspaper, *the world's greenhouse gas emissions must be brought close to zero by the end of the century, at the latest.* (That was in 2015, when I was still alive; I wasn't even starving or choking!) But whether that zeroing was practical, or even possible, we preferred not to start working toward it just yet. In a statement on "Greenhouse Gas Management," Chevron explained: *We strive to both manage our GHG emissions and support the growth of the global economy.* Meanwhile, an **Alaska Senator Wants to Fight Climate Change, but Drill in Arctic,** because *for the people who live and work and raise their families in a very small population state, we don't think that we are the problem*—which merely afflicted *some ecosystem somewhere* where unknown entities burned Alaskan oil. We'd have our cake and eat it; we'd keep everything and avoid pain. Among our majority stood India's former minister of the environment—quite the realist, who firmly explained that his nation *must continue to grow at 7.5 to 8 percent a year for the next 15 years**—because, as a famous Japanese actor had been made to say in a 2009 advertisement from the Nuclear Power Plant Environment Maintenance Organization (I quote the charming translation of Mrs. Keiko Golden), the goal was *to continue the life with electricity to the grandchild generations.*—Whoever could vote against that?—It might require Indians to burn three or four times more coal than before—after all, how else could they "continue their path of economic development?"†—Mr.

---

* In 15 years, if we all reduced carbon emissions considerably, *even the most disruptive climate change* would bestow dire poverty on a mere 3 million Indians. If we kept flicking our lights switches and the lights came on, 42 million Indians would achieve that situation. That would surely be for the best, because in the immortal words of Senator Manchin, "We don't understand what devastation is going to come when we don't have a coal industry to supply the country in the next 30 years."

† From the U.S. Energy Information Administration: "Primary energy consumption in India has more than doubled between 1990 and 2012, reaching an estimated 32 quadrillion British thermal units (Btu) . . . At the same time, India's per capita energy consumption is one-third of the global average,

Dunham, the retired CEO of Conoco, had called it right when he told me: "It's those 15 to 25% of the people that have no electricity today, they're gonna rapidly move in our direction regardless of what we want."*—And so the Australian government prognosticated:

> India is a likely candidate to be the next main driver of world energy consumption . . . It has a population of around 1.3 billion people, many of whom still do not have access to electricity, and is already investing heavily to address the issues in its electricity markets. Furthermore, its economy is starting to exhibit robust growth rates with the recently elected Modi government providing a substantial lift in business sentiment.

Meanwhile, in Vietnam, *the government has projected coal-fired energy will rise to 49% by 2020, up from 25% in 2014.*[†]

Back in our time we loved business sentiment because we loved a good party. We laughed all the way to the bank. Iran was *targeting India as its main destination for crude* oil, in part because *in a cutthroat market, Iran may . . . find it difficult to sell its oil, as the heavy grades it mostly offers are in low demand*—so sell them to India, just as South Africa sold inefficient coal to Bangladesh!—and also because *in China and India, stronger passenger car sales are fueling gasoline demand.* In addition to buying more cars, China was still building more coal plants than ever—speaking of which, do you remember good old Joe Manchin from West Virginia? He surely did love coal! *"The world's going to burn more of it than ever before,* Manchin said. *"This gives us a ray of hope that other decisions will be looked upon as unreasonable. If they are unattainable they are unreasonable."* By "unreasonable" he meant: decisions imposed upon American states by the executive branch in hopes of decreasing carbon emissions. As he so patriotically put it, "You can't regulate what we didn't legislate."—Hurrah, hurrah! *He* wouldn't

---

according to the International Energy Agency (IEA), indicating potentially higher energy demand in the long term as the country continues its path of economic development."

* See above, p. 456.

[†] It might not have been Vietnam whose nationals would deserve the prize for pushing us over that somewhat arbitrary 2° Celsius "tipping point." Maybe the straw that broke the camel's back was laid down not by Modi-ists at all—because just last Saturday the newspaper ran: **Until Recently a Coal Goliath, Energy-Hungry India Is Rapidly Turning Green**. Shortly after *Carbon Ideologies* was published, a neighbor yawned quickly through this second volume and in 42 seconds had reached this page, whereupon he triumphantly said: "See? You didn't write anything about the recession in India; their smokestacks are idle! And you forgot to consider the natural gas boom in Antarctica. That just goes to show you can't predict humanity's effect on the climate."—I defended myself by saying that it didn't matter who did it, but he shook his finger and said: "Wrong is wrong." In other words, business is business.

legislate any traitorous carboniferous zeroing! Instead, he contented himself with what was immutable: "They turn their lights on and they come on," said he. "They turn on their air conditioners and they come on." Such was our greatness.

And you, my unknown reader from the future, do you absolve me now? Would you agree at last that since even my own President* was nearly powerless to reduce carbon emissions (his predecessor and successor both actively *declined* to do so), and since China undid whatever Germany might wish to attempt, I ought to be held harmless? (When this book went to press, China was importing less coal—while Trump was pimping it out.) I think I can almost see you toiling in your radioactive fields, plowing your top 15 centimeters of dirt deep under, to keep fallout products well below the roots of whatever spindly crop you can still grow in between droughts and megastorms; you must be munching on insects, covering up against the sun, pollinating by hand, breathing smoke, keeping your children cool in caves and holes, collecting acid rain to drink and looking over your shoulder for the next catastrophe. Perhaps you are deploying bags of powdered zeolite at your sluice gates, to sequester more cesium from the irrigation water—although I once read that after Chernobyl the Swedes fertilized the soil with potassium, which plants will ingest in preference to cesium. Which is your procedure?

I'd rather imagine that you are happily, healthily burning carbon, just as I used to do. In the words of "a father and grandfather with 40 years of experience in the energy industry" (he was also "an expert in leadership"):

Climate change? Keep winter clothes handy

We are starting to see more scientists coming forward to question the global warming theory. An article in the most recent *West Virginia Executive Magazine* talks about the case against climate change . . . Even an experienced weatherman in Orlando, Florida thinks cooler oceans are the reason we are having less [*sic*] hurricanes . . .

---

* Obama, who remained in place like a pasteboard figure throughout most of my fieldwork for *Carbon Ideologies*.

FASTER THAN THE WORST-CASE SCENARIO

Beauty awakens. Set your shades in motion at sunrise, sunset and anytime in-between—automatically . . . Shades . . . move to schedules you create.

Advertisement, 2016

Fukushima, as anti-nuclear scientists readily grant, *was a well-defended nuclear plant by accepted standards, with robust, redundant layers of protection. But all of Fukushima's defensive barriers failed for the same reason. Each had a limit that provided too little safety margin to avert failure.**

Thus the story of humanity.

In the 1830s, John Audubon observed a killing of passenger pigeons in Green River, Kentucky. He wrote: *Two farmers from the vicinity of Russelville, distant more than a hundred miles, had driven upwards of three hundred hogs to be fattened on the pigeons which were to be slaughtered . . . The noise which they [the pigeons] made, though distant, reminded me of a hard gale at sea . . . Thousands were soon knocked down by the pole-men. The birds continued to pour in . . . Here and there the perches gave way under the weight with a crash . . . I found it quite useless to speak, or even to shout to those persons who were nearest to me . . . I was aware of the firing only by seeing the shooters reloading . . . Persons unacquainted with these birds might naturally conclude that such dreadful havock [sic] would soon put an end to the species. But I have satisfied myself, by long observation, that nothing but the gradual diminution of our forests can accomplish their decrease, as they not unfrequently quadruple their numbers yearly, and always at least double it.—* Audubon was wrong, of course. Fifty years later, even though large swathes of forest remained, the species was in catastrophic decline. Extinction came in 1914.

When I was alive, we often felt surprised and even mournful when something went wrong. We made plans convenient to our desires. Our safety margins were affordably optimistic. You see, the worst case would never happen, at least not in our time.

---

* And here I should repeat what that former Tepco worker said to me in Kawauchi (I:368): "It was only a matter of cost that prevented them from building a dike or siting the emergency power generator in a higher place."

From the 1990s until at least 2014, when this item was published, our oceans rose 3.4 millimeters per year. *This value aligns with the worst-case scenario from the IPCC report in 1990 . . . It appears that the I[ntergovernmental] P[anel on] C[limate] C[hange] underestimated the scale of the change.* I predict that they kept rising.

In California a wildfire that should have taken a week to burn 22,000 acres did the job in five hours. A fire captain said: *I've got 30 years in, and in the last 10 I have seen fire behavior that I had never seen in my entire career.* Three months later, in Western Australia, another firefighter warned: *It's going to be a horror summer. I've never seen conditions like this.**

Meanwhile our polar icecaps kept trickling, cracking and breaking. *The actual melting is faster than the worst-case scenario coming out of those models.* Logically enough, the shrinkage of glaciers was also *historically unprecedented.* (In Alaska, whose *chief source of revenue* was oil, a politician fretted: "I'm just hoping we don't get blamed for the fact that our glaciers are melting." How silly! How could anyone be to blame?)—The West Antarctic ice sheet had now begun breaking up, evidently unstoppably, *and a rise in sea level of 10 feet or more may be unavoidable in coming centuries.* About the latter *historically unprecedented* mishap a man who had been Chairman of the Council on Environmental Quality from 1979 to 1981 wrote in to the newspaper: *This possibility was foreseen more than three decades ago. It was . . . one of the warnings in the 1981 climate change assessment.* But really, reader, shouldn't we try to make you hate us less by pretending that we never could have predicted it?

Anyhow, perhaps the oceans would rise only six feet. That would harm a mere 13 million Americans, a number *three times the most current estimates.*[†]

The strongest hurricane to ever assault the Western Hemisphere now struck Mexico. An American atmospheric scientist said: *None of the models we use to forecast intensity, including my own, got this right.* The winds reached 165 miles per hour; there was talk of adding higher categories to the current measurement scale. But the hurricane fizzled out, so why not be optimistic again?

At the approach of the Christmas season a New York furrier lamented: "I've been in the business 20 years, and I haven't seen a December like this." Who wanted to wear mink coats in this weird warm weather? His sales of luxury pelts

---

* A climatologist consoled us that "this pattern—longer fire seasons, more burned acres of forest—is likely to continue as long as there is enough fuel to burn, but that there will come a point, probably in the middle of the century, when there are not enough trees left to sustain wildfires."

† "The complete loss of the Greenland ice sheet is not inevitable because this would take a millennium or more; if temperatures decline before the ice sheet has completely vanished, the ice sheet might regrow. However, some part of the mass loss might be irreversible."—A millennium, eh? I'll interpret that as a hundred years . . .

had decreased by 30%. Of course global warming remained a leftwing myth, like the predictions of climate change researchers that within 200 years the Persian Gulf might be afflicted by humid heat waves *so severe that simply being outside for several hours could threaten human life.*—Actually, no: That improvement was now scheduled to arrive within 85 years. "This is truly shocking," said one of the scientists.

In California the peculiar lack of rainfall was getting ominous; it was already *a historic drought,** and I am quoting no less august a source than the Hilton Checkers in Los Angeles. Fortunately, the staff offered two grand remedies: In the restaurant, and the same went for room service, *servers cannot bring out water unless customers ask.* Moreover, *hotel guests must get a chance to decline fresh towels and sheets.* Those measures would save us, no doubt—and they did, because two winters later, record-shattering rains struck the state, proving that climate change was a hoax!

In West Virginia, U.S. Representative Evan Jenkins had already presented his own solution to global warming. He had been touring a half-built hospital in Madison, and somehow the subject of coal came up—specifically, the War on Coal. It always did. "We are using the power of the purse strings," explained Mr. Jenkins. "The EPA had money budgeted for a theatrical performance on climate change and we said no." And then Donald Trump became President.† We became great again. *Scientists and environmentalists reacted with fear this week as the Trump administration purged nearly all mention of climate change programs from the White House and State Department websites . . .* So we walked out of the performance and shut out the light; I mean we kept the lights on.

MORE HEADLINES FROM 2016‡

Resettling the First American "Climate Refugees" . . . A $48 million
grant for Isle de Jean Charles, La. . . . is the first allocation of federal tax dollars to move an entire community struggling with . . . climate change.

---

* "Most revolutions in society have not power to interest, much less alarm us," wrote Thoreau; "but tell me that our rivers are drying up, or the genus pine dying out in the country, and I might attend." The rivers were drying in California, and the pine trees dying in Montana.

† "The 2016 presidential election was one of the most intense and unpredictable in U.S. history. In the aftermath, investors may be left with lingering questions . . . The answer, based on Vanguard research into decades of historical data, is that presidential elections typically have no long-term effect on market performance."

‡ I finished this book in 2017. When you from the future read these headlines, you will probably laugh, because I had it so good.

[Global] Warming Seen As Lit Match in Northern Forests . . . The near-destruction of a Canadian city last week by a fire that sent almost 90,000 people fleeing for their lives . . . is grim proof [of climate change] . . . In addition, winds are sometimes carrying soot from the northern fires onto the [Greenland icecap], darkening the surface and causing it to absorb more of the sun's heat . . . Limited evidence from Alaska suggests that fires in at least part of that state are at their worst in 10,000 years.

Flooding forces museums in Paris to shut down: TORRENTIAL RAIN: Thousands evacuated; at least nine killed in Germany . . . President François Hollande said Thursday that the rainfall and floods were "very serious" and linked them to global warming.

Temperatures Are on Course for Another Record Year: NASA Blames El Niño and Greenhouse Gases . . . Average temperatures for the first six months of this year were . . . "quite close" to 1.5 degrees Celsius above preindustrial levels . . . At the Paris climate treaty in December, the world agreed to aim to limit the increase . . . to that amount . . .

[Hawaiian islands.] Almost Unbearable: Warmer ocean temperatures, rain behind rising humidity level . . . The outlook calls for warmer than average temperatures . . . into the fall.

Thousands Displaced in Storm-Drenched Louisiana: At Least 14 Are Dead in Record-Breaking Flooding That Officials Warn May Spread . . . "The simple fact of the matter here is we're breaking records," Gov. John Bel Edwards said . . .

A Parasite Deadly to Fish Shuts a Montana River: Boating and Fly-Fishing Grind to a Halt as Officials Try to Protect Yellowstone Park . . . The white bodies of thousands of dead fish litter many parts of the river . . . Part of the reason for the outbreak is near-low record flows and warm water temperatures, which stress fish . . . Montana rivers are changing because of a warmer climate . . .

Climate Scientists Forecast More Floods Like Louisiana's.

Climate Policy Faces Reversal by New Leader . . . If Mr. Trump makes good on his campaign promises . . . the world . . . may have no way to avoid the most devastating consequences of global warming.

Carbon's Casualties: The Sweltering Sahel . . . The men and boys of West Africa . . . leave home . . . because the rains have become so fickle, the days measurably hotter, the droughts more frequent and more fierce, making it impossible to grow enough food . . .

Rising Seas Turn Coastal Houses into a Gamble: Fears of Market Crash . . . Politicians are more focused on keeping developers calm and reassuring people that technological solutions will save the day . . .

Shoreline Gentry Are Fake Climate Victims. The fake news problem is getting serious. An example appeared on the front page of the *New York Times* under the headline "Rising Seas Turn Coastal Houses into a Gamble" . . . Estimates vary, but sea levels may have risen at two millimeters a year over the past century. Meanwhile, tidal cycles along the U.S. east coast range from 11 feet every day (in Boston) to two feet (parts of Florida) . . . Background sea-level rise is a non-factor . . . Climate reporters aren't expected to understand the science or its limits—understanding is actively discouraged. Just memorize the word "consensus" and fling "denier" at anyone who proves inconvenient.

My fellows seem to me to be shortsighted, stupid, incapable of unanimity, and really not satisfied to wait for destruction: they seem bent on self-destruction, in their systematic unwillingness to face facts. Not I, of course: I am a rather sensible and level-headed fellow at all times.

John Hersey, 1950

Long ago a man fell gravely ill, and . . . he knew in his heart that he was to die . . . This I always knew / but yesterday or today . . . / no! never had I thought it.

Arihara no Narihira[?], 10th cent.

THAT SUBTLY SUBTLE BOTTLE NECK

Without denying the possibility, or even the probability, of the establish-ment of the fact of secular changes, there is yet no sufficient warrant for believing in considerable permanent changes over large areas . . . A change in climate has not been proved.

Encyclopaedia Britannica, 1911

While you in the hot dark future eke out what Pasolini called *remnants of a life that mirrors its shadow,* I lie quietly on a cool spring night in 2015, enjoying electric light over my left shoulder as I type these words on a laptop computer, whose battery will soon need to drink more electric-ity from the wall; and although I have written this book with good faith, dili-gence and I hope even imagination, my imagination often fails to uphold my knowledge that my generation and our luxuries are truly, as you know, *dead.*— Once upon a time the devil appeared to a man in a novel, proposing the usual bargain. The man of course expressed certain anxieties. Sooner or later, whatever the devil lent must reach its end, and then your hot dark future would devour him. But the devil consoled him that the future was still a good long way off! *And here silence is but natural, yet not twixt us nor over time—not when the hour-glass has been turned, not when the red sand has begun to run through that subtly subtle bottle neck . . . What lies below is still nothing in comparison with the quan-tity above—we give time, abundant, immeasurable time . . .*—So it seems to me, even now. I write this book unable to comprehend in my bones that someday all our choices will probably run out.

Transmission towers, Ruwais

Acknowledgments

Carbon Ideologies could never have achieved whatever limited competence it offers without the help of many people. (As any generic acknowledgments section runs, "Responsibility for errors is most definitely my own.")

One of this book's many faults is inadequate representation of the pro-fuel side. Although I tried my utmost to accomplish "due diligence" in this respect, the *no comment* of the *regulated community* repeatedly foiled me. For this reason I feel all the more grateful to those few adherents who took the trouble to educate me about the carbon ideologies they knew so well. I wish to thank them first and foremost.

Mr. Archie Dunham was a sharp-sighted, gracious gentleman who spoke his mind without compromise. Time will prove him right or wrong; I admire him for standing fast in his position. I also like him as a person. Mr. Sam Hewes said some of the most resonant things in this book. He patiently endured my pushy impositions; his political cynicism delighted me. Both men took pride in their petroleum careers, and shared that pride with me. The environmental consultant Mr. John Mahoney was another who felt proud of the oil companies, and helped me to understand why. Mr. Chris Hamilton, then Vice-President of the West Virginia Coal Association, treated me with kindness. I agreed with much of what he said, such as this: "Rightfully or wrongfully, we're so dependent on electricity, and there's only a handful of things used to make it." His interview, broken into bits, seasons several parts of *Carbon Ideologies*. All these men expressed the "can do" technological optimism toward which I cherish fond feelings, in memory of my father, my two machinist grandfathers (one of whom, Mr. Gilbert Vollmann, gave me his weatherbeaten 1958 *Mechanical Engineers' Handbook,* whose electron-volts and BTUs became the basis of this book's tables), and my own mid-20th-century American childhood. It goes without saying that technological optimism is something in which I still long to believe. Thank you, all.

"Can do" is not the best description of Tokyo Electric Power Company's Fukushima Daiichi Nuclear Plant—all the more reason for me to express my appreciation to Tepco for meeting with me to address my written questions. The interviewees were Mr. Sakakibara Kohji, Group Leader; Mr. Togawa Satoshi, Deputy Manager, International Public Relations Group, Corporate Communications

Department; Mr. Hitosugi Yoshimi, Section Manager, Corporate Communications Department; and a woman who did not give her name. Let me here express my sincere gratitude for their trouble on my behalf. If only Noble Energy in Colorado, or Alpha Resources in West Virginia, or Breitling Energy, or any refrigerator company whatsoever, or the Chamber of Commerce in Cushing, Oklahoma, had been as willing to comment as Tepco! *Domo arigato gozaimasu.*

Next let me thank the experts whose contributions to *Carbon Ideologies* had more to do with issues than with any particular location. (Other authorities will be thanked below, in their place-specific sections.) Dr. Jacqueline Agesa, Professor of Financial and Insurance Economics at Marshall University, checked a passage on GDP and electric demand. Peter Bradford, Board of Trustees, Union of Concerned Scientists, shared his expertise with me on the occasion of the Fukushima crisis. Mr. Scott Burnell, who was Public Affairs Officer at the Nuclear Regulatory Commission, responded kindly and fully to a quoted challenge regarding NRC policy on containment vents. Mr. Christopher Capra, Sr., Public Information Specialist for the Sacramento Municipal Utility District, answered questions about power plant efficiencies. Wai Cheng, Ph.D., Director of the Sloan Automotive Lab and Professor of Mechanical Engineering at MIT, kindly corrected and improved my discussion of internal combustion engine efficiencies and rolling resistance. Richard Crownover, M.D., Ph.D., Professor and Residency Director, Department of Radiation Oncology, University of Texas Health Science Center, San Antonio (he asked to be listed as "a physicist and a physician"), corrected, improved and verified several parts of the nuclear section. J. Eric Dietz, Ph.D., P.E., Professor, Purdue Polytechnic Institute, answered two questions about batteries. Mr. Robert Finkelman at the U.S. Geological Survey took the trouble to improve my understanding of coal HHVs. Dr. Canek Fuentes-Hernandez, Senior Research Scientist, at Georgia Tech, brought my ignorant and outdated chapter on solar energy up to date, then submitted to more questions. Professor Timothy Gutowski (another luminary of MIT) vastly improved my approach to the big questions of waste, thermodynamic ceilings and losses, manufacture and recycling. My many requests for help abused his goodheartedness, so let me just say that without him this book would have been much narrower. Dr. Edwin Lyman of the Union of Concerned Scientists (who told me to call him Ed) picked up my phone calls over and over, year after year, so that I could ask him the same stupid questions about cesium isotopes. The book he co-wrote about Fukushima was quite helpful to *Carbon Ideologies.* I greatly respect him. Mr. Matsumoto Jun verified various calculations rapidly and efficiently in personal letters and in his own publications. Professor Anna Mummert and Mr. Ben Coleman, both then of Marshall University, checked several tables and most of the definitions and conversions section, thereby finally reassuring me that I understood the difference between kilowatts and kilowatt-hours. Professor David Spiegelhalter at the Centre for Mathematical Science in Cambridge, U.K., was kind enough to answer my query about micromorts, which concept I uselessly hoped could be applied to this project. Dr. Pieter Tans, Chief, Carbon Cycle Greenhouse Gases Group, National Oceanic and Atmospheric Administration, made time for my written climate questions on at least half a dozen occasions. I have never met him, but come away with a sense of his intelligent benignity. Some of the most fundamental issues in this study, such as the atmospheric lifetime of CO_2 emissions, were addressed for me by him, precisely and concisely. Between 2014 and 2017 Mr. Yamasaki Hisataka of the NGO No Nukes Plaza helped me several times, both in person and via translated communications, addressing the Fukushima accident, power plant efficiencies and climate change. My thanks to all of you.

Now for people whom my mind links to certain places.

The pages devoted to Japan would be blank without Mr. Abe Masahiko; Mr. Tom Colligan (who had nothing directly to do with that country but whose skippering of the *Water Witch* at Hanford, Washington, deepened my appreciation of that Japanese-occurring vitamin, plutonium); Ms. Janis Heple; Mr. Endo Kazuhiro, the head of one community of Tomioka evacuees; Mr. and Mrs. Hamamatsu Koichi; Ms. Hotsuki Minako and her mother-in-law Ms. Hotsuki Keiko; Mr. Kawanami Shugoro; Mr. Kawama Tatsuhiko of the National Railway Mito Motive Power Union; Mr. Kida Shoichi, decontamination specialist in the Nuclear Hazard Countermeasure Division, Iwaki, and his deputy, the jovial Mr. Kanari Takahiro; Mr. Kojima of the organization Isshin-Juku; Ms. Kuwahara Akiyo of the Tomioka Life Recovery Support Center; Mrs. Ito Yukiko; "Source ML"; Mr. M.Y. and his mother; "Michiko"; Professor Morimoto Motoko; the Murakami family (including Takuto, his mother Kaoru, his grandmother Fumiko, his father, his other grandmother and two of the other grandsons); Nancy from UCS; the late Dr. Jean Pouliot, Vice-Chair of the Radiation Oncology Department at Mount Zion Hospital in San Francisco, and his graduate student assistant Ms. Josephine Chen; Mr. Sato Yoshimi; Mr. Shigihara Yoshitomo, the head of Nagodoro Subdistrict, Iitate, and his kindly wife; Mr. Suda Giichiro, Secretariat of Fukushima Kyodo Clinic, for his long letter of reply to me about long-term health effects of the nuclear disaster; Mr. Suzuki Hisatomo and his unnamed colleague, both of whom escorted me through the Okuma red zone; Mr. Takamitsu Endo; Mr. Tazawa Norio; the Utsumi family (of whom I met Takehiro, Yuya and their mother Yoshie); the eternally sweet Ms. Tochigi Reiko; Mrs. Yoshida; Mr. Yoshikawa Aki, founder of Appreciate FUKUSHIMA Workers (and valiant drinking partners);

and various anonymous nuclear refugees, anonymous decontaminators and former reactor workers, and brave unnamed taxi drivers.

Ms. Kawai Takako, with whom I have now been working for many years, showed her accustomed helpful bravery in facilitating trips to radiocontaminated areas, and in interpreting for me there, with and without a mask. It always did my heart good to see her completing a one-minute timed count for me with that pancake frisker while I finished loading another roll of film or scribbled a paragraph about sewer gratings. She never said no to anything. If she gets cancer from those excursions, I certainly hope to do the same. Her quiet, industrious compassion is my inspiration. Believing in this project, and sorrowing for the victims of Fukushima, she prepared, arranged, translated and, not least, witnessed, year after year, almost without compensation. Takako, I will never forget what you have done for me.

Ms. Hannah Scott and Ms. Miriam Feuerle of the Lyceum Agency made it possible for me to use the pancake frisker in Singapore. Thank you, sweet ladies.

For the Appalachian stretch I would like to thank the indispensable Mr. Atkins; Mr. Tim Bailey, who took time from his law practice upon very little notice to discuss coal politics with me, thoughtfully and at length; Ms. Jean Battlo; Pastor Bob Blevins, whose interview is one of my favorites; Ms. Tatiana Castro; the anti-MTR activist Mr. Chad Cordell, for repeated efforts and courtesies; Ms. Karen Elkins; Barney and Jackie Frazier, not just for an interview but for hospitality; Ms. Kelley J. Gillenwater, Communications Director, WVDEP, for explaining the mining permitting process; Professor Tom Jones of the Department of Integrated Science and Technology at Marshall University (whom I have never met, but whom Professor Laura Michele Diener [thanked below] interviewed partially on my behalf; his remarks greatly enrich *Carbon Ideologies*); Mary; Mr. Mark Mooney; Ms. April Mounts; Mr. Daniel Phillips, who let me interrupt his class; Ms. Wendy E. Radcliff, Environmental Advocate, WVDEP, for commenting on Barney Frazier's description of the mining permitting process; the brave Ms. Mary Rahall, for her corrections and her statement regarding the stand she took on fracking in Fayette County; Ms. Rhonda Reed and her family; Mr. Avril Richmond; Robin; Ms. Vivian Stockman of the Ohio Valley Environmental Coalition (OVEC); the former miner and inspector and current activist Mr. Stanley Sturgill; Ms. Jackie Wheeler; Mrs. Patricia Ann McNeely Wheeler; Mr. Bill White; Mr. Dustin White; Mr. and Mrs. Harry White; Mrs. Glenna Wiley; Mr. Arvel Wyatt; unnamed miners, security guards, housewives and others.

Professor Laura Michele Diener, in between chairing the Department of Women's Studies, teaching classes on everything from ancient Egypt to Norse sagas, researching medieval hair styles and raising money, some of it her own, to help impoverished women and girls in both Huntington and McDowell County, somehow found time to help me turn my vague hopes of an Appalachian coal chapter into what finally appeared in *Carbon Ideologies*. In a mere paragraph I cannot begin to enumerate my debt to Laura Michele for her devoted toil, which vampirized four years of her life. Sometimes she did something for this book every day. She saved big bundles of local newspapers for me. Whenever one of those articles made me want to know more, Laura Michele would do the research, transmit my questions electronically to interviewees and print out their answers (or damning *no comments*), arrange real-time meetings and take me to and from them in her car. She explored the coal country with me (getting rearended in the process), all the while never failing to be fair and kind with everyone we met. At the last possible editorial minute I was still telephoning her to ask the spelling of this name or the contents of that Senate bill. The only trouble I ever had with her was in persuading her to accept a token payment, which might have compensated her trouble at five cents an hour. Her selflessness will always inspire me. Should *Carbon Ideologies* turn out to be a failure, it will hardly be her fault.

What I learned in Bangladesh came courtesy of Dr. K. Anis Ahmed, who not only paid my way to his country but also put two of his staff and one of his vehicles at my disposal (Anis, I earnestly hope to someday be able to do something for you); Ms. Pushpita Alam, also known as Señorita Pita de Push, who fixed and interpreted, then refused to accept a single taka; our driver, who might be safer unnamed; Mr. Mohammed Aminul Islam Bablu [sometimes spelled Babu], who was then Chairman of the National Committee to Protect Oil, Gas, Mineral Resources, Power and Ports; Ms. Rehena Begom; Mr. R.C.; my generous friend Mr. Chris Heiser, whose recommendation to an old college roommate commenced this chapter; Mr. Mohammed Maksud Hossain, proprietor of Messgrs SYC Brick; a certain Mr. Islam; Mr. Sayed Saiful Islam Juuel, National Committee, Convener, Phulbari Committee; the great Mr. Nakeeb Nazmun, aka the living corpse; Mr. Rabiul Islam Rabi, who was then president of the Barapukuria Workers' Union (whose other members named and unnamed I also thank here); Mr. Muhammad Anishur Rahman, Deputy Manager (Personnel) at the Barapukuria Coal Mining Company Limited; Mr. Idu Sarker; Mr. David Shook, who like Chris is now a good friend; the activist Mr. Shahriar Sunny; Mr. Mohammad Sana Ullah, Manager (Admin.), Barapukuria Coal Mining Company Limited; and the mine worker Mr. Zhang Wen.

For the Colorado part let me thank the activist Ms. Sara Barwinski; the activist Ms. Sharon Carlisle; Mr. Eric Ewing, whose account of his family's ordeal may be the most effective part of this chapter; Mr. Brad Mueller, the Community Development Director of Greeley, who provided one of this book's few American municipal perspectives on carbon ideologies (and gave this out-of-towner a break just before a national holiday); Mr. Sam Schabacker, then of Food & Water Watch in Denver; the Wellington, Colorado, town clerk; and lionhearted Mr. Bob Winkler, whose "Tour of Destruction" gave me context.

Let me now thank the following people from Poza Rica, Mexico: Sr. Jonathan Carrillo Espinosa, who "took the contamination from wells"; Omar the watchman, the "subsidiary workers" at the refinery; not to mention the airport official in Poza Rica who when I asked for a hand inspection of my high-speed film, the request also being written out in my interpreter's impeccable Spanish, found herself quite at a loss, and turned the films round and round in her sweet young hand, then finally asked me, with worry in her guileless face, whether these items were truly not explosives, and when I looked into her eyes and promised that they were not, she smiled, trusted me and allowed me to take my film away.

The Oklahoma pages are much indebted to Mr. Vann Bighorse, director of the Wah Zha Zhi Cultural Center in Pawhuska, Oklahoma; Mr. Doug Hayes of the Sierra Club, for three rounds of review and correction of my text regarding the Alberta Clipper Project; Jackie; Keith and Deanna Lambert; Mr. David Shook (again), for introductions; and Mr. Roger Taylor, the former Mayor of Pawhuska.

In the Emirates I learned much from Mr. Baljeet (throughout this paragraph I employ the honorific per local usage); Mr. Gilbert; Mr. Iwahar; Mr. Kamal; Mr. Gafar Khan; Mr. Shavan Kumar; Mr. Mahiveer; Mr. Paul; Mr. Absay Pratap; Mr. Marat Sagimbayev, refinery medical coordinator; Mr. Sahabudin; Mr. Rana Saqib; my friend Mr. Priyank Srivastava, refinery safety auditor, who never tired of answering my queries; and Mr. Utkir Umarov, refinery senior process safety engineer. The *Harper's Magazine* fact-checker Ms. Rachel Poser corrected several errors and challenged some citations. Most of all, I thank my friend and fixer Ravindra. Thank you, everyone.

A more miscellaneous crew: Herr Georg Bauer, for the book *Gazprom City;* the nurse Ms. Cathy Behr; Ms. Patricia Broderick, for kindly supplying the *Pocket Dictionary of Terms Used to Describe Glass and Glassmaking;* Mr. Jeff Cox; my old friend and classmate Mr. Jacob Dickinson, for quoted remarks on cars and global warming; Ms. Ansel Elkins; Mr. Mats Engström; Empress Keiko Golden and her husband David, for translations, introductions, opinions, scientific verifications and iodine tablets; Mr. Brian Hewes; Mr. Steve Jones; Ms. Priscilla Juvelis; Mr. Kent Lacin; Mr. Chris Leonard, for sending oil-related newspaper clippings and friendly encouragement; my kind dentist, Dr. Jeffrey Light, and his X-ray tech, the lovely Amanda Ellis, for allowing me to measure the output of their X-ray machine, using my dosimeter and pancake frisker; Ms. Heidi Lehrman; the Samson-like Mr. Rob MacAulay; Mr. Mark Merin, for innumerable helps and kindnesses;

Mr. George R. Minkoff; Mr. L. Jackson Newell and his beautiful wife Linda, for their friendship (of course) but also for their work on L. L. Nunn, from one passage of which I repeatedly quote; Mr. Christopher Porter, who furnished me with documents pertaining to nuclear-powered-socialism; Ms. Vanessa Renwick; the late David Roberts and his widow, Ms. Elizabeth Allakariallak, for information on internal combustion engines and friendships at cold temperatures; Dr. Janice K. Ryu, for nuclear information, number questions and many electronic searches; Platinum Goddess Mary Swisher—and her husband Bob (who agreed with me about millisieverts); and Ms. Elisabeth K. Vollmann, whose existence prevents me from giving up on our future.

Unimpeded by a "new arrival" in the family, Mr. Jordan Rothacker, my paid research assistant (and also my friend), cheerfully and with immense energy busied himself on my behalf. Whenever I worried that I might not have done enough homework on internal combustion engines, fracking in eastern Europe or veganism's potential amelioration of climate change I would call Jordan and he would be on the case, with a DVD or memory stick soon arriving in my mail box—with baby photos included at no charge. And, speaking of money, I have the distinct feeling that after awhile he began to bill his time at approximately zero. Whenever I would ask him whether I owed him anything, he'd say, "Don't worry about it." Many of the experts in the *Carbon Ideologies* gratitude list—and nearly all of the priceless *no comments*—come courtesy of this tireless fellow. His work has benefitted the book's abstracter expositions of greenhouse gases, thermodynamic work, etcetera. Without the information he gathered for me, I would have had to content myself with reportage. Jordan, I am sincerely grateful.

I used to pride myself on turning in relatively immaculate copy. No more. Pushing 60, growing scatterbrained and fighting losing battles with my computer (whose occasional inexplicable doublings of words are nearly as inconspicuous to me as radiation), I was quite foolish to embark on this numbers-heavy work. One reason I decline to have a checking account is to avoid balancing my checkbook. Over and over while I was writing *Carbon Ideologies,* my mind got boggled. I will not go so far as to say that what I gave Viking was a hateful mess, but it was, as the production editor, Mr. Bruce Giffords, assured me, even more work than my previous book, *The Dying Grass.*

Although I hired number checkers here and there (and have thanked them in these notes), doubts of my own arithmetical acumen continued to trouble me even after both volumes had been line edited and copyedited. What if 47,000,000,000,000 BTUs should have been 4,700,000,000,000—or, worse yet, 0.047? Dr. Teresa A. McFarland, who used to make her living with budgetary mathematics, accordingly came to my rescue. "Soon I, too, will be gamboling through HHVs and exajoules," she sighed. Reader, was I happy to share that pain! *Carbon Ideologies* has benefitted from her labors in several ways. My Colorado, Poza Rica and Oklahoma researches might have gone nowhere without her. My long divisions and Spanish translations helped themselves to her native intelligence. And she refused to take a cent. Thank you so much for all your care, Teresa. I cannot express how much I owe you.

Now for my Viking colleagues.

Let me begin by thanking Mr. Bruce Giffords. He dealt with my smeary pencil-scrawls on what might well be Viking's last snail-mail-and-hard-copy edit; he (together with his wife Reiko) verified and corrected my transliterations of Japanese land-units and vegetable-words. He understood why I might prefer one spelling or punctuation to another. He cared about words. I cannot imagine all the pains he took to get this book into shape. [Bruce to WTV, very tactfully: "If you think you could write more clearly with a standard black graphite pencil, go ahead and use it. Legibility trumps other considerations."] In my 30-odd years in this business I have never worked with a more self-effacingly exemplary colleague.

The copy editor, Mr. Roland Ottewell, accomplished a sickening amount of drudgery to save me from misidentifying a certain Oklahoma highway and from misspelling (over and over!) the names of two cited and epigraphed authors and three corporations. [A sample of his diligence: "Since you don't count Sahabudin's salary, isn't it more accurate to divide by 5?"] Looking over the many pages of source-notes which he had to correlate and shuffle, I find myself in awe of his patience.

Ms. Nancy Resnick, the designer, never failed to be cheerful and agreeable when I called her up about this or that. (When I promised her that my next book would be shorter, she burst out laughing.) She must have been put to as hideous a labor as Bruce and Roland; thank goodness I cannot envision the weariness I caused her.

Mr. Paul Slovak, my editor, advocated for this long book, and worked tirelessly to improve it. And *Carbon Ideologies* needed improvement. I spent so much effort educating myself about the science and mathematics of it that I sometimes failed to distance myself sufficiently from the argument to detect stylistic infelicities. In a word, I had to reread, and rewrite, so many times that I could no longer see it fresh. Paul hacked at the typographical undergrowth of my submitted draft, so that Bruce, Roland and Nancy could create parklands. I struggled as best as I could, but never did any of my manuscripts need as much help as this one. Thank you, Paul, for the huge difference that you have made.

Bruce, Roland, Nancy and Paul: Many thanks and apologies, my friends.

Paul's assistant, Ms. Haley Swanson, did her own cutting, pasting and I don't know what. She was always nice to me on the phone.

My agent, Ms. Susan Golomb, sold this book. In another life, she and I will make each other millions! In the meantime, thank you, dear Susan, and thank you also to her former assistant, the much adored Ms. Soumeya Bendimerad Roberts.

Finally, to my readers: Thank you for continuing to support these projects.

WTV

 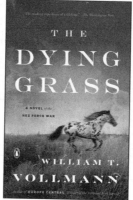 

NO IMMEDIATE DANGER

*Volume One of
Carbon Ideologies*

THE DYING GRASS

*Volume Five of Seven Dreams:
A Book of North American
Landscapes*

**LAST STORIES AND
OTHER STORIES**

IMPERIAL

EUROPE CENTRAL

ARGALL

*Volume Three of Seven Dreams:
A Book of North American
Landscapes*

THE ROYAL FAMILY

THE ATLAS

THE RIFLES

*Volume Six of Seven Dreams:
A Book of North American
Landscapes*

WHORES FOR GLORIA

FATHERS AND CROWS

*Volume Two of Seven Dreams:
A Book of North American
Landscapes*

THE ICE-SHIRT

*Volume One of Seven Dreams:
A Book of North American
Landscapes*

THE RAINBOW STORIES

**YOU BRIGHT AND RISEN
ANGELS**

PENGUIN BOOKS